DISCARDED
FROM
UNH LIBRARY

MOLECULAR VIBRATIONS
IN CRYSTALS

Exploring the diamond lattice (Peter Pearce and Synestructics, Inc.).

McGRAW-HILL
INTERNATIONAL
BOOK COMPANY

New York
St. Louis
San Francisco
Bogotá
Auckland
Düsseldorf
Johannesburg
London
Madrid
Mexico
Montreal
New Delhi
Panama
Paris
São Paulo
Singapore
Sydney
Tokyo
Toronto

JOHN COURTNEY DECIUS
Professor of Chemistry
Oregon State University

and

ROBERT MAURICE HEXTER
Professor of Chemistry
University of Minnesota

Molecular Vibrations in Crystals

This book was set in Times New Roman Series 327

Chem
QC
454
.V5
D4

Library of Congress Cataloging in Publication Data

Decius, J. C.
Molecular vibrations in crystals.

Includes bibliographical references and index.
1. Crystals—Spectra. 2. Vibrational spectra.
I. Hexter, Robert Maurice, 1925– joint author.
II, Title.
QC454.V5D4 535′.842 77-23259
ISBN 0-07-028615-9

MOLECULAR VIBRATIONS IN CRYSTALS

Copyright © 1977 McGraw-Hill Inc. All rights reserved.
Printed in the United States of America. No part of this publication may be reproduced, stored in a retrieval system, or transmitted, in any form or by any means, electronic, mechanical, photocopying, or otherwise, without the prior written permission of the publisher.

1 2 3 4 MHMH 7 9 8 7

Printed and bound in the United States of America

To our wives

ANNE AND NORMA

and to our mentors

E.B.W., R.S.H., D.F.H., AND P.C.C.

CONTENTS

Preface xi

1 Introduction 1

1-1 Gas-phase vibrational spectroscopy 1
1-2 Types of crystalline solids 3
(a) Metallic crystals 4
(b) Van der Waals crystals 5
(c) Covalent and ionic crystals 5
(d) Hydrogen-bonded crystals 6
1-3 The perturbation approach and the nature of solid state effects 7
1-4 Summary of motivation for crystal studies 11
(a) Studies of intermolecular forces 11
(b) Studies of molecular ions 11
(c) Special crystal problems: rotation of molecules in solids; order-disorder transitions, ferroelectricity; acentric crystals; energy transfer and relaxation 12

2 Symmetry and normal vibrations of isolated molecules 15

2-1 Symmetry point groups 15
2-2 Representations of point groups 23
2-3 Normal modes of vibration 33
(a) Normal coordinates 33
(b) Quantum mechanics of the normal modes 35
(c) Transformation between internal and normal coordinates; determination of frequencies 35
(d) Symmetry and internal coordinates 38
(e) The example of planar $XY_3(D_{3h})$; redundancy 40

2-4 Selection rules 44
(a) Direct product representations 45
(b) Activity representations 49
(c) State and transition representations 49
2-5 Isotope effects 53
(a) Simple **G** matrix elements 54
(b) Isotopic substitution without change of symmetry 56
(c) Isotopic substitution with change of symmetry 57
(d) Order of magnitude of isotopic shifts 59
(e) Rayleigh's Principle 59
2-6 Normal modes of simple symmetric molecules 60
(a) $XY_2(D_{\infty h})$ 60
(b) $XYZ(C_{\infty v})$ 61
(c) $XY_2(C_{2v})$ 61
(d) $XY_3(D_{3h})$ 62
(e) $XY_3(C_{3v})$ 62
(f) $XYZ_2(C_{2v})$ 63
(g) $XY_4(T_d)$ 64
(h) $XY_3Z(C_{3v})$ 64
2-7 Effects due to anharmonicity 65
2-8 Line and band shapes, widths and intensities 67

3 Vibrations of a linear lattice 75
3-1 The linear continuum 76
3-2 The linear monatomic lattice 77
3-3 Linear diatomic lattices 85
(a) Types of linear diatomic lattices 85
(b) The diatomic linear lattice with symmetry 86
(c) The diatomic linear lattice without symmetry 93
3-4 Linear polymers 94

4 Vibrations of crystals composed of monatomic units 101
4-1 Translational symmetry of crystals 102
4-2 Crystallographic space groups 105
4-3 The Brillouin zone 110
4-4 Symmetry species at $\boldsymbol{\kappa} = 0$ 113
4-5 Symmetry species at $\boldsymbol{\kappa} \neq 0$ 123
(a) Transformation of translationally symmetrized coordinates 123
(b) Symmetrization of coordinates under $H(\boldsymbol{\kappa})$ 129
(c) The body-centered lattice 129
(d) The rocksalt, diamond, and fluorite lattices 130
4-6 The secular determinant as a function of $\boldsymbol{\kappa}$ 133
(a) The simple cubic (sc) lattice 138

(b) The face-centered cubic (fcc) lattice 140
(c) The body-centered cubic (bcc) lattice 143
(d) A lattice with a basis: CsCl 149
4-7 Selection rules 152
(a) One phonon transitions 154
(b) Multiphonon transitions 154
(c) Density of states and critical points 156
4-8 Directional and polarization properties in single crystals 167
4-9 Longitudinal and transverse modes 170

5 Optical properties of single crystals 179
5-1 The dielectric constant 179
5-2 Plane wave propagation in anisotropic crystals 181
5-3 Separation of the dielectric constant and refractive index into real and imaginary parts 182
5-4 Reflectivity at a single crystal surface 183
(a) Normal incidence 184
(b) Oblique incidence 186
(c) Overlapping modes 188
(d) Reflectivity from an arbitrary face 188
5-5 Transmission through a slab 190
5-6 Polarization of Raman scattering 193
(a) Right angle scattering 193
(b) Example of a T_d unit cell 194
(c) Transverse and longitudinal modes in the Raman spectrum 195
(d) Errors in relative intensities for polarized Raman spectra 198
(e) Raman scattering at other angles 199
5-7 Microscopic theory of transverse and longitudinal modes 200
5-8 Microscopic theory of the dielectric constant 214

6 Molecular crystals 218
6-1 Internal and external coordinates 218
6-2 Symmetry species in molecular crystals by correlation 221
(a) $HCN(C_{\infty v})$, space group $C_{2v}^{20} = Imm2$ 222
(b) $CO_2(D_{\infty h})$, space group $T_h^6 = Pa3$ 223
(c) $CHI_3(C_{3v})$, space group $C_6^6 = P6_3$ 227
(d) NaN_3, space group $D_{3d}^5 = R\bar{3}m$ 228
(e) Calcite, space group $D_{3d}^6 = R\bar{3}c$ 229
(f) $NaClO_3$, space group $T^4 = P2_13$ 230
6-3 Selection rules and intensities: fundamentals 231
(a) $HCN(C_{\infty v})$, space group $C_{2v}^{20} = Imm2$ 233
(b) $CO_2(D_{\infty h})$, space group $T_h^6 = Pa3$ 234
(c) $CHI_3(C_{3v})$, space group $C_6^6 = P6_3$ 236
(d) NaN_3, space group $D_{3d}^5 = R\bar{3}m$ 237

(e) $CaCO_3$, space group $D_{3d}^6 = R\bar{3}c$ 238
(f) $NaClO_3$, space group $T^4 = P2_13$ 241
6-4 Exciton and phonon methods of analyzing intermolecular couplings
(a) One molecule per unit cell 246
(b) Several molecules per unit cell 248
6-5 Analysis of multiplets and quantitative theories of intermolecular potentials 249
6-6 Combinations and overtones 267
(a) Selection rules 267
(b) Dispersion curves 270
(c) Density of one- and two-phonon states 271
(d) Anharmonicity 273
(e) Examples 280

7 Solid solutions, matrix isolation, and impurity phenomena 289
7-1 Impurity theory 290
7-2 Isotopic solid solutions 298
7-3 Inert gas matrices 303
7-4 Ionic matrices 307
(a) The diatomic species OH^- and CN^- 308
(b) Linear triatomic anions 309
(c) NO_2^- and NO_3^- 310
(d) NH_4^+ and BH_4^- 310
(e) The tetrahedral oxyanions, XO_4^{-n} 311

Appendices
I Character tables, symmetry species of translation, rotation and forms of the dipole ($\mu' = \partial\mu/\partial Q$) and polarizability ($\alpha' = \partial\alpha/\partial Q$) derivatives for the crystallographic point groups 312
II Symmetry analysis in cartesian coordinates 318
III Harmonic oscillator state functions: creation and annihilation operators 322
IV Correlation between point groups, etc. 326
V The 230 space groups 336
VI Factor group theorems 342
VII List of sites in the 230 space groups 345
VIII Reciprocal lattices and space groups; reciprocity of sites and special points 356
IX Brillouin zones and special points 361
X Irreducible representations for some selected Brillouin zone points in non-symmorphic groups 364
XI Evaluation of lattice sums 371
XII Rotational motion in crystals 378

Author index 381
Subject index 385

PREFACE

This book has been written in response to the needs of research workers concerned with the vibrational spectra of molecular crystals—using the adjective "molecular" in the broad sense to cover all crystalline solids in which there are discernible molecular units, whether these be neutral, as in solid O_2, CO_2, C_6H_6 or charged as in NH_4Cl, $CaCO_3$, etc.

As chemists, the authors were enabled to embark upon their own studies by a mature corpus of theory developed by many workers for the analysis of the vibrational spectra of gaseous molecules. Symmetry plays a central role in such theories and, in the case of crystalline solids, the extension of the symmetry theory involves the addition of translational symmetry to that enjoyed by isolated molecules. Fortunately, this additional symmetry almost compensates for the vast increase in the number of degrees of freedom of a crystal as contrasted with a single molecule. Many of the new features of the symmetry theory have been developed by solid state physicists, and an important goal of the present work is to make the representation theory for the space groups available to the structural chemist. We have been distressed by the conflicting demands for completeness and for reasonable brevity, and have attempted to resolve this conflict by contenting ourselves with accounts of important examples, criteria which we hope will alert experimentalists to those situations in the non-symmorphic space groups where degeneracies higher than those to which the molecular chemist is accustomed will appear, and by reference to the useful existing tables of characters and representatives of all the 230 space groups such as those by Love and the splendid monograph by Bradley and Cracknell.

After an introductory chapter, we review the vibrational theory for isolated molecules (Chapter 2) drawing heavily upon the methods developed by E. Bright

Wilson, Jr., and described in the monograph *Molecular Vibrations* by E. B. Wilson, Jr., J. C. Decius, and P. C. Cross (McGraw-Hill, 1955). Then, in Chapter 3, we introduce the concept of translational symmetry in one dimension, proceed to a discussion of real crystals in three dimensions composed of monatomic units (Chapter 4), and before giving an account of the main theme of this work (Chapter 6) provide a chapter devoted to the unique aspects of optical studies of single crystals, which emphasizes the rather detailed information that can be obtained either from the infrared or Raman spectra of oriented samples using polarization.

The spectral effects peculiar to crystals include the phenomena of multiplets arising from a single (possibly degenerate) isolated molecular mode. Models of the potential function used to treat such phenomena have proved only partly successful. We hope that the account given in chapter 6 will stimulate further quantitative work in this direction.

A concluding chapter is devoted to the vibrations of isolated molecules in matrices. Extensive tabular and specialized material is contained in twelve appendices.

Although we have illustrated the important quantitative concepts with examples, no attempt is made to survey the rapidly growing experimental literature *in toto*. The examples chosen were those most familiar to us, and we hope we offend none of our colleagues by omissions.

CHAPTER
ONE
INTRODUCTION

1-1 GAS-PHASE VIBRATIONAL SPECTROSCOPY

The study of vibrational energy levels of gaseous molecules is carried out principally with the aid of infrared absorption and inelastic light scattering measurements (Raman effect). Of these two techniques, infrared absorption is the more venerable, since in the years following the discovery of infrared radiation by Herschel (1800), a great many pioneer researchers established the phenomenon of selective absorption by simple gaseous substances, so that by 1914 it was possible for Bjerrum to discuss the vibrational motion and force constants in carbon dioxide, 14 years prior to the experimental discovery by Raman of the effect which bears his name. Especially in symmetric molecules, the information obtained from the Raman effect is complementary to that available from infrared studies.

The advent of quantum mechanics in the mature form developed by Heisenberg (1925) and Schrödinger (1926) stimulated a thorough theoretical analysis of molecular dynamics, to which particularly distinguished contributions were made by C. Eckart, D. M. Dennison, E. B. Wilson, and H. H. Nielsen, together with their several collaborators, during the decade prior to the Second World War. These studies revealed the theory whereby the complex interaction between vibrational and rotational motions of isolated molecules could be interpreted in the light of the data from the experiments whose spectroscopic resolution was gradually increasing.

Then after the Second World War, commercially manufactured infrared spectrophotometers, and more recently Raman spectrophotometers, became available and were put to use in rapidly increasing numbers. The new instruments offered great convenience and, at least for observations at modest resolution—say 1–10 cm^{-1}—sufficient speed of recording so that the observation of vibrational spectra, not only of pure compounds in the gaseous phase, but also in the liquid and solid phases, became a routine task in chemistry laboratories everywhere.

In gases, the detailed band structure at high resolution, or the band envelope at lower resolution is governed by the rotational transitions which accompany a given vibrational transition. In condensed phases, particularly in solids, the band shape is considerably modified owing to the elimination of free rotational motion. Fine structure arising from new causes frequently appears in the crystal phase and will be discussed in increasing detail since it is one of the principal themes of this book.

A gas phase molecule composed of n atoms possesses $3n - x = f$ degrees of vibrational freedom, where $x = 5$ for a linear molecule with its three translations and two rotations, and $x = 6$ for a nonlinear molecule which has three rotational degrees of freedom. The positions of the vibration–rotation band origins are used to determine the vibrational energy levels which may be characterized by a vector $\mathbf{v} = (v_1, v_2, \ldots, v_f)$ whose components are integers specifying the number of quanta for each degree of vibrational freedom or normal mode of vibration. The simplest dynamical model is that of independent harmonic oscillators which according to quantum mechanics will have energy levels

$$E = E^\circ + h\boldsymbol{\nu}'\mathbf{v} \tag{1-1-1}$$

where $\boldsymbol{\nu}$ is a vector whose components are the frequencies of normal modes. This model is a consequence of the assumption that the potential energy is a quadratic function of the vibrational coordinates. Experiment shows that this is a fairly satisfactory approximation, but that higher order terms in the potential energy do have readily observable effects on the energy levels, typically leading to anharmonic corrections of the order of 1–5 percent on the lowest excited vibrational states, i.e., the fundamental levels in which $|v\rangle = |100\ldots0\rangle$ or $|010\ldots0\rangle$ or $|001\ldots0\rangle$, etc. The theory then indicates (quantum mechanical perturbation methods are used to deal with cubic and quartic terms in the potential energy) that the quantized vibrational energy is given by

$$E = E^\circ + h(\boldsymbol{\nu}'\mathbf{v} + \mathbf{v}'\mathbf{X}\mathbf{v} + \mathbf{1}'\mathbf{X}\mathbf{v}) \tag{1-1-2}$$

in which the elements of $\mathbf{X}$ (a square symmetric matrix) are the anharmonicity constants. The symbol $\mathbf{1}$ is a vector all of whose components are unity. The theory does not attempt to predict the parameters of this expression, but rather to relate $\boldsymbol{\nu}$ and $\mathbf{X}$ to the parameters of the potential energy of the molecule.

As will be shown in Chapter 2, the fact that a number of fundamental modes less than $3n - x$ is frequently observed in gas phase infrared or Raman studies of symmetric molecules may readily be understood with the aid of group theory. In part, this situation arises due to *degeneracy*, i.e., the fact that two or more fundamental frequencies of a molecule will necessarily be identical when the equilibrium structure of the molecule possesses a threefold or higher axis of symmetry. Symmetry also imposes certain restrictions on the forms of vibration which may give rise to dependence of the dipole moment or polarizability upon the normal coordinates, with the consequence that those normal modes which do not satisfy these requirements will be inactive in either (or both) the infrared spectrum or the Raman spectrum, whose mechanism of coupling with the electromagnetic field requires respectively a normal coordinate dependent dipole or polarizability.

Many of the techniques, both experimental and theoretical, which have been developed for the interpretation of gas-phase vibrational spectra can also be employed for the investigation of molecular vibrations in crystals. This book is devoted to and will subsequently review the vibrational theory for free molecules (Chapter 2) and augment this subject with the extensions needed to deal with such vibrations in crystals (Chapters 3, 4, 5). First, however, in the following section, we shall review the various classes of crystals in terms of the types of binding forces, since this subject governs the sort of intermolecular effects which we can expect to observe in spectroscopic experiments.

1-2 TYPES OF CRYSTALLINE SOLIDS: METALLIC, IONIC, COVALENT, VAN DER WAALS, HYDROGEN-BONDED

The several types of crystalline solids are thoroughly described and contrasted in treatises devoted to the chemistry and physics of the solid state.[1-4] We shall only present a short review of the subject which will emphasize those aspects which are of importance to vibrational spectroscopy.

The early infrared spectroscopists recognized the importance of the covalent bond by insisting that for spectroscopic activity due to vibrational motions, it was necessary that the material system be, at least approximately, a system of atomic masses held together by a discrete array of springs. In this array, each spring connects precisely two atoms and is permitted only two modes: longitudinal and transverse to its axis. Since homonuclear diatomic molecules in the gas phase (such as N_2) had no infrared spectrum, it was also argued that in general all elementary molecular systems both in the gas phase and when condensed, would be spectroscopically inactive. On this basis both P_4(g) and diamond were supposed to be transparent in the infrared.

We now know both of these conjectures to be erroneous. Any material system, even one with a continuous distribution of mass (like a drum head), can acquire vibrational energy when disturbed from equilibrium and the rules of spectroscopic activity, as indicated in the previous section, while grounded in the symmetry

1. J. C. Slater, *Quantum Theory of Molecules and Solids,* McGraw-Hill Book Co., New York. Volume 2, 1965, and Volume 3, 1967.
2. F. Seitz, *The Modern Theory of Solids,* McGraw-Hill Book Co., New York, 1940.
3. C. Kittel, *Introduction to Solid State Physics,* John Wiley & Sons, Inc., New York, 1971, 4th ed.
4. W. C. Hamilton and J. A. Ibers, *Hydrogen Bonding in Solids,* W. A. Benjamin, Inc., New York, 1968.

of the system, are more involved than those based on the chemical identity of all nuclei.

Since our fundamental interest here is in molecular vibrations we shall, for the most part, stress those crystals wherein molecular identities are retained, as much as possible. Indeed, it is this principle which enables us to distinguish quantitatively between the several kinds of crystals for it is in fact the extent of *loss* of molecular identity which forms the basis of the classification scheme presented below. In each case, in contrast to the preconceived notions of our predecessors, we shall indicate, qualitatively, the kinds of vibration and spectroscopic activity each may enjoy.

(a) Metallic Crystals

Metallic crystals are examples of extreme loss of molecular identity. These systems consist of crystalline arrays of positively charged nuclei held together by a three-dimensional electron cloud. The fundamental idea underlying the Born–Oppenheimer principle, that is, the extreme mass difference between electrons and stable nucleons, persists in metals. The solution of the many-body Schrödinger equation for such a system yields as the ground state one which is absolutely lower than that of the free atom. Thus, although it costs energy to ionize the collection of nuclei and although each electron and each charged nucleus repels its replicas, there is a net gain in the energy of the system when the charged nuclei are imbedded in the sea of electrons. This gain is nothing more than the same exchange energy which stabilizes the hydrogen molecule compared to its separated atoms, except that in the many-electron problem the gain is relatively larger per atom than in most small covalent systems. Hence the molar heats of sublimation of many metals are frequently greater than the heats of dissociation of small (covalent) molecules.

The elementary excitations of metals correspond to a collective motion of the electron cloud in the field of all the nuclei. The several states of this system, unlike the situation in the isolated atom, are no longer discrete *energy levels* but rather *energy bands*. A band results from the fact that there is a variation in the energy with the phases of the couplings between the many isolated atom wave functions which may be visualized as the basis set from which the final wave functions of the system are formed. (In Chapter 4 we shall discuss in greater detail the construction of energy bands in solids.) In this model, metals are those systems which have one of two qualitative characteristics: (1) Half-filled "conduction" bands or (2) overlapping bands. An electronic current can result from the application of an arbitrarily small electric field. On the other hand, an insulator is characterized by filled bands and a "forbidden" gap, which separates the highest filled and lowest empty bands. There is a close analogy between metals and unsaturated molecular systems, on the one hand, and between insulators and saturated molecules on the other.

For the same reason that metals conduct electricity with ease, they are totally absorbing in the infrared. Still, the vibrations of the lattice of nuclei are capable of interaction with a thermal radiation field; unfortunately, due to the peculiar nature of the electronic excitations, these vibrations are ordinarily inaccessible to study by either absorption or ordinary inelastic light scattering. Nevertheless, they may be investigated by neutron scattering, a technique which has some very important advantages, particularly for the study of the lower energy excitations in crystals, as compared with optical methods with which this book is principally concerned.[5,6] In summary, metals represent systems in which a major chemical change has occurred in producing them from their parent, isolated atoms.

(b) Van der Waals Crystals

Van der Waals crystals represent the other extreme in our scheme of crystal categories. In these crystals, the prototype of which is illustrated by crystals of the rare gases, the first ionization potentials of the free molecules are so high that energy gain due to intermolecular delocalization of the electrons, as in metals, has little chance to take place—the electrons are just too tightly bound to the molecules. Energy banding is thus slight, also the bands are ordinarily filled.

The vibrations of van der Waals crystals are dominated by the internal vibrations of their constituent molecules (see Chapter 6). As opposed to metals, the loss of molecular identity is very small in van der Waals crystals and the internal vibrations, to a good approximation, are little changed upon condensation. There are, however, weak forces between the molecules—otherwise the substances would never crystallize. These forces are indeed of the van der Waals type and hence the name given to these crystals. Because of the existence of these intermolecular forces there are, in addition, lattice vibrations which take place in these solids. These lattice motions can arise either from quasi-translational or quasi-rotational (librational) motions of the molecular units. They are usually low in frequency, quite separated from the internal vibrations. The general rules of infrared or Raman activity mentioned in Section 1-1 govern their appearance just the same as they do the "internal modes." The understanding of these motions is of considerable importance as it can illuminate the subject of intermolecular forces.

(c) Covalent and Ionic Crystals

Covalent and ionic crystals form the middle ground between the two extremes just discussed. In both cases, electronic delocalization takes place, but only

5. B. N. Brockhouse, in *Phonons and Phonon Interactions,* W. A. Benjamin, Inc., New York, 1964, T. A. Bak, Ed., page 221.
6. G. Dolling, in *Proceedings of the International School of Physics,* "Enrico Fermi," Course XV, "Lattice dynamics and intermolecular forces," S. Califano, Ed., Academic Press, New York, 1975, page 176.

moderately. In contrast to metals, where the delocalization is complete, in covalent crystals electron pair bonds are established between each pair of neighboring nuclei. Should one of these nuclei have a considerable electron affinity as compared to its neighbor, the electron pair bond will have proportionate polarity. In extreme cases, *both* electrons are quite localized on one member of the pair; thus the pair is best represented by the configuration M^+X^-. We then say an ionic bond has been formed and the crystal compounded from such pairs is said to be ionic. In such crystals, one ion is then surrounded by a "shell" of nearest neighbors of opposite charge. The number of ions in each shell and its symmetry fall into a small number of categories. Thus we readily distinguish between the rock salt structure, the cesium chloride structure, the zinc blende structure, etc.

We are thus able to conceive of a complete spectrum from the covalent to the ionic cases and there are indeed crystals whose ionicity is difficult to establish, e.g., InAs. *Both* types will necessarily be insulators and of quite considerable binding energy. Elementary crystals composed of monatomic units (such as C, Si, etc.) are the solid analogs of the homonuclear diatomic molecules (H_2, Cl_2, etc.); they are always covalent. On the other hand, as with the heteronuclear diatomics, binary compounds can be either ionic or covalent.

Since both types of crystals contain no molecules as their basic units, the only vibrations manifested by these systems are lattice vibrations. In this respect, these crystals resemble the metals. Symmetry dictates a number of restrictions on the infrared and Raman activity of these vibrations. For example, in monatomic elementary crystals (C, Si, etc.), there is no infrared spectrum due to single excitation of one lattice mode. (This rule is the analog of a similar one for homonuclear diatomic molecules.) In lattices with centers of symmetry, such as that of rock-salt, the analogous Raman process is forbidden; similarly, the double excitation of one lattice mode (an overtone) is Raman active but is not infrared active at all. These selection rules, as well as others, will be derived in Section 4-7.

There is another category of ionic crystal we must cite before leaving this subsection. We refer here to crystals containing *complex ions*, e.g., NH_4Cl, $NaNO_3$, or NH_4NO_3. Structurally and physically these are very similar to the simple ionic crystals, such as NaCl. The complex ions they contain, however, are *polyatomic units*, bonded together by covalent forces. These remain essentially unchanged when these solids are precipitated from solution, or from a melt. As such, these crystals also demonstrate internal modes of vibration, which are probably little changed from what they would be in the gas phase. To date, none has been so studied; they have been studied in solution and in their melts, although always with some difficulty.

(d) Hydrogen-bonded Crystals

There is one other category of crystal which does not conveniently fit the scheme outlined above. The anomalous category is that of hydrogen-bonded crystals. Our other categories were all related, in some way, to the formation of more or less extensive chemical bonds which were either two-center or many-center, with few other possibilities. There is another possibility which is of considerable importance in chemistry and physics: the three-center bond, which involves a proton and two other nuclei besides the bonding electrons.

Hydrogen bonds are well-appreciated in solution and even gas-phase spectroscopy. Carboxylic acid dimers, liquid water, and even such complicated systems as solutions of DNA have had their spectra interpreted quite reasonably on the basis of the existence of bonds such as these.

The importance of the hydrogen bond is at least as great in the solid state. For example, several forms of ice owe their complexity to the various ways H_2O molecules can be H-bonded together to form regular, three-dimensional arrays.

As far as vibrational spectra are concerned, however, there are primarily two respects in which hydrogen-bonded systems differ from other crystals. Often, as in, say, KHF_2, the hydrogen-bonded species may be interpreted as another complex ion, FHF^-, which is convenient to study in the solid state as it is impossible to produce in the gas phase and cannot be obtained simply (and pure) in solution. In the solid state, crystals containing it may be studied with polarized light, so that certain directional properties of the H-bond may be ascertained. However, in this regard, the ion is in no way different from numerous other complex ions.

One major difference between H-bonded crystals and others is demonstrated by any of the ice structures. In the gas phase, water molecules are quite free and in no way polymerized via H-bonds. Thus ice represents a major modification of the isolated molecule. Accordingly H-bonded crystals should be recognized as systems which are considerably different from van der Waals crystals.

The other difference arises in those systems where the H-bonding also takes place in the melt. Again we may use water as our example. In the liquid state, there are a large variety of H-bonds formed. They vary in length, number formed per water molecule, geometry, etc. Furthermore, they are probably all involved in dynamic equilibria. In the several forms of ice, for reasons of symmetry, the possible forms of H-bonds are very much more limited in number. Furthermore, several of the known phase changes, from one form of ice to another, have to do with a change from one H-bonding scheme to another. In some ferroelectric crystals, such as KDP, the ferroelectric effect is a cooperative one in which, under the influence of an external electric field, one distribution of H-bonds can be made to dominate another.

1-3 THE PERTURBATION APPROACH AND THE NATURE OF SOLID STATE EFFECTS

For the one class of crystalline solids described in Section 1-2 which is available for observations in both the gas and crystal phases, namely the van der Waals solids, abundant experimental evidence shows that the changes in the vibrational energy levels on condensation are quite small. In Table 1-1 we have collected some vibrational frequencies which are known for a number of diatomic and simple polyatomic molecules. The conventional notation defining the numbering of the normal modes will be introduced in Chapter 2—particularly in Section 2-6. Table 1-1 indicates that the shifts in frequency on condensation are quite small, usually less than one percent.

This same table also indicates other types of changes. For example, the fundamental of HCl and the bending modes of HCN and CO_2 are all doublets. Two explanations of such splittings are available: (i) a degenerate mode of a gas molecule may be split owing to the influence of a less symmetrical environment in the crystal and (ii) multiplets may also occur, even in the case of non-degenerate vibrations, owing to the fact that the unit cell of the crystal contains two or more molecules whose vibrations then couple with different phase relations and different frequencies in consequence of terms in the potential energy of the crystal of the form $Q_{k1}Q_{k2}$ where Q_{k1} and Q_{k2} are respectively the normal coordinates of the kth vibration mode in molecule 1 and in molecule 2. Translationally equivalent molecules, i.e., equivalent molecules in different unit cells, may also produce couplings, but in this case the different frequencies are not in general observable since the selection rules in good approximation limit the spectroscopically active modes to those which involve in-phase motion for all unit cells.

The sorting out of the two types of splitting has been the subject of a number of investigations which will be described in Chapters 6 and 7; without entering into details at this stage we may remark that a mode known to be non-degenerate which exhibits a multiplet can only be explained by mechanism (ii) and, more generally, the presence of more multiplet components than the degeneracy can also be taken as positive evidence of intermolecular coupling.

These two effects have often been described in the literature as *site* symmetry splittings versus *correlation field* couplings and may also be thought of as static versus dynamic crystal field effects.

Not only are there changes in the vibration frequencies, but also there occur changes in the intensities. The most conspicuous of these are the appearance with finite intensity of certain modes which are inactive in the corresponding gas-phase spectra. All these effects lend themselves to theoretical treatment by the methods of perturbation theory, which simply splits the Hamiltonian of the dynamical system into two parts

$$H = H^0 + H^1$$

where it is assumed that the wave functions and energy levels of the zero-order part (gas phase), H^0, are known. H^1 is the perturbation (first-order part of the Hamiltonian) which arises from the intermolecular potential energy in the crystal and standard methods which have been known almost since the beginnings of quantum mechanics are available for the calculation of the shifts in the energy levels and of the first-order corrections to the wave functions. These first-order solutions are also valuable for the discussion of the intensity changes.

Although the perturbation method is particularly powerful for dealing with

Table 1-1 Fundamental frequencies of some simple molecules in gas and crystalline phases

Molecule	Crystal symmetry	Temperature (°K)	Number of molecules per unit cell	Mode	Frequencies: Crystal Infrared	Frequencies: Crystal Raman	Gas (cm^{-1})	Shift (cm^{-1})
N_2	T^4 (approx. T_h^6)	14	4			2329	2330.7	−1.7
		40				2328		
O_2	D_{3d}^5		1			1553	1556.2	−3.2
CO	T^4		4		2138.3	2240?	2143.3	−5.0
HCl	C_{2v}^{12}		2		2705, 2748		2886	−181, −138
CO_2	T_h^6		4	ν_1		1277	1285	−8
				$2\nu_2$		1385	1388	−3
				ν_2	653, 660		667	−12, −7
				ν_3	2344		2349	−5
N_2O	T^4		4	ν_1	1293	1292	1285	+8
				ν_2	591		589	+2
				ν_3	2238	2229	2224	+14, +5
HCN	C_{2v}^{20}		1	ν_1	2097		2097?	0
				ν_2	828, 838		713	+115, +125
				ν_3	3132		3311	−180

problems of this type where the actual shifts in energy levels are small compared with the spacings between the unperturbed levels, it is also possible to formulate the problem of molecular vibrations in crystals in such a way that exact solutions are obtained provided one limits the treatment to the approximation in which the molecules are treated as coupled harmonic oscillators. These topics will be fully developed in Chapter 6.

At this point it should be remarked that there are important cases, namely ionic and covalent crystals, in which the gas phase cannot be used as the reference, i.e., zero-order state. The vibrational spectrum of ammonium ions for example has not been observed in the gas state, and that of diamond "molecules" certainly will never be observed in such a condition. Nevertheless important information can of course be obtained from the spectra of such types of crystals, and comparisons of NH_4X where X = various halides, and moreover of crystalline phase transitions of a given ammonium salt have proved interesting and have shed new light on problems of crystal structure. Indeed the opportunity which is available to study infrared and Raman spectra of several classes of crystals over a wide range of temperatures, particularly down to a few degrees Kelvin, is one of the especially interesting aspects of such researches.

In all types of molecular crystals, there will be *external* modes of translational (T) and rotational (R) types in addition to the internal modes which we designate by ν_k. Examples for the molecular ionic crystals $NaNO_3$ and $CaCO_3$ are given in Table 1-2. In this case, there is a wide energy gap between the internal and external modes which greatly simplifies the characterization of the vibrations. The internal modes numbered ν_k, $k = 1$ through 4, are essentially those of the molecular units, NO_3^-, CO_3^{2-} (and also of ClO_3^- though the external modes are not given for $NaClO_3$ in Table 1-2). Note that ν_3 and ν_4 give rise to several multiplet components in these three cases.

In concluding this section it must be remarked that there are a few exceptions to the generalization that the vibrational energy differences between gas and

Table 1-2 Internal and external frequencies for some ionic crystals (all in cm^{-1})

	$NaClO_3(T^4 = P2_13)$		$NaNO_3(D_{3d}^6 = R\bar{3}c)$		$CaCO_3(D_{3d}^6 = R\bar{3}c)$	
	IR	R	IR	R	IR	R
ν_1	937	936, 937		1068		1088
ν_2	624	618, 623	838		871	
ν_3	966, 987	957, 965, 988	1353	1385	1410	1432
ν_4	482	482, 489	727	724	712	714
R_-				185		283
T_-				98		156
$T_\pm$			205		303	
R_-			51		92	
$T_\pm$			204		297	
T_+			175		223	
R_-			87		102	

crystal are small and may be regarded as small perturbations. These include (i) hydrogen bonding, where frequencies shifts may be as large as 1000 cm^{-1}; (ii) dimerization or polymerization; and (iii) such obvious changes in molecular structure as occur, for example, when neutral, covalent molecules such as N_2O_5, PCl_5, and PBr_5 are transformed respectively into ionic lattices composed respectively of $NO_2^+NO_3^-$, $PCl_4^+PCl_6^-$, and $PBr_4^+Br^-$. Spectroscopic methods are particularly rapid and convenient for the detection of such phenomena although in most cases they were preceded by X-ray crystallographic studies.

1-4 SUMMARY OF MOTIVATION FOR CRYSTAL STUDIES

The contrast of Sections 1-1 and 1-3, together with the types of solid state spectra discussed in Section 1-2 immediately suggests a large number of possible investigations, most of which have had a distinguished history and will be thoroughly discussed, particularly in Chapters 5–7. A few of these will now be enumerated, for the purpose of summarizing the motivation of studies of vibrational spectra of molecular crystals.

(a) Studies of Intermolecular Forces

We have already indicated (Section 1-2 (b)) that the study of the lattice vibrations of van der Waals crystals affords the opportunity of understanding weak intermolecular forces. Since the lattice motions in these systems correspond to whole-molecule translations and librations, the lattice vibrational energies correspond to the allowed energies dictated by the intermolecular force field. The translatory lattice modes thus offer detailed information on the radial dependence of the intermolecular potential energy function and the librations furnish similar information concerning the angular parts of that function. Thus, in principle, these studies can yield rather complete information about the intermolecular potential energy surface. With the recent advent of commercial instrumentation for the far infrared, as well as laser Raman equipment, quantitative investigations toward these ends are only just beginning.

Other kinds of intermolecular potentials are those which couple internal vibrations in van der Waals crystals, as well as in crystals containing complex ions. As indicated in Section 1-3, these potentials are responsible for shifts, removal of degeneracies and further multiplicities of spectroscopic transitions in the infrared. Much effort has been devoted to this subject during the past decade and we shall review the major results.

(b) Studies of Molecular Ions

Solid state spectroscopy of crystals containing complex ions offers the experimentalist one of the few opportunities to study such chemical systems. In general, these species decompose long before they sublime, so vapor-phase studies are precluded. Because of their ionic nature, they are insoluble in non-polar solvents such as CCl_4 and CS_2, the standard infrared solvents, and although most are water soluble, the infamous opacity of water prevents their investigation in that medium. On the other hand, the same solubility in water facilitates their study via the Raman effect. However, since most of these ions are centro-symmetric, the Raman and infrared spectra are mutually exclusive; thus the Raman spectra do not provide us with a complete set of vibrational frequencies. In contrast, the infrared spectrum of the crystal is relatively easy to obtain. Most complex ionic crystals are relatively easy to grow from aqueous solution, although such crystals are often hydrates of the desired compound. When anhydrous forms are absolutely necessary, modern methods of growth from the melt are quite common.

The desirable size of such crystals varies widely. In general, for infrared work, it is currently possible to work with samples as small as 1×5 mm^2. On the other hand, the thicknesses may have to be as little as 2–20 μm in order to achieve any light transmission. Indeed, as we shall demonstrate later (Section 5-5) true transmission spectra can only be observed with samples having thicknesses considerably less than the radiation wavelength in the medium, so that an analysis of reflection spectra is often a valuable adjunct to transmission. For Raman work, 1 cm^3 crystals are quite convenient (although with lasers, they can be 1 mm^3). Since many complex ion crystals are naturally occurring in these sizes as mineral specimens, they have been the subject of much investigation and a bibliography of such studies is available. Such samples must be viewed with caution, however, as they are notoriously impure, aesthetic though they may be. The impurities can imperil the infrared results and they can devastate Raman investigations as they are often fluorescent and the strongest scattered signal recorded may be due more to the impurity than to the sample.

The vibrational information resulting from these studies can, in the usual manner, help in determining the structure of the complex ion and the orientation in the crystal, although, as with most spectroscopic studies, this is usually better obtained by means of X-ray diffraction techniques.

(c) Special Crystal Problems

i. Rotation of molecules in solids Since the early 1930s, it has been repeatedly suggested that high-temperature, usually cubic crystalline phases of symmetrical molecules were the consequence of the fact that in these phases the molecules had achieved rotational (but not translational) freedom. Much thermodynamic argument has been marshalled on behalf of this conjecture, most of which simply restates the basic datum that most of the heat of vaporization of these materials has already been supplied to them by the time they reach their high temperature phases. Indeed, a simple thermodynamic parameter can be used to select, without any knowledge of crystal structure, those systems which fall into this category. One has only to evaluate the ratio of the normal boiling to the normal melting point of the substance; if this parameter is close to unity (1.1–1.3), the crystal is very likely one which has a high temperature, cubic phase and is called "plastic."

Unfortunately, among van der Waals crystals of this category, vibrational spectroscopy has failed to reveal any evidence whatsoever of regular, free rotation by the constituent molecules, with one exception: solid hydrogen. In all other systems, the high-temperature transition to the cubic phase seems to be due instead to an order-disorder transition. An example of such a system is solid CH_4. The crystal structure of the high-temperature phase of this compound is still not known with certainty but it can nevertheless serve for an example of the type of order-disorder transition which most likely takes place in plastic crystals. We may imagine that in this phase, no matter what the exact structure may be, one CH bond of each molecule is parallel to some fixed axis of the crystal but the others are organized randomly about a small set of equilibrium configurations. In such a state it is easily seen that the crystal becomes a three-dimensional gear box. As the temperature is raised, the gears begin to oscillate about their equilibrium orientations, the amplitudes of such oscillations increasing with temperature. Eventually one or another gear surmounts the barrier which isolates different equilibrium orientations and "snaps" into another. At a still higher temperature a greater number of gears "snap" until the lower temperature phase is no longer stable and the entire system undergoes a change to the new set of equilibrium orientations, and thus does so cooperatively. Transitions such as these can explain all the data that have led some crystallographers, magnetic resonance spectroscopists, and thermodynamicists to conceive of "free rotation" for these systems. For example, quantitative calculations based on a model such as the one presented here has accounted for the NMR line-narrowing observed in solid CH_4 at $\sim 80°K$.[7] The model also accounts for the fact that no rotational fine structure is observed in the spectrum of any plastic crystal other than solid H_2 and it is this desideratum of the spectrum upon which the concept of free rotation is either established or not.

On the other hand, *hindered rotation* of some complex ion impurities or of molecular impurities in other ionic or van der Waals crystals is well established. Examples are those of NO_2^- in alkali halides or CH_4 in solid Kr. These studies have illustrated much rotational fine structure which has been interpreted so as to yield information on the barriers hindering free rotation in systems. As such, this subject is another source of intermolecular potential energy information.

ii. Order-disorder transitions, ferroelectricity In the previous subsection we have discussed one type of order-disorder transition amenable to some investigation by infrared or Raman spectroscopy. In Section 1-2(d) we mentioned another which is intimately connected with the appearance of ferroelectricity in certain hydrogen-bonded crystals, such as KDP. Vibrational spectroscopy has been of some value in locating these transitions and in some instances, in characterizing them.

7. M. Bloom and H. S. Sandhu, *Can. J. Phys.*, **40**, 289 (1962).

iii. Acentric crystals It has come to be appreciated relatively recently that Raman scattering by crystals *lacking* centers of symmetry is quite different from that due to centered crystals. The subject is an important one, particularly now that laser Raman spectroscopy is coming to the fore, in that the directional aspects of the Raman effect can now be more accurately and quickly carried out. We shall therefore treat this subject in some detail, in Section 5-6(c).

iv. Energy transfer and relaxation We close this section with some mention of what may be the most important of all problems in solid state, vibrational spectroscopy. The question is: how does a crystal dispose of its vibrational excitation? The various channels for relaxation available to an excited molecule need to be discussed, and quantitatively, for it is of course the relaxation, or equilibration, process which determines band shapes and widths in solids, just as in any other phase.

In the gas phase, the fundamental width of a vibrational transition is determined by its rotational envelope. Each vibration–rotation line is, to a good approximation, Lorentzian, the damping constant arising from rotational–translational relaxation. We have already observed, however, that in the solid state, rotation is usually quenched. Even so, solid-state spectral lines are normally broad: they are at least as wide as the instrumental slit. Where does this width arise? We shall discuss some of these questions further in Section 6-6.

We thus see that there is abundant motivation for vibrational spectroscopic study of molecular crystals. In the following chapters we shall first present the theoretical basis of these studies, followed by some experimental examples, in considerable detail.

CHAPTER

TWO

SYMMETRY AND NORMAL VIBRATIONS OF ISOLATED MOLECULES

In this chapter we shall present, in a very much condensed fashion, some of the salient features of the topic announced in the chapter heading. It is not our purpose here to give such detail as will enable the reader to become expert in vibrational analysis of isolated molecules. For this purpose, there are a number of monographs available.[1] Instead, it is our desire to remind the reader familiar with this topic of those elements of the subject we shall be putting to use when we come to discuss the vibrations of monatomic and molecular lattices. There are quite a few aspects of molecular vibrations which have close correspondencies in crystal vibrations and we shall stress these. We shall state, but not derive, certain theorems, lemmas, and rules. Unless he desires to pursue the subject in all of its mathematical elegance and rigor, the theoretical apparatus offered to the reader will be self-contained; that is to say, there will be no necessity to make reference to other, more exhaustive, treatments in order to continue with the subject matter of this book.

2-1 SYMMETRY POINT GROUPS

Most isolated molecules are characterized by having one equilibrium configuration. There are some which have several equivalent equilibrium configurations, with relative ease of interconversion among each. An example of this category is

1. One example is E. B. Wilson, Jr., J. C. Decius, and P. C. Cross, *Molecular Vibrations*, McGraw-Hill Book Co., New York, 1955, hereinafter referred to as WDC.

ammonia, in which there are two equilibrium configurations; this substance has an equal number of both types at all times, in rapid dynamic equilibrium. In this interconversion, an ammonia molecule must either overcome or penetrate a low potential barrier which opposes the process.

There are some molecules, e.g., dimethylacetylene, in which the height of the analogous barrier (in this case, to free internal rotation) is vanishingly small. When this happens, the molecule has *no* equilibrium configuration with respect to this coordinate. There are a few other examples of this phenomenon. For the most part, however, isolated molecules belong to the first category—they possess equilibrium configurations.

By a configuration, we refer to a three-dimensional arrangement of all *nuclei.* In any problem of interest in our subject, fundamentally we must solve the appropriate Schrödinger equation, $H\psi = E\psi$, for the system of nuclei and electrons under consideration, be it atom, molecule, or crystal. The Hamiltonian operator H consists in general of two parts: a kinetic energy operator and the total potential energy of the system, which is itself made of two parts—the coulombic terms between nuclei and electrons plus internuclear repulsion terms. Both the operator H and the wave function ψ are functions of the coordinates chosen for their expression.

Hence, we shall restrict our interest to the set of all orthogonal transformations which leave the *nuclear framework* unchanged, i.e., those which preserve all internuclear distances and bring the framework into coincidence with itself. The transformations which can accomplish this differ from one molecule to another but are always of three types—rotation about some axis, reflection in some plane, or a combination of the first two. There can be more than one of each of these possible transformations appropriate to a given molecule. If so, they all intersect in at least one point. Those points of intersection always remain fixed in space as we carry out the several transformations admitted by each molecule. Did they not, the full meaning of coincidence would not be achieved; neither could we guarantee the invariance of the complete Hamiltonian under the transformation.

In any molecule, the set of transformation thus defined is found to form a *group*[2] which, since it leaves at least one point fixed in space, is called a *point group.* Each transformation, sometimes called a "covering operation," can be represented by a matrix, the basis of which is a $3n$-dimensional vector. The components of this vector are the cartesian (or other) coordinates of the n-nuclei of the molecule in its equilibrium configuration, relative to an arbitrary, space-fixed orthogonal system of axes. Thus each matrix tells in detail how the coordinates of each nucleus are transformed such that each is brought into coincidence with itself or with another, identical to it in the untransformed molecule—a *symmetrically equivalent nucleus.* It is a necessary requirement upon a point group operation, that it accomplish such coincidence for the entire molecule. That is, *all* nuclei within *each* set of symmetrically equivalent nuclei must be permuted by the operation and no non-identical nuclei may be interchanged.

2. A. Speiser, *Die Theorie der Gruppen von Endlichen Ordnung,* Dover, New York, 1943, page 85.

It is not necessary, however, to work with the matrix representation of each transformation. Rather, it is more convenient to visualize each covering operation as a rotation or reflection of the molecule in a purely conceptual way. For this purpose, we can associate with each transformation a *symmetry element* which symbolizes the fact that a transformation matrix can be derived for the particular case. As an example, the 1,1-dibromoethylene molecule is one which has two sets of symmetrically equivalent nuclei—the two gem-protons and the two gem-bromine nuclei. The carbon nuclei are not symmetrically equivalent. The coordinates of one member of each set are brought into coincidence with themselves or with those of the other by only four operations. One of these operations is easily seen to be that corresponding to the reflection of the molecule in its own plane. Another corresponds to reflection in a plane perpendicular to the first one, the intersection of the two being the carbon–carbon bond of the molecule. Indeed, rotation of the molecule by 180° about an axis through that bond is a third operation of this type. Finally, another operation which is trivial but always appropriate, is the complete rotation of the molecule about any axis. The latter operation is called the *identity* and is usually symbolized E (for *einheit*). All operations can be represented by matrices defined in the manner stated above. Each matrix may be in block form in which each block relates the "old" cartesians of a particular atom to "new" coordinates, i.e., after the transformation. All other blocks will be null matrices. If we define the z-axis of the molecule as coincident with the C=C bond, the molecule lying in the yz plane, and write the components of the basis serially by atom and in the order $(x_1, y_1, z_1, \ldots)$, each diagonal block of E will be the unit matrix whereas each such block of the reflection operation will be of the form

$$\mathrm{R}(\sigma_{yz}) = \begin{pmatrix} -1 & 0 & 0 \\ 0 & 1 & 0 \\ 0 & 0 & 1 \end{pmatrix} \qquad (2\text{-}1\text{-}1)$$

The full matrix so constructed is denoted by $\boldsymbol{\sigma}_{yz}$ in which the symbol $\boldsymbol{\sigma}$ is used to represent the reflection operation and the subscript yz refers to the fact that the reflection is in the (vertical) plane of the molecule. It is readily recognized that writing down the full 18×18 matrices is tiresome and, as indicated previously, unnecessary. Instead, we simply visualize the *process* of reflection, for example, associate with it the plane of the reflection, and denote that *symmetry element*: σ_{yz}.

To illustrate these matters with other types of operations which do in fact exchange equivalent nuclei, we consider the water molecule. Here there is only one set of symmetrically equivalent nuclei. The two protons are brought into coincidence with themselves by two operations and are exchanged by another two. The first two are again the identity (E) and the reflection in the molecular plane (σ_{yz}) (all planar molecules have these two symmetry elements in common). Each of the four transformation matrices is 9×9. As before, E and σ_{yz} have only 3×3 blocks along the principal diagonals.

Since they exchange proton coordinates and not those of the oxygen nucleus, the remaining transformations have two off-diagonal blocks and one diagonal block. Thus each full 9×9 matrix has the form

$$\mathbf{Q} = \begin{pmatrix} 0 & \mathbf{R} & 0 \\ \mathbf{R} & 0 & 0 \\ 0 & 0 & \mathbf{R} \end{pmatrix} \tag{2-1-2}$$

where $\mathbf{R}$ is a 3×3 block whose form is different for each operation. One of the two operations turns out to be another reflection, which we designate as (σ_{xz}); its form is therefore similar to that characteristic of σ_{yz}. It is

$$\mathbf{R}(\sigma_{xz}) = \begin{pmatrix} 1 & 0 & 0 \\ 0 & -1 & 0 \\ 0 & 0 & 1 \end{pmatrix} \tag{2-1-3}$$

The remaining operation is in fact a rotation and is different in form from $\mathbf{R}(\sigma_{yz})$ and $\mathbf{R}(\sigma_{xz})$:

$$\mathbf{R}(C_2^z) = \begin{pmatrix} -1 & 0 & 0 \\ 0 & -1 & 0 \\ 0 & 0 & 1 \end{pmatrix} \tag{2-1-4}$$

The planes of symmetry σ_{yz} and σ_{xz}, whose intersection is the C_2^z axis, are frequently designated as "vertical planes," σ_v and $\sigma_{v'}$. The $\mathbf{R}(C_2^z)$ operation rotates the molecule by π radians to interchange the symmetrically equivalent nuclei. The symmetry element C_2 is then said to be a twofold axis of symmetry and in this case it is the z-axis of the molecule.

With this distinction between symmetry transformations and symmetry elements now understood, we shall proceed to list the various symmetry elements out of which the several point groups are formed. Frequently, as part of the definition of some symmetry element, we shall state an identity between one symmetry element and the product of several others. These are group theoretical statements which may be verified either by performing the implied matrix multiplication or by successively carrying out the implied operations in a conceptual way. The several operations are as follows:

C_n: An n-fold rotation axis, where n is always integral. $C_n^n = E$. There can be more than one of these in the molecule. For example, in C_2H_4 there are three C_2 axes, mutually perpendicular to each other. In C_2H_6 there is a C_3-axis along the CC bond and three C_2 axes perpendicular to it and with 120° between each. In cubane, there are four C_3 axes, etc. The operation C_n^k corresponds to the rotation of the cartesian coordinates by the angle $2\pi k/n$ about the axis of symmetry.

In terms of matrix operators, the effect of C_n on the cartesian coordinates of any one member of a set of equivalent nuclei is given by the (block) matrix (here z is chosen as the axis of C_n)

$$\begin{pmatrix} \cos 2\pi/n & -\sin 2\pi/n & 0 \\ \sin 2\pi/n & \cos 2\pi/n & 0 \\ 0 & 0 & 1 \end{pmatrix}$$

σ: A reflection in *some* plane. $\sigma^2 = E$, always. If σ is perpendicular to C_m, where m is the greatest n, attach the subscript "h" to σ. If σ contains C_m, call it σ_v.

S_n: A rotation-reflection, $S_n = C_n\sigma_\perp = \sigma_\perp C_n$, where $\sigma_\perp$ is the plane perpendicular to C_n.

When n is even, $S_n^n = E$.
When n is odd, $S_n^n = \sigma_\perp$.

The matrix operator for S_n is

$$\begin{pmatrix} \cos 2\pi/n & -\sin 2\pi/n & 0 \\ \sin 2\pi/n & \cos 2\pi/n & 0 \\ 0 & 0 & -1 \end{pmatrix}$$

There is one important special case, called the inversion:

$$S_2 = C_2\sigma_h = i$$

These operations or symmetry elements may be compounded with one another in various ways to form the point groups. For example, the cyclic groups C_n each consist of the n pure rotational operations $C_n^1, C_n^2, \ldots, C_n^{n-1}, C_n^n = E$. Thus the order of a cyclic group is n.* The dihedral groups, D_n, are of order $2n$ and consist of the cyclic group C_n augmented by n C_2 operations each perpendicular to the principal symmetry axis C_n. Thus partially rotated C_2H_6 belongs to the point group D_3.

The groups C_{nh} are formed by joining the reflection σ_h to the groups C_n. The new element $S_n = C_n\sigma_h$ is called an alternating, or rotation–reflection, axis. This is demonstrated in Fig. 2-1 (page 20) with the group C_{3h}.

There are some instances where the S_n operations generate their own groups. For example, the group S_2 (often called C_i) consists simply of E and $S_2 = i$. There are two other cases: S_6, with the elements S_6, $S_6^2 = C_3$, $S_6^3 = i$, $S_6^4 = C_3^2$, S_6^5 and $S_6^6 = E$, and S_4 with elements S_4, $S_4^2 = C_2$, S_4^3 and $S_4^4 = E$. Note that there is no such group as S_3.

The groups C_{nv} are generated by augmenting the cyclic groups C_n with n vertical planes of symmetry *through* the axis C_n.

Four more groups are obtained by adjoining the horizontal reflection σ_h to the dihedral groups D_n. The new groups are called D_{nh} and are of order $4n$. Besides

* In this section we shall restrict our attention to the point groups which can occur in crystals. This restriction allows only the values of $n = 1, 2, 3, 4$, and 6 for C_n and S_n operations and as we shall see limits the total number of point groups to 32. Free molecules can also exhibit other symmetries, for example linear molecules (discussed at the end of this section), and such cases as D_{4d}, I_h, etc.

Figure 2-1 The group C_{3h}

the $2n$ elements of D_n and the n rotation–reflections $S_n^k = C_n^k \cdot \sigma_h$, they also consist of the reflections in the vertical reflection planes $\sigma_v', \sigma_v'', \sigma_v'''$, etc. That these are generated may be appreciated by means of the identity $C_2 \cdot \sigma_h = \sigma_v$ where the equality is interpreted to mean "has the same effect as." Each σ_v plane *contains* a dihedral (C_2) axis.

In contrast to this last situation, there are two other groups of order $4n$ which are formed by the addition of a *vertical* reflection plane to the groups D_n, such that this plane *bisects* the angle between two neighboring dihedral axes. As soon as one such plane is added, the twofold rotations generate a total of $n - 1$ others. Such groups are generally called D_{nd}, but there are only two possible: $n = 2$ or 3. The $4n$ operations consist of the $2n$ rotations of D_n and the n vertical planes (now called σ_d to distinguish them from σ_v's, which *contain* dihedral axes) plus n rotation–reflections about the principal axis, each denoted S_{2n}^{2k+1}. These groups may be confused with the groups S_n; to appreciate the difference, the reader should contrast the two possible structures of S_4N_4 illustrated in Fig. 2-2. One of these structures (Fig. 2-2(*a*)) is supported by diffraction results whereas the other corresponds to the group S_4.

Finally, we wish to discuss the groups defined by the covering operations of the regular polyhedra: the tetrahedron, the cube, and the octahedron. Euler's theorem also admits of the dodecahedron and the icosahedron; however, since the group which describes both of these figures has six fivefold axes, they are not permitted crystallographically. We shall therefore not discuss them. The remaining polyhedra have considerable crystallographic importance.

Tetrahedra, cubes, and octahedra are closely related in their symmetry. We begin by restricting the permitted elements to pure rotations (no reflections) and further only to those rotations which are threefold or less. In a limited way, we can thus describe certain tetrahedral and cubic molecules, e.g., partially rotated neopentane. Such a molecule has only threefold and twofold axes, four of the former along each C—C bond, and three of the latter, each of which bisects a CCC bond angle. When the carbon framework of this molecule is inscribed in a cube, the threefold axes parallel the body diagonals of the cube and the twofold axes bisect the faces. The complete set of the operations of this group, called T, may be listed as E, $C_3(4)$, $C_3^2(4)$ and $C_2(3)$. Thus the order of $T = 12$.

The simplest octahedral group is similarly generated by two-, three-, *and* four-fold rotation axes. The fourfold axes bisect the faces of the cube, the threefold axes are as before with the T group, and the twofold axes bisect opposite edges of a cube. The result is the group O: E, $C_2(6)$, $C_3(8)$, $C_4(6)$, $C_4^2(3)$ of order 24.

Each of the symbols used in the last two examples of point groups denotes a *class* of operations, and the number in parentheses indicates the number of group elements in the class. The class of the operation R_i is the set R_j where $R_j = X^{-1}R_iX$ and X is any element of the group. If R_k is a group operation not in the class of R_i, a second set can be generated by $X^{-1}R_kX$, and repetition of this process will decompose the entire group into mutually exclusive classes.

By adjoining T or O with the inversion, the new groups T_h and O_h are respectively obtained, each having twice the order as their non-centric subgroups. The group O_h demonstrates the full symmetry of either the cube or an octahedron.

In forming T_h out of $T \times i$ reflection planes parallel to cube faces are created ($C_2 \cdot i = \sigma_h$). The last identity suggests an alternative method of generating T_h: T may be augmented with reflection planes parallel to the faces of the cube.

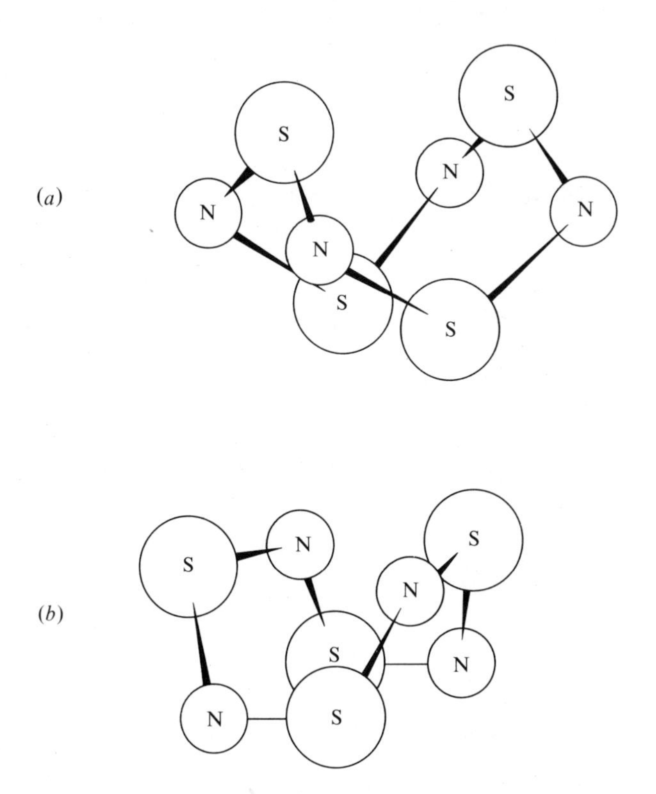

Figure 2-2 Two possible structures for the S_4N_4 molecule: (*a*) D_{2d}; (*b*) S_4

A final group is constructed by augmenting T with reflection planes which contain opposite edges of the cube. It is called T_d and exhibits the full symmetry of a tetradedral molecule, such as CH_4.

In Table 2-1, we list all 32 point groups. Each group is identified using both the Schönflies as well as the Maugin–Hermann (or International) notation. The first system, in common usage among spectroscopists, has been the one used in this text. The International System is more commonly used by

Table 2-1

Crystal class	Schönflies symbol (order)	International
Triclinic	C_1 (1)	1
	$C_i \equiv S_2$ (2)	$\bar{1}$
Monoclinic	C_{2h} (4)	$2/m$
	C_2 (2)	2
	C_s (2)	m
Orthorhombic	D_{2h} (8)	mmm
	$D_2 \equiv V$ (4)	222
	C_{2v} (4)	$mm2$
Trigonal	D_{3d} (12)	$\bar{3}m$
	D_3 (6)	32
	C_{3v} (6)	$3m$
	S_6 (6) $\equiv C_{3i}$ (6)	$\bar{3}$
	C_3 (3)	3
Tetragonal	D_{4h} (16)	$4/mmm$
	D_4 (8)	422
	C_{4v} (8)	$4mm$
	C_{4h} (8)	$4/m$
	D_{2d} (8)	$\bar{4}2m$
	S_4 (4)	$\bar{4}$
	C_4 (4)	4
Hexagonal	D_{6h} (24)	$6/mmm$
	D_6 (12)	622
	C_{6v} (12)	$6mm$
	C_{6h} (12)	$6/m$
	D_{3h} (12)	$\bar{6}m2$
	C_{3h} (6)	$\bar{6}$
	C_6 (6)	6
Cubic	O_h (48)	$m3m$
	O (24)	432
	T_d (24)	$\bar{4}3m$
	T_h (24)	$m3$
	T (12)	23

crystallographers. Inasmuch as our purpose is to serve both groups, we shall introduce the latter system.

The cyclic groups are denoted by their order. A symmetry plane is denoted by m. If the plane is perpendicular to a symmetry axis, the order of the axis and the symbol m are separated by a solidus. Thus, for example C_{2v} is $m2$ but C_{2h} is $2/m$. The presence of a center of symmetry is indicated by a bar over the group order symbol. Thus S_2, S_4, and S_6 become $\bar{1}$, $\bar{4}$, and $\bar{3}$ (not $\bar{6}$). If the index n of the group $\bar{n}$ is twice an odd integer, the group contains the plane m and has already been included in the category of n/m. The reader should carefully note that because of certain arbitrary conventions, in the International notation $D_3 = 32$ while $T = 23$ and $C_{3v} = 3m$ while $T_h = m3$.

This completes the enumeration of the finite point groups. In addition to these, molecules also assume the symmetries of two "continuous groups," denoted $C_{\infty v}$ and $D_{\infty h}$. Continuous groups admit rotations by any angle, $0 < \varphi < 2\pi$. In fact, if the group operations are restricted to such rotations about a single axis, we generate the group C_{∞}, which is called the two-dimensional pure rotation group. These rotations are carried out about some fixed (say, z) axis but the z coordinates of all nuclei are clamped. The matrix form of each operation, R, is therefore two-dimensional:

$$R = \begin{pmatrix} \cos\varphi & -\sin\varphi \\ \sin\varphi & \cos\varphi \end{pmatrix} \tag{2-1-5}$$

but φ is a continuous variable in the range $0 \leq \varphi < 2\pi$. Each time R is applied, the coordinates of the symmetrically equivalent nuclei are brought into coincidence, as with any symmetry operation. The difference is that since φ can be arbitrarily small, R can be repeated indefinitely. Thus the order of this group is infinite.

$C_{\infty v}$ is simply C_{∞} augmented by σ. The presence of the C_{∞} operations implies that the class population of σ is infinite. A molecular example is CO.

$D_{\infty h}$ may be formed from C_{∞} in a variety of ways—by adding i or σ_h or an infinite class of dihedral C_2 axes, as the group symbol suggests. The cyanogen molecule is an example.

2-2 REPRESENTATIONS OF POINT GROUPS

We have already seen that we may represent an individual covering operation of a molecule with a matrix, the basis of which was a $3n$-dimensional vector, whose components were the cartesian coordinates of the n-nuclei of the molecule. The set of such matrices, one for each covering operation in the point group, is called a *representation,* and in this case the basis of the representation was the equilibrium configuration of the molecule.

In that we are dealing with vibrations and vibrational spectroscopy, we must also consider configurations of the molecule other than that it possesses at equilibrium, although it is true that we usually restrict our interest to only small departures from that configuration. Any displaced configuration of a molecule (away from equilibrium) can form a basis of a representation of the group. For example, consider a non-symmetrical water molecule, say with one of its O—H bonds lengthened and the other shortened. (This motion must be accompanied by a displacement of the oxygen atom transverse to the bisector of the equilibrium bond angle in order to keep the center of gravity fixed.) Let us see what happens to this figure as we apply the several operations of C_{2v}, the point group appropriate to water.

E and σ_v (the reflection in the plane of the molecule) of course change nothing. The C_2 operation interchanges the positions of the long and short bonds —it is therefore an "antisymmetry operation." σ_v' does the same. Let us now represent each of these operations by a matrix which gives us this same information. One possible form of this representation is:

E	C_2	σ_v	σ_v'
$+1$	-1	$+1$	-1

where the $+1$ matrices denote no change, and the -1 matrices are associated with those operations which interchange the long and short O—H bonds. If we multiply the several matrices thus formed we find that $E \cdot R = R$, where R is any one of the matrices and E is the unit matrix, while $C_2 \cdot \sigma_v = \sigma_v \cdot C_2 = \sigma_v'$; $C_2 \cdot \sigma_v' = \sigma_v' \cdot C_2 = \sigma_v$; and $\sigma_v \cdot \sigma_v' = \sigma_v' \cdot \sigma_v = C_2$, where in these last identities, C_2, σ_v, and σ_v' are the matrices just tabulated. Thus we reproduce the multiplication table of the abstract group C_{2v}. We therefore have a *matrix representation*, suitable to the description of a non-equilibrium H_2O molecule, or a "displacement coordinate."

As a second, slightly more illuminating example, we consider some deformations of a square molecule, which we simply call X_4. The symmetry group is D_{4h}. Consider the deformations illustrated in Fig. 2-3, (*a*) and (*b*).

Test these with respect to the effects of some of the operations of D_{4h}. Clearly

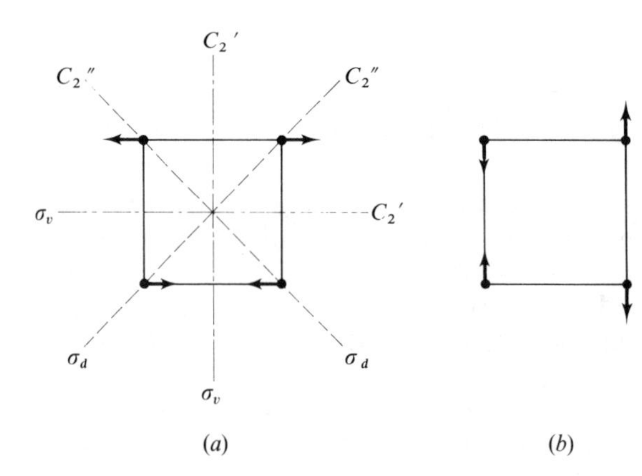

Figure 2-3 Orthogonal components of a degenerate deformation mode of a square molecule, X_4. (*a*) and (*b*). One choice of the basis.

each is invariant under E, σ_h, and, in each case, *one* of the σ_v's and C_2's. Note that neither (a) nor (b) is invariant under C_4^n, S_4^n, i, σ_d, or C_2''. Finally we observe that $C_4^3(a) = C_4^{-1}(a) = (b)$ and herein is the clue to the problem. Neither figure by itself is a suitable basis for a representation of D_{4h} but together they form the orthogonal components of a *two-dimensional basis* that is suitable. To demonstrate this, we conceive of (a, b) as a column vector, apply the operations of D_{4h} to it and note the results.

$$C_4\begin{pmatrix} a \\ b \end{pmatrix} = \begin{pmatrix} -b \\ a \end{pmatrix} \tag{2-2-1}$$

To achieve this result,

$$C_4 = \begin{pmatrix} 0 & -1 \\ 1 & 0 \end{pmatrix} \quad \text{or} \quad \begin{pmatrix} 0 & 1 \\ -1 & 0 \end{pmatrix} \tag{2-2-2}$$

In the same way we find

$$C_2 = \begin{pmatrix} -1 & 0 \\ 0 & -1 \end{pmatrix} \tag{2-2-3}$$

$$C_2'(x) = \begin{pmatrix} -1 & 0 \\ 0 & 1 \end{pmatrix} \quad \text{or} \quad \begin{pmatrix} 1 & 0 \\ 0 & -1 \end{pmatrix} \tag{2-2-4}$$

$$C_2''(xy) = \begin{pmatrix} 0 & 1 \\ 1 & 0 \end{pmatrix} \quad \text{or} \quad \begin{pmatrix} 0 & -1 \\ -1 & 0 \end{pmatrix} \tag{2-2-5}$$

The invariances previously denoted demand

$$E = \begin{pmatrix} 1 & 0 \\ 0 & 1 \end{pmatrix} = \sigma_h \tag{2-2-6}$$

although the last equality is not general. Similarly, in this example, $i = C_2$, $\sigma_v = C_2'$, $\sigma_d = C_2''$, and $S_4 = C_4$ (also $C_4^3 = S_4^3$). The reader will please note that when alternative forms of the transformation matrices are given these refer to the two members of the class under consideration. Thus the two matrices labelled C_2' refer to rotations about the x- or y-axes Note that the traces (the sums of the diagonal elements) of all operations in each class are the same. This is a general rule which may be easily proven algebraically.

The selection of the particular diagram (Fig. 2-3) was quite arbitrary. Given one part of a diagram—such as Fig. 2-3 (a)—that is, one component of a basis—it is necessary that the other be orthogonal to it, as Fig. 2-3 (b) is. Furthermore, orthogonal linear combinations of Figs. 2-3 (a) and (b) can also serve as suitable bases. The reader may test this proposition by testing Figs. 2-3 (c) and (d), which correspond to $(a) \pm (b)$. The matrices generated by the transformation properties of (c) and (d) may be different from those generated by (a) and (b) but they will still satisfy the group multiplication rules as well as some other rules to be discussed shortly.

Among the many possible linear combinations of (a) and (b) we can also

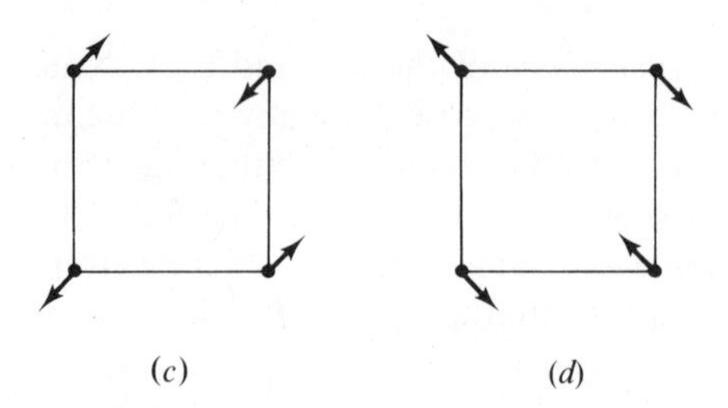

Figure 2-3 Orthogonal components of a degenerate deformation mode of a square molecule, X_4. (*c*) and (*d*). Alternative basis to that of Fig. 2-3 (*a*) and (*b*).

include (*a*) $\pm$ *i*(*b*). This complex basis has two interesting properties. First, it may be represented by the column vector

$$\begin{pmatrix} \rho\, e^{i\varphi} \\ \rho\, e^{-i\varphi} \end{pmatrix}$$

where $\rho = (a^2 + b^2)^{1/2}$ and φ is the phase angle $\tan^{-1}(b)/(a)$. Second, this manner of writing the basis shows that all possible bases, including (a, b), (c, d), etc., can be generated by letting the angles the displacement vectors make with, say, the x-axis as origin, sweep out any angle φ. Indeed when φ becomes $5\pi/4$, (a, b) will have become (c, d) and so forth. Furthermore, when φ becomes $\pi/2$, (a, b) will have become $(b, -a)$. We thus see that there are not two distinct deformations here but really only one—a so-called *doubly degenerate* mode. Figures 2-3 (*a*) and (*b*), etc., are real representations of the two components of this mode; there is an infinity of such pairs. The complex form is actually closer to the actual motion when account is taken of vibration–rotation interaction, or certain anharmonic effects which occur in real molecules. In this representation each atom moves in a *circular* orbit about its "equilibrium position." (The *direction*, or *sense*, of this motion is indeterminate; hence the degeneracy implied by the sign ambiguity $e^{\pm i\varphi}$ remains.) Note from Fig. 2-2 (*a*) and (*b*) (as well as from (*c*) and (*d*)) that in this mode each atom is constantly $+\pi/2$ radians out-of-phase with its neighbor as it goes about its orbit. This is also true for the complex form (*a*) $\pm$ *i*(*b*).

In the complex basis, the several matrices are

$$C_4 = \begin{pmatrix} e^{+\pi i/2} & 0 \\ 0 & e^{-\pi i/2} \end{pmatrix} = \begin{pmatrix} i & 0 \\ 0 & -i \end{pmatrix} \tag{2-2-7}$$

$$C_2 = \begin{pmatrix} e^{\pi i} & 0 \\ 0 & e^{-\pi i} \end{pmatrix} = \begin{pmatrix} -1 & 0 \\ 0 & -1 \end{pmatrix} \tag{2-2-8}$$

$$C_4^3 = \begin{pmatrix} e^{+3\pi i/2} & 0 \\ 0 & e^{-3\pi i/2} \end{pmatrix} = \begin{pmatrix} -i & 0 \\ 0 & i \end{pmatrix} \tag{2-2-9}$$

$$C_2' = \begin{pmatrix} 0 & -1 \\ -1 & 0 \end{pmatrix} \tag{2-2-10}$$

$$C_2'' = \begin{pmatrix} 0 & i \\ -i & 0 \end{pmatrix} \tag{2-2-11}$$

Again E and σ_h are unit matrices, $i = C_2 = -E$, and the other equalities stated earlier are unchanged. The second C_2' matrix is the same in the new basis, but the other C_2'' is given by

$$C_2'' = \begin{pmatrix} 0 & -i \\ i & 0 \end{pmatrix} \tag{2-2-12}$$

Note that in going to the complex basis, the traces of each matrix do not change. This will have already been discovered by the reader who formed the basis (c, d). This invariance of the trace is a general property of any *similarity transformation*. After a coordinate transformation,

$$x' = ax \tag{2-2-13}$$

any matrix R, corresponding to a linear operator,* such as one of the group operations, becomes

$$R' = aRa^{-1} \tag{2-2-14}$$

The matrix a which takes the real $\begin{pmatrix} x \\ y \end{pmatrix}$ basis into the complex $\frac{1}{\sqrt{2}}\begin{pmatrix} x + iy \\ x - iy \end{pmatrix}$ basis is obviously

$$a = \frac{1}{\sqrt{2}}\begin{pmatrix} 1 & i \\ 1 & -i \end{pmatrix} \tag{2-2-15}$$

Its inverse is

$$a^{-1} = \frac{1}{\sqrt{2}}\begin{pmatrix} 1 & 1 \\ -i & i \end{pmatrix} \tag{2-2-16}$$

If (2-2-14) is applied, using (2-2-15) and (2-2-16) to (2-2-7)–(2-2-12), the matrices (2-2-2)–(2-2-6) will be recovered.†

* A linear operator R is one which satisfies:

$$R(\mathbf{x} + \mathbf{z}) = R\mathbf{x} + R\mathbf{z}$$

and

$$R(\alpha\mathbf{x}) = \alpha R\mathbf{x}$$

† This transformation is quite general and will take any rotation matrix such as

$$C_n^k(2\pi k/n) = \begin{pmatrix} e^{+2\pi ik/n} & 0 \\ 0 & e^{-2\pi ik/n} \end{pmatrix}$$

into the form

$$C_n^k(2\pi k/n) = \begin{pmatrix} \cos 2\pi k/n & -\sin 2\pi k/n \\ \sin 2\pi k/n & \cos 2\pi k/n \end{pmatrix}$$

Note that the trace of each matrix is $2 \cos 2\pi k/n$.

We have thus generated a particular two-dimensional representation of the point group D_{4h} and shown that it can be written in a large number of ways, each of which is related to the other by a similarity transformation.

Each point group can have an indefinite number of representations, matrices or otherwise, which serve to reproduce the properties of the group. It can be shown, however, that an arbitrary representation can be written as a linear combination of a small and finite number of representations, appropriate to each point group.

Each of these is called an *irreducible representation* and can be viewed as a set of matrices, one corresponding to each member of the group. For example, consider the following matrix representation of the group C_{3v},

$$\underset{E}{\begin{pmatrix} 1 & 0 & 0 & 0 \\ 0 & 1 & 0 & 0 \\ 0 & 0 & 1 & 0 \\ 0 & 0 & 0 & 1 \end{pmatrix}} \underset{C_3}{\begin{pmatrix} e^{-2\pi i/3} & 0 & 0 & 0 \\ 0 & 1 & 0 & 0 \\ 0 & 0 & 1 & 0 \\ 0 & 0 & 0 & e^{2\pi i/3} \end{pmatrix}} \underset{C_3^2}{\begin{pmatrix} e^{2\pi i/3} & 0 & 0 & 0 \\ 0 & 1 & 0 & 0 \\ 0 & 0 & 1 & 0 \\ 0 & 0 & 0 & e^{-2\pi i/3} \end{pmatrix}}$$

$$\underset{\sigma_v}{\begin{pmatrix} 0 & 0 & 0 & 1 \\ 0 & 0 & 1 & 0 \\ 0 & 1 & 0 & 0 \\ 1 & 0 & 0 & 0 \end{pmatrix}} \underset{\sigma_v''}{\begin{pmatrix} 0 & 0 & 0 & e^{2\pi i/3} \\ 0 & 0 & 1 & 0 \\ 0 & 1 & 0 & 0 \\ e^{-2\pi i/3} & 0 & 0 & 0 \end{pmatrix}} \underset{\sigma_v'}{\begin{pmatrix} 0 & 0 & 0 & e^{-2\pi i/3} \\ 0 & 0 & 1 & 0 \\ 0 & 1 & 0 & 0 \\ e^{2\pi i/3} & 0 & 0 & 0 \end{pmatrix}}$$

The basis of this representation is the vector

$$\begin{pmatrix} e^{i(\phi_1+\phi_2)} \\ e^{i(\phi_1-\phi_2)} \\ e^{-i(\phi_1-\phi_2)} \\ e^{-i(\phi_1+\phi_2)} \end{pmatrix}$$

in the same sense that $\begin{pmatrix} e^{i\phi} \\ e^{-i\phi} \end{pmatrix}$ was a basis for the doubly degenerate representation of D_{4h} discussed previously. We start to make a coordinate transformation which, like forming $x \pm iy$, is simply a re-mixing of the basis functions.

As a result, we will still have a four-dimensional representation, each matrix of which will have the same trace as in that just exhibited. The new basis will be of the form

$$\begin{pmatrix} e^{i(\phi_1+\phi_2)} + e^{-i(\phi_1+\phi_2)} \\ e^{i(\phi_1+\phi_2)} - e^{-i(\phi_1+\phi_2)} \\ e^{i(\phi_1-\phi_2)} + e^{-i(\phi_1-\phi_2)} \\ e^{i(\phi_1-\phi_2)} - e^{-i(\phi_1-\phi_2)} \end{pmatrix}$$

If we let $(x + iy) \sim e^{i\phi}$, this can be shown to be equivalent to

$$\begin{pmatrix} x_1x_2 - y_1y_2 \\ x_1y_2 + y_1x_2 \\ x_1x_2 + y_1y_2 \\ y_1x_2 - x_1y_2 \end{pmatrix}$$

In this basis, the four-dimensional representation is ($\varepsilon = e^{-2\pi i/3}$)

$$\begin{matrix} E & C_3 & C_3^2 \\ \begin{pmatrix} 1 & 0 & 0 & 0 \\ 0 & 1 & 0 & 0 \\ 0 & 0 & 1 & 0 \\ 0 & 0 & 0 & 1 \end{pmatrix} & \begin{pmatrix} \varepsilon^* & 0 & 0 & 0 \\ 0 & \varepsilon & 0 & 0 \\ 0 & 0 & 1 & 0 \\ 0 & 0 & 0 & 1 \end{pmatrix} & \begin{pmatrix} \varepsilon & 0 & 0 & 0 \\ 0 & \varepsilon^* & 0 & 0 \\ 0 & 0 & 1 & 0 \\ 0 & 0 & 0 & 1 \end{pmatrix} \end{matrix}$$

$$\begin{matrix} \sigma_v & \sigma_v' & \sigma_v'' \\ \begin{pmatrix} 0 & 1 & 0 & 0 \\ 1 & 0 & 0 & 0 \\ 0 & 0 & 1 & 0 \\ 0 & 0 & 0 & -1 \end{pmatrix} & \begin{pmatrix} 0 & \varepsilon & 0 & 0 \\ \varepsilon^* & 0 & 0 & 0 \\ 0 & 0 & 1 & 0 \\ 0 & 0 & 0 & -1 \end{pmatrix} & \begin{pmatrix} 0 & \varepsilon^* & 0 & 0 \\ \varepsilon & 0 & 0 & 0 \\ 0 & 0 & 1 & 0 \\ 0 & 0 & 0 & -1 \end{pmatrix} \end{matrix}$$

Note that any of these matrices can be written in "block form"

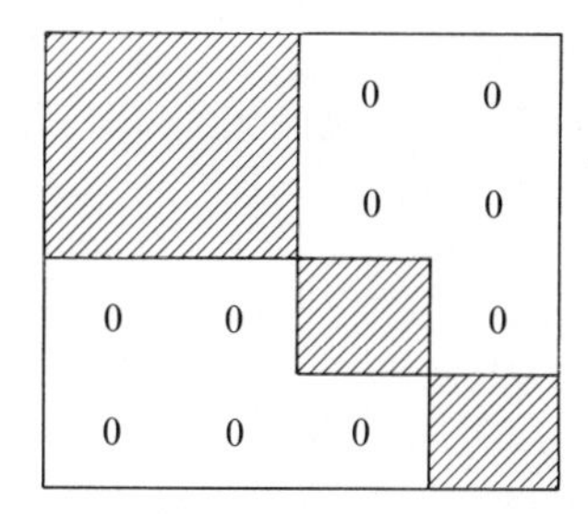

such that the sets of six matrices composed from each of the separately shaded blocks will itself be a representation of C_{3v}. Alternatively, we can say that the four-dimensional space spanned by the original basis functions contains three subspaces, two of which are one-dimensional and the other of which is two-dimensional, each of which serves perfectly well as a basis for a representation of this group. Notice that one of the one-dimensional blocks consists of the number 1 for each operation. The basis of this representation, $x_1x_2 + y_1y_2$, is in fact the scalar product $\mathbf{r}_1 \cdot \mathbf{r}_2$ of two vectors $\mathbf{r}_1$ and $\mathbf{r}_2$ in the xy plane with components (x_1, y_1) and (x_2, y_2). Since it is a scalar it must be invariant under all group operations. On the other hand, the other one-dimensional basis, $y_1x_2 - x_1y_2$, is the z-component of the cross product $\mathbf{r}_1 \times \mathbf{r}_2$, which behaves as a pseudo-vector parallel to $\mathbf{z}$; hence it is not invariant under all operations. It is not affected by a rotation about an axis parallel to that of the pseudo-vector but it is antisymmetric with respect to reflections in planes containing **z**, which is a peculiar property of an axial (pseudo) vector, as opposed to a polar vector.

The remaining functions are not identifiable in their present form. However, because they are in no way mixed with the other two functions (note the zeros in all the matrices), they can be mixed with each other (another coordinate transformation) which accomplishes the following

$$\begin{pmatrix} x_1x_2 - y_1y_2 \\ x_1y_2 + y_1x_2 \end{pmatrix} \rightarrow \begin{pmatrix} (x_1 + iy_1)(x_2 + iy_2) \\ (x_1 - iy_1)(x_2 - iy_2) \end{pmatrix}$$

We have, as far as the two-dimensional subspace is concerned, rotated back to

$$\begin{pmatrix} e^{i(\phi_1+\phi_2)} & \\ & e^{-i(\phi_1+\phi_2)} \end{pmatrix}$$

a basis with which we have some familiarity. No further coordinate transformations can be found which will simultaneously eliminate the off-diagonal terms in these 2×2 blocks for all group operations. We say we have *reduced* the original reducible representations to two one-dimensional *irreducible* representations and one two-dimensional irreducible representation. *If the two-dimensional representation were reducible at all, a coordinate transformation could be found which would project two one-dimensional representations out of it. In fact, no such transformation can be found: this two-dimensional representation of C_{3v} is also irreducible, as was that used in the discussion of the X_4 molecule.*

There are a number of algebraic theorems concerning irreducible representations which we shall make use of in later sections. We shall state them here, without proofs. These may be found in any standard text devoted to group theory.[3] The ijth element of the matrix representation of the operation R in the μth irreducible representation is denoted $\mathscr{R}_{ij}^{(\mu)}$. These elements satisfy the orthogonality relation

$$\sum_R \mathscr{R}_{ij}^{(\mu)} \mathscr{R}_{lm}^{(\nu)*} = \frac{g}{d_\mu} \delta_{\mu\nu} \delta_{il} \delta_{jm} \tag{2-2-17}$$

where g is the order of the group and d_μ is the dimension of the irreducible representation $\mathscr{R}^{(\mu)}$. Equation (2-2-17) tells us that for each μ, i, and j, the quantities $\mathscr{R}_{ij}^{(\mu)}$ are components of vectors in a g-dimensional space, orthogonal to each other.

It is useful to restate Eq. (2-2-17) in terms of the *character* of R in the μth irreducible representation $\mathscr{R}^{(\mu)}$,

$$\chi^{(\mu)}(R) = \chi_R^{(\mu)} = \sum_i \mathscr{R}_{ii}^{(\mu)} \tag{2-2-18}$$

the quantity usually called the *trace* of the matrix $\mathscr{R}^{(\mu)}$. Thus the character is the same for all members of a class. A representation which differs from another only by a coordinate transformation is indeed a different representation but it has

3. M. Hamermesh, *Group Theory*, Addison-Wesley Publishing Co., Reading, Mass., 1962.

the same set of characters. It is called an *equivalent representation.* Hence, the characters of the reducible representation of C_{3v} discussed previously are

$$\chi(E) = 4; \quad \chi(C_3) = 1; \quad \chi(\sigma_v) = 0.$$

Those of the two-dimensional irreducible representations of D_{4h} used as an example are

$$\chi(E) = 2; \quad \chi(C_4) = 0; \quad \chi(C_2) = 2; \quad \chi(C_2') = \chi(C_2'') = 0$$
$$\chi(i) = -2; \quad \chi(S_4) = 0; \quad \chi(\sigma_h) = 2; \quad \chi(\sigma_v) = \chi(\sigma_d) = 0$$

Equation (2-2-17) also states that each vector with components $\mathscr{R}_{ij}^{(\mu)}$ is orthogonal to another with components $\mathscr{R}_{lm}^{(\nu)}$, provided μ and ν are *inequivalent.* Equation (2-2-17) restated in terms of the characters is

$$\sum_R \chi^{(\mu)}(R)\chi^{(\nu)}(R)^* = g\delta_{\mu\nu} \tag{2-2-19}$$

Since the characters of all g_i elements of the class K_i are the same, we may restate Eq. (2-2-19) as a summation over the classes:

$$\sum_i g_i\chi_i^{(\mu)}\chi_i^{(\nu)*} = g\delta_{\mu\nu} \tag{2-2-20}$$

The number of possible non-equivalent irreducible representations is fixed. It can be shown to equal the number of classes in the group.* Furthermore, the dimensions of the representations must satisfy

$$\sum_\nu d_\nu^2 = g \tag{2-2-21}$$

There will always be one irreducible representation for which $\chi_i^{(\mu)} = 1$, for all i. This representation, called the totally symmetric representation, is frequently denoted Γ_1.

The operation E is always in a class by itself, since

$$R^{-1}ER = E \tag{2-2-22}$$

for any R. The character of the identity $\chi^{(\mu)}(E)$ is always the dimension d_μ of the representation, irreducible or not. For the irreducible representations of a group, this rule, combined with Eq. (2-2-21) leads to

$$\sum_\nu [\chi^{(\mu)}(E)]^2 = g \tag{2-2-23}$$

The characters of the irreducible representations may be arrayed in a table, each column of which is conventionally a different class, beginning with E, and the

* There are some point groups, notably the pure rotation groups C_n, the groups C_{nh}, S_n, plus D_2, C_{2v}, and D_{2h}, in which each operation is in a class by itself: $g_i = 1$ for all i. This circumstance arises when group multiplication is commutative as well as associative, for all group elements. Such groups are said to be *Abelian* and have as many irreducible representations as there are operations g in the group.

rows are then the several irreducible representations, beginning with Γ_1. Equation (2-2-20) indicates the orthogonality between rows, considered as vectors; there is a similar relationship between the columns:

$$\sum_{\nu=1}^{r} \chi_i^{(\nu)} \chi_j^{(\nu)*} = \frac{g}{g_j} \delta_{ij} \tag{2-2-24}$$

For each point group there are, accordingly, well-known *character tables,* which are to be found in Appendix I. The derivation of these characters can be accomplished in various ways; for example, see Appendix X of WDC. It has become conventional in spectroscopy to denote certain types of irreducible representations systematically. Thus, the totally symmetric representation is always labelled *A*. A representation which is antisymmetric to C_n is called *B*. Groups which have threefold or higher axes of symmetry will always have irreducible representations which are more than one-dimensional. Two-dimensional, three-dimensional, or higher dimensional representations are labelled, respectively, *E, F* (sometimes called *T*), *G, H,*.... These symbols are frequently decorated with primes, double primes, numerical subscripts as well as the letter subscripts *g* (for *gerade*) or *u* (for *ungerade*). *g* and *u* distinguish between representations which are respectively symmetric or antisymmetric with respect to *i*. Similarly, primes and double primes differentiate between representations which are symmetric or antisymmetric under σ_h. When further distinct irreducible representations occur, they are distinguished by numerical subscripts which the reader can best comprehend by a careful scrutiny of Appendix I.

We shall now outline the algebraic technique commonly used for the reduction of a representation. Instead of the laborious procedure of finding a coordinate transformation that will accomplish this, we shall use a procedure which is based upon the fact that a reducible representation has a *structure**

$$\Gamma(R) = \sum_{\nu} \eta^{(\nu)} \Gamma^{(\nu)}(R) \tag{2-2-25}$$

where the $\eta^{(\nu)}$ are the number of times the irreducible representation $\Gamma^{(\nu)}(R)$, or its equivalent, appears in $\Gamma(R)$. In terms of the characters, Eq. (2-2-25) may be written

$$\chi_i = \sum_{\mu} \eta^{(\mu)} \chi_i^{(\mu)} \tag{2-2-26}$$

Because of Eq. (2-2-20), it then follows that

$$\eta^{(\mu)} = \frac{1}{g} \sum_{i} g_i \chi_i^{(\mu)*} \chi_i \tag{2-2-27}$$

We shall refer to Eq. (2-2-27) as the decomposition formula.

* The summation which appears in (2-2-25) is to be understood in the sense that the change of basis has the effect of putting the original set of transformation matrices in the diagonal block form exemplified by the four-dimensional representation of C_{3v} which was reduced to one two-dimensional plus two one-dimensional blocks.

We close this section with another prescription, also based upon the transformation of vector bases under symmetry operations of point groups, which is useful in generating functions which may serve as bases of representations. In general, such functions are called *symmetry adapted,* although in molecular vibration theory they are often referred to as "symmetry coordinates." The prescription is

$$\psi_i^{(\mu)} = \mathcal{N} \sum_R \mathscr{R}_{ii}^{(\mu)*} \mathbf{R}\psi \tag{2-2-28}$$

in which ψ is *any* function, $\mathbf{R}\psi$ is the function obtained by operating on ψ by the symmetry operation $\mathbf{R}$, and the other quantities have all been previously defined. If $\mathscr{R}^{(\mu)}$ is non-degenerate, $\mathscr{R}_{ii}^{(\mu)}$ may be replaced by $\chi_R^{(\mu)}$.

On the other hand, if $\mathscr{R}^{(\mu)}$ is degenerate, there will always exist $d_\mu - 1$ "partner functions," generated by

$$\psi_l^{(\mu)} = \frac{d_\mu}{g} \sum_R \mathscr{R}_{li}^{(\mu)*} \mathbf{R}\psi_i^{\mu} \tag{2-2-29}$$

which may also be desired in some particular problems. The operator,

$$\mathbf{W}_l^{(\mu)} = \sum_R \mathscr{R}_{li}^{(\mu)*} \mathbf{R} \tag{2-2-30}$$

is known as the "Wigner Projection Operator." The reader will note that (2-2-29) is not normalized. The normalization constant $\mathcal{N}$, in a given case, depends upon the choice of ψ and is easily found.

2-3 NORMAL MODES OF VIBRATION

The simplest dynamic model of a polyatomic molecule neglects the coupling between vibration and rotation, and regards the motion as a combination of harmonic oscillation and rigid rotation. Such a model yields a reasonable description of the vibration–rotation bands for non-degenerate vibrations, but is spectacularly incorrect in accounting for the rotational structure of bands arising from degenerate vibrations. Since the intent of this chapter is to lay the foundations for the discussion of vibrations in crystals, where with very few exceptions, nothing approximating free rotational motion occurs, we shall not discuss the rotation–vibration couplings, and proceed immediately to an account of the vibrational motion alone.

(a) Normal Coordinates

In order to justify the quantum mechanical energy formula (1-1-1), it is necessary to express both kinetic and potential energy as sums of dynamically independent degrees of freedom, i.e., to give the vibrational Hamiltonian as a sum such as

$$H = \sum_{k=1}^{f} H_k \tag{2-3-1}$$

where

$$H_k = \tfrac{1}{2}(P_k^2 + \omega_k^2 Q_k^2) \tag{2-3-2}$$

This form of the vibrational Hamiltonian expressed in terms of normal coordinates Q_k, and their conjugate momenta P_k, will be related to other coordinate systems so that the ω_k can be calculated in terms of molecular properties in Section 2-3(c). It immediately permits the following elementary, but important deduction about the transformation of the Q_k (or P_k) under the symmetry operations. Since the Hamiltonian is an invariant, and since the geometrical point group operations do not mix coordinates with momenta, the kinetic and potential energies must be individually invariant to such operations. In matrix notation, the potential energy V is

$$V = \tfrac{1}{2}\mathbf{Q}^\dagger \mathbf{\Omega}^2 \mathbf{Q} \tag{2-3-3}$$

Let the transformation of the normal coordinates by the symmetry operation R be represented by the *unitary*[4] matrix $\mathbf{R}$, so that if $\mathbf{Q}$ is the normal coordinate vector before the operation, its value after the operation is denoted by $\mathbf{RQ}$. Since $(\mathbf{RQ})^\dagger = \mathbf{Q}^\dagger \mathbf{R}^\dagger$, the potential energy after the operation is

$$V = \tfrac{1}{2}\mathbf{Q}^\dagger \mathbf{R}^\dagger \mathbf{\Omega}^2 \mathbf{RQ} \tag{2-3-4}$$

But since V is an invariant,

$$\mathbf{R}^\dagger \mathbf{\Omega}^2 \mathbf{R} = \mathbf{\Omega}^2$$

or

$$\mathbf{\Omega}^2 \mathbf{R} = (\mathbf{R}^\dagger)^{-1} \mathbf{\Omega}^2 = \mathbf{R}\mathbf{\Omega}^2 \tag{2-3-5}$$

where the last step follows from the definition of a unitary matrix. We see that the invariance of $\mathbf{\Omega}^2$ implies that it commutes with all the transformation matrices $\mathbf{R}$ of the group. Having found this result, we examine its meaning in terms of individual elements of the matrix products $\mathbf{\Omega}^2\mathbf{R}$ and $\mathbf{R}\mathbf{\Omega}^2$ and find that the k, k' element is

$$\omega_k^2 R_{kk'} = R_{kk'} \omega_{k'}^2$$

or

$$(\omega_k^2 - \omega_{k'}^2) R_{kk'} = 0 \tag{2-3-6}$$

First, we observe that if ω_k is non-degenerate, i.e., not equal to any other $\omega_{k'}$, then $R_{kk'}$ must vanish for all $k' \neq k$. This clearly implies that Q_k is not mixed with any other $Q_{k'}$ by any symmetry operation, and that it is the basis of a one-dimensional and hence necessarily irreducible representation of the symmetry group. The matrix elements R_{kk} must correspond to one of the rows of the character table for a non-degenerate species, which, as has been pointed out above, can be derived purely from the combinatory properties of the symmetry operations without reference to the specific molecule.

4. Unitary matrices, $\mathbf{U}$, have the property that the transpose conjugate is equal to the inverse: $\mathbf{U}^\dagger = \mathbf{U}^{-1}$

Second, suppose that ω_k belongs to a set of d equal ω's: if $\omega_{k'}$ is another member of this set so that $\omega_k = \omega_{k'}$, (2-3-6) shows that $R_{kk'}$ need not equal zero and we must admit that the part of the **R** matrix corresponding to the degenerate Q_k's will no longer be a single diagonal term $R_{kk} \neq 0$ but a $d \times d$ block. This situation may occur because of "accidental" degeneracy, but much the most common cause is that the set of degenerate Q's arise because of symmetry, in particular because two-dimensional (or higher) irreducible representations are included in the representation by the full set of vibrational coordinates. In the rare case of accidental degeneracy the $d \times d$ block is reducible, and appropriate linear combinations of the initial set of Q's can be found which do not modify the set of eigenvalues ω_k^2 but which correspond precisely with the bases of irreducible representations.

(b) Quantum Mechanics of the Normal Modes

The value of (2-3-1) and (2-3-2) lies in the evident facts that (i) the Hamiltonian is separated so that the wave function (in the Dirac notation) becomes a *product* of factors

$$|\mathbf{v}\rangle = |v_1\rangle\,|v_2\rangle \cdots |v_f\rangle$$

and the energy becomes a sum of independent terms; (ii) the form of each separate contribution to the Hamiltonian (2-3-2) is that of the harmonic oscillator, whose energies

$$E_k = h\nu_k(v_k + \tfrac{1}{2}) = \hbar\omega_k(v_k + \tfrac{1}{2}) \qquad \text{(2-3-7)}$$

with

$$\nu_k = \omega_k^{1/2}/2\pi \qquad \text{(2-3-8)}$$

are the solutions of one of the best known problems in elementary quantum mechanics. Thus

$$\begin{aligned} E &= \sum_{k=1}^{f} h\nu_k(v_k + \tfrac{1}{2}) = (h/2)\sum \nu_k + h\sum \nu_k v_k \\ &= E^\circ + h\boldsymbol{\nu}'\mathbf{v} \end{aligned} \qquad \text{(2-3-9)}$$

as stated in chapter 1.

(c) Transformation between Internal and Normal Coordinates; Determination of Frequencies

Returning to the justification of the vibrational Hamiltonian (2-3-1) and (2-3-2), we first observe that this is only the *harmonic approximation,* in which coordinate terms in the potential energy of order greater than two are neglected (see Section 2-7 for a resumé of anharmonicity effects). In this approximation, the potential energy in terms of a set of *internal* coordinates r_t, $1 \leq t \leq f$, is conveniently written as

$$2V = \mathbf{r}^\dagger \mathbf{F} \mathbf{r} \qquad \text{(2-3-10)}$$

where $\mathbf{r}$ is the f-dimensional vector of the internal coordinates and $\mathbf{F}$ is a corresponding square matrix whose elements are the force constants. The definition of internal coordinates requires that, *when supplemented by rigid translations and rotations,* they be capable of describing any configuration of the molecule, and further that if the atoms are displaced from their equilibrium positions according to any combination of the rigid motions, that all $r_t = 0$. The most commonly utilized types include bond stretches, valence angle bends, out-of-plane bends (e.g., of CH bonds in benzene), twisting around bonds such as C—C in ethane, and are described in considerable detail in standard references.[5]

What now of the kinetic energy? One always knows its definition in terms of cartesian coordinates of the individual particles, either in terms of velocities or in terms of the momenta conjugate to such coordinates. Moreover in any system of coordinates, the momentum vector $\mathbf{p}$ is related to the velocity vector $\dot{\mathbf{q}}$ by

$$\mathbf{p} = \frac{\partial T}{\partial \dot{\mathbf{q}}} \tag{2-3-11}$$

Then, by the laws of transformation between a cartesian system x, and the above mentioned internal system r, one has

$$\mathbf{p}_x^\dagger = \frac{\partial T^\dagger}{\partial \dot{\mathbf{x}}} = \frac{\partial T^\dagger}{\partial \dot{\mathbf{r}}} \frac{\partial \dot{\mathbf{r}}}{\partial \dot{\mathbf{x}}} = \mathbf{p}_r^\dagger \frac{\partial \mathbf{r}}{\partial \mathbf{x}} = \mathbf{p}_r^\dagger \mathbf{B} \tag{2-3-12}$$

where the last expression introduces $\mathbf{B}$, a rectangular $f \times 3n$ matrix which defines the linear transformation between the $3n$ cartesian displacement coordinates and the f internal coordinates:

$$\mathbf{r} = \mathbf{B}\mathbf{x} \tag{2-3-13}$$

Inasmuch as the kinetic energy in the cartesian basis is

$$2T = \mathbf{p}_x^\dagger \mathbf{m}^{-1} \mathbf{p}_x \tag{2-3-14}$$

where $\mathbf{m}^{-1}$ is a (diagonal) matrix whose elements are the reciprocal masses of the atoms (each repeated three times), one sees in the light of (2-3-12) and (2-3-14) that

$$2T = \mathbf{p}_r^\dagger \mathbf{B} \mathbf{m}^{-1} \mathbf{B}^\dagger \mathbf{p}_r = \mathbf{p}_r^\dagger \mathbf{G} \mathbf{p}_r \tag{2-3-15}$$

in which the "effective inverse mass matrix," $\mathbf{G} = \mathbf{B}\,\mathbf{m}^{-1}\mathbf{B}^\dagger$, is defined simply by the atomic masses together with the transformation matrix from cartesian displacement to internal coordinates. The computation of the elements of $\mathbf{G}$ is facilitated by a number of simplifications and tabulations,[6] and can be accomplished with great convenience employing computers[7] whose requisite input is just the equilibrium molecular geometry, the masses, and the coordinate types (bond stretches, angle bends, etc.). The calculation of elements of $\mathbf{G}$ is carried out below

5. See footnote 1 on page 15.
6. WDC, Appendix VI.
7. J. H. Schachtschneider, Tech. Report No. 57–65, Shell Development Co., Emeryville, CA, 1964.

in Section 2-5. At this stage we may write the vibrational Hamiltonian as:

$$H = \tfrac{1}{2}(\mathbf{p}_r^\dagger \mathbf{G}\mathbf{p}_r + \mathbf{r}^\dagger \mathbf{F}\mathbf{r}) \tag{2-3-16}$$

But we want to write it as:

$$H = \tfrac{1}{2}(\mathbf{P}^\dagger \mathbf{P} + \mathbf{Q}^\dagger \boldsymbol{\Omega}^2 \mathbf{Q}) \tag{2-3-17}$$

where Ω (and hence Ω^2) is a diagonal matrix. To achieve this, assume a linear, non-singular transformation

$$\mathbf{r} = \mathbf{L}\mathbf{Q} \qquad (\mathbf{r}^\dagger = \mathbf{Q}^\dagger \mathbf{L}^\dagger) \tag{2-3-18}$$

which implies that

$$\mathbf{P}^\dagger = \mathbf{p}_r^\dagger \mathbf{L} \qquad (\mathbf{P} = \mathbf{L}^\dagger \mathbf{p}_r) \tag{2-3-19}$$

Therefore one finds that

$$\mathbf{P}^\dagger \mathbf{P} = \mathbf{p}_r^\dagger \mathbf{L}\mathbf{L}^\dagger \mathbf{p}_r \tag{2-3-20}$$

and

$$\mathbf{r}^\dagger \mathbf{F}\mathbf{r} = \mathbf{Q}^\dagger \mathbf{L}^\dagger \mathbf{F}\mathbf{L}\mathbf{Q} \tag{2-3-21}$$

so that by comparison with (2-3-16) and (2-3-17) one may conclude that

$$\mathbf{L}\mathbf{L}^\dagger = \mathbf{G} \tag{2-3-22}$$

and

$$\mathbf{L}^\dagger \mathbf{F}\mathbf{L} = \boldsymbol{\Omega}^2 \tag{2-3-23}$$

To complete this phase of the discussion, it is only necessary to point out that (2-3-22) implies $\mathbf{L}^\dagger = \mathbf{L}^{-1}\mathbf{G}$, consequently from (2-3-23) one has

$$\mathbf{L}^{-1}\mathbf{G}\mathbf{F}\mathbf{L} = \boldsymbol{\Omega}^2 \tag{2-3-24}$$

i.e., the similarity transformation $\mathbf{L}^{-1}$ [ ] $\mathbf{L}$ acting upon the matrix product **GF** produces the matrix $\boldsymbol{\Omega}^2$ whose diagonal elements (off-diagonal elements are by definition zero) are the required parameters of (2-3-2), and are called the eigenvalues (or characteristic values) of the matrix **GF**. Their determination, given **G** and **F**, can in principle be accomplished by algebraic methods, but except for very small values of f, the number of rows or columns of **G** or **F**, numerical methods are greatly preferable, especially with modern computers.

Despite the ease with which such computations can be effected, it requires great discretion to achieve physically meaningful results from such numerology. Granted even that one knows the molecular geometry (and hence **G**), in the general case, **F**, which is a *symmetric* matrix meaning that $F_{tt'} = F_{t't}$), contains $f(f+1)/2$ unknown parameters, whereas there are only f items of data (the ω_k^2). Thus except for the diatomic molecule, the problem appears to be indeterminate. The great advances which have been achieved in vibrational molecular spectroscopy would not have been possible without further principles most of which stem from symmetry.

(d) Symmetry and Internal Coordinates

To introduce these principles, let us consider the example of the XY_3 planar molecule of D_{3h} symmetry (the boron trihalides are gas phase examples). In such molecular structures, $f = 3 \cdot 4 - 6 = 6$, so that our initial estimate of the number of force constants would be $6 \cdot (7/2) = 21$ parameters. But this number (as well as the number $f = 6$ identified with the number of observables) is an overestimate and does not allow for certain relations between the force constants and also between the eigenvalues ω_k^2 implied by the symmetry. In fact there are only four distinct eigenvalues ω_k^2 or frequencies ν_k and only five distinct force constants. Since the gap between the observables and the parameters is so drastically narrowed, it is possible that the use of additional information obtained from the spectra of isotopic molecules (see Section 2-5) can lead to a fully determinate solution. Even more favorable cases are known such as CO_2 or P_4 where the symmetry actually reduces the number of parameters to equality with the number of observables.

The way such situations arise can now be explained either in an unsophisticated intuitive way, or by appeal to group theoretical arguments. We shall explore and employ the latter route throughout the rest of this volume, but first we give the elementary argument for CO_2 or similar molecules.

In CO_2 (symmetry $D_{\infty h}$) one first notes that $f = 4$ and that a satisfactory (i.e., linearly independent) internal coordinate set is composed of two bond stretches and two angle bendings, the latter occurring in mutually perpendicular planes intersecting at the equilibrium molecular axis. Since there is nothing to distinguish these deformations, we may assume that the force constants associated with them are identical. As to the two bond stretches, again we note that their force constants must be identical. Then, denoting the two bond stretches by r_1 and r_2, the two angle bends by α_x and α_y, we assert that the complete expression for the force constant matrix is:

$$\mathbf{F} = \begin{pmatrix} F_r & F_r' & 0 & 0 \\ F_r' & F_r & 0 & 0 \\ 0 & 0 & F_\alpha & 0 \\ 0 & 0 & 0 & F_\alpha \end{pmatrix} \tag{2-3-25}$$

Although one additional parameter, namely F_r', has crept into our formulation, we note that there are only three, not ten force constants. Note particularly that there are no interaction force constants between the two bendings, nor between either bond stretch or angle bend. An elementary proof that such constants must vanish can be given by recalling that the potential energy must be invariant to the point symmetry operations; thus if there were a term say of the form $F_\alpha' \alpha_x \alpha_y$, reflection through the xz plane would yield the following transformations $\alpha_x \to \alpha_x$, $\alpha_y \to -\alpha_y$ and consequently $F_\alpha' \alpha_x \alpha_y \to -F_\alpha' \alpha_x \alpha_y$ which is contrary to the requirement of invariance and therefore implies $F_\alpha' = 0$. The reader can easily supply the analogous argument which prohibits the occurrence of terms such as $r\alpha_x$, etc.

Since the symmetry transformations affect the momentum components in a manner strictly analogous to the coordinate components, one concludes that

$$\mathbf{G} = \begin{pmatrix} G_r & G'_r & 0 & 0 \\ G'_r & G_r & 0 & 0 \\ 0 & 0 & G_\alpha & 0 \\ 0 & 0 & 0 & G_\alpha \end{pmatrix} \tag{2-3-26}$$

Now one further simplification is possible: by introducing linear combinations of the bond stretching coordinates of the form $\mathsf{S}_\pm = (2)^{-1/2}(r_1 \pm r_2)$, the potential and kinetic energies can be expressed with the aid of the F and G matrices:

$$\mathsf{F} = \begin{pmatrix} F_r + F'_r & 0 & 0 & 0 \\ 0 & F_r - F'_r & 0 & 0 \\ 0 & 0 & F_\alpha & 0 \\ 0 & 0 & 0 & F_\alpha \end{pmatrix}; \quad \mathsf{G} = \begin{pmatrix} G_r + G'_r & 0 & 0 & 0 \\ 0 & G_r - G'_r & 0 & 0 \\ 0 & 0 & G_\alpha & 0 \\ 0 & 0 & 0 & G_\alpha \end{pmatrix} \tag{2-3-27}$$

One now recognizes (i) that the Hamiltonian is separated into four independent terms and (ii) that for the two terms associated with the bending, the coefficients appearing in the Hamiltonian are identical. The latter statement implies an example of two-fold degeneracy.

Actually we can immediately obtain the eigenvalues of GF, which are nothing more nor less than the four quantities $(G_r + G'_r)(F_r + F'_r) = \omega_1^2$, $(G_r - G'_r)(F_r - F'_r) = \omega_3^2$, $G_\alpha F_\alpha = \omega_{2x}^2 = \omega_{2y}^2$ appearing along the diagonal.* In other words, no further calculation is required because all off-diagonal terms vanish. The present example, then, is an exceptionally simple and infrequently encountered case in which the number of distinct frequencies (3) is identical with the number of force constants.

It should be noted that a special type (sans serif) has been employed to designate G and F when expressed as above in terms of the four coordinates, $\mathsf{S}_+ = (2)^{-1/2}(r_1 + r_2)$, $\mathsf{S}_- = (2)^{-1/2}(r_1 - r_2)$, α_x and α_y. Such coordinates are designated as "symmetry coordinates," and although we have introduced them in an intuitive manner, they can always be found by application of (2-2-30). In fact they transform like one or another of the irreducible representations of the point group for CO_2 ($D_{\infty h}$), in particular, $(2)^{-1/2}(r_1 + r_2)$ like Σ_g^+, $(2)^{-1/2}(r_1 - r_2)$ like Σ_u^-, and the degenerate pair, α_x and α_y like Π_u. Application of Eq. (2-2-30) to r_1 and α_x or α_y will yield just these functions.

The great value of symmetry coordinates in molecular vibration problems is due to the simplification of the secular determinant; there exist no terms in the determinant coupling symmetry coordinates which transform like different irreducible representations, or indeed, like different members (α_x and α_y) of a degenerate irreducible representation. In the spectroscopic literature the term "species" is often used to signify the nature of a quantity which transforms like a particular irreducible representation. Thus we say that CO_2 has one vibrational coordinate of species Σ_g^+, one of species Σ_u^-, and a degenerate pair of species Π_u.

We can now readily understand that the introduction of symmetry coordinates

* The numbering of modes, which is obviously not identical with the order of rows or columns in (2-3-27) corresponds to a long accepted convention.

brings about the maximum possible simplification in the F and G matrices owing to an extension of the argument given above, to wit: if S_k and S_l are two symmetry coordinates belonging to different species, there exists at least one symmetry transformation with respect to which S_k and S_l behave differently, so that the product $S_k S_l$ cannot appear as a term in the invariant quantity V. An analogous argument holds pertaining to the terms P_k and P_l in the kinetic energy. Thus one anticipates that in a basis of symmetry coordinates, F and G will assume a form with non-vanishing blocks along the diagonal but with zeros in the off-diagonal blocks corresponding to rows associated with one species of symmetry coordinates and columns associated with any other species of symmetry coordinate. In addition, some blocks may be repeated identically twice or more times if symmetry coordinates which are twofold or more degenerate occur.

(e) The Example of Planar XY_3 (D_{3h}); Redundancy

The carbon dioxide example is unusual in the sense that no more than one symmetry coordinate (or one degenerate pair) occurs in any symmetry species, so that once the dynamical problem is cast in terms of symmetry coordinates there is no further algebraic labor required to find the eigenvalues. In the following example we shall encounter a case in which for one of the symmetry species there exists a 2×2 non-vanishing block. The XY_3 molecule is an example of D_{3h} symmetry, for which the character table, as well as further information appropriate to XY_3 is given in Table 2-2.

Now a planar molecule possesses $2n - 3$ in-plane and $n - 3$ out-of-plane modes, i.e., 5 in- and 1 out-of-plane modes. It is natural to choose XY bond stretches and YXY bond angles as in-plane coordinates, and the out-of-plane coordinate can be defined in a number of ways, such as the distance of X measured perpendicular to the plane of the three Y's. Alternatively, one might select for the in-plane coordinates the XY distances and the (non-bonded) YY distances. However, for either alternative there is a minor dilemma: the various symmetry

Table 2-2 Characters of the irreducible representations (D_{3h}) and of the internal coordinates for XY_3

D_{3h}	E	$2C_3$	$3C_2$	σ_h	$2S_3$	$3\sigma_v$
A_1'	1	1	1	1	1	1
A_2'	1	1	-1	1	1	-1
E'	2	-1	0	2	-1	0
A_1''	1	1	1	-1	-1	-1
A_2''	1	1	-1	-1	-1	1
E''	2	-1	0	-2	1	0
χ_r	3	0	1	3	0	1
χ_α	3	0	1	3	0	1
χ_p	1	1	-1	-1	-1	1

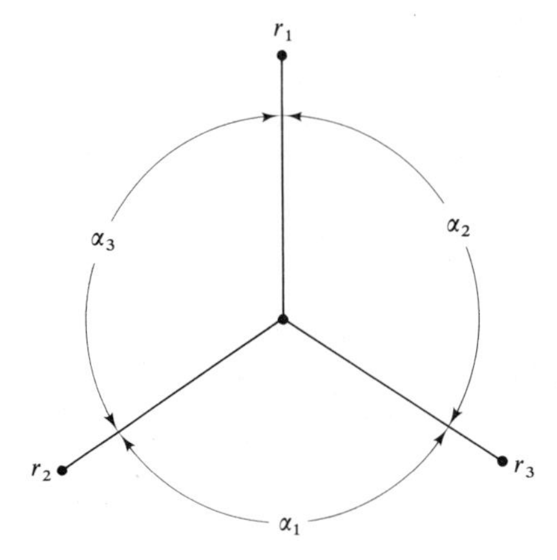

Figure 2-4 Coordinate numbering in the XY_3 molecule

operations of D_{3h} clearly interchange all three XY's among themselves and similarly interchange *all* three YXY's or *all* three YY's among themselves. Thus whichever choice of internal coordinates we make, in order to analyze the transformation properties of the internal coordinates, it appears necessary to introduce six in-plane coordinates, of which only five can be independent. This is an example of the *redundancy* problem, which has been the motivation of a number of discussions in the literature, some of which are vacuous if not actually erroneous.

It is convenient to carry the redundant coordinate through the preliminary stages of the analysis and to eliminate it at a later stage. In Section 2-4 (Selection Rules) it will be shown that the symmetry species occurring in a vibrational problem are uniquely determined by the character of the coordinate representation. For XY_3 in the XY, YXY representation it is clear first that adoption of the coordinate numbering exhibited in Fig. 2-4 renders the three-dimensional representations by the three bond stretches and by the three in-plane bends identical. Further, to calculate the character it is only necessary to subtract the number of members of the set which are sent into their negatives from the number which are sent into themselves by the operation in question, since the elements of the transformation matrix are either ± 1 or 0. For example, the twofold rotation C_2 about bond number 2 is represented by the matrix

$$\begin{pmatrix} 0 & 0 & 1 \\ 0 & 1 & 0 \\ 1 & 0 & 0 \end{pmatrix}$$

Thus the two rows below the character table proper (Table 2-2) labelled χ_r and χ_α have been calculated in a most elementary manner.* The out-of-plane coordinate p is the only member of its set and its character is easily reckoned by visualizing whether the direction of the displacement if the X atom is changed by the symmetry operation.

* Note that C_2, which interchanges bonds 1 and 3, leaves the change in bond angle α_2 unaffected and does not change its sign.

At this stage, the problem is to discover what linear combination of the rows in the table proper (A_1' through E'') yields the rows labelled χ_r, χ_α, and χ_p. For χ_p it is obvious that A_2'' corresponds exactly. For χ_r it is easy to determine the answer by inspection, but the decomposition formula, Eq. (2-2-27),

$$\eta^{(\mu)} = \frac{1}{g} \sum_i g_i \chi_i^{(\mu)*} \chi_i^{(r)} \tag{2-3-28}$$

yields $\eta_r^{(A_1)} = 1$, $\eta_r^{(E)} = 1$, and $\eta_r^{(\mu)} = 0$ for all other species. We therefore conclude that it is possible to form one linear combination (symmetry coordinate) of the r_t which transforms like A_1' and one pair of linear combinations which transform like E'. Thus for all seven coordinates of the sets r, α, and p, one would find symmetry coordinates of the species:

$$2A_1' + 2E' + A_2''$$

Were we now to write out the GF matrix we would expect to find a 2×2 block corresponding to A_1', another 2×2 repeated once identically (E'), and one 1×1 block (A_2''). However, we must remember the redundancy lurking in the algebraic thicket. In the present case it is easy to recognize where it lurks, since the totally symmetric (A_1') symmetry coordinates are necessarily simple sums of the form:

$$S_r(A_1') = (r_1 + r_2 + r_3)(3)^{-1/2}$$

$$S_\alpha(A_1') = (\alpha_1 + \alpha_2 + \alpha_3)(3)^{-1/2}$$

[$(3)^{-1/2}$ is the normalizing constant]. The former of these is a genuine coordinate, but the latter is the redundancy for the simple reason that the sum of the three angle changes must vanish identically.* Had we employed the three non-bonded YY distances in place of the α's, they would similarly have contributed $A_1' + E'$ to the set of symmetry coordinates. The redundancy again is in the A_1' species, not because the totally symmetric sum vanishes identically, but because it prescribes a motion which is linearly dependent upon the sum of the three XY stretches; each symmetry coordinate (XY and YY) would in fact correspond to a simple scaling (breathing) mode in which X is stationary and the Ys move in and out along the bond directions with equal phase and amplitude. We have therefore reached the conclusion that the genuine symmetry coordinates, and *a fortiori* the normal coordinates of XY_3 are of species $A_1' + A_2'' + 2E'$; the normal coordinates are conventionally numbered ν_1, in A_1', ν_2 in A_2'', and ν_3 and ν_4 in E'. The conventional numbering systems assign numbers in order of decreasing frequency within a given species. Since ν_3 and ν_4 fall in the same species we may expect that the corresponding normal coordinates are mixtures of the symmetry coordinates, but that, since stretching frequencies are usually higher than bending frequencies, Q_3 is mainly the E' symmetry coordinate formed from the XY set and Q_4 is mainly the E' symmetry coordinate formed from the YXY set.

* For in-plane displacements, the sum of the three angles is always 360°, so that the sum of the changes from their equilibrium values is zero.

Since the choice of internal coordinates and the recognition and removal of redundancies has some pitfalls for the inexperienced worker, it is fortunately possible to deduce the symmetry properties of the normal modes entirely in terms of the $3n$-dimensional coordinate representation afforded by cartesian displacements. The procedure is defined and illustrated in Appendix II, where for the XY_3 example presently under discussion it is shown that after removal of the three translations and three rotations the genuine vibrational representation consists precisely of $A_1' + A_2'' + 2E'$ as stated above.

In concluding this section we now wish to examine the nature and number of the force constants. In symmetry coordinates, we have

$$\begin{array}{ccccccc} A_1' & & A_2'' & E' & & E' & \end{array}$$

$$\mathsf{F} = \begin{pmatrix} \mathsf{F}_{11} & \mathsf{F}_{1,1a} & 0 & 0 & 0 & 0 & 0 \\ \mathsf{F}_{1,1a} & \mathsf{F}_{1a,1a} & 0 & 0 & 0 & 0 & 0 \\ 0 & 0 & \mathsf{F}_{22} & 0 & 0 & 0 & 0 \\ 0 & 0 & 0 & \mathsf{F}_{33} & \mathsf{F}_{34} & 0 & 0 \\ 0 & 0 & 0 & \mathsf{F}_{34} & \mathsf{F}_{44} & 0 & 0 \\ 0 & 0 & 0 & 0 & 0 & \mathsf{F}_{33} & \mathsf{F}_{34} \\ 0 & 0 & 0 & 0 & 0 & \mathsf{F}_{34} & \mathsf{F}_{44} \end{pmatrix}$$

Here we have indicated the redundant coordinate with the subscript $1a$; note also that since F is symmetric, a 2×2 block involves three (not four) parameters, and that the E' factor is repeated identically. In general if there are $\eta^{(\gamma)}$ symmetry (or normal) coordinates of a given species, counting each degenerate species *only once*, there will be

$$\frac{1}{2}\sum_{\gamma} \eta^{(\gamma)}(\eta^{(\gamma)} + 1)$$

distinct force constants, which appears to be seven in the present case. However, the redundant coordinate $1a$ in species A_1' either vanished identically (YXY angle) or is linearly dependent upon the coordinate 1 (YY stretch), so that there is only one independent parameter contained in the A_1' factor. Thus the true number of independent parameters is not seven but five, which simply means that $\eta^{(\gamma)}$ must be interpreted in terms of the actual degrees of freedom after elimination of redundancies.

Just as in the case of CO_2, the observation of the frequencies ν_1 and ν_2 permits the deduction of F_{11} and F_{22} (when the appropriate G matrix elements are known), but unlike CO_2, the observation of ν_3 and ν_4 in the E' species still leaves the problem indeterminate. We shall discuss the use of isotopes as a means of resolution of this type of problem in Section 2-5, where a few further details of such calculations will be outlined. First, however, in the following section we turn to the problem of identifying the species of the *experimentally* observed frequencies, i.e., the *assignment* of the modes. This is accomplished in part through a detailed study of the *selection rules,* which restrict the allowed transitions in infrared absorption and the Raman effect respectively.

2-4 SELECTION RULES

A spectrum is a two-dimensional array of information—it conveys both the position, on an energy scale, of a spectroscopic transition as well as its strength, or intensity.

All spectroscopic transitions do not occur with equal probability. In that the density of vibrational states (levels per cm^{-1}) of a polyatomic molecule becomes very large above, say, 10 000 cm^{-1}, were this the case, all matter would be opaque to visible light. There are, then, *selection rules* which state when a transition may take place. In some instances, they may also state when a transition cannot take place, but then statements are usually conditional, as we shall see.

Inasmuch as we shall be primarily interested in vibrational transitions in crystals, we shall assume that the translational and rotational motions of the molecule have been quenched and that the electronic part of the complete wave function of the molecule remains constant. We shall therefore concentrate on *vibrational selection rules.*

We shall only consider the selection rules and intensity theory in the electric dipole approximation, since intensities due to magnetic dipole or electric quadrupole coupling with the radiation field are known to be at least five orders of magnitude weaker than those arising from elastic dipole coupling. The coupling operator between the molecule and the radiation field is thus of the form

$$H' = -\mathbf{p} \cdot \mathscr{E} \tag{2-4-1}$$

where $\mathbf{p}$ is the total molecular dipole moment and $\mathscr{E}$ the field vector of the radiation. It is convenient to separate the dipole into an intrinsic (field free) part $\boldsymbol{\mu}$ and additional terms which represent the induced part as a power series in the applied field:

$$\mathbf{p} = \boldsymbol{\mu} + \boldsymbol{\alpha}\mathscr{E} + \boldsymbol{\beta}\mathscr{E}^2 + \cdots \tag{2-4-2}$$

In (2-4-2), $\boldsymbol{\alpha}$, called the polarizability, is a second rank symmetric tensor, $\boldsymbol{\beta}$, the hyperpolarizability is a third rank tensor, etc. Substitution of such an expression for $\mathbf{p}$ in (2-4-1) evidently will give rise to terms in H' which are linear, quadratic, cubic in $\mathscr{E}$. The linear terms give rise to single photon creation or annihilation, i.e., ordinary emission or absorption; the quadratic terms lead to several two-photon effects, of which the most important for this work is Raman scattering, consisting of the scattering of an incident photon of one energy into a second photon of a different energy; the cubic terms lead to a still richer variety of processes, including the hyper-Raman effect in which two incident photons are scattered into a single photon. In each case, of course, the conservation of energy applies, so that in the first case (infrared) the photon energy gives the molecular transition energy directly, in the second case (Raman effect), the difference between the incident and scattered photon energies equals the molecular transition energy, and in the third case (hyper-Raman), the difference between the scattered photon energy and the sum of energies of the two incident photons is the transition energy.

Time dependent perturbation theory[8] then shows that the rates of transitions between states $|\mathbf{v}\rangle$ and $|\mathbf{v}'\rangle$ are proportional to the squared magnitude of the appropriate transition matrix elements, i.e.,

$$\begin{aligned} &|\langle \mathbf{v}'|\boldsymbol{\mu}|\mathbf{v}\rangle|^2 \text{ for infrared} \\ &|\langle \mathbf{v}'|\boldsymbol{\alpha}|\mathbf{v}\rangle|^2 \text{ for Raman} \qquad (2\text{-}4\text{-}3) \\ &|\langle \mathbf{v}'|\boldsymbol{\beta}|\mathbf{v}\rangle|^2 \text{ for hyper-Raman} \end{aligned}$$

Since the last mentioned effect is very weak and has not yet led to any appreciable body of data, we will drop it from further discussion, referring the reader to a few works suggesting its possible uses.[9]

The matrix elements themselves are *invariants,* which means that they either vanish or transform like totally symmetric representations of the molecular group. There is a temptation to regard such matrix elements as quantities which may change under the application of symmetry operations because they bear the labels of a component of the dipole vector or the polarizability tensor. *This temptation should be resisted:* it is the *operators* $\boldsymbol{\mu}$ and $\boldsymbol{\alpha}$ which may transform in a non-totally symmetric manner, not the numbers which arise from evaluation of quantum mechanical matrix elements.

Each of the three factors defining a transition matrix element is itself the basis for a representation of the molecular group, so we must next examine the nature of the representation whose basis is a product of bases, i.e., $|\mathbf{v}\rangle$, $\boldsymbol{\mu}$ or $\boldsymbol{\alpha}$, and $\langle \mathbf{v}'|$. It is simplest to start with a binary product.

(a) Direct product representations Let $\psi_{k\alpha}$, $\alpha = 1, 2, \ldots, d_k$ and $\psi_{l\beta}$, $\beta = 1, 2, \ldots, d_l$ be any two sets of basis functions for representations of the molecular group. The character of the respective representations are

$$\chi_R(k) = \sum_\alpha R^{(k)}_{\alpha\alpha}$$

and

$$\chi_R(l) = \sum_\beta R^{(l)}_{\beta\beta}$$

Then consider the set of product functions $\psi_{k\alpha}\psi_{l\beta}$: if the sets are distinct, there are $d_k d_l$ functions in such a set, but if the two sets are identical, there are only $d_k(d_k - 1)$ basis functions; in the latter case, one refers to the representation as a *symmetrized direct product.*

It is very easy to evaluate the character in the first case. The direct product matrix has diagonal terms of the form $R^{(k)}_{\alpha\alpha} R^{(l)}_{\beta\beta}$ so that the character of the direct

8. This topic is presented in all the standard textbooks on quantum mechanics; see, for example, F. L. Pilar, *Elementary Quantum Chemistry,* McGraw-Hill, New York, 1968, Chapter 5.

9. The effect was discovered by R. W. Terhune, P. D. Maker, and C. M. Savage, *Phys. Rev. Letters,* **14,** 681 (1965). The selection rules have been reported by S. J. Cyvin, J. E. Rauch, and J. C. Decius, *J. Chem. Phys.,* **43,** 4083 (1965).

product representation is

$$\chi_R(k \times l) = \sum_\alpha \sum_\beta R_{\alpha\alpha}^{(k)} R_{\beta\beta}^{(k)} = \chi_R(k) \cdot \chi_R(l) \tag{2-4-4}$$

This result can clearly be extended to a product of three or more factors. Thus we have a simple theorem: the character of a direct product representation is the product of the characters.

The evaluation of the character for a symmetrized direct product is not quite so simple. By defining the basis functions as ψ_α^2 (all α) and $\psi_\alpha\psi_\beta$ ($\alpha < \beta$), one finds that the character of the binary symmetrized direct product is

$$\chi_R(2) = \sum_\alpha R_{\alpha\alpha}^2 + \sum_{\alpha<\beta} (R_{\alpha\alpha}R_{\beta\beta} + R_{\alpha\beta}R_{\beta\alpha})$$

Then since

$$\sum_{\alpha<\beta} (R_{\alpha\alpha}R_{\beta\beta} + R_{\alpha\beta}R_{\beta\alpha}) = \tfrac{1}{2} \sum_{\alpha\neq\beta} (R_{\alpha\alpha}R_{\beta\beta} + R_{\alpha\beta}R_{\beta\alpha})$$

the character of the symmetrized direct product representation may be written as

$$\begin{aligned}\chi_R(2) &= \tfrac{1}{2}\sum R_{\alpha\alpha}^2 + \tfrac{1}{2}\sum_{\alpha\neq\beta} R_{\alpha\alpha}R_{\beta\beta} + \tfrac{1}{2}\sum_\alpha R_{\alpha\alpha}^2 + \tfrac{1}{2}\sum_{\alpha\neq\beta} R_{\alpha\beta}R_{\beta\alpha} \\ &= \tfrac{1}{2}[(\sum_\alpha R_{\alpha\alpha})^2 + \sum_\alpha\sum_\beta R_{\alpha\beta}R_{\beta\alpha}] \\ &= \tfrac{1}{2}[(\chi_R)^2 + \chi_{R^2}]\end{aligned} \tag{2-4-5}$$

since

$$\sum_\alpha\sum_\beta R_{\alpha\beta}R_{\beta\alpha} = \sum_\alpha (R^2)_{\alpha\alpha} = \chi_{R^2}$$

A generalization of formulas of this type for symmetrized direct products of any order, based upon two- and three-dimensional representations is given in WDC.[10]

We are now in a position to evaluate the characters of the representations whose bases are the transition matrix elements appropriate either to infrared or Raman processes. In general, such representations are reducible, but we know how to ascertain whether they contain the totally symmetric representation: all that is necessary is to apply the decomposition formula (2-2-27). Thus, if the appropriate characters for the transition matrix elements are simply designated as χ_R(infrared) or χ_R (Raman), the reduction of the respective representation contains the totally symmetric representation precisely $\eta^{(1)}$ times where

$$\begin{aligned}\eta^{(1)}(\text{infrared}) &= \frac{1}{g}\sum_R \chi_R^{(1)}\chi_R(\text{infrared}) \\ &= \frac{1}{g}\sum_R \chi_R(\text{infrared})\end{aligned}$$

10. WDC, page 153.

and

$$\eta^{(1)}(\text{Raman}) = \frac{1}{g}\sum_R \chi_R(\text{Raman})$$

In practice it is more convenient to separate the triple direct product representation defined by the transition matrix elements into two binary products. First we consider the *transition representation* based simply upon the initial and final states:

$$\chi_R(\mathbf{v}'^* \times \mathbf{v}) = \chi_R^*(\mathbf{v}')\chi_R(\mathbf{v})$$

Note the appearance of the complex conjugate sign in consequence of the quantum mechanical rules involving bra and ket functions. We then consider the product of the transition representation with the *activity representation*, i.e., the representation based upon $\boldsymbol{\mu}$ or $\boldsymbol{\alpha}$. It is desirable actually to express all representations as sums (technically *direct sums*) of irreducible representations. We thus may use the symbol Γ to designate a representation in general and express it as a sum of irreducible representations by

$$\Gamma = \Sigma \eta^{(\gamma)}\Gamma^{(\gamma)}$$

where $\Gamma^{(\gamma)}$ stands for an irreducible representation. Thus we could now write

$$\Gamma(\mathbf{v}'^* \times \mathbf{v}) = \sum_\gamma \eta^{(\gamma)}\,(\text{transition})\,\Gamma^{(\gamma)}$$

$$\Gamma(\boldsymbol{\mu}) = \sum_{\gamma'} \eta^{(\gamma')}(\boldsymbol{\mu})\Gamma^{(\gamma')}$$

$$\Gamma(\boldsymbol{\alpha}) = \sum_{\gamma'} \eta^{(\gamma')}(\boldsymbol{\alpha})\Gamma^{(\gamma')}$$

From what was said above, we know that a process is *allowed* if the direct product of the transition representation with the activity representation contains the totally symmetric representation, *forbidden* otherwise. But this question is easily settled with the aid of the following simple theorem: a direct product of two irreducible representations contains the totally symmetric representation if and only if the two irreducible representations are identical. The proof is elementary: the character of the direct product is simply

$$\begin{aligned}\chi_R(\gamma \times \gamma') &= \chi_R^{(\gamma)}\chi_R^{(\gamma')}\\ &= \sum_{\gamma''} \eta^{(\gamma'')}\chi_R^{(\gamma'')}\end{aligned}$$

and to find $\eta^{(\gamma'')}$ for $\gamma'' = 1$ we use the decomposition formula (2-2-27) and find

$$\begin{aligned}\eta^{(\gamma''=1)} &= \frac{1}{g}\sum_R \chi_R^{(1)}\chi_R(\gamma \times \gamma')\\ &= \frac{1}{g}\sum \chi_R(\gamma \times \gamma') = \frac{1}{g}\sum_R \chi_R^{(\gamma)}\chi_R^{(\gamma')} \qquad \text{(2-4-6)}\\ &= \delta_{\gamma\gamma'}\end{aligned}$$

This means that the issue of the selection rules is settled by comparing the irreducible representations contained in $\Gamma(\mathbf{v}'^* \times \mathbf{v})$ with those contained in the activity representation. For infrared processes we have

$$\Gamma(\mathbf{v}' \times \mathbf{v}) \times \Gamma(\boldsymbol{\mu}) = \sum_{\gamma}\sum_{\gamma'} \eta^{(\gamma)}\,(\text{transition})\,\eta^{(\gamma')}(\boldsymbol{\mu})\Gamma^{(\gamma)} \times \Gamma^{(\gamma')}$$

so that an infrared process is allowed if the transition representation has species (irreducible representations) in common with $\Gamma(\boldsymbol{\mu})$, forbidden otherwise. For the Raman effect, one compares the transition species with those of $\boldsymbol{\alpha}$. The irreducible representations appearing in $\boldsymbol{\mu}$ and $\boldsymbol{\alpha}$ can be tabulated once and for all for each point group: this is done in Appendix I. For example, in D_{3h} one finds

$$\Gamma(\boldsymbol{\mu}) = A_2'' + E'$$

and

$$\Gamma(\boldsymbol{\alpha}) = 2A_1 + E' + E''$$

The irreducible representations involved in the state functions $|\mathbf{v}\rangle$ and $|\mathbf{v}'\rangle$ and hence in the transition representation depend in a detailed way upon the modes and quantum numbers for a particular molecule. Presently we shall give an account of such analysis, but a simple example may help the reader's understanding at this stage. For the XY_3 molecule, we may designate the ground state as $|0000\rangle$, the state with one quantum of v_2 as $|0100\rangle$, and that with one quantum of v_3 as $|0010\rangle$. These states have the respective symmetry species:

$$|0000\rangle \qquad A_1'$$
$$|0100\rangle \qquad A_2''$$
$$|0010\rangle \qquad E'$$

The transition representations are then

$$\langle 0100|0000\rangle \qquad A_2'' \times A_1' = A_2''$$
$$\langle 0010|0000\rangle \qquad E' \times A_1' = E'$$
$$\langle 0010|0100\rangle \qquad E' \times A_2'' = E''$$

The reduction of the several direct products is readily accomplished using the character methods established above, but can be read very rapidly from tables.[11] Note that $\Gamma^{(1)} \times \Gamma^{(\gamma)} = \Gamma^{(\gamma)}$ since $\chi_R(1 \times \gamma) = \chi_R^{(1)} \times \chi_R^{(\gamma)} = \chi_R^{(\gamma)}$. The first of the three transitions is allowed in the infrared (A_2'' appears in $\Gamma(\boldsymbol{\mu})$) but forbidden in the Raman. The second transition is allowed in both types of spectra. The third transition is allowed only in the Raman effect.

11. WDC, pages 331–333.

(b) Activity representations Since $\boldsymbol{\mu}$ is an ordinary cartesian vector, its character is

$$\chi_R(\boldsymbol{\mu}) = \pm 1 + 2\cos\theta_R \tag{2-4-7}$$

where the sign choice is + for rotations and − for operations involving a reflection (σ, i, S), and θ_R is the angle of rotation. The reduction of this three-dimensional representation, which is the same as that of a translation, is indicated in Appendix I.

The polarizability has a character which may be found in a number of ways. Its six components may be regarded as a symmetrized direct product formed from the cartesian representation, so that (2-4-5) and (2-4-7) combined yield

$$\begin{aligned}\chi_R(\boldsymbol{\alpha}) &= \tfrac{1}{2}[(\pm 1 + 2\cos\theta_R)^2 + (1 + 2\cos 2\theta_R)] \\ &= 2\cos\theta_R(\pm 1 + 2\cos\theta_R)\end{aligned} \tag{2-4-8}$$

This representation, like that of $\boldsymbol{\mu}$, can be reduced once and for all in each group. Appendix I goes a stage further: with the aid of the WPO (2-2-30) it is possible to find the appropriate linear combinations of the individual elements of $\boldsymbol{\mu}$ and of $\boldsymbol{\alpha}$ which belong to the several irreducible representations appearing in the respective representations. In fact, for subsequent use in the analysis of crystal spectra, Appendix I displays detailed information about $\boldsymbol{\mu}$ and $\boldsymbol{\alpha}$ by tabulating the non-vanishing derivatives of the components of $\boldsymbol{\mu}$ and of $\boldsymbol{\alpha}$ with respect to normal modes of each species. For the D_{3h} example, the entry n in the column headed μ'_z indicates that μ_z transforms like the A''_2 species, and the identical entries l in the μ'_x and μ'_y columns show that μ_x and μ_y transform respectively like the two basis functions for the E' species. This is equivalent to the statement

$$\Gamma(\boldsymbol{\mu}) = A''_2 + E'$$

made above for this example. Similarly, one finds that $\partial\alpha_{xx}/\partial Q = \partial\alpha_{yy}/\partial Q$ when Q is a totally symmetric coordinate, and that α_{zz} also transforms like the totally symmetric species, but that $\partial\alpha_{zz}/\partial Q \neq \partial\alpha_{xx}/\partial Q$. Thus the diagonal elements $\alpha_{xx}+\alpha_{yy}$ and α_{zz} give rise to $2A'_1$. Further entries (c) show that $\alpha_{xx} - \alpha_{yy}$ and α_{xy} are of species E' and finally that α_{yz} and α_{xz} (d) belong to E'', so

$$\Gamma(\boldsymbol{\alpha}) = 2A'_1 + E' + E''$$

(c) State and transition representations It remains to explain the method of deducing the representations associated with a general vibrational state indicated by $|\mathbf{v}\rangle$. The ground state, $|0\rangle$, is obviously non-degenerate, and so necessarily is the basis for some one-dimensional irreducible representations. In fact the ground state is totally symmetric, as can be shown by detailed examination of the harmonic oscillator wave function.

Any excited state can be obtained by application of suitable products of the creation operators. Omitting an irrelevant constant factor, with the aid of Appendix III one sees that

$$|v_1 v_2 v_3 \ldots\rangle = (\text{constant})(a_1^+)^{v_1}(a_2^+)^{v_2}(a_3^+)^{v_3}\ldots|000\ldots\rangle \tag{2-4-9}$$

From (III-9) it is evident that a_k^+ must transform like Q_k, therefore it is only necessary to form the direct product

$$\Gamma^{v_1}(Q_1) \times \Gamma^{v_2}(Q_2) \times \Gamma^{v_3}(Q_3) \times \cdots \Gamma^{(1)} \tag{2-4-10}$$

in order to specify the representation of any molecular vibration state. If Q_k is a degenerate normal mode and $v_k > 1$, the symmetrized direct product must be used.

The most important excited states are those in which a single $v_k = 1$ and all other v's are zero: the so-called *fundamental excitations*. For such states we have simply

$$\Gamma(\mathbf{v}) = \Gamma(Q_k)$$

which we used in the XY_3 example. The selection rules for transitions from the ground state evidently can be stated in the simple form: the transition is allowed if the symmetry species of Q_k is one of the species of $\boldsymbol{\mu}$ (infrared) or $\boldsymbol{\alpha}$ (Raman).

Overtones of *non-degenerate* modes obey the particularly simple rule

$$\begin{aligned} \Gamma^{v_k}(Q_k) &= \Gamma^{(1)} && \text{if } v_k \text{ is even} \\ &= \Gamma(Q_k) && \text{if } v_k \text{ is odd} \end{aligned} \tag{2-4-11}$$

We conclude with two examples of combinations and overtones involving degenerate states, in particular $|0011\rangle$ and $|0020\rangle$, where Q_3 and Q_4 both belong to E'. In the first case

$$\Gamma(\mathbf{v}) = \Gamma(Q_3) \times \Gamma(Q_4) = \{E' \times E'\} = A_1' + A_2' + E'$$

so transition from the ground state is allowed in the Raman effect because of the appearance of A_1' and E' and in the infrared, because of E'. In contrast, the state $|0020\rangle$ has the symmetry species $[E']^2 = A_1' + E'$ (we use $[E']^2$ to indicate the symmetrized direct product) and is again allowed.

The selection rules as stated above are clearly of a very general nature; only that part of the analysis which describes the symmetry of states in terms of the symmetry of normal modes is limited by the assumption of the harmonic oscillator. For real molecules, anharmonicity shifts the energy levels, but the specification of state symmetries is unaffected, since inclusion of anharmonicity can only mix states of identical symmetry. The ground state is still totally symmetric, and excited states, as described by harmonic oscillator quantum numbers in zeroth order, also conserve their symmetry, although their more accurate description may require small admixture with other harmonic oscillator states. Although this complicates the quantitative discussion of the intensities, it is important to have available a framework for the discussion of intensities, since intermolecular coupling in crystals involves the infrared transition integrals.

We have seen that, subject to the symmetry based selection rules, the transition integrals of the form $\langle v|\mu|v'\rangle$ govern the infrared intensity. Let us continue to assume that the molecule is mechanically harmonic and now investigate such integrals a little further. We may expand the expression for the dipole moment as a power series in the normal coordinates, retaining only constant, linear, and quadratic terms:

$$\boldsymbol{\mu} = \boldsymbol{\mu}_e + \sum_k \left(\frac{\partial \boldsymbol{\mu}}{\partial Q_k}\right)_e Q_k + \frac{1}{2}\sum_k \sum_l \left(\frac{\partial^2 \boldsymbol{\mu}}{\partial Q_l\, \partial Q_k}\right)_e Q_k Q_l \tag{2-4-12}$$

where the subscripts e refer to the equilibrium configuration of the molecule (all $Q_k = 0$). The first term, the equilibrium dipole $\boldsymbol{\mu}_e$, cannot give rise to any vibrational transition, since

$$\langle \mathbf{v} \,|\, \boldsymbol{\mu}_e \,|\, \mathbf{v}' \rangle = \boldsymbol{\mu}_e \langle \mathbf{v} \,|\, \mathbf{v}' \rangle = \boldsymbol{\mu}_e \prod_k \delta_{v_k v_k'}$$

The second term, linear in Q_k, permits transitions of the type $\Delta v_k = \pm 1$, $\Delta v_l = 0$, $l \neq k$, i.e., the fundamental transitions, *provided* $(\partial \boldsymbol{\mu}/\partial Q_k)_e \neq 0$. A moment's reflection shows that this is equivalent to the general rule of symmetry, which requires that the activity representation $(\boldsymbol{\mu})$ have species in common with the transition representation, which in this case is identical with the representation with Q_k as a basis. Combination and overtone transitions similarly arise from the fact that if $(\partial^2 \boldsymbol{\mu}/\partial Q_l\, \partial Q_k)_e$ is non-vanishing, the transition representation, for which $\Delta v_k = \pm 1$, $\Delta v_l = \pm 1$, must have species in common with $\boldsymbol{\mu}$. An analogous expansion of the polarizability $\boldsymbol{\alpha}$ leads to similar conclusions.

For the fundamental transition of frequency ν_k, the matrix element is easily evaluated and is

$$\langle \mathbf{v}' | \left(\frac{\partial \boldsymbol{\mu}}{\partial Q_k}\right) Q_k \,|\, \mathbf{v} \rangle = \left(\frac{\partial \boldsymbol{\mu}}{\partial Q_k}\right) \langle 1 \,|\, Q_k \,|\, 0 \rangle = \frac{\partial \boldsymbol{\mu}}{\partial Q_k} \left(\frac{\hbar}{2\omega_k}\right)^{1/2} \tag{2-4-13}$$

However, one must remember that all transitions in which $\Delta v_k = 1$ occur at the same frequency for the harmonic oscillator, so that it is necessary to consider all the matrix elements of the form

$$\langle v_k + 1 | \left(\frac{\partial \boldsymbol{\mu}}{\partial Q_k}\right) Q_k \,|\, v_k \rangle = \left(\frac{\partial \boldsymbol{\mu}}{\partial Q_k}\right) \left(\frac{\hbar}{2\omega_k}\right)^{1/2} (v_k + 1)^{1/2} \tag{2-4-14}$$

There are several measures of the strength of a transition, including the Einstein coefficients for spontaneous emission, $A_{\mathbf{vv}'}$, and of induced emission or absorption, $B_{\mathbf{vv}'}$ or $B_{\mathbf{v}'\mathbf{v}}$. These are related to one another by the expressions

$$A_{\mathbf{vv}'} = \frac{8\pi h \nu_{\mathbf{vv}'}^3}{c^3} B_{\mathbf{vv}'}; \qquad B_{\mathbf{vv}'} = B_{\mathbf{v}'\mathbf{v}} \tag{2-4-15}$$

and to the dipole transition matrix element by

$$B_{\mathbf{vv}'} = \frac{8\pi^3}{h^2} |\langle \mathbf{v}' \,|\, \mu_x \,|\, \mathbf{v} \rangle|^2 \tag{2-4-16}$$

assuming that x-polarized radiation is employed. For randomly oriented molecules in a fluid phase it is appropriate to replace (2-4-16) by

$$B_{\mathbf{vv}'} = \frac{8\pi^3}{3h^2} \{|\langle \mathbf{v}' \,|\, \mu_x \,|\, \mathbf{v} \rangle|^2 + |\langle \mathbf{v}' \,|\, \mu_y \,|\, \mathbf{v} \rangle|^2 + |\langle \mathbf{v}' \,|\, \mu_z \,|\, \mathbf{v} \rangle|^2\} \tag{2-4-17}$$

Although the transition rate theory leads to selection rules as described above, a more quantitative description of the actual spectrum requires consideration of the rotational and translational degrees of freedom. The rotational quantization of free gaseous molecules leads of course to a rich fine structure,[12,13] but if this is neglected and the molecules are assumed to be isolated, the vibrational lines would be very narrow, having a width determined by the radiative lifetimes and by Doppler broadening. Most studies are conducted under conditions such that *pressure broadening* must be considered. This means that collisions influence the lifetimes of initial and final states and one must appeal to the methods of statistical mechanics for a satisfactory theory. As a consequence of a very general principle, the so-called "fluctuation-dissipation" theorem due mainly to the work of Kubo,[14] it is possible to establish a connection between the infrared spectrum and the *dipole autocorrelation function.* For a vibration transition $\mathbf{v} \rightarrow \mathbf{v}'$ this relation takes the form[15]

$$\alpha_{\mathbf{vv}'}(\omega + \omega_0)[(\omega + \omega_0)(1 - e^{-\hbar(\omega+\omega_0)/kT}]^{-1} = (4\pi^2 N/3\hbar c)\left[\frac{1}{2\pi}\int_{-\infty}^{+\infty} e^{-i\omega t}\langle \boldsymbol{\mu}_{\mathbf{vv}'}(0) \cdot \boldsymbol{\mu}_{\mathbf{vv}'}(t)\rangle \, dt\right] \quad (2\text{-}4\text{-}18)$$

which shows that the absorption coefficient $\alpha_{\mathbf{vv}'}$ is essentially a Fourier transform of the correlation function for the transition dipole moment $\boldsymbol{\mu}_{\mathbf{vv}'} = \langle \mathbf{v}' | \boldsymbol{\mu} | \mathbf{v} \rangle$. The brackets in (2-4-18) indicate an *ensemble average.* Fourier inversion enables one to deduce the time behavior of the correlation function from the experimental absorption coefficient

$$\frac{2\pi}{3\hbar c} N\langle \boldsymbol{\mu}_{\mathbf{vv}'}(0) \cdot \boldsymbol{\mu}_{\mathbf{vv}'}(t)\rangle = \int_{\text{band}} e^{i\omega t} \alpha_{\mathbf{vv}'}(\omega + \omega_0)[(\omega + \omega_0)(1 - e^{-\hbar(\omega+\omega_0)/kT}]^{-1} \, d\omega \quad (2\text{-}4\text{-}19)$$

As carefully pointed out by Gordon, the current use of these elegant relations requires care in the selection of the band center frequency ω_0; several interesting examples of the time behavior of such correlation functions and extension to the Raman case for condensed phase samples of CO and CH_4 are given in Gordon's work.

As noted above, the frequencies of the fundamental transitions $v_k \rightarrow v_k + 1$ all coincide and must be added for the harmonic oscillator. When this is done, one replaces $\alpha_{\mathbf{vv}'}$ by $\sum_{\mathbf{v}} \alpha_{\mathbf{vv}'} = \alpha_k$ and the summation of the transition dipoles

12. G. Herzberg, *Infrared and Raman Spectra of Polyatomic Molecules,* Van Nostrand, New York, 1945.
13. H. C. Allen, Jr. and P. C. Cross, *Molecular Vib-Rotors,* Wiley, New York, 1963.
14. R. Kubo, *Rep. Prog. Phys.,* **29,** 255 (1966).
15. R. G. Gordon, *J. Chem. Phys.,* **43,** 1307 (1965).

is weighted by

$$\rho_{v_k} = (1 - e^{-h\omega_k/kT})\, e^{-\varepsilon_{v_k}/kT} \tag{2-4-20}$$

Thus one finds

$$\sum \rho_{v_k} \boldsymbol{\mu}_{v_k v_k'}(0) \cdot \boldsymbol{\mu}_{v_k' v_k}(t) = \frac{\partial \boldsymbol{\mu}}{\partial Q_k} \cdot \frac{\partial \boldsymbol{\mu}}{\partial Q_k} \sum \rho_{v_k} [Q_k(0)]_{v_k v_k'} [Q_k(t)]_{v_k' v_k}$$

In most situations, $1 - e^{-h\omega_k/kT}$ is nearly unity, as is the factor $1 - e^{-h(\omega+\omega_0)/kT}$ at all frequencies for which α_k is appreciable ($\omega_0 = \omega_k$). Then putting $t = 0$ in (2-4-19), using (2-4-14) for the v_k dependence of the transition matrix elements $[Q_k(0)]_{v_k v_k'}$, and explicitly summing, one obtains an important special case, first given by Crawford:[16]

$$(N\pi/3\omega_k c)\, \frac{\partial \boldsymbol{\mu}}{\partial Q_k} \cdot \frac{\partial \boldsymbol{\mu}}{\partial Q_k} = \int_{\text{band}} (\alpha_k(\omega)/\omega)\, d\omega$$

which has been widely used for the determination of the molecular parameter, $\partial\mu/\partial Q_k$. The experiments require careful precautions to avoid instrumental distortion if the individual rotational lines are too narrow compared with the spectrometer slit; pressurizing with "inert" gases is commonly used.[17]

2-5 ISOTOPE EFFECTS

The comparative study of the vibrational spectra of chemically identical molecules containing different isotopes has been of value for a number of reasons, principally as an aid in making or confirming the assignments of the frequencies and by providing further data from which the force constants may be evaluated. Both these desiderata are dependent upon the assumption, amply demonstrated by the results of diatomic vibrational spectroscopy, that force constants are invariant to isotopic substitution. The theoretical basis of this approximation is of course the Born–Oppenheimer separation of electronic and nuclear motion.

Amongst the consequences of the assumed invariance to isotopic substitution are various rules relating the frequencies and intensities of isotopic molecules, of which the earliest and most useful is the frequency *product rule,* first elucidated by Redlich and by Teller. Other relations exist, including a sum rule and more intricate interrelations. The reader must here be referred to other sources of more detailed information.[18] Our purpose here is simply to illustrate the magnitude of isotope effects and their use in resolving the indeterminancy of potential functions.

16. B. L. Crawford, Jr., *J. Chem. Phys.*, **29,** 1042 (1958).
17. E. B. Wilson, Jr. and A. J. Wells, *J. Chem. Phys.*, **14,** 578 (1946).
18. WDC, Section 8.5. Also, see footnote references, pages 186 and 191, WDC.

(a) Simple G Matrix Elements

Although we shall refer the reader to WDC for full details on the evaluation of G matrix elements, it will be illuminating to provide the basic information needed for their evaluation so that the subject matter of the present and certain later sections can be put on a more quantitative basis. Recall that in (2-3-15) we defined $\mathbf{G} = \mathbf{B}\,\mathbf{m}^{-1}\mathbf{B}^{\dagger}$ where $\mathbf{m}^{-1}$ is a diagonal matrix with each reciprocal atomic mass repeated thrice (x, y, z) and $\mathbf{B}$ is the rectangular matrix which defines the linear (first order) transformation from the $3n$ cartesian displacement (external) to the $3n - f$ internal coordinates of our choice. An element of G is therefore defined by

$$G_{tt'} = \sum_{j=1}^{3n} B_{tj}(m^{-1})_j B^*_{t'j} \tag{2-5-1}$$

It was shown by Wilson[19] that it is convenient to collect the elements of B for a fixed t and for the three values of j corresponding to the cartesian coordinates of a given atom, α, and to regard such elements as a vector $\mathbf{s}_{t\alpha}$: then (2-5-1) becomes

$$G_{tt'} = \sum_{\alpha=1}^{n} \mu_\alpha \mathbf{s}_{t\alpha} \cdot \mathbf{s}_{t'\alpha} \tag{2-5-2}$$

where $\mu_\alpha = (m_\alpha)^{-1}$.

For a bond stretch, $\mathbf{s}_{t\alpha}$ is an outward directed unit vector along the bond for the two atoms defining the bond and vanishes for all other atoms. Indeed, for any coordinate t and atom α, $\mathbf{s}_{t\alpha}$ is a gradient vector for the (scalar) coordinate with respect to vectorial displacement of atom α. In the case of an angle bending coordinate, these gradient directions are, for the terminal atoms, perpendicular to the bonds, but $\mathbf{s}_{t\alpha}$ for the central atom in the general unsymmetrical case is a little more complicated. However, suppose we write

$$S_t = \sum_{\alpha=1}^{n} \mathbf{s}_{t\alpha} \cdot \boldsymbol{\rho}_\alpha$$

where S_t is any internal coordinate and $\boldsymbol{\rho}_\alpha$ is the displacement vector of atom α. Suppose $\boldsymbol{\rho}_\alpha = \mathbf{t}$ where $\mathbf{t}$ is a constant independent of α, i.e., a translation. For such a motion, any internal coordinate must vanish. Therefore

$$S_t = 0 = \mathbf{t} \cdot \sum_{\alpha=1}^{n} \mathbf{s}_{t\alpha}$$

which implies that

$$\sum_{\alpha=1}^{n} \mathbf{s}_{t\alpha} = \mathbf{0}$$

19. E. B. Wilson, Jr., *J. Chem. Phys.*, **9**, 76 (1941).

since **t** is arbitrary. For a bending motion, only three atoms are involved in the definition, so that once we know the $\mathbf{s}_{t\alpha}$'s for the terminal atoms, that for the central atom is simply the negative sum of the other two.

In the numerical treatment of vibrational problems by this so-called *FG* system it is convenient to keep *G* (and *F*) homogeneous in physical dimensions. Thus although the magnitude of the $\mathbf{s}_{t\alpha}$ for terminal atoms in a bending coordinate would be the reciprocal of the bond length in order to define an internal coordinate S_t in radians, it is more customary to replace a bending angle in radians by the product of some distance by this same angle in order that all coordinates may have the dimensions of distance. When the two bonds defining the angle are equivalent, this scaling distance is taken as the equilibrium bond length; less symmetric situations require a careful definition of the working coordinates in order to avoid confusion.

Figure 2-5 gives diagrams illustrating the $\mathbf{s}_t$ vectors corresponding to the internal coordinates $r_1, r_2, \alpha_x, \alpha_y$ and $r_1, r_2, r_3, \alpha_1, \alpha_2, \alpha_3, p$ introduced in Section 2-3 for the CO_2 and XY_3 molecules. Because of the scaling convention (angular coordinates are multiplied by the equilibrium bond length) all vectors in these diagrams are of unit length, except the central atom vectors, which for α in CO_2 is of length 2, for α in XY_3 is of length $(3)^{1/2}$, and for p (out-of-plane mode) in XY_3 is of length 3. In less symmetrical molecules the situation would of course be less simple.

From these diagrams, using the definition (2-5-2) one can obtain the following non-vanishing **G** elements for CO_2:

$$G_r = G_{r_1r_1} = G_{r_2r_2} = \mu_C + \mu_O$$

$$G'_r = G_{r_1r_2} = -\mu_C$$

$$G_\alpha = 4\mu_C + 2\mu_O$$

$$G_r + G'_r = \mu_O$$

$$G_r - G'_r = 2\mu_C + \mu_O$$

and for XY_3:

$$G_r = G_{r_1r_1} = G_{r_2r_2} = G_{r_3r_3} = \mu_X + \mu_Y$$

$$G'_r = G_{r_1r_2}, \text{ etc.} = -\tfrac{1}{2}\mu_X,$$

$$G_\alpha = 3\mu_X + 2\mu_Y$$

$$G'_\alpha = -\tfrac{1}{2}(3\mu_X + 2\mu_Y)$$

$$G_{r\alpha} = (3)^{1/2}\mu_X$$

$$G'_{r\alpha} = -\frac{(3)^{1/2}}{2}\mu_X$$

$$G_p = 9\mu_X + 3\mu_Y$$

all others vanishing. We have also exhibited the simple combinations $G_r \pm G'_r$

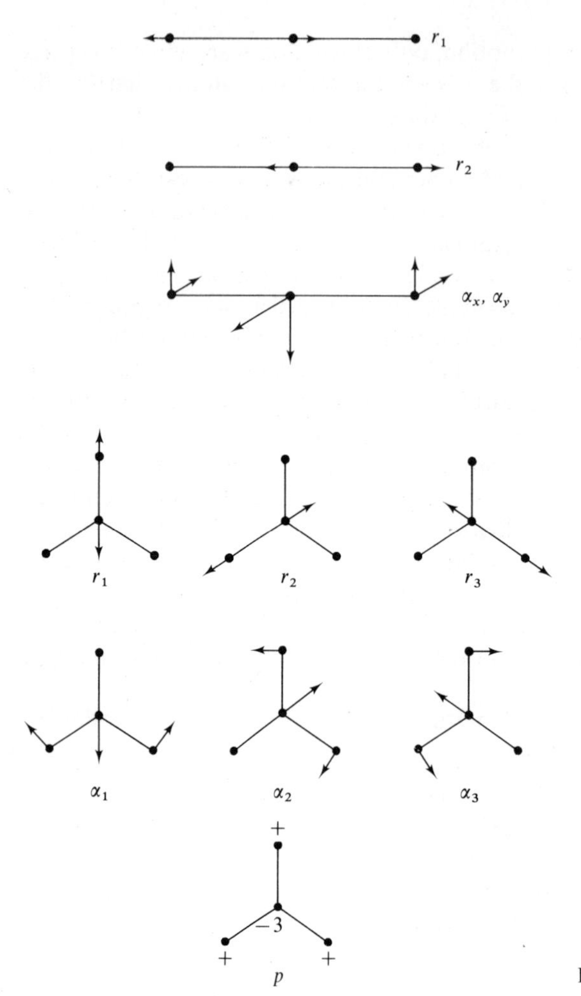

Figure 2-5 s vectors for CO_2 and XY_3

appropriate for the $\Sigma_g^+(\nu_1)$ and $\Sigma_u^-(\nu_3)$ modes of CO_2, [cf. (2-3-27)]; note particularly that the frequency of ν_1 is independent of the central mass.

(b) Isotopic Substitution without Change of Symmetry

In this situation, which is exemplified by the comparison $^{12}CO_2$ with $^{13}CO_2$, H_2O with D_2O, $^{10}BF_3$ with $^{11}BF_3$, or any substitution in linear N_2O, the symmetry factoring described in Section 2-3 will be undisturbed. Thus in the harmonic approximation and in a case like carbon dioxide where each frequency mode is the sole member of its symmetry species, no new information about the force constants is available, since the invariance of the potential function implies that the square of each frequency ($\lambda = 4\pi^2\nu^2$) is proportional to the appropriate individual G matrix element. However, in CO_2 and molecules of similar structure and symmetry, the totally symmetric mode cannot involve displacement of the carbon atom, since such a normal coordinate would not transform like Σ_g^+, so that one concludes without detailed calculation of the corresponding G element that it must be independent of the mass of the central atom and that ν_1 is unshifted by the substitution of isotopic carbon. This is confirmed by the calculation of $G_r + G_r' = \mu_O$. Thus even in the absence of the distinction between Raman and infrared selection rules one has evidence which aids in the assignment, since ν_2 and ν_3 are shifted by such substitution.

In the XY_3 (D_{3h}) example, we found in Section 2-3 that the vibrational representation comprised the symmetry species $A_1' + A'' + 2E'$. The modes $\nu_1(A_1')$ and $\nu_2(A_2'')$ are, each, the only one of their respective symmetry types and hence give rise to one-dimensional blocks in the secular determinant. Thus, the isotopic shifts can be predicted directly from the G matrix elements. The two modes ν_3 and ν_4 of symmetry species E' give rise to a 2×2 factor, and the calculation of individual isotopic frequency shifts would require solution of this determinant. However, in this situation the *product rule* comes into play. The product rule is based upon the algebraic theorem that the product of all eigenvalues of a matrix is equal to its determinant. Hence for any symmetry species, one has

$$\prod_{\text{(species)}} \omega_k^2 = |\mathbf{GF}| = |\mathbf{G}|\,|\mathbf{F}| \tag{2-5-3}$$

where (species) means that k runs over the indices of modes in a single symmetry species. The last form of (2-5-3) in which the product of determinants of **G** and of **F** occurs is crucial to the argument and is a further consequence of the algebra of matrices. By taking the ratio of frequencies ω_k^{*2} of an isotopically substituted molecule (2-5-3) to those of the parent molecule ω_k^2, one may cancel $|\mathbf{F}|$ and obtain

$$\prod_{\text{(species)}} \left(\frac{\omega_k^{2*}}{\omega_k^2}\right) = \frac{|\mathbf{G}^*|}{|\mathbf{G}|} \tag{2-5-4}$$

For the XY_3 molecule, it shows us explicitly that for X^*Y_3 and XY_3 we have

$$\left(\frac{\nu_1^*}{\nu_1}\right)^2 = \frac{\mathsf{G}_{11}^*}{\mathsf{G}_{11}} \qquad \left(\frac{\nu_2^*}{\nu_2}\right) = \frac{\mathsf{G}_{22}^*}{\mathsf{G}_{22}}$$

$$\left(\frac{\nu_3^*\nu_4^*}{\nu_3\nu_4}\right) = \frac{\begin{vmatrix} \mathsf{G}_{33}^* & \mathsf{G}_{34}^* \\ \mathsf{G}_{34}^* & \mathsf{G}_{44}^* \end{vmatrix}}{\begin{vmatrix} \mathsf{G}_{33} & \mathsf{G}_{34} \\ \mathsf{G}_{34} & \mathsf{G}_{44} \end{vmatrix}}$$

(c) Isotopic Substitution with Change of Symmetry

Comparison of molecules such as $^{16}O^{12}C^{16}O$ and $^{18}O^{12}C^{16}O$, NH_3 and NH_2D, or CH_3D and CH_2D_2, even though still based upon the invariance of the potential function, in particular of the F matrix, poses a new problem in that the symmetries of the two molecules are different. For carbon dioxide, $^{18}O^{12}C^{16}O$

has no center of inversion and is an example of the $C_{\infty v}$ rather than the $D_{\infty h}$ point group. Similarly, NH_3 is C_{3v} but NH_2D only C_s, and the symmetries of CH_3D (C_{3v}) and of CH_2D_2 (C_{2v}) are not even related in the sense that one is a supergroup and the other a subgroup of its partner, although both deuterated methanes have symmetry which is a subgroup of that of the parent molecule (T_d).

Lowering of symmetry by isotopic substitution affords examples qualitatively similar to the phenomena described in Chapter 1, in which lowering of the symmetry of a gas phase molecule by subjecting it to a less symmetric crystal field in the condensed phase can give rise to splitting of degenerate modes or activation of frequencies which are forbidden by the gas phase selection rules. Thus in $^{18}O^{12}C^{16}O$, ν_1 becomes allowed in the infrared, or in $B^{35}Cl_2{}^{37}Cl$, the modes ν_3 and ν_4, which are doubly degenerate (species E') in the D_{3h} $B^{35}Cl_3$ molecule, are split, and all modes are allowed in both infrared and Raman spectra under the less stringent selection rules appropriate to the molecule of C_{2v} symmetry. Molecules such as methane, which have triply degenerate modes, exhibit partial lifting of the degeneracy (two component frequencies) in CH_3D, or complete lifting of the degeneracy (three component frequencies) in CH_2D_2.

The analysis of cases of these sorts provides an introduction to the use of *correlation tables*, which provide an even more valuable technique for the crystal (Chapters 4 and 6).

We consider first the cases in which the isotopically substituted molecule has a symmetry which is a subgroup of the symmetry of the parent molecule, for instance C_{2v} for $B^{35}Cl_2{}^{37}Cl$ compared with D_{3h} for $B^{35}Cl_3$. Now the irreducible representations of D_{3h} are representations in the subgroup C_{2v}, which may either be irreducible or reducible in C_{2v}. Since C_{2v} has no degenerate species, the E' species in D_{3h} must be reducible in the smaller group C_{2v}. All that is needed to define the relations between the species of a group G and its subgroup H, is the decomposition formula

$$h_{\gamma}^{(\eta)} = \frac{1}{h} \sum_{R \subset H} \chi_R^{(\eta)*} \chi_R^{(\gamma)} \tag{2-5-5}$$

which gives the number of times that the ηth irreducible representation of H appears in the γth irreducible representation of G. Since these relationships are entirely independent of particular molecular examples, they can be developed solely from the character tables, and the results for all relevant symmetries can be, and are in fact tabulated in Appendix IV. Under the heading of D_{3h} one finds in that Appendix IV (and can easily verify using the individual D_{3h} and C_{2v} tables with Eq. (2-5-5)) that the symmetry species of D_{3h} are *correlated* to those of C_{2v} as follows:

$$A_1' \rightarrow A_1, \quad A_2' \rightarrow B_2, \quad E' \rightarrow A_1 + B_2, \quad A_1'' \rightarrow A_2, \quad A_2'' \rightarrow B_1, \quad E'' \rightarrow A_2 + B_1$$

These correlations imply that the normal modes of the boron trihalide are $(A_1' + A_2'' + 2E') \rightarrow (A_1 + B_1 + 2A_1 + 2B_2)$. Since the species A_1, B_1, B_2 in C_{2v} are active in both Raman and infrared, all modes are now allowed in both spectra; moreover, the degeneracies of ν_3 and ν_4 are split, and these modes each give rise to distinct frequencies ν_{3A} and ν_{3B}, ν_{4A} and ν_{4B}. We shall henceforth refer to (2-5-5) as the *correlation theorem.*

(d) Order of Magnitude of Isotope Shifts

Thus far in this section no mention of the magnitude of isotope effects has been made. In a diatomic molecule XY the G matrix is simply $\mu_X + \mu_Y$ (μ's are reciprocal atomic masses). For a hydride, the relative change in mass on substitution of D for H is very large; moreover, if X = hydrogen and Y = some element from the second or later row of the periodic table, $G \approx \mu_X$, so that the ω^*/ω ratio approaches 0.7071. Conversely, the frequency ratio for HY*/HY where, say $Y = {}^{35}Cl$ and $Y^* = {}^{37}Cl$ is very close to unity (0.999249) and the frequency shift in this example will be only -2.2 cm^{-1}. In molecules other than hydrides the frequency shifts are quite small, but in favorable cases in solids, where the line widths are small and uncomplicated by large amplitude motions of a rotational character, a large number of isotopic frequencies are observed, e.g., for ν_3 of the *cyanate ion* NCO^-, the frequencies of a number of isotopic species are given in Table 2-3.

(e) Rayleigh's Principle

It is of considerable value to have bounds upon the isotope shifts produced by the substitution of a single heavier (or lighter) atom, both in the free molecule case and as we shall see later in Chapter 7, for the case of an isotopic "impurity" in a solid. Such bounds are provided by a theorem due originally to Lord Rayleigh which may be stated for our purposes in the following way. The increase (decrease) of mass of any atom will lead to a new set of frequencies which are unchanged or decreased (increased) by an amount which is no greater than the interval to the next unperturbed frequency. Such a principle is perhaps more cogent in cases where a large relative mass change is made, but may be useful in understanding changes in spectra where there exist a number of frequencies in a relatively narrow interval.

Quantitative calculations of frequency shifts in cases where the relative mass

Table 2-3 The frequencies of the mode ν_3 in the NCO^- ion as a substitutional impurity in KCl(^{14}N, ^{15}N, ^{12}C, ^{13}C, ^{16}O, ^{18}O)

Isotopic molecule			Frequency (cm^{-1})	Isotopic molecule			Frequency (cm^{-1})
14	12	16	2181.8	14	13	16	2124.7
14	12	18	2172.7	14	13	18	2117.6
15	12	16	2165.3	15	13	16	2106.9
15	12	18	2157	15	13	18	2097

change is small may conveniently be performed with the aid of a perturbation technique.[20]

2-6 NORMAL MODES OF SIMPLE SYMMETRIC MOLECULES

It is now possible to summarize the results of the application of group theory to the analysis of the normal modes of a number of simple molecular units which will be the main objects of interest in Chapter 6. In this section the following structures will be discussed: XY_2 ($D_{\infty h}$), XYZ ($C_{\infty v}$), XY_2 (C_{2v}), XY_3 (D_{3h}), XY_3 (C_{3v}), XYZ_2 (C_{2v}), XY_4 (T_d), and XY_3Z (C_{3v}). In each case we shall indicate the numbering of normal modes in their various symmetry species, their infrared or Raman activity, and we shall also give expressions for the symmetry coordinates associated with each species. If there is only a single symmetry coordinate in a non-degenerate species, or a single pair in a doubly degenerate species, etc., such a symmetry coordinate is, aside from a scale factor, also a normal coordinate. But when two or more such coordinates (or sets of degenerate pairs or triplets) occur within a given species, the normal coordinates will in general be linear combinations of the symmetry coordinates whose precise form can only be determined by solution of the secular determinant. Nevertheless, it often happens that the mixing is not so extensive as to prevent us from using the symmetry coordinates as a good initial approximation to the description of the normal mode; this will be true when the several symmetry coordinates involve motions which are widely different in frequency, as for example the CH stretch compared with the HCH bend or the C—Cl stretch in methyl chloride.

We shall now discuss in detail the normal modes of these molecules. The reader may wish to refer to Reference 12, which illustrates all of them.

(a) XY_2 ($D_{\infty h}$)

The case of the linear symmetric triatomic molecule, of which CO_2, CS_2, NO_2^+, N_3^-, BO_2^- are significant examples, has been discussed in earlier sections. Symmetry coordinates are

$$
\begin{aligned}
S_1 &= 2^{-1/2}(r_1 + r_2)\\
S_{2x} &= \alpha_x\\
S_{2y} &= \alpha_y\\
S_3 &= 2^{-1/2}(r_1 - r_2)
\end{aligned}
$$

where r_i is a bond stretch and α_x an angle bend in the xz plane.

20. WDC, pages 190–92. The theorem is originally due to Lord Rayleigh, *Theory of Sound*, Vol. I, see pages 88, 92a, Dover, New York, 1945.

The normal coordinates (in this case identical with the symmetry coordinates except for a scale factor) are distributed as follows:

		Infrared activity	*Raman activity*
Σ_g^+	ν_1		$\alpha_{xx} + \alpha_{yy}; \alpha_{zz}$
Σ_u^-	ν_3	μ_z	
Π_u	ν_2	μ_x, μ_y	

(b) XYZ ($C_{\infty v}$)

The linear triatomic molecule without a center of symmetry differs from the previous example in the respect that ν_1 and ν_3 now fall in the same species, and that it is no longer possible to form symmetric and antisymmetric combinations of the bond stretches. Thus we may as well write

$$S_1 = r_1$$

$$S_{2x} = \alpha_x$$

$$S_{2y} = \alpha_y$$

$$S_3 = r_2$$

The symmetry classification is:

		Infrared activity	*Raman activity*
Σ^+	ν_1, ν_3	μ_z	$\alpha_{xx} + \alpha_{yy}; \alpha_{zz}$
Π	ν_2	μ_x, μ_y	$(\alpha_{yz}; \alpha_{zx})$

It is apparent that all three fundamental frequencies are allowed in both infrared and Raman spectra.

(c) XY_2 (C_{2v})

The nonlinear triatomic molecule with symmetry includes a great many familiar cases, e.g., H_2O, SO_2, etc. The symmetry coordinates appropriate in this case are:

$$S_1 = 2^{-1/2}(r_1 + r_2)$$

$$S_2 = \alpha$$

$$S_3 = 2^{-1/2}(r_1 - r_2)$$

The bending mode (ν_2) is no longer degenerate. The symmetry species are:

		Infrared activity	*Raman activity*
A_1	ν_1, ν_2	μ_z	$\alpha_{xx}; \alpha_{yy}; \alpha_{zz}$
B_1	ν_3	μ_x	α_{zx}

In classifying ν_3 as B_1, we employ the convention that the molecular plane is xz.

(d) XY_3 (D_{3h})

We have already discussed this example at length in an earlier section and merely indicate here that

$$
\begin{aligned}
S_1 &= 3^{-1/2}(r_1 + r_2 + r_3) \\
S_2 &= p \\
S_{3a} &= 6^{-1/2}(2r_1 - r_2 - r_3) \\
S_{3b} &= 2^{-1/2}(r_2 - r_3) \\
S_{4a} &= 6^{-1/2}(2\alpha_1 - \alpha_2 - \alpha_3) \\
S_{4b} &= 2^{-1/2}(\alpha_2 - \alpha_3)
\end{aligned}
$$

and

		Infrared activity	*Raman activity*
A_1'	ν_1		$\alpha_{xx} + \alpha_{yy}; \alpha_{zz}$
A_2''	ν_2	μ_z	
E'	ν_3, ν_4	μ_x, μ_y	$(\alpha_{xx} - \alpha_{yy}, \alpha_{xy})$

(e) XY_3(C_{3v})

The pyramidal, as distinct from the planar molecule of this type with threefold symmetry, includes such well-known examples as NH_3, PCl_3, ClO_3^-, etc. The symmetry coordinates are the same as for XY_3 (D_{3h}) except that the sum of the three angle bends, $\alpha_1 + \alpha_2 + \alpha_3$, is no longer a redundancy but a genuine coordinate, and that we drop the out-of-plane mode p and use $S_2 = 3^{-1/2}(\alpha_1 + \alpha_2 + \alpha_3)$. The distribution of modes under the regime of C_{3v} is:

		Infrared activity	*Raman activity*
A_1	ν_1, ν_2	μ_z	$\alpha_{xx} + \alpha_{yy}; \alpha_{zz}$
E	ν_3, ν_4	μ_x, μ_y	$(\alpha_{xx} - \alpha_{yy}, \alpha_{xy}); (\alpha_{yz} - \alpha_{zx})$

Here, as in XYZ linear, all modes are allowed in both spectra. For detailed calculations involving force constants and frequencies it is especially important to observe bond and angle numbering similar to that indicated in Fig. 2-4.

(f) XYZ_2 (C_{2v})

A planar molecule such as formaldehyde or phosgene provides an illustration of this sort of structure. Defining s as the XY stretch, r_1, r_2 as XZ stretches, and α_1, α_2 as XYZ angle bends, a set of symmetry coordinates may be defined as follows:

$$\begin{aligned}
S_1 &= s \\
S_2 &= 2^{-1/2}(r_1 + r_2) \\
S_3 &= 2^{-1/2}(\alpha_1 + \alpha_2) \\
S_4 &= 2^{-1/2}(r_1 - r_2) \\
S_5 &= 2^{-1/2}(\alpha_1 - \alpha_2) \\
S_6 &= p
\end{aligned}$$

The out-of-plane mode p is again required as with XY_3 (D_{3h}). The species and spectral activities of these modes are:

		Infrared activity	*Raman activity*
A_1	ν_1, ν_2, ν_3	μ_z	α_{xx}; α_{yy}; α_{zz}
B_1	ν_4, ν_5	μ_x	α_{zx}
B_2	ν_6	μ_y	α_{yz}

Note that unlike any of the previous examples, this molecular model has two modes in the A_1 species which are both of a bond stretching type. In previous cases, we encountered at most one stretch and one bend in a given species, making it plausible to describe the first numbered mode as primarily stretch and the second numbered mode as primarily bend, such as ν_3 and ν_4, respectively, in the planar or pyramidal XY_3 molecules. Here we must be more careful. Thus in $O{=}CCl_2$, our $S_1 = s$ or OC stretch motion will be the main contributor to ν_1, by definition the highest frequency mode in A_1. This is so because (i) the CO bond is of higher order than the CCl bond and consequently has a larger force constant and (ii) because the G matrix element involving the OC bond is also larger than that involving the CCl bond. However, if the example under discussion were formaldehyde, $O{=}CH_2$, the mass effect would be reversed, i.e., the G element for CH stretch would be (considerably) larger than for CO stretch, and would in fact dominate over the force constant effect, leading to a situation in which ν_1 is largely $S_2 = 2^{-1/2}(r_1 + r_2)$, i.e., the symmetric CH stretch, and ν_2 is mainly $S_1 = s$. Clearly as the number of modes of a given species increases, one must be very cautious about the description of normal coordinates in terms of symmetry coordinates, as the opportunity for mixing becomes more favorable and more intricate.

(g) XY_4 (T_d)

Our first example of a pentatomic molecule includes the important examples of methane, carbon tetrachloride, and in solids, such examples as the ammonium, perchlorate, and sulfate ions. The symmetry coordinates here appropriate are

$$S_1 = 4^{-1/2}(r_1 + r_2 + r_3 + r_4)$$

$$S_{2a} = 12^{-1/2}(2\alpha_{12} - \alpha_{13} - \alpha_{14} - \alpha_{23} - \alpha_{24} + 2\alpha_{34})$$

$$S_{2b} = 4^{-1/2}(\alpha_{13} - \alpha_{14} - \alpha_{23} + \alpha_{24})$$

$$S_{3a} = 4^{-1/2}(r_1 + r_2 - r_3 - r_4)$$

$$S_{4a} = 2^{-1/2}(\alpha_{12} - \alpha_{34})$$

$$S_{3b} = 4^{-1/2}(r_1 - r_2 - r_3 + r_4)$$

$$S_{4b} = 2^{-1/2}(\alpha_{14} - \alpha_{23})$$

$$S_{3c} = 4^{-1/2}(r_1 - r_2 + r_3 - r_4)$$

$$S_{4c} = 2^{-1/2}(\alpha_{13} - \alpha_{24})$$

The mode distribution in the species of T_d is:

		Infrared activity	*Raman activity*
A_1	ν_1		$\alpha_{xx} + \alpha_{yy} + \alpha_{zz}$
E	ν_2		$(\alpha_{xx} + \alpha_{yy} - 2\alpha_{zz}, \alpha_{xx} - \alpha_{yy})$
F_2	ν_3, ν_4	(μ_x, μ_y, μ_z)	$(\alpha_{xy}, \alpha_{yz}, \alpha_{zx})$

Here it will be noted that all four fundamental frequencies are allowed in the Raman effect, but only the triply degenerate modes, ν_3 and ν_4, are permitted in the infrared.

(h) XY_3Z (C_{3v})

Here we have three equivalent XY stretches r_1, r_2, r_3, a single XZ stretch s, three YXY bends $\alpha_1, \alpha_2, \alpha_3$, and three ZXY bends $\beta_1, \beta_2, \beta_3$: the set of ten internal coordinates contains a redundancy which arises from the fact that an appropriate sum over the angular changes must vanish to the first order. It is thus a matter of taste whether we abandon the symmetry coordinate $3^{-1/2}(\alpha_1 + \alpha_2 + \alpha_3)$ or $3^{-1/2}(\beta_1 + \beta_2 + \beta_3)$; we shall abandon the latter and thus obtain the set:

$$
\begin{aligned}
S_1 &= 3^{-1/2}(r_1 + r_2 + r_3)\\
S_2 &= s\\
S_3 &= 3^{-1/2}(\alpha_1 + \alpha_2 + \alpha_3)\\
S_{4a} &= 6^{-1/2}(2r_1 - r_2 - r_3)\\
S_{4b} &= 2^{-1/2}(r_2 - r_3)\\
S_{5a} &= 6^{-1/2}(2\alpha_1 - \alpha_2 - \alpha_3)\\
S_{5b} &= 2^{-1/2}(\alpha_2 - \alpha_3)\\
S_{6a} &= 6^{-1/2}(2\beta_1 - \beta_2 - \beta_3)\\
S_{6b} &= 2^{-1/2}(\beta_2 - \beta_3)
\end{aligned}
$$

The species classification is:

		Infrared activity	*Raman activity*
A_1	ν_1, ν_2, ν_3	μ_z	$\alpha_{xx} + \alpha_{yy}; \alpha_{zz}$
E	ν_4, ν_5, ν_6	(μ_x, μ_y)	$(\alpha_{xx} - \alpha_{yy}, \alpha_{xy}); (\alpha_{yz}, \alpha_{zx})$

Although the totally symmetric coordinates are listed in the order of XY stretch (S_1), XZ stretch (S_2), and YXY bend (S_3), this ordering may not correspond to ν_1, ν_2, ν_3. For example, in methyl chloride, bromide, and iodide, ν_1 is almost exclusively CH stretch (S_1), but ν_2 is mainly bend, i.e., our S_3 above, and ν_3 mainly our S_2. However, in methyl fluoride, S_2 and S_3 are more extensively mixed, since the C—F stretch mode in isolation is not so much lower in frequency than the symmetric methyl group bend as in the case of C—Cl, C—Br, and C—I.

2-7 EFFECTS DUE TO ANHARMONICITY

We have already mentioned that the harmonic oscillator model developed in the earlier sections of this chapter is only an approximation, and have stated that the actual energy levels are better represented by a formula which includes quadratic as well as linear terms in the vibrational quantum numbers:

$$E = E^\circ + h\boldsymbol{\nu}'\mathbf{v} + h\mathbf{v}'\mathbf{X}\mathbf{v} + h\mathbf{1}'\mathbf{X}\mathbf{v}$$

Indeed, precise experimental studies of small molecules and theory are in accord that still higher terms are actually required in the energy formula, namely cubic, quartic, etc., expressions in the quantum numbers.

The effects of anharmonicity include not only a modification of the energy levels, but they also are one of the sources of a relaxation of the harmonic oscillator selection rule, $\Delta v_k = \pm 1$ for a single mode k, which would restrict the observable transitions to the fundamental frequencies ν_k. It must be emphasized, however, that the transitions allowed under the more permissive conditions of anharmonicity must still conform to the restrictions imposed by symmetry which require that $\Gamma_v \times \Gamma_{\mu \text{ or } \alpha} \times \Gamma_{v'}$ contains Γ_1. A characteristic feature of the spectrum of the anharmonic molecule is the appearance of *overtone* transitions, e.g., $\Delta v_k = \pm 2, \pm 3$, etc., *combination* transitions, e.g., $\Delta v_k = \pm 1$ simultaneously with $\Delta v_{k'} = \pm 1$, together with still more complicated transitions involving two or more quantum numbers.

One feature of gas-phase vibrational absorption spectra which assumes perhaps greater importance in the case of crystals, is the phenomenon of *hot bands*. In the simplest case (diatomic molecule) one means the transitions $v = 1 \rightarrow 2, v = 2 \rightarrow 3$, etc., which in the harmonic approximation would all coincide with the $v = 0 \rightarrow 1$ (fundamental) frequency. The anharmonicity constant X in this case leads to a series of slightly different frequencies, namely $\nu + (2v + 1)X$ for the transition $v \rightarrow v + 1$. The intensities of the absorptions which originate in excited states ($v > 0$) are naturally progressively diminished by the Boltzmann factor which specifies the relative populations in these higher states under conditions of thermal equilibrium, and when $h\nu > kT$, such hot bands may be of negligible intensity. In crystals one has the opportunity to vary kT over a wide range and thereby to exert a strong control on the relative intensities of hot bands, all of which are practically completely quenched at liquid helium (and perhaps much higher) temperatures.

In polyatomic molecules, the role of hot bands is clearly more complex owing to the existence of a variety of initial states corresponding to low excitations of the various fundamentals, with various $X_{kk'}$ entering into the expressions for the frequencies of the transitions.

In addition to these two effects, modification of the energy levels and of the selection rules, other more subtle but at least equally striking phenomena occur in consequence of anharmonicity. One of these is *Fermi resonance,* first noted in the Raman spectrum of CO_2, and subsequently detected in many other simple molecules. In the classic example it happens that the overtone of the bending fundamental $2\nu_2$ is nearly coincident with the symmetric stretch fundamental ν_1. Moreover, the level $2\nu_2$, of symmetry species $\Sigma_g^+ + \Pi_g$, includes a component (the totally symmetric Σ_g^+) which is identical with the species of ν_1. Thus the stage is set for an interaction between these two states via higher order (i.e., cubic, quartic, etc.), terms in the potential energy. Such an interaction reveals itself (a) through a marked shift in the energy of the level "$2\nu_2$" from its expected value in addition to the effect of the X_{22} term, (b) through a borrowing of intensity by "$2\nu_2$" at the expense of "ν_1." This typically quantum mechanical perturbation will be discussed in connection with anharmonicity effects in crystal spectra (Section 6-6).

The anharmonic molecule is readily described by appending to the quadratic part of the potential energy terms of the third, fourth, and higher degrees in the Q_k so that

$$V = \tfrac{1}{2}\sum_k \lambda_k Q_k^2 + \tfrac{1}{6}\sum\sum\sum g_{kk'k''}Q_kQ_{k'}Q_{k''} + \tfrac{1}{24}\sum\sum\sum\sum h_{kk'k''k'''}Q_kQ_{k'}Q_{k''}Q_{k'''}\cdots$$

In the presence of this formidable array of new parameters $g_{kk'k''}$ and $h_{kk'k''k'''}$ one can draw a little consolation from the reflection that the number of such parameters is restricted by the basic requirement that V is an invariant, hence $g_{kk'k''}$ must vanish unless $\Gamma(k) \times \Gamma(k') \times \Gamma(k'')$ contains Γ_1 and similarly $h_{kk'k''k'''}$ must vanish unless $\Gamma(k) \times \Gamma(k') \times \Gamma(k'') \times \Gamma(k''')$ contains Γ_1. Thus in the $XY_3(D_{3h})$ molecule, there are eight cubic and fifteen quartic constants.

Perturbation theory is employed to relate the anharmonic constants g and h to the phenomenological parameters X_{kl}.

2-8 LINE AND BAND SHAPES, WIDTHS, AND INTENSITIES

In the gas phase, a vibrational transition is nearly always accompanied by a change in the rotational energy of the molecule. Gas-phase infrared spectroscopists are therefore accustomed to encounter vibration–rotation *branches,* or at least their envelopes. At low resolution or with large molecules, the fundamental width of a vibrational transition is determined by its rotational envelope. At high resolution and with small molecules, individual ro-vibronic lines may be resolved. If these are unperturbed by other vibrational transitions—as in the ground electronic states of diatomic molecules—the shapes and widths of individual lines may be the subject of important investigation, for the entire mechanism of rotational relaxation is convoluted into the moments of a rotational line.

In the solid state, with the absence of molecular rotation, the intensity, width, and shape of a *vibrational* line are in principle the bearers of information concerning the process by means of which the vibrational excitation becomes relaxed. Before beginning the subject of crystal spectra, we shall therefore discuss the general principles which govern these parameters of individual spectral lines.

The semiclassical theory of radiation gives us, by way of a perturbation calculation, the transition probability of induced absorption. The inverse process of induced emission has an identical probability. The induction of each is due to an external radiation field, but an accelerated charge, or charge distribution (like a molecule) may also radiate in the absence of an external field. The process is called *spontaneous emission,* and it is ultimately responsible for the natural lifetime of any quantum state. The ratio of the transition probabilities for spontaneous to induced emission is given by that of the Einstein coefficients for the two processes,

$$\frac{A_{nm}}{B_{mn}} = \frac{8\pi h\nu^3}{c^3} \tag{2-8-1}$$

where

$$B_{mn} = \frac{8\pi^3}{3h^2}|\langle m|\boldsymbol{\mu}|n\rangle|^2$$

Equation (2-8-1) is ordinarily derived with the assumption of thermal equilibrium; however, A_{nm} and B_{mn} are *molecular* constants. Both are independent of temperature, and (2-8-1) is always valid.

The integrated absorption coefficient is defined by

$$\int \alpha(\nu)\, d\nu$$

where

$$\begin{aligned} \alpha &= \frac{1}{l} \ln \frac{I_0}{I} \\ &= \frac{h\nu_{nm}}{c} B_{nm}(N_m - N_n) \end{aligned} \tag{2-8-2}$$

is the absorption cross-section. In the above, I_0 is the radiant power incident and I is that transmitted by a sample of length l. The units of I are watts $\mathrm{cm}^{-2}\,\mathrm{str}^{-1}$.

If we consider a molecule in an excited state, in the absence of an external radiation field, the number of molecules in the state $|n\rangle$ will decrease according to

$$-\frac{dN_n}{dt} = N_n \sum_m A_{nm} = N_n \gamma_n \tag{2-8-3}$$

so that there is an exponential decay of the excited state density,

$$N_n = N_n^{\circ} e^{-\gamma_n t} \tag{2-8-4}$$

where

$$\gamma_n = \sum_m A_{nm} = \tau_n^{-1} \tag{2-8-5}$$

is the reciprocal of the mean lifetime of the molecule in the state $|n\rangle$.

The state $|m\rangle$ to which the excited state decays may itself have a lifetime τ_m comparable with τ_n. Hence, the decay constant characteristic of the transition $|n\rangle \rightarrow |m\rangle$ is given by

$$\begin{aligned} \gamma_{nm} &= \tau_n^{-1} + \tau_m^{-1} \\ &= \gamma_n + \gamma_m \end{aligned} \tag{2-8-6}$$

In the classical theory of emission by an elastically bound electron, the equation of motion relates the inertial force on the electron to the sum of an (elastic) restoring force plus one other, a dissipative, or radiation damping, force, assumed to be proportional to the velocity, due to which radiation is emitted, either as heat or light. Without the latter, the motion would be strictly harmonic; in quantum mechanics, the system would remain in a stationary state. With damping, the amplitudes of the oscillations die off such that they are bounded by the function $e^{-\gamma t}$. Likewise, the electric field $\mathscr{E}(t)$ resulting from the oscillations has an additional time dependence of the same form.

Since we are interested in the absorption coefficient as a function of frequency, it is necessary to Fourier analyze $\mathscr{E}(t)$. The result can be shown[21] to have the Lorentzian form

$$\alpha(\nu) = \frac{8\pi^3}{3ch} |\langle m | \boldsymbol{\mu} | n \rangle|^2 \frac{\gamma_{nm}}{4\pi^2} \frac{1}{(\nu - \nu_0)^2 + (\gamma_{nm}/4\pi)^2} \tag{2-8-7}$$

It is also possible to derive the complex index of refraction of a gas of simple (one-electron) oscillators. Use the continuum model (Section 5-8) is made together with the definition of the polarizability

$$\mathbf{p} = \alpha\mathscr{E}$$

and the electric displacement

$$\mathscr{D} = \varepsilon\mathscr{E} = \mathscr{E} + 4\pi\mathscr{P}$$

where $\varepsilon = n^2$ is the dielectric constant and n is the index of refraction. Together with these relationships, the electric field and oscillator amplitude which result from the equation of motion described earlier, the form of (2-8-7) can be easily derived.[22]

In quantum mechanics, the derivation of (2-8-7) proceeds in a similar way to that used to derive the Fermi Golden Rule for the transition probability per unit time,

$$w = \frac{d}{dt} \sum_n |c_n(t)|^2 = \frac{2\pi}{\hbar} |\langle n | V | m \rangle|^2 \rho(\nu_{mn}) \tag{2-8-8}$$

with one change. In this derivation, a perturbation [a photon field of density $\rho(\nu_{mn}) = I(\nu_{mn})/c$] acts to mix the initial and final states, and the mixing coefficient is proportional to the matrix element $V_{nm} = \langle n | V | m \rangle$. The time dependence of the mixing coefficient of a state, $|n\rangle$, is given by

$$i\hbar \frac{dc_n}{dt} = \sum_m V_{nm} c_m e^{i\omega_{nm} t} \tag{2-8-9}$$

where

$$\omega_{nm} = \frac{E_n^\circ - E_m^\circ}{\hbar}$$

and

$$E_n^\circ = \langle n | H^\circ | n \rangle$$

21. W. Kauzmann, *Quantum Chemistry*, Academic Press, New York, 1957, Chapter 15.
22. L. H. Aller, *Astrophysics*, Ronald Press, New York, 1963, Chapters 4 and 7.

It is normally assumed that $V_{nm} \ll H^\circ$, and that the time period of interest is so short that $c_n(t) \cong 1$ and $c_m(t) \cong 0$. In fact, exponential decay (2-8-4) will not follow from these assumptions. Instead, (2-8-9) implies transitions both to and from $|n\rangle$. However, *the several transitions occur with different phases.* The resulting interference gives rise to the exponential decay.

Integration of (2-8-9) yields the time development of the mixing coefficient c_n. Since the resulting distribution of transition probabilities in the frequency domain is desired, the integration is carried out by the use of the Fourier transform

$$f(\omega) = \frac{1}{2\pi} \int c_n(t)\, e^{i\omega t}\, dt \tag{2-8-10}$$

Not only is (2-8-7) the result, but it can be shown that $\gamma = \hbar w$, where w is the rate of transition given by the Fermi Golden Rule.[23]

In the presence of a radiation field, the occupation number of $|n\rangle$ may change by way of several channels:

1. Spontaneous emission to *lower* levels, $|m\rangle$.
2. Induced emission to *lower* levels, $|m\rangle$.
3. Induced absorption from *lower* levels, $|m\rangle$.
4. Induced absorption to *higher* levels, $|n'\rangle$.

All of these contribute to the effective width, which becomes

$$\gamma_n = \sum_m A_{nm} + \sum_m B_{nm} I(\nu_{nm}) + \sum_{n'} B_{nn'} I(\nu_{nn'}) \tag{2-8-11}$$

Using the Planck Law, the principle of detailed balancing and (2-8-1), (2-8-11) becomes*

$$\gamma_n = \sum_m A_{nm}(1 - e^{-h\nu/kT})^{-1} + \sum_{n'} A_{n'n}(e^{h\nu/kT} - 1)^{-1} \tag{2-8-12}$$

Because of (2-8-6) and (2-8-12), we have the qualitative rule that strong transitions are broad. Equation (2-8-6) also enables one to describe the width of *levels* as narrow or broad. For example, since τ_G of a ground state is infinite, $\gamma_G = \tau_G^{-1}$ is small, i.e., a ground state is always sharp.

23. E. Merzbacher, *Quantum Mechanics,* John Wiley & Sons, Inc., New York, 1970, Chapter 18.

* If the transition is vibrational,

$$|\langle v|\boldsymbol{\mu}|v-1\rangle|^2 = \left(\frac{\partial \boldsymbol{\mu}}{\partial Q}\right)^2 \frac{v}{8\pi^2\nu} h$$

For most internal vibrations, the second sum in (2-8-12) is negligible compared to the first. Hence,

$$\gamma_v \cong A_{v,v-1} = \frac{8\pi^2\nu^2}{3c^3}\, v \left(\frac{\partial \mu}{\partial Q}\right)^2$$

γ_n is also related to the dimensionless *oscillator strength* of a transition

$$f_{nm} = \frac{2m_e\omega}{3\hbar} |\langle n|\mathbf{r}|m\rangle|^2 \tag{2-8-13}$$

which in turn is proportional to the integral of the absorption coefficient over the effective width of the spectroscopic transition. Vibrational fundamentals have oscillator strengths in the range 10^{-6}–10^{-4}, in contrast to allowed molecular electronic transitions, for which $f \sim 10^{-1}$. A typical vibrational band will have an "integrated absorption" S,

$$S = \int_{\text{band}} \alpha(\nu)\, d\nu$$

of 500 $\text{cm}^{-2}\,\text{atm}^{-1}$. Among infrared spectroscopists it is conventional to multiply S by the factor 22.415 darks cm^2 atm. Thus, a nominal vibration band intensity is $\sim$10 000 darks. The dipole derivative $(\partial\mu/\partial Q)$ is related to the intensity by

$$\left(\frac{\partial\mu}{\partial Q}\right)^2 = \frac{3c^2 \times 10^3}{\pi N_0} \frac{\text{darks}}{\text{degeneracy}} \tag{2-8-14}$$

where N_0 is the Avogadro number. The numerical factor in (2-8-14) has the magnitude 1.42468.*

It is also possible to relate the lifetime to the oscillator strength or the integrated absorption. The last relationship is

$$\tau = \frac{Nc}{8\pi^2 pS\nu^2} \tag{2-8-15}$$

Using (2-8-15), we find that a typical natural lifetime of an excited vibrational state is $\sim 10^{-2}$ sec at STP. Thus the relative *improbability* of a vibrational transition is reflected in the natural longevity of an excited vibrational state. Accordingly, if there is available some alternative to radiative decay, the molecule will certainly utilize it.

Since $\tau = \gamma^{-1}$, and since a Lorentzian line shape has a width at half peak intensity of 2γ, any relaxation mechanism which causes the state to have a shorter lifetime will *broaden* the spectral line.

There are two principal sources of broadening of vibration–rotation lines. One of these is a relativistic effect, and is due to the fact that gas molecules are far from stationary. The actual frequency absorbed or emitted by a molecule moving with a velocity v_x is

$$\nu = \nu_{nm}\left(1 - \frac{v_x}{c}\right) \tag{2-8-16}$$

* There are several other systems of units for vibrational intensities. Their interrelations have been summarized by I. M. Mills, *Ann. Repts. Chem. Soc.* (London), **55,** 55 (1958).

Hence, the Doppler shift of the frequency is

$$\frac{\Delta \nu}{\nu} = \frac{v_x}{c}$$

which results in a Gaussian line

$$I(\nu) \sim \exp\left[-mc^2(\Delta\nu)^2/2\nu^2kT\right] \tag{2-8-17}$$

The Doppler *half-width* b_D can be shown to be given by

$$b_D = \left(\frac{2kT \ln 2}{mc^2}\right)^{1/2} \frac{\nu_{nm}}{c} \tag{2-8-18}$$

Equation (2-8-18) can be used to conclude that Doppler widths at room temperatures are of the order of 10^{-4} cm^{-1} for molecules with molecular weight 100 and vibrational transitions of $\sim 10^3$ cm^{-1}.

Of greater importance in broadening a spectral line is the decay of excitation by way of collisional energy transfer, commonly referred to as collisional broadening. The intensity profile of a line subject to both natural and collisional broadening follows a dispersion formula similar to that for natural broadening alone (2-8-7), except that the damping constant is treated completely parametrically, and is called the line-shape parameter b, which is the sum of the damping constants due to both effects. A typical value of b for a vibration–rotation line is ~ 0.04 cm^{-1}. There are a number of theories which attempt to relate b to other physical parameters by way of particular models for the collisional deactivation process. In most of these, a collision is viewed as a process which more or less abruptly *interrupts* the radiative decay. The radiative damping constant γ is then replaced by the *time between collisions*, which may be calculated from the kinetic theory of gases. By this means optical collision diameters may be obtained. These are often compared with collision diameters obtained from transport process experiments, such as gas viscosities. The orders of magnitude of both parameters are the same, but the values are rarely equal. Nor should they be, for the types of energy being transferred are different—i.e., vibrational, as compared with translational.

In any theory, b is pressure-dependent, and it is an experimental fact that b increases with pressure. That it should be is qualitatively understandable from the model of collisional interruption of the radiation process; the time between collisions becomes shorter, so b grows. The effect on the overall line is mainly concentrated in the wings, thus leading to the *pressure broadening* of the entire vibration–rotation band. In some cases, using high but easily achieved pressures, the fine structure of a band which results from the details of the selection rules tends to be "washed out." As shown in Section 2-4, the *integrated* absorption coefficient of a vibration–rotation band, as previously defined, is not pressure dependent. Consequently, if the experimentalist is forced to use a spectrometer whose resolving power is not high enough to precisely map out a vibration–rotation *line*, it behooves him to pressure-broaden the complete band and measure its intensity, in order to achieve his ultimate goal, the determination of $(\partial\mu/\partial Q)$. Much time and effort has been devoted to both procedures, both with respect to experimental techniques and data handling.[24]

In the solid state, the collisional deactivation mechanism must clearly be modified, inasmuch as random translational motion is surely absent. Nevertheless, a theory of line shapes, widths and intensities is obviously necessary, since at most temperatures the line widths are far from natural.

As has been aptly summarized by R. G. Gordon,[25] the analysis of any spectrum in terms of its assignment to transitions between specific quantum states may be made difficult by three circumstances:

1. Especially in condensed phases, transitions are often so numerous and so close in energy that a continuous band is formed, rather than a line spectrum. Thus, a spectrum may be so rich in transitions as to make its analysis difficult.
2. In condensed phases, important portions of the intensity *versus* energy distribution we call a spectrum are the result of couplings between the motions of the molecules which make up the system. Strictly speaking, in order to begin to analyze the spectrum we should construct the appropriate many-body wave functions of the system and calculate the necessary off-diagonal matrix elements among them. Not only are these calculations difficult, but so are the *visualizations* of the wave functions and matrix elements.
3. As is well-known, classical mechanics can be derived from quantum statistics—that is, finding the *expectation value* in a given state of a quantum mechanical operator which corresponds to a classical variable, followed by an evaluation of a weighted average of all expectation values. For this reason, the simple, two-level system which was the basis of (2-8-3) lacks a realistic counterpart in classical mechanics.

Part of the problem underlying these observations is that our emphasis upon *energies* is a result of the use of the *Schrödinger representation* of quantum mechanics, the basis functions of which are time-dependent $[\psi(r,t) = \psi(r)\,e^{-iEt/h}]$ and are solutions of Schrödinger's equation, $H\psi = i\hbar\dot{\psi}$. In this equation, the operator H has no time dependence, and its eigenvalues E are the definite energy levels we ordinarily seek. Gordon has also pointed out that by relinquishing our usual preference for knowledge of definite energies we can instead gain information concerning the response of a system to an external perturbation, such as a radiation field, as a function of time. In order to do this, use is made of the *Heisenberg representation,* in which all of the time dependence is associated only with the operators, and the wave functions are time-independent. The transformation from one basis to another is unitary,

$$\begin{aligned}\psi'(r) &= e^{iHt/h}\,\psi(r,t)\\ O(t) &= e^{iHt/h}\,O\,e^{-iHt/h}\end{aligned} \tag{2-8-19}$$

24. S. S. Penner, *Quantitative Molecular Spectroscopy and Gas Emissivities*, Addison-Wesley, Reading, Mass., 1959.
25. R. G. Gordon, *J. Chem. Phys.*, **43**, 1307 (1965).

In the Schrödinger representation, the absorption intensity as measured by the left-hand side of (2-4-18) may be written as

$$\tfrac{1}{3}\sum_{i,f} \rho_i |\langle i|\boldsymbol{\mu}|f\rangle|^2 \, \delta(\omega_{if} - \omega) \tag{2-8-20}$$

where ρ_i is the Boltzmann factor for the initial state, and $\boldsymbol{\mu}$ is the transition dipole moment for a particular vibrational band.

Transformation to the Heisenberg representation involves taking the Fourier transform of the delta function

$$\delta(\omega) = \frac{1}{2\pi}\int_{-\infty}^{\infty} e^{i\omega t}\, dt \tag{2-8-21}$$

Thus

$$\begin{aligned} I(\omega) &= \frac{1}{2\pi}\sum_{i,f} \tfrac{1}{3}\rho_i \langle i|\boldsymbol{\mu}|f\rangle\langle f|\boldsymbol{\mu}|i\rangle \int_{-\infty}^{\infty} e^{i(E_f - E_i)t/h}\, e^{-i\omega t}\, dt \\ &= \frac{1}{2\pi}\int_{-\infty}^{\infty} e^{-i\omega t} \sum_{i,f} \tfrac{1}{3}\rho_i \langle i|\boldsymbol{\mu}|f\rangle\langle f|e^{iHt/h}\,\boldsymbol{\mu}\, e^{-iHt/h}|i\rangle\, dt \\ &= \frac{1}{2\pi}\int_{-\infty}^{\infty} e^{-i\omega t} \sum_{i} \tfrac{1}{3}\rho_i \langle i|\boldsymbol{\mu}\cdot(e^{iHt/h}\,\boldsymbol{\mu}\, e^{-iHt/h})|i\rangle\, dt \\ &= \frac{1}{2\pi}\int_{-\infty}^{\infty} e^{-i\omega t}\, \tfrac{1}{3}\langle \boldsymbol{\mu}(0)\cdot\boldsymbol{\mu}(t)\rangle\, dt \end{aligned} \tag{2-8-22}$$

where the *correlation function* $\langle \boldsymbol{\mu}(0)\cdot\boldsymbol{\mu}(t)\rangle$ corresponds classically to the projection of $\boldsymbol{\mu}(t)$ on $\boldsymbol{\mu}(0)$. According to Fourier's theorem,

$$\tfrac{1}{3}\langle \boldsymbol{\mu}(0)\cdot\boldsymbol{\mu}(t)\rangle = \int_{\text{band}} I(\omega)\, e^{i\omega t}\, d\omega \tag{2-8-23}$$

Using procedures similar to those outlined here, considerable progress has been made in relating infrared band shapes in liquids to the rotational diffusion of the component molecules. In Section 6-7(d) we relate the bandwidth parameter to the cubic term in the mechanically anharmonic potential energy function of a crystal by means of which one phonon decays into two others.

CHAPTER

THREE

VIBRATIONS OF A LINEAR LATTICE

A useful study preliminary to the consideration of real, three-dimensional crystals is that of the *linear lattice* which is the topic of this chapter. The theory herein developed is actually applicable, granted certain conditions and approximations, to polymeric substances. The chief distinctive feature of this model is the existence of translational symmetry, and the methods outlined in Chapter 2 can be developed and applied to the linear lattice, and are, in fact, the main mathematical tool for the solution of the vibrational problem.

In Chapter 2 we recognized as a central problem in the analysis of molecular vibrations the determination of a linear transformation of some initial set of coordinates, in terms of which it was convenient to express the kinetic and potential energies, to the normal coordinates which separated the Hamiltonian. Exactly the same problem recurs for the linear lattice, but now the number of atoms is very much larger. We need not, however, be intimidated by the prospect of an enormous secular determinant, because the translational symmetry permits the reduction of the problem to one of tractable size.

Because we wished to pass immediately to the quantum mechanical solution of the molecular vibration problem, no mention was made in Chapter 2 of the treatment of vibration as a problem in classical mechanics. In dealing with the linear lattice there is no need to discuss the classical mechanical problem, and we could concentrate as we did in Chapter 2 exclusively upon the algebraic problem of the secular determinant, but it is illuminating to develop the present problem in classical language, and this will be done in Section 3-1 for the case of the vibrating string, in order to emphasize certain physical characteristics of the solution of the vibrational problem for a lattice of discrete particles, which is given in Section 3-2 for a monatomic lattice, in 3-3 for diatomic lattices, and in 3-4 for linear polymers.

3-1 THE LINEAR CONTINUUM

In order to gain some insight into the form of elastic waves—which, in discontinuous media we shall come to call acoustic phonons—we shall quickly review the classical mechanics of the vibrations of a homogeneous linear system, often called a string. Most texts in elementary mechanics, and even in advanced acoustics, restrict the attention of the reader to the transverse vibrations of a line, inasmuch as these waves form the basis of operation of a very important class of musical instruments. For our purposes, however, it is somewhat more instructive to concentrate on the longitudinal vibrations of a string. If the latter has finite stiffness—or elasticity—such waves are of course propagated following a distortion, but they are observed with greater difficulty. To emphasize just what motions are of interest, we recall to the reader his early experiments with the "slinky" toy, in which longitudinal waves are very much in evidence following a localized distortion. In acoustics texts such motions are usually discussed under the heading of those of a bar, where they are treated separately from the transverse vibrations of the same system.

As in the description of the motion of any continuum, we begin by focussing our attention on an infinitesimal *line element* dx, located at coordinate x along the line, relative to the origin taken at one end of the line. Upon the application of a longitudinal force, one end of the line element is displaced by ξ and the other end, at $x + dx$, by $\xi + d\xi$. We can expand this last displacement in a Taylor's series:

$$\xi + d\xi = \xi + \left(\frac{\partial \xi}{\partial x}\right) dx \tag{3-1-1}$$

The *net* displacement of the line element is thus $(\partial\xi/\partial x)\,dx$. The *strain* S in the line element is defined as the ratio of net displacement to the original length. Thus

$$S = \left(\frac{\partial \xi}{\partial x}\right) \tag{3-1-2}$$

Since the strain is a function of both position (x) and time (t), a longitudinal wave motion results. The reason for this is that there are elastic forces which oppose the local strains. It is of course these forces which hold the material system together. According to Hooke's Law, these forces must be proportional to the strain and of opposite sign. The proportionality constant is the local force constant, or elastic stiffness; for a bar it is related to Young's modulus. We again consider matters at either end of the line element. At x, the restoring force is proportional to the strain $S(x)$ and at $x + dx$ it must be proportional to $[S(x) + (\partial S/\partial x)\,dx]$. The *net force* on the line element dx is then proportional to $(\partial^2\xi/\partial x^2)\,dx$. Letting the force constant be f, and using Newton's Second Law, we obtain a wave equation of motion for the displacement at any point:

$$\frac{\partial^2 \xi}{\partial x^2} = \left(\frac{\rho}{f}\right)\frac{\partial^2 \xi}{\partial t^2} \tag{3-1-3}$$

where ρ is the mass density of the line. The solutions of (3-1-3) are well-known; they have the form

$$\xi = \xi_{\pm}^{\circ} e^{i(\omega t \pm kx)} \tag{3-1-4}$$

that is, they are traveling waves with angular frequency ω and wave vector k. If the ends of the string are clamped, the boundary conditions restrict k:

$$k = \frac{n\pi}{l} \tag{3-1-5}$$

where l is the length of the line, n is integral, and they require that $\xi_{+}^{\circ} = -\xi_{-}^{\circ}$. The allowed wavelengths are $2\pi/k = 2l/n$ which is useful for picturing the allowed waves. The allowed frequencies are computed from $\nu\lambda = U$; thus

$$\nu = \frac{nU}{2l} \tag{3-1-6}$$

where U is the phase velocity of the wave, according to (3-1-3), given by

$$U^2 = \frac{f}{\rho} \tag{3-1-7}$$

From (3-1-6), U is seen to be independent of the frequency. Thus there is no dispersion suffered by these waves and a plot of ω *vs.* k will be one of constant slope, U. We emphasize this point for it is different from the situation to be encountered in the next section.

3-2 THE LINEAR MONATOMIC LATTICE

This model comprises a set of atoms all of mass m, spaced along a line at constant intervals a of total number N. If N is allowed to pass to the limit of infinity, it appears that each atom is equivalent, but for finite N they are not. Recognizing this fact, we nevertheless introduce the fiction that the first and last atoms are identical. In doing this we follow Born and von Kármán,[1] and gain the huge advantage that the kinetic and potential energies become invariant to symmetry operations of translation, which constitute a group. Justification for the neglect of end effects when N is large has been derived.[2]

It will be worthwhile to consider this translational symmetry group in some detail before embarking upon the vibrational problem. It is remarkably simple.

The generating element of the group is the operation of translation by one interatomic distance a. The general operation will then be indicated as a translation t, in which the generating element is repeated t times, i.e., a translation by the distance ta; t can then take on all values from 0 to $N - 1$, and $t = 0$ corresponds

1. M. Born and Th. von Kármán, *Phys. Z.*, **13,** 297 (1912); **14,** 15 (1913).
2. W. Lederman, *Proc. Roy. Soc. (London)*, **182A,** 362 (1944).

to the identify operation. The group is Abelian, i.e., the operations $t't$ and tt' are identical, both corresponding to a translation by the distance $(t + t')a$.

For such a group, the irreducible representations are all one-dimensional, and the characters must be complex roots of unity; we use the symbol $k = 0, \pm 1, \ldots, \pm N/2$ to identify the distinct irreducible representations,* and since $t^N = E$,

$$[\chi_t^{(k)}]^N = \chi_E^{(k)} = 1 \tag{3-2-1}$$

The N distinct solutions of (3-2-1) are

$$\chi_t^{(k)} = \exp(2\pi ikt/N) \tag{3-2-2}$$

Presently we shall employ (3-2-2) in the Wigner Projection Operator in order to generate symmetry coordinates and hence factor the secular determinant, but first we turn to a consideration of the energy.

We have our choice of a number of possible coordinate systems. The reader is referred to Appendix II for a discussion in terms of external coordinates, i.e., the cartesian components of the displacement of each atom from its equilibrium position. Here we shall use a set of internal coordinates, which are nothing more than an extension of the bond stretches r and angle bends α_x and α_y employed in the discussion of the triatomic linear molecule in Section 2-3.

Now the linear lattice possesses not only translational symmetry, but also point symmetry. Indeed, each atom occupies a position of $D_{\infty h}$ point symmetry, and just as in CO_2, the bond stretches, which we shall henceforth designate as *longitudinal* modes, belong to a different point symmetry species from the bond bends, which are doubly degenerate, and which hereafter will be termed *transverse* modes. For this reason we may: (i) treat the longitudinal and transverse modes separately and (ii) consider only one component of the transverse modes.

We consider first the longitudinal modes represented by the r_t. The potential energy is

$$2V = \mathbf{r}'\mathbf{F}_L\mathbf{r} \tag{3-2-3}$$

where the subscript L reminds us that we are looking at the longitudinal part of the force constant matrix. Similarly

$$2T = \mathbf{P}_r'\mathbf{G}_L\mathbf{P}_r \tag{3-2-4}$$

Now, although it would be tedious to write out $\mathbf{G}_L$ completely it has a quite simple form. Along the diagonal, the elements of $\mathbf{G}_L$ are 2μ where $\mu = 1/m$, and along the borders of the diagonal and in the off-diagonal corners, the elements are $-\mu$. In short,

* Strictly if N is even we omit one of the two values $\pm N/2$, say $-N/2$, and if N is odd we use k up to $\pm N/2$.

$$\mathbf{G}_L = \mu \begin{pmatrix} 2 & -1 & 0 & 0 & \dots & -1 \\ -1 & 2 & -1 & 0 & \dots & 0 \\ 0 & -1 & 2 & -1 & \dots & 0 \\ 0 & 0 & -1 & 2 & \dots & 0 \\ \dots & \dots & \dots & \dots & \dots & \dots \\ -1 & 0 & 0 & 0 & \dots & 2 \end{pmatrix}$$

This simple form is a consequence of the fact that G matrix elements vanish unless the internal coordinates corresponding to the row and column in question have at least one common atom; for diagonal terms, the G matrix element is the sum of the reciprocal masses of the atoms defining the bond, and for two adjacent bonds, the G matrix element is the reciprocal mass of the common atom times the scalar product of unit vectors projecting outwards from the respective bonds (see Section 2-5(a)).

Our next step will be to transform $\mathbf{G}_L$ to a symmetry coordinate basis. Recalling the Wigner Projection Operator (2-2-30) and utilizing (3-2-2) we find that

$$\begin{aligned} S_k &= \mathbf{W}^{(k)} r_0 \\ &= N^{-1/2} \sum_{t=0}^{N-1} [\exp -(2\pi ikt/N)] r_t \end{aligned} \qquad (3\text{-}2\text{-}5)$$

or in the matrix form

$$\mathbf{S} = \mathbf{U}\mathbf{r}$$

where

$$U_{kt} = N^{-1/2} \exp -(2\pi ikt/N) \qquad (3\text{-}2\text{-}6)$$

Since each symmetry coordinate, corresponding to values of k differing by integers between $-N/2$ and $+N/2$, belongs to a different irreducible representation, in such a basis both $\mathbf{F}_L$ and $\mathbf{G}_L$ will be diagonal; using the law of transformation,

$$\mathsf{G}_L = \mathbf{U}\mathbf{G}_L\mathbf{U}^\dagger \qquad (3\text{-}2\text{-}7)$$

it follows that

$$\begin{aligned} (\mathsf{G}_L)_{kk} &= N^{-1} \sum_{tt'} \exp[2\pi ik(t'-t)/N](G_L)_{tt'} \\ &= 2\mu - 2\mu \cos(2\pi k/N) \\ &= 4\mu \sin^2(\pi k/N) \end{aligned} \qquad (3\text{-}2\text{-}8)$$

The simplest possible potential function is expressed by a *diagonal* and *constant* force matrix $\mathbf{F}_L = F_0\mathbf{E}$ which remains unchanged on transformation to a symmetry coordinate basis. In this case the eigenvalues are given by

$$\lambda_k = \mathsf{F}_0(\mathsf{G}_L)_{kk} = 4\mathsf{F}_0\mu \sin^2(\pi k/N) \qquad (3\text{-}2\text{-}9)$$

so that the oscillator frequencies are simply ($\lambda = 4\pi^2\nu^2$)

$$\nu_k = \frac{(F_0\mu)^{1/2}}{\pi} \sin(\pi k/N) = \nu_{LM} \sin(\pi k/N) \tag{3-2-10}$$

where ν_{LM} is the maximum frequency for the longitudinal mode.

More complicated potential functions do not lead to any essential difficulty since the symmetry factoring is preserved. In fact if the F matrix contains F_0 on the diagonal, F_1 on the two strips bordering the diagonal, ... and F_τ is the interaction force constant between r_t and $r_{t+\tau}$, then

$$(\mathsf{F}_L)_{kk} = \sum F_\tau \exp(2\pi ik\tau/N) = F_0 + 2\sum F_\tau \cos(2\pi k\tau/N) \tag{3-2-11}$$

and

$$\lambda_k = 4\mu \sin^2(\pi k/N)\left\{F_0 + 2\sum_{\tau=1} F_\tau \cos(2\pi k\tau/N)\right\} \tag{3-2-12}$$

When N is of the order of Avogadro's number, or even very much less, it is convenient, indeed imperative to regard $\phi(k) = 2\pi k/N$ as a continuous rather than a discrete variable. In this way one is led to the concept of the *dispersion curve* in which $\nu(\phi)$ or $[\nu(\phi)]^2$ is plotted against ϕ, as is done in Fig. 3-1 for case (*a*) corresponding to the simple diagonal F matrix and case (*b*) for a more complicated potential function in which $F_1 = 0.1F_0$. Because of the degeneracy of the modes for $\pm k$, it will suffice to show only the interval $0 \leq k \leq N/2$ or $0 \leq \phi \leq \pi$.

We have called a figure such as Fig. 3-1 a *dispersion curve* (strictly this should be a plot of ν rather than ν^2 versus the k or ϕ parameter, but we will often plot ν^2 since it is more simply expressed in terms of F and G elements); why is this nomenclature appropriate? The answer is that we are dealing with a relation between frequency and (reciprocal) wavelength, and that for any sort of wave motion, such a function implies the specification of the phase velocity. A non-dispersive wave satisfies the simple relation $\nu = U(1/\lambda)$ where U is the phase velocity and is constant. In our monatomic lattice, the atoms separated by t unit cells, that is by a distance ta, move with a phase difference of $2\pi kt/N = (2\pi k/Na) \cdot ta$ in the kth mode according to (3-2-5). Thus the wavelength is Na/k, and plots of ν versus k (or ϕ) are essentially relations between frequency and reciprocal wavelength.

Examination of (3-2-12) indicates that in the case of more complicated potential functions, Fourier analysis of experimental values of λ_k divided by

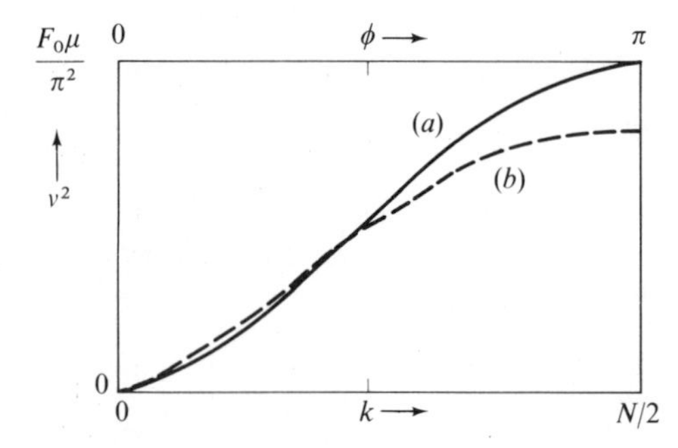

Figure 3-1 Dispersion of longitudinal mode.
(a) $F_0 = 1$, $F_\tau = 0$ $(\tau \geq 1)$
(b) $F_0 = 1$, $F_1 = 0.1$, $F_\tau = 0$ $(\tau \geq 2)$

$4\mu \sin^2 (\phi/2)$ is capable of yielding separately the parameters F_0, F_1, etc. Although spectroscopy is unable to provide knowledge of the complete $\lambda(\phi)$ curve because of the severe restrictions imposed by selection rules (*vide infra*), neutron inelastic scattering experiments have succeeded in providing such information, particularly for metals, and such Fourier analyses have in fact been carried out in real, three-dimensional cases, for example for Pb.[3]

One apparent defect of the foregoing treatment is not real. It might well be objected that the mode for $k = 0$ does not exist, for the following reason: from (3-2-5) we note that the $k = 0$ mode is the simple sum

$$S_0 = N^{-1/2} \sum r_t$$

Thus it follows that in this mode every atom must execute a displacement to the right in virtue of the bond of which it is the right-hand member, and a displacement to the left in virtue of the adjacent bond of which it is the left-hand member: since these displacements are equal in magnitude but opposite in sign, the net effect is no displacement on any atom. Thus S_0 resembles a redundancy rather than a genuine mode. Nevertheless, every mode of finite frequency has been correctly calculated, and if we had employed cartesian displacement coordinates, the result for $k = 0$ would necessarily have been the same, for in that case, this mode could have been described as a rigid translation of the whole linear lattice, which would also have had a zero frequency since the potential energy for such a motion is by definition zero.

Turning next to the transverse modes, we make use of the G matrix elements for bending which are easily derived with the aid of the $\mathbf{s}_{t\alpha}$ vectors introduced in Section 2-5(a). There it was shown that for a simple bending coordinate, the **s** vectors could be assigned unit values ($\perp$ to the bond) for terminal atoms and a magnitude of 2 of opposite direction for the central atom when the two bonds defining the angle wave were equivalent. Thus for the linear lattice, the bending coordinates centered at atoms t, $t + 1$, $t + 2$, $t + 3$ will have **s** vectors as in Fig. 3-2.

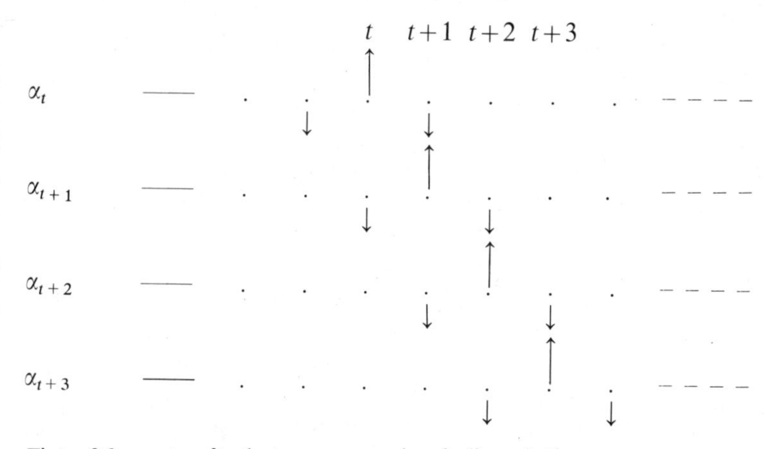

Figure 3-2 **s** vectors for the transverse modes of a linear lattice

3. B. N. Brockhouse, in *Phonons and Phonon Interactions,* T. A. Bak, Ed., W. A. Benjamin, Inc., New York, 1964, pages 253 and 268.

All masses being the same, we have

$$G_{t,t+\tau} = \mu \sum_{\alpha} \mathbf{s}_{t\alpha} \cdot \mathbf{s}_{t+\tau,\alpha}$$

and easily find that there is no overlap when $\tau > 2$. The non-vanishing elements are

$$G_{t,t} = 6\mu$$

$$G_{t,t+1} = -4\mu$$

$$G_{t,t+2} = \mu$$

The transformation to symmetry coordinates is exactly analogous to (3-2-8), but owing to the presence of $G_{t,t+2}$ terms, this time one finds a term in $\cos(4\pi k/N)$:

$$(\mathsf{G}_T)_{kk} = 6\mu - 8\mu \cos(2\pi k/N) + 2\mu \cos(4\pi k/N)$$

or written more simply as a continuous function

$$\mathsf{G}_T(\phi) = 2\mu(3 - 4\cos\phi + \cos 2\phi) \qquad (3\text{-}2\text{-}13)$$

It is obvious that even with the simplest possible potential function (a diagonal F_T), the dispersion curve for this, the transverse branch, will be a little more complicated owing to the two different periodicities in (3-2-13).

Although the highest craft of the chemist has proved inadequate for the synthesis of molecules such as $R_2C{=}C{=}C{=}{-}{-}{-}CR_2$ with more than 5 to 10 carbon atoms, we illustrate in Fig. 3-3 the spectrum, longitudinal and transverse branches together, for such a hypothetical linear lattice (the R groups at the ends are of course neglected), utilizing the simplest possible F matrix with $F_L = 10$ md/Å and $F_T = 0.6$ md/Å, which would be not inappropriate for allene with three carbons.

A characteristic feature of the dynamical system which we have been investigating in this section is the result that for both L and T branches $\nu \to 0$ as $\phi(k) \to 0$. We reiterate that this is a consequence of the elementary fact that the $k = 0$ modes correspond to rigid translations of the entire lattice, which by hypothesis are not opposed by any restoring forces. It is not until the following section, in which we first encounter a lattice containing more than one atom per *unit cell*, that we shall arrive at modes of finite frequency for $k = 0$. We have thus implicitly defined a unit cell as the minimal array of particles from which the entire lattice can be generated by application of the translational symmetry operations. In this section, there has been only a single particle per unit cell.

It is now appropriate to introduce the spectroscopic selection rules. In marked contrast to the case of a free molecule, the momentum conservation rule is now of dominant importance. For photons the quantized energy is just $h\nu$, so that the energy is proportional to frequency by the same constant (Planck's) which is

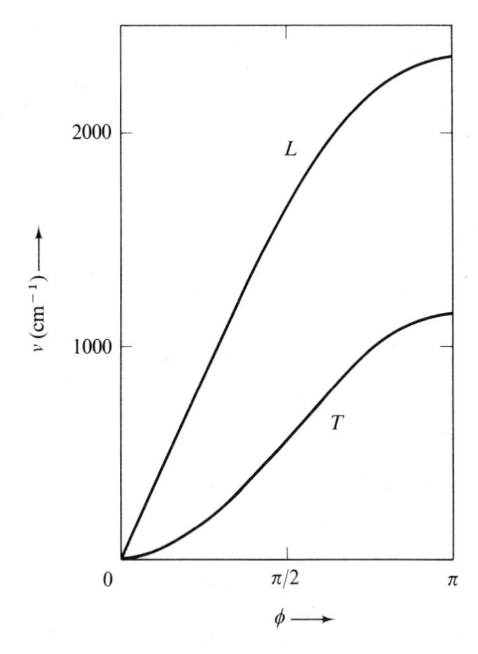

Figure 3-3 Dispersion of the longitudinal and transverse modes of an infinite linear carbon chain: $F_L = 10$ md/Å, $F_T = 0.6$ md/Å

appropriate for the material vibrations in the lattice. On the other hand, de Broglie's relation $p = h/\lambda$, similarly shows that momentum is proportional to reciprocal wavelength via the same universal constant. Thus we see that a plot of frequency versus reciprocal wavelength is equivalent to a plot of energy versus momentum. This implies that if we superpose such a curve for photons on our curve for vibrational modes in the lattice, the conservation of energy and of momentum require that *such curves intersect for an allowed transition.*

In a diagram where ϕ is the independent variable, we have $\phi = 2\pi k/N = 2\pi a(1/\lambda)$,* so that since for photons (in vacuo), $\nu = c(1/\lambda)$, the frequency plotted against ϕ would be a straight line of slope $c/2\pi a$, which is of the order of 10^{18} sec^{-1}. Indeed, such a (photon) line will not intersect the dispersion curve for the lattice oscillations at all. For the longitudinal mode, according to (3-2-10) one finds

$$\frac{d\nu}{d\phi} = \frac{\nu_{LM}}{2} \cos(\phi/2) \tag{3-2-14}$$

which has its maximum at the origin ($\phi = 0$). Since ν_{LM} is of the order of 10^{13} sec^{-1} it should be clear that the photon line cannot possibly intersect the longi-

* The reader will note that in view of well-established conventions, we employ the symbol λ to specify wavelength as well as $\lambda = 4\pi^2\nu^2$. This introduces a possibility for confusion, particularly because of the relation between these quantities, but with a little care such confusion can be avoided.

tudinal branch and will be practically indistinguishable from the vertical axis in Fig. 3-1. The same (no intersection) holds for the transverse branch.

The corresponding dispersion curves for three-dimensional lattices, even though spectroscopically inactive, have nevertheless physical importance, for example in determining the specific heat and other thermal properties. They are termed *acoustic branches,* and are the same in number, i.e., three, regardless of the complexity of the crystal. Moreover, since

$$\frac{d\nu}{d\phi} = \frac{1}{2\pi a}\frac{d\nu}{d(1/\lambda)} = \frac{1}{2\pi a}U$$

we find that for very low frequencies in the longitudinal branch that

$$U_L = \pi a \nu_{LM} \tag{3-2-15}$$

This velocity, U_L, is in fact the velocity of (longitudinal) sound waves (hence the name "acoustic" branch), with a plausible order of magnitude, namely 10^5 cm sec^{-1}. Transverse, or shear waves, will have a significantly lower velocity in our model.

There is at least one defect in the previous treatment which supposes that $\nu = c(1/\lambda)$ for photons: in real crystals we shall have to concern ourselves with the well-known phenomenon that the velocity, or wavelength for radiation of a given frequency, is not the same "inside" a material medium as in a vacuum. This complication will be considered at length in Chapter 5.

For many purposes, it is useful to describe the *density of modes,* either for a single or for all the branches of the lattice spectrum (i.e., dispersion curves). The distinct degrees of freedom are uniformly distributed along the ϕ (or k) axis, and in fact if $g(\nu)\,d\nu$ is the number of modes lying between frequency ν and $\nu + d\nu$, one has

$$g(\nu)\,d\nu = \frac{dk}{d\nu}\,d\nu = \frac{N}{2\pi}\cdot\frac{d\phi}{d\nu}\,d\nu$$

so that by simply inverting (3-2-14), we have

$$\begin{aligned} g_L(\nu) &= \frac{N}{2\pi}\cdot\frac{2}{\nu_{LM}}\cdot\frac{1}{\cos(\phi/2)} \\ &= \frac{N}{\pi}(\nu_{LM}^2 - \nu^2)^{-1/2} \end{aligned} \tag{3-2-16}$$

The singularity in this distribution function at $\nu = \nu_{LM}$ follows of course from the fact that ν versus ϕ is horizontal at $\phi = \pi$ (Fig. 3-3). We show in Fig. 3-4 the individual and composite, $g(\nu) = g_L(\nu) + 2g_T(\nu)$, densities of modes corresponding to the spectrum of Fig. 3-3.

A quantum of energy in any of the modes which we have discussed in this section is called a *phonon,* more particularly, as the case may be, a longitudinal acoustic (LA) phonon, transverse acoustic (TA) phonon. In the next section we shall encounter optic phonons.

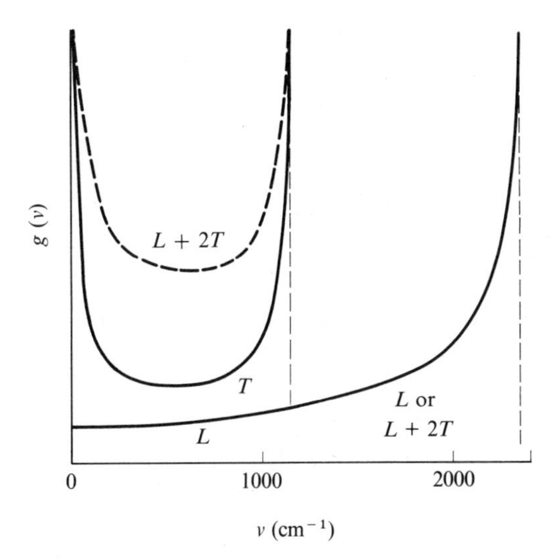

Figure 3-4 Density of modes corresponding to dispersion of Fig. 3-3

3-3 LINEAR DIATOMIC LATTICES

(a) Types of Linear Diatomic Lattices

As heralded in the previous section, the distinctive feature of the diatomic lattice is the existence of a *basis*—a unit cell containing more than one atom, in this case, two. It is this distinction, we shall see, which gives rise to a phenomenological difference: namely, optical phonons, which have finite frequencies as $k \rightarrow 0$.

We restrict our interest, as elsewhere in this chapter, to the one-dimensional crystal. In Section 3-1, for reasons of their greater physical significance, we emphasized only longitudinal waves. In the subsequent section, we explored both longitudinal acoustic (LA) and transverse acoustic (TA) phonons, in order to contrast their different dispersion relations and mode densities. In discussing linear diatomic lattices we shall again concentrate upon longitudinal modes since we wish to artificially restrict the problem to only one dimension.

There are only two possible kinds of linear diatomic lattices. In each of these we shall refer to the two different kinds of atoms as those with mass m_a and m_b, respectively. We can be even more precise and state that there are but two possible space (or line) groups in one dimension.* The distinction between the two possible line groups is the presence (or absence) of a point symmetry element which is in addition to the translational symmetry and which exchanges like nuclei. This

* A number of spectroscopists have studied polymeric systems and have described their motions in terms of what are loosely called "line groups." Strictly, these should be called "border groups" as they are in fact ornamental strips which exhibit translational symmetry, as well as certain point symmetries which include glide planes and screw axes, but always restrict these to the plane of the strip. Speiser has shown that there are a total of seven border groups in contrast to the mere pair of line groups. (A. Speiser, *Die Theorie der Gruppen von Endlicher Ordnung,* Dover, New York, 1943, pages 81–83.)

symmetry element may be viewed as a mirror plane, as a twofold rotation axis perpendicular to the line of atoms, or, as a center of symmetry. We arbitrarily choose the last convention and hence refer to the two line groups as C_1 (line), which has only translational operations, and C_i (line). Several examples are given of these groups in Fig. 3-5.

In Fig. 3-5, all three lattices are arrays of two different kinds of atoms. In the first of these, however, the lattice is also characterized by two different "bond lengths"—it is a "two-parameter lattice." It obviously lacks a center of symmetry. The second lattice is centered and indeed all atoms are located on centers of symmetry. It necessarily is a one-parameter lattice. The last lattice is also centered but it is a two-parameter lattice. Note that it may be viewed as a head-to-head lattice of diatomic "molecules" as opposed to the first example, in which the "molecules" are head-to-tail. These two lattices are therefore prototype molecular crystals whereas the second is a prototype ionic lattice. As $m_a \to m_b$, the last case resembles a covalent crystal.

In the first two cases, which are the only ones we shall discuss, there are always two atoms per unit cell. (In the last case, $Z = 4$.) Hence in the two $Z = 2$ cases there are altogether six degrees of freedom. However, because we have restricted our considerations to only longitudinal motions, there are only two degrees of freedom so polarized. Accordingly in either case, in an internal coordinate representation, there are two "bond stretches" per unit cell. An internal coordinate representation sometimes leads to certain artificialities in the nature of the symmetry coordinates derived from them. We encountered one in the last section and another similar to the last occurs with the diatomic lattice. Nevertheless, the convenience of the internal coordinate representation, especially in deriving the energy matrices, is a great advantage; and the aforementioned artificialities are easily recognized.

Each lattice has translational symmetry: the unit cell, composed of one heteronuclear diatomic unit (or basis) is repeated indefinitely by the translation t. Because it is more fundamental we shall treat lattice (b) of Fig. 3-5 first.

(b) The Diatomic Linear Lattice with Symmetry

In this lattice, as we have remarked earlier, the generating element is the inversion operation i, together with the translation t. The only other member of the line group, C_i (line) is the identity.

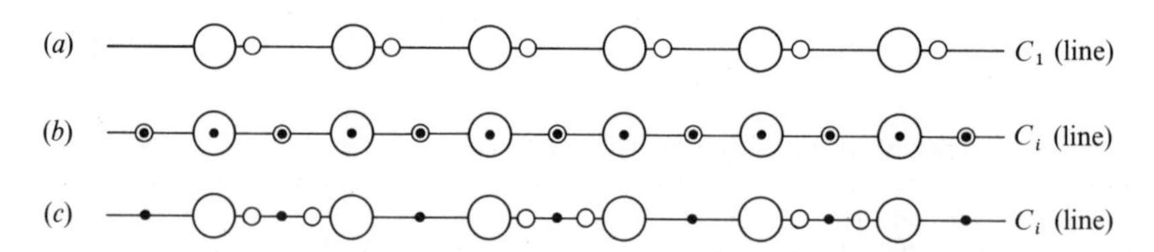

Figure 3-5 Three linear diatomic lattices. A center of symmetry is indicated by a solid dot

Even with an element of point symmetry associated with each point in the lattice, the translational symmetry persists and the lattice is invariant under the operations of the translational group. The overall, or space, symmetry is described by both point and translational symmetry—the lattice is invariant under the operations of both kinds, the complete set of these constituting the elements of the space group. However, it is possible to imagine another group, in which the translational subgroup of the space group behaves like an identity element and the remaining elements of this group consist of the *cosets*, R_1T, $R_2T, \ldots$, where R_1 is one of the point operations and T is the translational subgroup. This new group is called the *factor group* of the space group and is isomorphous with some one of the 32 crystallographic point groups. In the present example, of course, there is only one R_i and that is the inversion operation. (See Appendix VI.)

In the previous example of the monatomic lattice, C_i point symmetry also existed; there was no need, however, to take advantage of it. In the present example, we shall later take advantage of this point symmetry, but first we consider the symmetry classification of modes with respect to the translational group. Again we can index the representations with the symbol $-N/2 < k \leq N/2$. (Because of the presence of the center of symmetry, in this lattice N is required to be even.) When $k = 0$, the representations of the translational subgroup are all $+1$ and any further factoring of the energy matrices is dictated by the point symmetry. Examination of (3-2-5) shows why this must be so: $k = 0$ always corresponds to the same local motion in all unit cells, i.e., all unit cells are in phase. Hence the translational symmetry is complete; $k = 0$ corresponds to the totally symmetric representation of the translational subgroup.

In the present example, when $k = 0$ we have available the point group C_i to describe the symmetry of the $k = 0$ motions. Hence we could use the characters of the point group C_i in the Wigner Projection Operator to form symmetry coordinates at $k = 0$. For $k \neq 0$, however, we can no longer use that character table associated with C_i symmetry. Lattice motions described by $k \neq 0$ may be pictured as waves which are no longer invariant under the operations of the point group, C_i. We therefore return to the dynamical problem.

Before constructing symmetry coordinates, we shall establish a notation for the masses and coordinates which differs a little from that used in the last section. This is done in Fig. 3-6, in which we recognize that there are two kinds of internal coordinates, which are actually equivalent but are not related by the translational symmetry.

We proceed to derive the G matrix in the internal coordinate representation. As with G_L of the last section, its diagonal elements are the sum of the reciprocals of the masses which define each "bond"—hence $\mu_a + \mu_b$ instead of 2μ. Similarly,

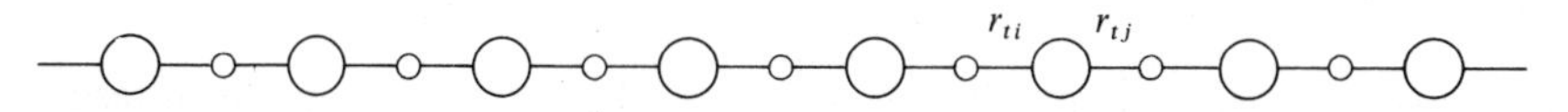

Figure 3-6 Coordinate numbering in the C_i (line) group. The first subscript denotes the "unit cell number," while the second identifies one or the other of each kind of coordinate in the unit cell

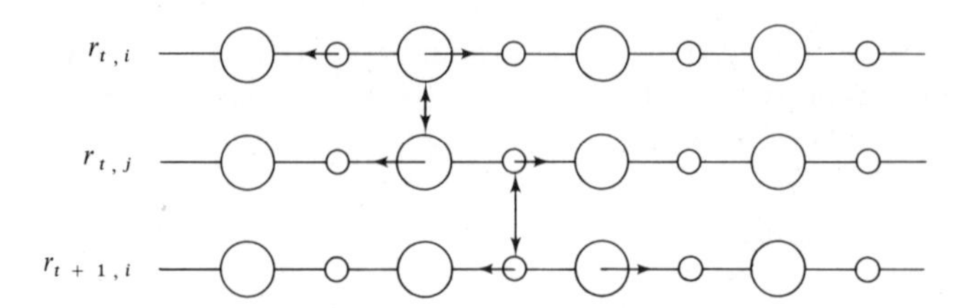

Figure 3-7 G_L matrix elements between successive internal coordinates in the C_i (line) group (common atoms are indicated by vertical arrows)

we may expect that the elements along the borders of the principal diagonal will each be the reciprocal masses of the common atoms, multiplied by the usual scalar product. The matter is illustrated in Fig. 3-7.

We see that for any two *neighboring* coordinates there is but one common atom, always indicated by the vertical double-headed arrow, that only neighboring coordinates have a G matrix element connecting them and that this is so whether two coordinates are in the same or different unit cells. Finally, we see that $\mathbf{s}_{ti} \cdot \mathbf{s}_{tj} = \mathbf{s}_{tj} \cdot \mathbf{s}_{t+1,i} = -1$. According to our (arbitrary) choice of unit cell origin in Fig. 3-6, $G_{ti,tj} = -\mu_a$ and $G_{tj,t+1,i} = G_{ti,t-1,j} = -\mu_b$. All other G matrix elements vanish.

Since all internal coordinates r_{ti} and r_{tj} are respectively related by the translational symmetry of the lattice, even for $k \neq 0$ we can construct symmetry coordinates S_i^k and S_j^k making use of the translational symmetry alone. Thus

$$S_i^k = N^{-1/2} \sum_{t=0}^{N} r_{ti} \exp(-2\pi ikt/N)$$

and (3-3-1)

$$S_j^k = N^{-1/2} \sum_{t=0}^{N} r_{tj} \exp(-2\pi ikt/N)$$

At the point $k = 0$ the appropriate symmetry coordinates are linear combinations of S_i^0 and S_j^0 which are bases for the one-dimensional irreducible representations of the point group C_i. Elsewhere, at $k \neq 0$, we could also form similar linear combinations of S_i^k and S_j^k but they would be no better descriptions of the proper motions of the lattice than are S_i^k and S_j^k themselves—that is, they will not further factor the energy matrices. In the expressions (3-3-1) k is unsigned, since the modes for $\pm k \neq 0$ are degenerate. The consequence of (3-3-1) is that

$$U_{kt} = N^{-1/2} \exp(-2\pi ikt/N) \qquad (3\text{-}3\text{-}2)$$

independent of i, j. It follows that

$$G(k) = N^{-1} \sum_{t,t'} G_{ti;t'j} \exp\{2\pi ik(t' - t)/N\} \qquad (3\text{-}3\text{-}3)$$

Before applying (3-3-3) it is helpful to write out the form of G in the internal coordinate basis. It is convenient to order the columns and rows of the G matrix such that the coordinates r_{ti} and r_{tj} are segregated:

$$
G = \begin{array}{c|ccc|cccc}
 & r_{t,i} & r_{t+1,i} & r_{t+2,i}\ \cdots & r_{t,j} & r_{t+1,j} & r_{t+2,j} & \cdots r_{t-1,j} \\
\hline
r_{t,i} & \mu_a + \mu_b & 0 & 0 \ \cdots & -\mu_a & 0 & 0 & \cdots -\mu_b \\
r_{t+1,i} & 0 & \mu_a + \mu_b & 0 \ \cdots & -\mu_b & -\mu_a & 0 & \cdots\ 0 \\
r_{t+2,i} & 0 & 0 & \mu_a + \mu_b \cdots & 0 & -\mu_b & -\mu_a & \cdots\ 0 \\
\hline
r_{t,j} & & & & \mu_a + \mu_b & 0 & 0 & \cdots \\
r_{t+1,j} & & \text{(symmetric)} & & 0 & \mu_a + \mu_b & 0 & \cdots \\
r_{t+2,j} & & & & 0 & 0 & \mu_a + \mu_b & \cdots \\
\hline
\end{array}
$$

The double summation over the rows and columns of G indicated in (3-3-3) can be replaced by a summation over the diagonals, of which there are only five with finite elements. There is, first, the principal diagonal, summation over which yields

$$G_{ii}^{(k)} = G_{jj}^{(k)} = \mu_a + \mu_b \tag{3-3-4}$$

There remain only the diagonals found in the off-diagonal blocks which connect coordinates $r_{t,i}$ and $r_{t,j}$. Focussing our attention on the upper right-hand off-diagonal block, we note that for the principal diagonal of that block, $t - t' = 0$ while for the border diagonal just under the principal, $t - t'$ (row – column) $= 1$. Applying (3-3-3) we therefore find

$$G_{ij}^{(k)} = -\mu_a - \mu_b\, e^{-2\pi ik/N} \tag{3-3-5}$$

We see that $G_{ij}^{(k)}$ is complex; hence G must be Hermitian. Using the Hermitian property (or by explicit application of (3-3-3) to the lower left-hand off-diagonal block of $G_{tt'}$), we obtain

$$G_{ji}^{(k)} = -\mu_a - \mu_b\, e^{+2\pi ik/N} \tag{3-3-6}$$

Thus the complete form of $G(k)$ is

$$G(k) = \begin{pmatrix} \mu_a + \mu_b & -\mu_a - \mu_b\, e^{-2\pi ik/N} \\ -\mu_a - \mu_b\, e^{+2\pi ik/N} & \mu_a + \mu_b \end{pmatrix} \tag{3-3-7}$$

We proceed to form the secular determinant $|FG - \omega^2 E| = 0$ and solve for its roots. We again assume F to be diagonal. The roots at $k \neq 0$ are

$$\omega^2(k) = F_0\{(\mu_a + \mu_b) \pm [(\mu_a + \mu_b)^2 - 4\mu_a\mu_b \sin^2 \pi k/N]^{1/2}\} \tag{3-3-8}$$

A plot of the branches of $\omega^2(k)$ is illustrated in Fig. 3-8. Note that in Fig. 3-8, case (a), $m_a > m_b$, and there is a gap at $k = N/2$, while in case (b), $m_a = m_b$, there is no gap. The values at the special points, $k = 0$, $N/2$ may be found by direct substitution of those values in (3-3-8):

$$\omega^2(0) = 0, \quad \text{or} \quad 2F_0(\mu_a + \mu_b) \tag{3-3-9}$$

We see that at $k = 0$ we again have a mode of zero frequency but, in addition,

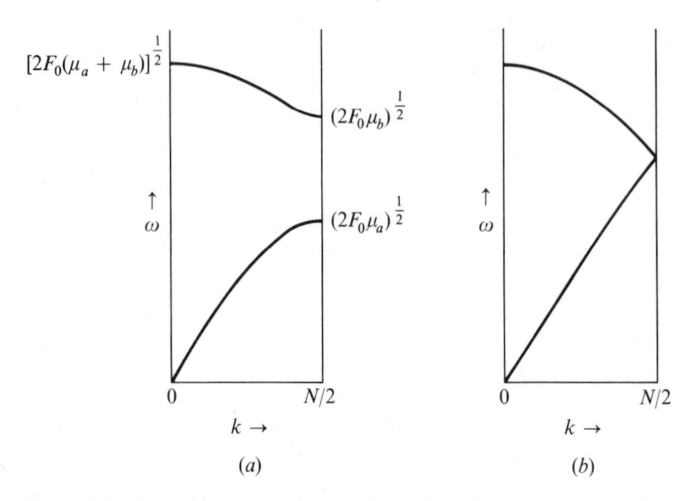

Figure 3-8 Dispersion curves of linear diatomic lattices: (a) $m_a > m_b$; (b) $m_a = m_b$

we also have one of finite frequency, a distinction from the example of the monatomic lattice.

Substitution of the value $k = 0$ into the $G(k)$ matrix gives its special form:

$$G(0) = (\mu_a + \mu_b)\begin{pmatrix} 1 & -1 \\ -1 & 1 \end{pmatrix} \tag{3-3-10}$$

The solution of the secular determinant formed using $G(0)$ of course gives the same results as (3-3-9) but this in no way gives any insight into the special nature of the $k = 0$ modes. The irreducible representations of the factor group are those of the point group C_i: introduction of symmetry coordinates corresponding to the A_g and A_u irreducible representations of this point group factors the secular determinant, as may be seen by forming linear combinations $2^{-1/2}[S_i(0) \pm S_j(0)]$.

$$\begin{aligned} S^g(0) &= (2N)^{-1/2} \sum_{t=0}^{N} (r_{ti} + r_{tj}) \\ S^u(0) &= (2N)^{-1/2} \sum_{t=0}^{N} (r_{ti} - r_{tj}) \end{aligned} \tag{3-3-11}$$

so that $G(0)$ can be factored using

$$U = 2^{-1/2}\begin{pmatrix} 1 & 1 \\ 1 & -1 \end{pmatrix} \tag{3-3-12}$$

which yields

$$G(0) = 2(\mu_a + \mu_b)\begin{pmatrix} 0 & 0 \\ 0 & 1 \end{pmatrix} \tag{3-3-13}$$

From (3-3-13) we will again find that $\omega^2(0) = 0$ or $2F_0(\mu_a + \mu_b)$ but the form of G demonstrates that the zero frequency root belongs to the mode $S^g(0)$, and the

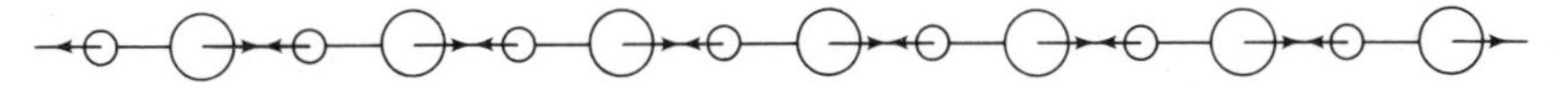

Figure 3-9 The optical mode, $S^u(0)$, of the linear diatomic lattice

finite frequency belongs to $S^u(0)$. The form of $S^g(0)$ is essentially identical with that of $S(0)$ of the monatomic linear lattice and the comments made in Section 3-2 concerning that mode are equally valid for $S^g(0)$. The mode $S^u(0)$ is new, however, and it is rewarding to use (3-3-11) to sketch it. Figure 3-9 illustrates the result.

Since the mode is seen to involve the motion of unlike atoms in opposition to each other, excitation of this motion by dipole radiation is allowed; this result also follows from the fact that $S^u(0)$ transforms like A_u, the activity representation of C_i. Because of this, this mode is commonly called the *optical mode.*

According to (3-3-13) $S^g(0)$ and $S^u(0)$ are not coupled. From the form of $G(k)$, however, we see that S_i^k and S_j^k are always coupled except at $k = 0$ and $\pm N/2$, and there exists no transformation of the coordinates that will decouple them. Also, neither S_i^k nor S_j^k go continuously into $S^g(0)$ or $S^u(0)$ as $k \to 0$. It is possible, however, to make the same linear combination $2^{-1/2}(S_i^k \pm S_j^k)$ and achieve the latter result:

$$S_{\pm}^k = (2N)^{-1/2} \sum_{t=0}^{N} (r_{ti} \pm r_{tj}) \exp(-2\pi ikt/N) \tag{3-3-14}$$

Transformation of $G(k)$ to this basis yields

$$\mathsf{G}(k) = \begin{pmatrix} \mu_b(1 - \cos 2\pi k/N) & -\mu_b \sin 2\pi k/N \\ \mu_b \sin 2\pi k/N & 2\mu_a + \mu_b(1 + \cos 2\pi k/N) \end{pmatrix} \tag{3-3-15}$$

which leads to the roots obtained with $G(k)$. Now $\mathsf{G}(k) \to \mathsf{G}(0)$ as $k \to 0$; hence $S_{\pm}^k \to S^{g,u}(0)$ at the same limit. Indeed, $S_{\pm}^k$ could have been obtained by using the Wigner Projection Operator. The lengthier derivation, however, has illuminated a number of points which would otherwise have been obscured.

Because of the correlations, $S_+^k \leftrightarrow S^g(0)$ and $S_-^k \leftrightarrow S^u(0)$, $\omega_+^2(k)$ and $\omega_-^2(k)$ are commonly referred to as the acoustical and optical branches, respectively. Once again because of the magnitude of the velocity of light, only the optical branch intersects with the dispersion curve for photons. The discussion of this problem is completed by one of the energy matrices, modes, and frequencies at the other limit of $k = \pm N/2$. At this point we find from (3-3-7):

$$G(\pm N/2) = \begin{pmatrix} \mu_a + \mu_b & -\mu_a + \mu_b \\ -\mu_a + \mu_b & \mu_a + \mu_b \end{pmatrix} \tag{3-3-16}$$

$$\omega^2(N/2) = 2F_0\mu_a \quad \text{or} \quad 2F_0\mu_b \tag{3-3-17}$$

As for the form of the modes, we note that at $k = \pm N/2$ the phase factors, $e^{2\pi ikt/N}$ in S_i^k, S_j^k and in $S_{\pm}^k$ all become $e^{\pi it}$, which tells us to begin with the modes S_i^0, S_j^0 at $k = 0$ and then proceed to reverse the senses of motion of atoms successively separated by one translational unit. The result is illustrated in Fig. 3-10.

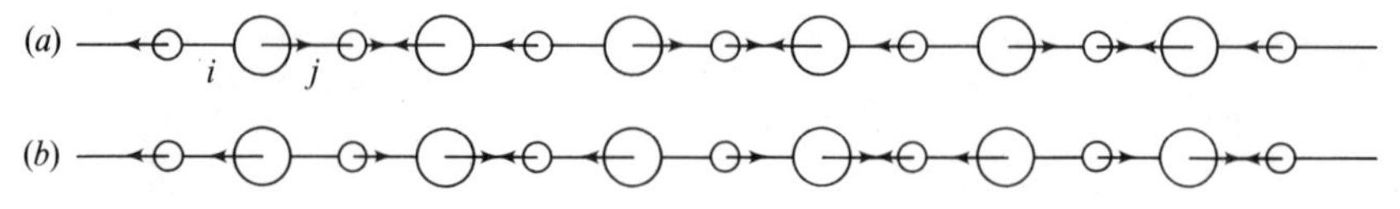

Figure 3-10 The modes $S_i^{N/2}$ and $S_j^{N/2}$: (a) $S_i^{N/2}$; (b) $S_j^{N/2}$

Upon inspection of Fig. 3-10 it is evident that both modes have "optical" character to them, thus indicating the mode-mixing which occurs at all $k \neq 0$. It is only as $k \to 0$ that $S_+^k \to S^g(0)$. On the other hand, (3-3-17) tells us that each mode involves the motion of one or the other of the masses m_a or m_b, which certainly is not evident in Fig. 3-10. The reason for this is that $G(\pm N/2)$, with which ω^2 was calculated, was in the S_i, S_j representation. That this is so is also evident in Fig. 3-10. In (*a*) of that figure, the reader will note that only coordinates r_{ti} are exercised; in (*b*), they are replaced by the r_{tj}. Hence, there must be linear combinations of S_i and S_j which diagonalize G. Indeed, they are the same linear combinations we used to form $S^g(0)$ and $S^u(0)$ as well as the $S_\pm^k$. The result is

$$\mathsf{G}(\pm N/2) = \begin{pmatrix} 2\mu_b & 0 \\ 0 & 2\mu_a \end{pmatrix} \tag{3-3-18}$$

The linear combinations, $S_\pm^{N/2}$, are easily pictured from Fig. 3-10. The results are displayed in Fig. 3-11. Figure 3-11 demonstrates the result indicated both by (3-3-17) and by the form of $\mathsf{G}(N/2)$, namely, that each mode involves the motion of one or the other of the masses m_a or m_b.

That $G(\pm N/2)$ may be factored to yield $\mathsf{G}(N/2)$ indicates that at the limiting values of $k = \pm N/2$, just as at the point $k = 0$, point symmetry of the system has been reestablished. In general, inversion changes the sign of k. But when $k = N/2$, this transformation leaves the phase factor, $\exp(2\pi ikt/N)$, unchanged. Therefore, by (3-3-1), $S^{N/2} \equiv S^{-N/2}$. In accordance with this result, we find that the energy matrices and the energies themselves are identical at $k = \pm N/2$.

The factor group of the space group at $k = N/2$ is thus isomorphous with C_i. The representations at $k = N/2$, however, are not identical with those of the point group C_i. We call these representations $\Gamma_+(N/2)$ and $\Gamma_-(N/2)$. The matrices and their characters of $\Gamma_+(N/2)$ are $e^{\pi it}$ for both the identity and inversion operations; those of $\Gamma_-(N/2)$ are respectively $e^{\pi it}$ and $-e^{\pi it}$, where t is the translational part of the symmetry operation, as can be appreciated from the forms of $S_+^{N/2}$ and $S_-^{N/2}$ [see (3-3-14) and Fig. 3-11].

All of the factorization of the energy matrices we have achieved up until now have been based upon the use of spatial symmetry. This raises the question as to whether or not there is any additional symmetry, the presence of which may help to achieve further factorization. Since the factorization was complete at $k = 0$, $\pm N/2$, the only value of additional symmetry would be at the remaining values of k. Because the Hamiltonian must be invariant to time reversal ($t \to -t$), and because $k \sim 1/\lambda \sim p$, the time reversal operation is realized by changing $k \to -k$. Inspection of (3-3-7) shows that the operation of time reversal effects

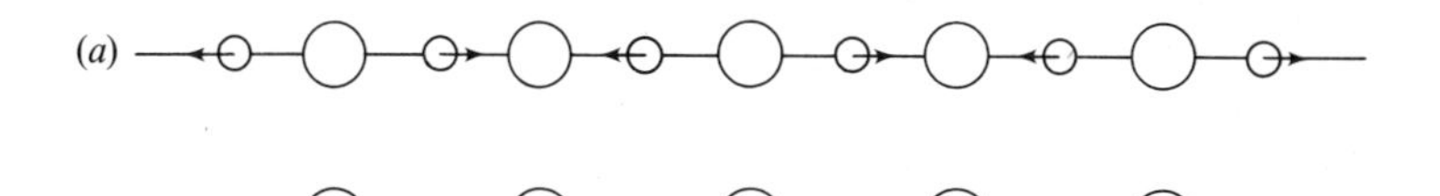

Figure 3-11 The modes $S_{\pm}^{N/2}$: (a) $S_{+}^{N/2}$; (b) $S_{-}^{N/2}$

no further factorization. All that we can say in this case is that $\omega^2(k) = \omega^2(-k)$, which result also followed from the general spatial symmetry operation, $ik = -k$. The equivalence of the spectrum at k and $-k$ is always valid; it is for this reason that dispersion curves need only be presented for $k > 0$.

(c) The Diatomic Linear Lattice without Symmetry

We complete this section with a short discussion of the remaining diatomic lattice —the one without any rotational symmetry. Earlier we called this the two-parameter lattice; the line group is C_1 (line). The lattice is illustrated in Fig. 3-5 (*a*). Although the only symmetry element is translational, we can construct symmetry coordinates exactly according to (3-3-1), since those, too, utilized only the translational symmetry of the lattice C_i (line). Furthermore, the G matrix (in either the internal coordinate or symmetry coordinate representations) is identical to that of the last problem. A difference enters, because of the loss of inversion symmetry, in that it is no longer reasonable to assume that F is diagonal. Instead, in the internal coordinate representation, we shall set all $F_{ti;ti} = F_i$ and all $F_{tj;tj} = F_j$ and all other elements as zero. Then the F matrix, written in same order as the G matrix, has the form:

	$r_{t,i}$	$r_{t+1,i}$	$r_{t+2,i}$	$\cdots$	$r_{t,j}$	$r_{t+1,j}$	$r_{t+2,j}$	$\cdots r_{t-1,j}$
$r_{t,i}$	F_i	0	0		0	0	0	0
$r_{t+1,i}$		F_i	0			0	0	0
$r_{t+2,i}$			F_i				0	
$r_{t,j}$					F_j	0	0	
$r_{t+1,j}$		symmetric				F_j	0	
$r_{t+2,j}$							F_j	

Using (3-3-2), in the (S_i^k, S_j^k) representation this becomes

$$F(k) = \begin{pmatrix} F_i & 0 \\ 0 & F_j \end{pmatrix} \qquad \text{(3-3-19)}$$

If we now form the secular determinant $|FG - E\lambda| = 0$ and solve for its roots, the quadratic in $\lambda = \omega^2$ is

$$\lambda^2 - 2\mu F\lambda + 4F_iF_j\mu_a\mu_b \sin^2(\pi k/N) = 0 \tag{3-3-20}$$

where $\mu = \mu_a + \mu_b$, $F = (F_i + F_j)/2$.

This case lends itself to consideration of a simple, one-dimensional molecular crystal by stipulating $F_i \gg F_j$, so that at all k one of the two roots will be much larger than the other. If this is true, we can make an approximate separation of the high and low frequency roots, as is often done in molecular vibrations (see WDC, Section 4-8 and Appendix IX). (This separation tacitly assumes that $S_i(k)$ and $S_j(k)$ are decoupled.) The "old" $G(k)$ matrix (3-3-7) then factors into $\mu_a + \mu_b$ for the high frequency part and into

$$\begin{aligned} G_l(k) &= \frac{4\mu_a\mu_b}{\mu}\sin^2(\pi k/N) \\ &= \frac{4}{M}\sin^2(\pi k/N) \end{aligned} \tag{3-3-21}$$

where $M = m_a + m_b = \mu/\mu_a\mu_b$, for the low frequency portion. Thus the two roots are

$$\lambda_h = \mu F_i$$

and

$$\lambda_l = \frac{4F_j}{M}\sin^2(\pi k/N)$$

Furthermore, λ_h has no k dependence; hence it has no dispersion. On the other hand λ_l behaves normally: in the vicinity of $k = 0$ it depends on Fk^2/M, it vanishes at $k = 0$ and rises according to the $\sin^2 \pi k/N$ function to its maximum value at $k = N/2$. It is interesting to note from the dependence on M^{-1} that this mode is similar to that of a monatomic lattice, since M is the total mass of the diatomic unit, which moves as a whole in this mode, at least near $k = 0$. The mode whose frequency is $(\mu F)^{1/2}$ is clearly optical in nature.

The addition of an off-diagonal interaction constant to (3-3-19) leads to a straightforward modification of (3-3-20) which illustrates the multiple periodicity discussed in Section 3-2.

3-4 LINEAR POLYMERS

Linear polymers provide a great diversity of examples to which the theory of the linear lattice is applicable, but with complications arising from large numbers of atoms per unit cell. Strictly, one cannot neglect the interactions between adjacent polymer chains in a real crystal, but such effects can be treated as a small perturbation superimposed upon the results of the following analysis.

The unit cell structure can range between very wide extremes, for example,

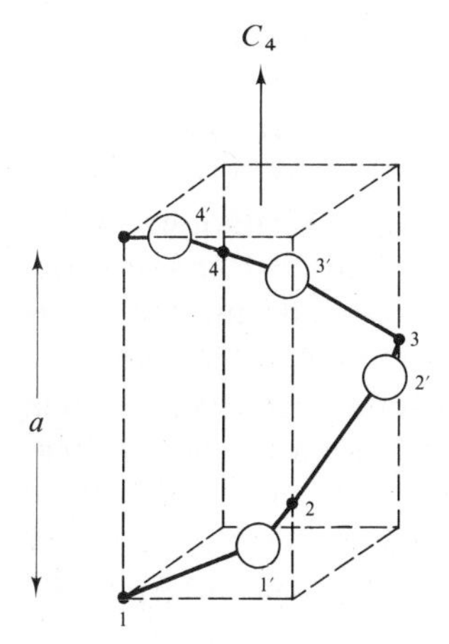

Figure 3-12 Helical chain of C_4 symmetry: the operation $\{C_4 | a/4\}$ transforms atom 1 to 2, 1′ to 2′, etc.

from polyethylene with two CH_2 groups per cell to synthetic DNA in which the complementary helices are constituted of all adenine in one strand and all thymine in the other. We shall concentrate upon the simpler examples.

At the outset one recognizes the value of the maximum use of symmetry, both of the translational and point operations. Fortunately, these can be treated independently, although some care is needed in the interpretation of cases in which there are apparently no point group operations. Thus in the simple helical coil shown in Fig. 3-12, it is clear that the generating element from which all covering operations can be created by repetition is a rotation by $2\pi/4$ accompanied by a translation by $a/4$. This is an example of a screw axis. One cannot really express all these so-called *line group* operations as products of point group operations by translational operations, but it is nevertheless important to recognize that all symmetry operations admitted by the structure are expressible as products of one of four elements consisting of $\{E|0\}$, $\{C_4|a/4\}$, $\{C_4^2|a/2\}$, $\{C_4^3|3a/4\}$, by the ordinary translational group consisting of translations by $\{E|ta\}$, where t is any integer. A line group operation, $\{R|\tau_R\}$, consists of a point group operation (R), followed by a fractional translation (τ_R). The four listed operations comprise a factor group of the full line group by the translational subgroup, more particularly, a unit cell group.[4,5,6] It is isomorphous, though not identical, with the point group C_4. By isomorphous, one means that the operations combine with one another in exactly the same way as do the simple rotational operations E, C_4, C_4^2, and C_4^3.

The special value of the unit cell group consists in the facts that (i) its character table is well-known, i.e., identical with that of C_4, and (ii) since to an excellent approximation the only spectroscopically active modes are those with $k = 0$, i.e.,

4. F. Seitz, *Z. Krist.*, **91,** 336 (1935).
5. D. F. Hornig, *J. Chem. Phys.*, **16,** 1063 (1948).
6. M. C. Tobin, *J. Chem. Phys.*, **23,** 891 (1955); *J. Mol. Spectry.*, **4,** 349 (1960).

totally symmetric under the translational group, we can carry out most of the necessary analysis using C_4.

Thus our unit cell with eight atoms has $f = 24$ degrees of freedom, of which we may anticipate that three will comprise acoustic modes and have vanishing frequency at $k = 0$. However, a new element has entered the situation which did not occur in the strictly linear structure described in Sections 3-2 and 3-3. The rotation of the whole lattice around its axis is now a real motion since the atoms no longer lie exclusively on the axis of the lattice. It is only in accord with our neglect of any external forces to assume that such a motion has vanishing frequency (at $k = 0$), so that in effect we have a fourth acoustic mode. Rigid rotations of the whole structure about axes perpendicular to the axis of the lattice cannot arise because of the boundary conditions. These complications make it very desirable to analyze the normal modes using both a cartesian basis (see Appendix II) and one constructed from internal coordinates. The cartesian analysis of the C_4 unit cell is particularly simple, since only the identity leaves any atoms unmoved; the character of a vector for the identity being 3, the cartesian character is the number of unmoved atoms (8) times 3 or 24 for E and zero for the other operations. The analysis of this simple character reveals a total of six modes in each species, but because R_z (as well as T_z) falls in A, we conclude that there exist only four optic modes in A plus two acoustic modes, etc., as shown in Table 3-1.

The problem may be treated entirely with four coordinates (per unit cell) of each of the six following types: r_i, s_i = stretches, α_i, β_i = bends, and σ_i, τ_i = twists (this last type of internal coordinate describes the change in the dihedral angle between planes defined by atoms at positions such as 1, 2, 3 and 2, 3, 4). Proof of the kinematic completeness for such a set in the case of finite molecules has been given.[7] Table 3-1 shows the symmetry analysis, which is particularly elementary inasmuch as the characters are identical for each coordinate set, $r, s, \alpha, \beta, \sigma$, and τ, and have the values four for E, 0 for all other operations, since such operations leave no coordinate unchanged. Although the analysis reveals the presence of six symmetry coordinates of each species (six degenerate pairs in E, which is called *separably degenerate*), one must realize that redundancies occur because of the use of the internal coordinate representation. In the present case the redundancies are

Table 3-1 Symmetry analysis of the fourfold simple helical chain

C_4	E	C_4	C_4^2	C_4^3	$n_r = n_\phi = n_T$		N Optic	N Acoustic
A	1	1	1	1	1	T_z; R_z; μ_z; $\alpha_{xx}+\alpha_{yy}$; α_{zz}	4	2
B	1	-1	1	-1	1	$\alpha_{xx}-\alpha_{yy}$; α_{xy}	6	0
E	$\begin{Bmatrix}1\\1\end{Bmatrix}$	$\begin{matrix}i\\-i\end{matrix}$	$\begin{matrix}-1\\-1\end{matrix}$	$\begin{matrix}-i\\i\end{matrix}$	1	(T_x, T_y); (μ_x, μ_y); $(\alpha_{yz}, \alpha_{zx})$	5	1
$\chi_r =$	4	0	0	0	$(= \chi_s = \chi_\alpha = \chi_\beta = \chi_\sigma = \chi_\tau)$		20	4

7. J. C. Decius, *J. Chem. Phys.*, **17**, 1315 (1949).

in one-to-one correspondence with the acoustic modes, i.e., two in A and one (pair) in E. In more complicated examples, this one-to-one correspondence may not exist, making it imperative to carry through the cartesian analysis, or to use the correlation methods which will be developed in Chapter 6. Nevertheless, kinematically complete internal coordinate sets of the type used here give a completely correct basis for the symmetry treatment of vibrations for $k \neq 0$, and provide a convenient framework for the employment of physically meaningful potential functions.

In the present instance the complete dynamical analysis can readily be carried out with the aid of the irreducible representations of the full line group. It is abelian, i.e., every pair of operations commute. Thus again we expect to encounter characters which are complex roots of unity. The order of the group is $4N$, where N is the number of ordinary translational unit cells in the lattice. Thus the character for the generating element must be $\chi_R^{(k)} = \exp(2\pi ik/4N)$, where, for example one may choose k as any integer in the range $-2N \leq k < 2N$. Then the character for R^p is $[\chi_R^{(k)}]^p = \exp(2\pi ikp/4N)$, and when one is considering those modes which are totally symmetric with respect to translation, it is easily seen that since $[\chi_R^{(k)}]^4 = 1$ we must have $4k/4N = m$ where m is an integer; thus $\chi_R^{(k)} = \exp(2\pi im/4)$, i.e., the special values of k selected by the requirement of translational symmetry being multiples of N, we discover that the only possible characters for the (unit cell) (factor) group are precisely those of C_4. The full dynamic analysis, that is the computation of the frequencies of all nine optic and all three acoustic branches, for all allowed k values can in principle be carried out with the aid of symmetry coordinates constructed with the aid of the Wigner projection operator and the *general* line group character, $\chi_R^{(k)} = \exp(2\pi ik/4N)$, employing techniques entirely analogous to those illustrated in section 3-2.

A different example which is very well known experimentally is that of the extended form of polyethylene[8,9] in which all carbons lie in a single plane (Fig. 3-13). In this instance, the unit cell group is D_{2h}. Some of the operations of this unit cell group are ordinary point group operations, namely, E, $C_2(z)$, $\sigma(xy)$, and i constituting the point group C_{2h}; note that according to Fig. 3-13 we choose the lattice axis in the x direction, the carbon plane as the xy plane, and the mid-point of the C—C bond as the origin. However, further reflection planes and twofold axes become covering operations when they are combined with translations by half the unit cell length, $a/2$. Containing six atoms, the unit cell must have 18 degrees of freedom, 4 acoustic and 14 optic in nature. The 18 degrees of freedom are describable with the aid of four hydrogen stretches r_i, two CC stretches s_j, two HCH bends ϕ_j, eight HCC bends η_h, and two torsions around CC bonds, τ_j. That such a set is kinematically complete follows from arguments given in the reference made during the discussion of the fourfold helical chain molecule. However, as will be seen from the details of the symmetry

8. S. Krimm, C. Y. Liang, and G. B. B. M. Sutherland, *J. Chem. Phys.*, **25**, 549 (1956).
9. R. G. Snyder, *J. Mol. Spectry.*, **4**, 411 (1960); **7**, 116 (1961).

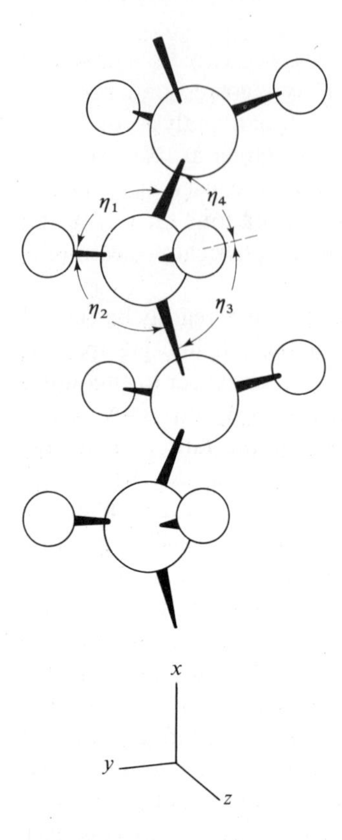

Figure 3-13 The extended form of polyethylene

analysis given in Table 3-2, the reduced representation based on this set of internal coordinates does not correspond to the reduction of the cartesian representation; we shall discuss this point at greater length below. Actually the η coordinates (HCC bends) are not so convenient as certain linear combinations. Using the numbering in Fig. 3-13, it may be seen that the linear combinations

$$\gamma(t) = \tfrac{1}{2}(\eta_1 - \eta_2 + \eta_3 - \eta_4) \quad \text{(twist)}$$

$$\gamma(w) = \tfrac{1}{2}(\eta_1 - \eta_2 - \eta_3 + \eta_4) \quad \text{(wag)}$$

$$\gamma(r) = \tfrac{1}{2}(\eta_1 + \eta_2 - \eta_3 - \eta_4) \quad \text{(rock)}$$

have the forms of a twist of the CH_2 group around its twofold axis, a wagging of the CH_2 group in the direction of the lattice axis, and a rocking of the CH_2 group perpendicular to the carbon plane, respectively. The fourth linear combination

$$\gamma = \tfrac{1}{2}(\eta_1 + \eta_2 + \eta_3 + \eta_4)$$

is equivalent to a combined bending of the CH_2 group (ϕ) plus a bending of the CCC angle in the carbon plane. The internal coordinate representation will accordingly be carried out in terms of a set of $4r_i$, $2s_j$, $2\phi_j$, $2\gamma_j$, $2\gamma(t)_j$, $2\gamma(w)_j$, $2\gamma(r)_j$, and $2\tau_j$.

Table 3-2 Symmetry analysis of polyethylene

D_{2h}	$\{E\|0\}$	$\{C_2^z\|0\}$	$\{C_2^y\|a/2\}$	$\{C_2^x\|a/2\}$	$\{i\|0\}$	$\{\sigma^{xy}\|0\}$	$\{\sigma^{zx}\|a/2\}$	$\{\sigma^{yz}\|a/2\}$		n	$n_{opt.}$	r	s	ϕ	γ	$\gamma(t)$	$\gamma(w)$	$\gamma(r)$	τ	$n_{red.}$
A_g	1	1	1	1	1	1	1	1	$\alpha_{xx}; \alpha_{yy}; \alpha_{zz}$	3	3	1	1	1	1	0	0	0	0	1
B_{1g}	1	1	−1	−1	1	1	−1	−1	α_{xy}	2	2	0	1	0	0	0	1	0	0	0
B_{2g}	1	−1	1	−1	1	−1	1	−1	α_{zx}	1	1	0	0	0	0	1	0	0	0	0
B_{3g}	1	−1	−1	1	1	−1	−1	1	$\alpha_{zy}; R_x$	3	2	1	0	0	0	0	0	1	0	0
A_u	1	1	1	1	−1	−1	−1	−1		1	1	0	0	0	0	1	0	0	1	1
B_{1u}	1	1	−1	−1	−1	−1	1	1	$T_z; \mu_z$	3	2	1	0	0	0	0	0	1	1	1
B_{2u}	1	−1	1	−1	−1	1	−1	1	$T_y; \mu_y$	3	2	1	0	1	1	0	0	0	0	1
B_{3u}	1	−1	−1	1	−1	1	1	−1	$T_x; \mu_x$	2	1	0	0	0	0	0	1	0	0	0
u_R	6	0	2	0	0	2	0	6												
$\pm 1 + 2\cos\alpha_R$	3	−1	−1	−1	−3	1	1	1												
χ (cartesian)	18	0	−2	0	0	2	0	6												
$\chi(r)$	4	0	0	0	0	0	0	4												
$\chi(s)$	2	2	0	0	2	2	0	0												
$\chi(\phi)$	2	0	2	0	0	2	0	2												
$\chi(\gamma)$	2	0	2	0	0	2	0	2												
$\chi[\gamma(t)]$	2	0	2	0	0	−2	0	−2												
$\chi[\gamma(w)]$	2	0	−2	0	0	2	0	−2												
$\chi[\gamma(r)]$	2	0	−2	0	0	−2	0	2												
$\chi(\tau)$	2	2	0	0	−2	−2	0	0												

There are a number of precautionary points which ought to be made in connection with the deductions which are carried out in Table 3-2. In the first place, it is now really imperative to carry through the cartesian analysis; this is done in the first three rows below the character table proper, using $\chi_R = u_R(\pm 1 + 2\cos\alpha_R)$ (see Appendix II) where u_R is the number of unshifted atoms and α_R is the rotation angle involved. The analysis of the cartesian character gives the total number of optic and acoustic modes as shown in the second column to the right of the character table proper. After subtracting the acoustic modes, T_x, T_y, T_z, and R_z, the number of optic modes is shown in the next column.

A second point of caution is the observation that the count of unshifted atoms by the operation R requires the following interpretation. If the factor group operation moves an atom to a translationally equivalent atom in another unit cell, the atom is counted as unshifted. Exactly the same remark applies in the examination of the transformation of internal coordinates. For example, the $\{C_2^y \mid a/2\}$ operation is interpreted as carrying one carbon atom into the equivalent one removed by one unit cell, and the other carbon into itself, hence $u_R = 2$. Similarly the $\{\sigma^{yz} \mid a/2\}$ operation carries one HCH angle into an equivalent one in the next unit cell, and restores the other HCH to its original position in virtue of the $a/2$ translation. Subsequently, the symmetry analysis of the coordinate sets $r =$ CH, $s =$ CC, $\phi =$ HCH, $\gamma(s) =$ HCH and CCC bend, $\gamma(t) = CH_2$ twist, $\gamma(w) = CH_2$ wag, $\gamma(r) = CH_2$ rock, and $\tau =$ twist of carbon skeleton is carried out. One notes that the redundancies are distributed as follows: $A_g + A_u + B_{1u} + B_{2u}$, which does not correspond with the acoustic modes in $B_{3g} + B_{1u} + B_{2u} + B_{3u}$. However, the internal coordinates serve *inter alia* the useful purpose of indicating that the infrared spectrum should consist of one z-polarized CH stretch (transverse to the C plane and therefore locally antisymmetric), one methylene rock, and apparently one skeletal torsion which, however, turns out to be the B_{1u} redundancy. The latter statement is easily verified by construction of the G matrix. The y-polarized modes comprise a (locally symmetric) CH stretch, a bend of the HCH type, but the third mode is again a redundancy. The sole longitudinal infrared active mode is clearly a methylenic wag. Similar interpretation of the Raman active modes is possible. One of the two remaining redundancies occur in A_g, where construction of the G matrix easily shows that there is a linear relation between s, ϕ, and γ, so that the polarized Raman modes consist of a CH stretch, an HCH bend, and a simple breathing mode of the carbon chain (CC stretch). The final redundancy (A_u) is purely τ; since both τ modes are redundant, they could well be omitted in the calculation of these, the $k = 0$ modes, but they would be necessary for all other values of $k \neq 0$.

CHAPTER

FOUR

VIBRATIONS OF CRYSTALS COMPOSED OF MONATOMIC UNITS

In this chapter we shall consider the theory for crystals in which there are either no molecular entities or in which the crystal as a whole may be regarded as a single giant molecule. In the former category one encounters purely van der Waals binding (but with a restriction to single atoms and therefore effectively the "inert" gases, He, Ne, etc.) and simple ionic compounds such as NaCl, CaF_2, etc. (but not ionic crystals of the types NH_4Cl or $CaCO_3$ in which the ions NH_4^+, $CO_3^=$ are *molecules*). In the second category one has such examples as diamond, silicon, boron nitride, etc., with obvious covalent binding extending throughout the entire lattice. Thus we postpone until the following chapters the central subject of this work in which we shall discuss small molecules in the crystalline phase. However, we should mention here that there are well defined types of solids which do not fall into either of the categories suggested above. Examples are selenium,[1] in which spiral chains of atoms are apparently covalently bonded, each chain however being bonded to other chains by weaker, van der Waals forces, and graphite, in which the covalency is two-dimensional, with much weaker van der Waals forces between layers. Obviously in these cases we cannot compare the solid spectrum with that of the vapor, yet we must allow for the existence of two classes of binding forces in the solid, one of which is similar to that encountered in small, gaseous molecules.

The subject of the present chapter has been extensively studied by solid state

1. R. W. G. Wyckoff, *Crystal Structures*, Interscience, New York, Volume I, 1963, page 36.

physicists and a number of excellent reviews are available.[2-5] The present treatment will be made as concise as possible, consistent with our aim to set the stage for the understanding of the following chapter.

The principal new feature to be revealed in the present chapter is, of course, the three-dimensional nature of the translational symmetry. We again encounter the existence of acoustic and optical "branches" of the spectrum, but these branches will now be functions of a three-dimensional independent variable instead of a single variable; i.e., $\mathbf{k}$ is a vector instead of a scalar. This will lead to an obvious difficulty in plotting the frequency spectrum as we did for the linear lattice; instead we shall often limit the discussion to certain directions of $\mathbf{k}$, i.e., of wave propagation, which have relatively high symmetry. Some understanding of the spectrum can be achieved by limiting the discussion to the $\mathbf{k} = \mathbf{0}$ modes. The concept of the factor (or unit cell) group will continue to be important as a means of discussing the symmetry and selection rules of the allowed fundamental modes.

4-1 TRANSLATIONAL SYMMETRY OF CRYSTALS

We now replace the concept of translational symmetry in one dimension comprising translations by ta where t is an integer and a is a lattice distance by the concept of vector translations by amounts

$$\mathbf{t} = \tau_1\mathbf{a}_1 + \tau_2\mathbf{a}_2 + \tau_3\mathbf{a}_3 \tag{4-1-1}$$

where $\mathbf{a}_1$, $\mathbf{a}_2$, $\mathbf{a}_3$ are primitive lattice vectors. The integers τ_1, τ_2, τ_3 each take on N distinct values: thus the order of the translational group is N^3. This group is again abelian, i.e., $tt' = t't$ because each expression has the logical significance of a translation by

$$\mathbf{t} + \mathbf{t}' = \mathbf{t}' + \mathbf{t} = (\tau_1 + \tau_1')\mathbf{a}_1 + (\tau_2 + \tau_2')\mathbf{a}_2 + (\tau_3 + \tau_3')\mathbf{a}_3$$

Then all irreducible representations will be one-dimensional. Each of the three degrees of freedom must separately satisfy the Born and von Kármán cyclic requirement, namely

$$\chi(\mathbf{a}_1)^N = 1; \quad \chi(\mathbf{a}_2)^N = 1; \quad \chi(\mathbf{a}_3)^N = 1$$

so that the characters of the generating elements $\mathbf{a}_1$, $\mathbf{a}_2$; $\mathbf{a}_3$ must be complex roots of unity. Thus for the general translation, $\mathbf{t}$, the character of an irreducible

2. J. de Launay, *Solid State Physics,* F. Seitz and D. Turnbull, Eds., Academic Press, New York, 1956, Volume 2, page 220.
3. G. Leibfried and W. Ludwig, *Solid State Physics,* F. Seitz and D. Turnbull, Eds., Academic Press, New York, 1961, Volume 12, page 276.
4. A. A. Maradudin, E. W. Montroll, G. H. Weiss, and I. P. Ipatova. *Theory of Lattice Dynamics in the Harmonic Approximation,* Academic Press, New York, 1971. (Supplement 3 to *Solid State Physics,* H. Ehrenreich, F. Seitz and D. Turnbull, Eds.)
5. J. L. Warren, *Rev. Mod. Phys.*, **40,** 38 (1968).

representation will involve three integers k_1, k_2, and k_3 and will assume the form

$$\chi_{\mathbf{t}}^{\mathbf{k}} = \exp(2\pi i k_1 \tau_1/N) \exp(2\pi i k_2 \tau_2/N) \exp(2\pi i k_3 \tau_3/N) \qquad (4\text{-}1\text{-}2)$$

It is very convenient mathematically and will prove helpful in forming the appropriate physical concepts to cast (4-1-2) in vectorial form. To achieve this, we introduce the concept of the *reciprocal lattice*.[6] One defines a set of vectors $\mathbf{b}_1$, $\mathbf{b}_2$, and $\mathbf{b}_3$ which are reciprocal to the primitive vectors of the original lattice in the sense that*

$$\mathbf{b}_i \cdot \mathbf{a}_j = \mathbf{b}_i'\mathbf{a}_j = \delta_{ij}. \qquad (4\text{-}1\text{-}3)$$

Solution of (4-1-3) leads to the following expression for the vectors $\mathbf{b}_i$:

$$\mathbf{b}_i = \frac{\mathbf{a}_j \times \mathbf{a}_k}{\mathbf{a}_i \cdot (\mathbf{a}_j \times \mathbf{a}_k)}$$

The definition (4-1-3) of the primitive vectors $\mathbf{b}_i$ of the reciprocal lattice allows the following possible definition of a vector $\mathbf{k}$:

$$\mathbf{k} = (k_1/N)\mathbf{b}_1 + (k_2/N)\mathbf{b}_2 + (k_3/N)\mathbf{b}_3 \qquad (4\text{-}1\text{-}4)$$

so that

$$\chi_{\mathbf{t}}^{(\mathbf{k})} = \exp 2\pi i(\mathbf{k} \cdot \mathbf{t}) \qquad (4\text{-}1\text{-}5)$$

Although the allowed values of the k_i, $i = 1, 2, 3$ in (4-1-4) are integers, one can for most purposes regard k_i/N as a continuous variable since N is so large. For this reason, and because it is convenient to suppress the ubiquitous 2π appearing in (4-1-5), we introduce the variables κ_1, κ_2, κ_3 defined by

$$\kappa_i = 2\pi(k_i/N) \qquad (4\text{-}1\text{-}6)$$

so that

$$2\pi\mathbf{k} = \kappa_1\mathbf{b}_1 + \kappa_2\mathbf{b}_2 + \kappa_3\mathbf{b}_3$$

In view of these definitions, one has

$$\chi_{\mathbf{t}}^{(\mathbf{k})} = e^{2\pi i \mathbf{k} \cdot \mathbf{t}} = e^{i\kappa \cdot \tau} = \chi_{\tau}^{(\kappa)}$$

which permits one to think of points in the lattice as represented by a vector of pure integers,

$$\tau = \begin{pmatrix} \tau_1 \\ \tau_2 \\ \tau_3 \end{pmatrix}$$

6. P. P. Ewald, *Z. Krist.*, **56,** 129 (1921).

* One can also express (4-1-3) in matrix form; if **A** is the matrix whose columns are the $\mathbf{a}$, and **B** is the matrix whose rows are the $\mathbf{b}_i$, then $\mathbf{B}'\mathbf{A} = \mathbf{E}$ or $\mathbf{B} = \mathbf{A}^{-1}$.

while reciprocal space is similarly described by the vector of pure numbers

$$\boldsymbol{\kappa} = \begin{pmatrix} \kappa_1 \\ \kappa_2 \\ \kappa_3 \end{pmatrix}$$

If our crystal possessed no symmetry other than this translational symmetry, we could proceed to construct symmetry coordinates with the aid of (4-1-5) and the Wigner Projection Operator with the assurance that the secular determinant would be factored in the sense that no interactions would occur between coordinates associated with vectors $\mathbf{k}$ and $\mathbf{k}'$ defined by the different sets of integers, k_1, k_2, k_3 and k'_1, k'_2, k'_3. However, in the case which will be of major interest, there will almost always be additional symmetry, and just as in the linear lattice we shall find that the recognition of a factor group is essential to our understanding and analysis of the normal modes.

Nevertheless, whether the crystal has point symmetry or not, we shall henceforth assume that the factoring made possible by the translational symmetry has been carried out as the initial step in the analysis of the dynamical problem. If the primitive unit cell contains n atoms, there will be $3n$ degrees of freedom per unit cell, or $3nN^3$ modes in all. The translational symmetry factoring reduces the problem back to the original size associated with a unit cell, but there are N^3 separate factors each in correspondence with one of the allowed values of $\mathbf{k}$. To be more specific, let us define a set of vibrational coordinates by $r_i(\boldsymbol{\tau})$ where $1 \leq i \leq 3n$ and $\boldsymbol{\tau}$ is a lattice vector which specifies the unit cell with which $r_i(\boldsymbol{\tau})$ is associated. Further, let the F and G matrices have elements denoted as follows:

$$F_{ij}(\boldsymbol{\tau}, \boldsymbol{\tau}'), \quad G_{ij}(\boldsymbol{\tau}, \boldsymbol{\tau}')$$

in which the notation is intended to indicate that $F_{ij}(\boldsymbol{\tau}, \boldsymbol{\tau}')$ is the coefficient of the term $r_i(\boldsymbol{\tau})r_j(\boldsymbol{\tau}')$ in the potential energy, etc.

Symmetry coordinates (with respect to translation) are then generated with the Wigner Projection Operator (WPO) and are simply:

$$s_i(\boldsymbol{\kappa}) = N^{-3/2} \sum_{\boldsymbol{\tau}} e^{-i\boldsymbol{\kappa}\cdot\boldsymbol{\tau}}\, r_i(\boldsymbol{\tau}) \tag{4-1-7}$$

When the transformation defined by (4-1-7) is used in the usual way to obtain F and G in the symmetry coordinate basis, and when one further notes that $F_{ij}(\boldsymbol{\tau}, \boldsymbol{\tau}')$ and $G_{ij}(\boldsymbol{\tau}, \boldsymbol{\tau}')$ are not really dependent upon the individual values of $\boldsymbol{\tau}$ and $\boldsymbol{\tau}'$ but only upon the (vectorial) difference, i.e., then

$$F_{ij}(\boldsymbol{\tau}, \boldsymbol{\tau}') = F_{ij}(\Delta\boldsymbol{\tau}) \quad \text{where} \quad \Delta\boldsymbol{\tau} = \boldsymbol{\tau}' - \boldsymbol{\tau}$$

$$F_{ij}(\boldsymbol{\kappa}) = N^{-3} \sum_{\boldsymbol{\tau}, \boldsymbol{\tau}'} e^{i\boldsymbol{\kappa}\cdot(\boldsymbol{\tau}'-\boldsymbol{\tau})}\, F_{ij}(\boldsymbol{\tau}, \boldsymbol{\tau}') \tag{4-1-8}$$

$$= \sum_{\Delta\boldsymbol{\tau}} F_{ij}(\Delta\boldsymbol{\tau})\, e^{i\boldsymbol{\kappa}\cdot\Delta\boldsymbol{\tau}}$$

In (4-1-8) we can adopt an arbitrary unit cell as the origin which means that the sum over $\Delta\boldsymbol{\tau}$ is a sum over $\boldsymbol{\tau}$ relative to this origin. Thus we find that

$F_{ij}(\boldsymbol{\kappa})$ and $F_{ij}(\boldsymbol{\tau})$ are essentially (finite) Fourier transforms. One can, of course, also write (4-1-8) in the form

$$F_{ij}(\boldsymbol{\kappa}) = F_{ij}(0) + 2 \sum_{\Delta\boldsymbol{\tau}} F_{ij}(\Delta\boldsymbol{\tau}) \cos(\boldsymbol{\kappa} \cdot \Delta\boldsymbol{\tau}) \tag{4-1-9}$$

4-2 CRYSTALLOGRAPHIC SPACE GROUPS

The set of all operations which transforms every point of a crystal into another point equivalent to it as well as every direction within a crystal into an equivalent one is called a *crystallographic space group.*

To appreciate the many possibilities offered by this definition, let us begin to examine an arbitrary, three-dimensional lattice of points, all interconnected by vectors which are the elements of the translation group T of the crystal, discussed in the previous section. The point symmetry operations which leave at least one point invariant consist of rotations, or of rotations combined with reflections. In the following, all such operations will be designated simply as rotational operations to distinguish them from the lattice translations. Let us examine the *rotational symmetry* of this three-dimensional network. If $\mathbf{t}$ is a member of T [Eq. (4-1-1)], so is $-\mathbf{t}$. Thus a lattice always has at least the rotational symmetry C_i. But a lattice can have greater rotational symmetry than C_i. Consider any two-dimensional lattice with fundamental lattice vectors $\mathbf{a}_1$ and $\mathbf{a}_2$. If we rotate this lattice about the normal to the plane defined by $\mathbf{a}_1$ and $\mathbf{a}_2$ subject to the condition that the lattice is carried into itself, the rotation matrix must necessarily be composed of integers. On the other hand, the rotation matrix must have the trace $2 \cos \varphi$, where φ is the rotation angle. Thus we conclude that $2 \cos \varphi$ is integral. This function can then only have the values -2, -1, 0, $+1$, $+2$. The corresponding values of φ are $2\pi/2$, $2\pi/3$, $2\pi/4$, $2\pi/6$, and 0. No other rotation angles are admissible. The abhorrence of nature for fivefold (and seven, eight, etc., fold) symmetry stems from this principle.

We have not considered the consequences of the fact that there are some three-dimensional lattices with $\mathbf{a}_3 \cdot \mathbf{a}_i \neq 0, i = 1$ or 2. The absence of $n = 5, 7, 8$, etc., is also valid in this case and can be shown to be a consequence of the "law of rational indices." It can also be shown[7] that when $n > 2$, a lattice is invariant under reflection in a plane containing the n-fold rotation axis.

Only seven point groups, C_i, C_{2h}, D_{2h}, D_{4h}, D_{3d}, D_{6h}, and O_h, have these properties; hence, all *point* lattices must demonstrate the rotational symmetry of one or another of these groups. We refer to each of these seven groups as a *system* and in the order of the list of groups just given, we call these systems, respectively, *triclinic, monoclinic, orthorhombic, tetragonal, trigonal, hexagonal,* and *cubic.*

7. C. J. Bradley and A. P. Cracknell, *The Mathematical Theory of Symmetry in Solids,* Clarendon Press, Oxford, 1972, page 40.

Within each *system,* we can sometimes distinguish among several lattice *types.* For example, in the monoclinic system, there are two *types* of lattices. One of these, called *primitive* (i.e., has points only at the corners of the lattice) and symbolized Γ_m, satisfies the condition that its set of primitive lattice vectors $\{\mathbf{a}_1, \mathbf{a}_2, \mathbf{a}_3\}$ is invariant under the operations of the point group C_{2h}. [Note that $\sigma_h \cdot \mathbf{a}_3 = -\mathbf{a}_3$, which is considered a member of the set.] In the other type of monoclinic *type,* called *base-centered* and symbolized Γ_m^b, the invariant vector set is $\{\mathbf{a}_1, \mathbf{a}_2, (2\mathbf{a}_3 - \mathbf{a}_2)\}$.

In Γ_m, $\mathbf{a}_1$, $\mathbf{a}_2$, and $\mathbf{a}_3$ define a parallelepiped, while in Γ_m^b $\mathbf{a}_3$ (and only $\mathbf{a}_3$) terminates at the *center* of one of the faces of the parallelepiped bounded by $\mathbf{a}_1$, $\mathbf{a}_2$, and $(2\mathbf{a}_3 - \mathbf{a}_2)$. It is for this reason that Γ_m^b is called base-centered. It still is invariant under C_{2h}.

There are 14 possible lattice types, frequently called Bravais lattices. That they are limited to precisely 14 is deduced in monographs devoted to X-ray crystallography, for example, that authored by Zachariasen.[8] The definitions of the lattices, as well as their distribution among the crystal *systems* is given in Table 4-1. The Bravais lattices are illustrated in Fig. 4-1.

Table 4-1

System	Bravais lattice			Basic vectors
Triclinic	C_i	Primitive	$\Gamma_t(P)$	arbitrary
Monoclinic	C_{2h}	Primitive	$\Gamma_m(P)$	$\mathbf{a}_3 \perp \mathbf{a}_1, \mathbf{a}_2$
		Base-centered	$\Gamma_m^b(B \text{ or } C)$	$(\mathbf{a}_3 - \frac{1}{2}\mathbf{a}_2) \perp \mathbf{a}_1, \mathbf{a}_2$
Orthorhombic	D_{2h}	Primitive	$\Gamma_0(P)$	$\mathbf{a}_1 \perp \mathbf{a}_2 \perp \mathbf{a}_3 \perp \mathbf{a}_1$
		Base-centered	$\Gamma_0^b(C)$	$(\mathbf{a}_3 - \frac{1}{2}\mathbf{a}_2) \perp \mathbf{a}_1, \mathbf{a}_2;\ \mathbf{a}_1 \perp \mathbf{a}_2$
		Body-centered	$\Gamma_0^v(I)$	$[\mathbf{a}_3 - \frac{1}{2}(\mathbf{a}_1 + \mathbf{a}_2)] \perp \mathbf{a}_1, \mathbf{a}_2;\ \mathbf{a}_1 \perp \mathbf{a}_2$
		Face-centered	$\Gamma_0^f(F)$	$\mathbf{a}_1 \perp (\mathbf{a}_2 - \frac{1}{2}\mathbf{a}_1) \perp (\mathbf{a}_3 - \frac{1}{2}\mathbf{a}_1) \perp \mathbf{a}_1$
Tetragonal	D_{4h}	Primitive	$\Gamma_q(P)$	$\mathbf{a}_1 \perp \mathbf{a}_2 \perp \mathbf{a}_3 \perp \mathbf{a}_1;\ a_1 = a_2$
		Body-centered	$\Gamma_q^v(I)$	$\mathbf{a}_1 \perp \mathbf{a}_2 \perp [\mathbf{a}_3 - \frac{1}{2}(\mathbf{a}_1 + \mathbf{a}_2)] \perp \mathbf{a}_1;\ a_1 = a_2$
Trigonal	D_{3d}	Primitive	$\Gamma_{rh}(R)$	$\mathbf{a}_1 \cdot \mathbf{a}_2 = \frac{2\pi}{3} a_1^2;\ \mathbf{a}_1 \perp [\mathbf{a}_3 - \frac{1}{2}(\mathbf{a}_1 - \mathbf{a}_2)] \perp \mathbf{a}_2$
Hexagonal	D_{6h}	Primitive	$\Gamma_h(P)$	$\mathbf{a}_1 \cdot \mathbf{a}_2 = \frac{\pi}{3} a_1^2;\ \mathbf{a}_1 \perp \mathbf{a}_3 \perp \mathbf{a}_2$
Cubic	O_h	Primitive	$\Gamma_c(P)$	$\mathbf{a}_1 \perp \mathbf{a}_2 \perp \mathbf{a}_3 \perp \mathbf{a}_1;\ a_1 = a_2 = a_3$
		Face-centered	$\Gamma_c^f(F)$	$\mathbf{a}_1 \perp \mathbf{a}_2 \perp [\mathbf{a}_3 - \frac{1}{2}(\mathbf{a}_1 + \mathbf{a}_2)] \perp \mathbf{a}_1;\ a_1 = a_3;$ $\lvert \mathbf{a}_3 - \frac{1}{2}(\mathbf{a}_1 + \mathbf{a}_2) \rvert = \frac{1}{\sqrt{2}} a_1$
		Body-centered	$\Gamma_c^v(I)$	$\mathbf{a}_1 \perp \mathbf{a}_2 \perp [\mathbf{a}_3 - \frac{1}{2}(\mathbf{a}_1 + \mathbf{a}_2)] \perp \mathbf{a}_1;\ a_1 = a_2;$ $\lvert \mathbf{a}_3 - \frac{1}{2}(\mathbf{a}_1 + \mathbf{a}_2) \rvert = \frac{1}{2} a_1$

8. W. H. Zachariasen, *Theory of X-ray Diffraction in Crystals,* Wiley, New York, 1945.

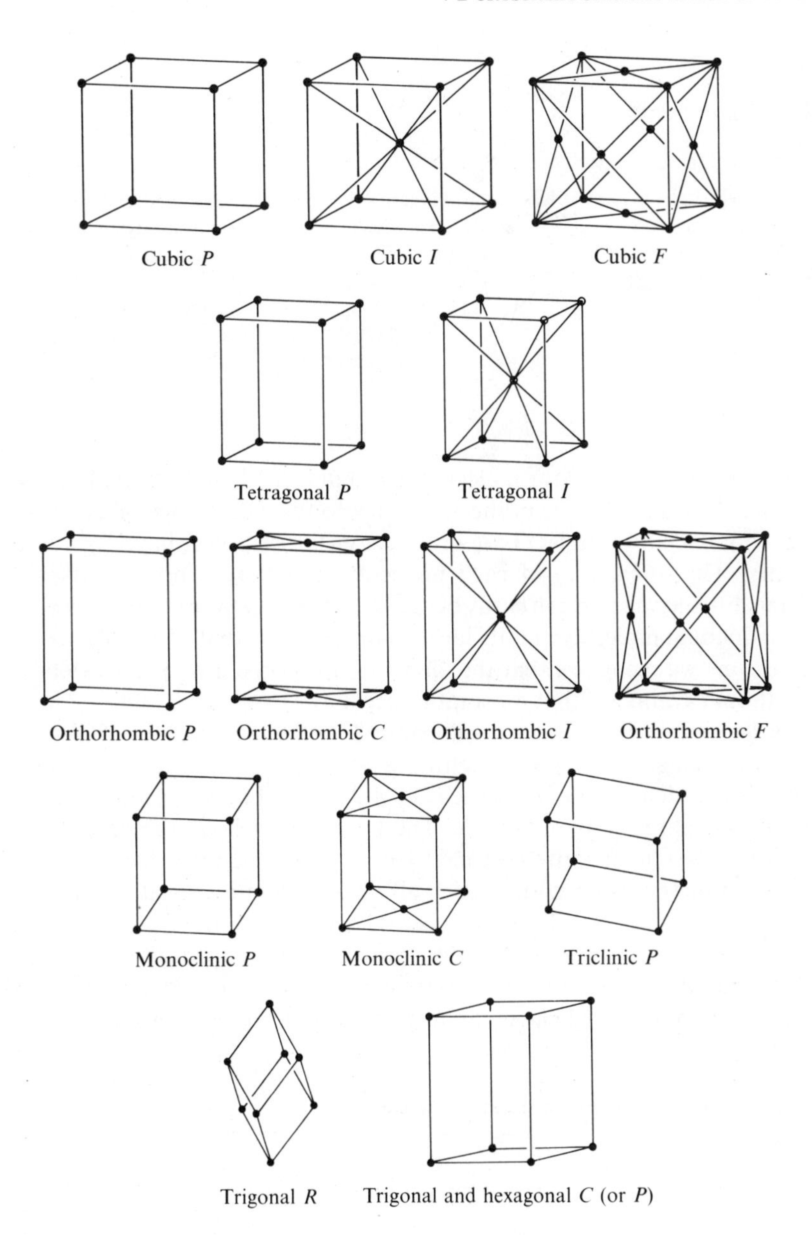

Figure 4-1 The 14 Bravais lattices. (After F. C. Phillips, *An Introduction to Crystallography*, 4th Edition, Oliver and Boyd, Edinburgh, 1971, page 253)

As opposed to the lattice of *points* which defines a crystal, we can examine the possible symmetry groups under which the several *directions* in a crystal may be invariant. For example, consider the crystal structure of HCl, recently shown

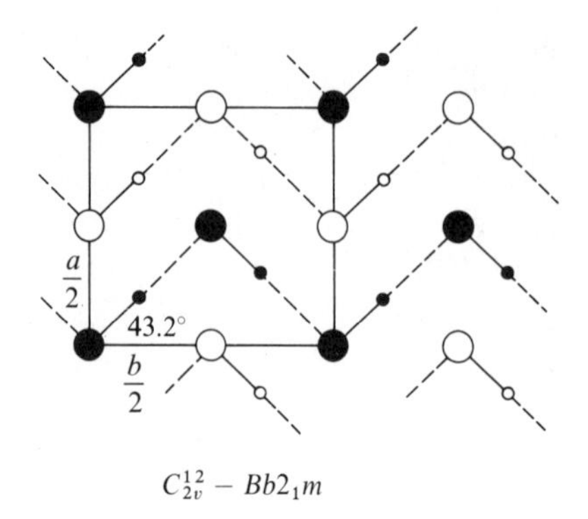

Figure 4-2 C_{2v}^{12} structure of solid orthorhombic HCl. The solid circles represent atoms lying in the plane of the diagram, while the open circles represent atoms displaced by the amount $c/2$ above and below the plane of the diagram. Hydrogen bonds are represented by the dashed lines. (After N. S. Gillis, J. C. Raich, L. B. Kanney, and A. Bickermann, *J. Chem. Phys.*, **64,** 2501 (1976))

to be orthorhombic with four molecules per unit cell.[9] It is illustrated in Fig. 4-2. The molecules are seen to lie in sheets parallel to the (001) plane. Since the crystal is orthorhombic and is in fact base-centered, its lattice points have D_{2h} rotational symmetry. The *orientation* of each molecule, however, is not invariant under inversion. Furthermore, it can easily be seen that the only point symmetry admitted by the individual molecule and its local environment is that of the plane (σ_h). For these reasons we recognize that the *directions* in a crystal may not exhibit all of the rotational symmetry that its point lattice may.

Another view of this matter is illustrated by the fact that the HCl crystal is really composed of several interpenetrating lattices, some of H atoms, others of Cl nuclei. We can define a point lattice (Fig. 4-2) of HCl *molecules* defined by, say, the center of charge of the molecule. The lattice of such charge centers exhibits D_{2h} rotational symmetry but the H and Cl atom lattices manifest lower symmetries.

The symmetry group under which all directions in a crystal are invariant is denoted the crystal *class*. These number 32 altogether—the 32 crystallographic point groups. For a given crystal, the *class* is a subgroup of the crystal *system*. For example, the monoclinic (C_{2h}) *system* contains the *classes* C_{2h}, C_2, and C_s. The distribution of crystal *classes* among the crystal *systems* is given in Table 4-2.

Table 4-2 Distribution of crystal classes among systems

System	Classes
Triclinic	C_1, C_i
Monoclinic	C_2, C_s, C_{2h}
Orthorhombic	D_2, C_{2v}, D_{2h}
Tetragonal	C_4, S_4, C_{4h}, D_4, C_{4v}, D_{2d}, D_{4h}
Trigonal	C_3, S_6, D_3, C_{3v}, D_{3d}
Hexagonal	C_6, C_{3h}, C_{6h}, D_6, C_{6v}, D_{3h}, D_{6h}
Cubic	T, T_h, O, T_d, O_h

9. N. Niimura, K. Shimaoka, H. Motegi, and S. Hoshino, *J. Phys. Soc. Japan*, **32,** 1019 (1972).

We now recall that a crystallographic space group consists of the set of operations (both translational and rotational) under which *both* points and directions remain unchanged. Thus a space group implies a crystal *system*, a lattice *type*, and a *class*. More exactly, to enumerate all possible *space groups*, we must compound the translational symmetry of a particular Bravais lattice with the rotational symmetry of a particular class. Although there are obviously certain prohibited combinations (for example, of D_{6h} with Γ_m^b), there are still 230 allowed combinations. The manner in which some of these are developed is illustrated below.

Consider the primitive monoclinic lattice Γ_m and the class C_s. There are two ways in which they can be combined. In one of these, the pure rotational operations of the group C_s (E and σ_h) are each carried out simultaneously with the translations of Γ_m. A typical space group operation would then be symbolized $\{R|\tau\}$, where R represents a point group operation and τ is *any* translation (not necessarily a member of the translation group, the primitive vectors of which define Γ_m).[10] The effect of $\{R|\tau\}$ on a vector $\mathbf{x}$ is defined by

$$\{R|\tau\} \cdot \mathbf{x} = R\mathbf{x} + \tau$$

consistent with which we represent group multiplication by

$$\{R|\tau_1\} \cdot \{S|\tau_2\} = \{RS|R\tau_2 + \tau_1\}$$

The identity is $\{E|0\}$ and the inverse is $\{R|\tau\}^{-1} = \{R^{-1}|-R^{-1}\tau\}$. Now, in the case under consideration, R and S belong to the point group (C_s) while τ_1 and τ_2, etc., are members of the translation group Γ_m. This space group then has $2N^3$ operations. N^3 of these are purely translational and N^3 consist of the "pure" reflection followed by one of the N^3 translations of Γ_m.

The other combination of Γ_m and C_s differs from the first in that whenever σ_h is applied it may be accompanied by a *non-primitive* translation of $\mathbf{a}_1/2$ besides integer multiples of the primitive translations $\mathbf{a}_1$, $\mathbf{a}_2$, and $\mathbf{a}_3$ (recall that the plane σ_h was parallel to the plane defined by $\mathbf{a}_1$ and $\mathbf{a}_2$). σ_h is said to be a *glide reflection* instead of a "pure" reflection. The first kind of space group is called *symmorphic* and the second *non-symmorphic*.

Similar possibilities exist in the combination of C_s with Γ_m^b.

In the combination of C_2 with Γ_m, there are again two choices. In the first of these the group is symmorphic and C_2 is always accompanied by one of the primitive translations. When the group is non-symmorphic and C_2 is applied, besides the primitive translations $\mathbf{a}_1$, $\mathbf{a}_2$, and $\mathbf{a}_3$, it may be accompanied by a non-primitive translation, parallel to $\mathbf{a}_3$ and of magnitude $a_3/2$. The operation $\{C_2|\mathbf{a}_3/2\}$ is called a *screw-rotation*.

In Γ_m^b, $\mathbf{a}_3 \cdot \mathbf{a}_2 \neq 0$. Since the C_2 axis is not parallel to the direction of $\mathbf{a}_3$, the structure is not invariant to an operation of the form $\{C_2|\mathbf{a}_3/2\}$. There is thus only one type of combination of C_2 with Γ_m^b which is possible, namely, $\{C_2|\mathbf{0}\}$ and the group is symmorphic.

10. F. Seitz, *Ann. Math.*, **37**, 17 (1936).

A continuation of this procedure results in the 230 space groups listed in Appendix V. Both the Schönflies and International Symbols for the space groups are given in the appendix. In the case of the space groups, there is little that is systematic in the use of the Schönflies system. In this system, all space groups are simply denoted by the class symbol and a superscript. If the superscript is unity, the group is always symmorphic; but there are other symmorphic groups for which the superscript is not unity. The International system is rational. The symbol begins with a capital letter which identifies the lattice type, but only to the extent that we are told whether the lattice is primitive, base-, face-, or body-centered. The symbols are, respectively, P, B (or C), F, and I. Occasionally, rhombohedral crystals are denoted by R. The capital letter is followed by a series of symbols which relate the nature of the axes or planes of symmetry. For example, $P\bar{6}2m$ is primitive, with a sixfold improper (S_6) axis, a twofold proper axis and an ordinary reflection plane; $P4_2/mmc$ is again primitive with a screw tetrad whose translational period is $\frac{1}{2}$ the unit cell length along the direction of the fourfold axis and there are three reflection planes, two ordinary and one a glide plane—in which the translation which accompanies this reflection is parallel to the c-axis.

It is shown in Appendix VI that the *factor group* constructed from the full translational subgroup T of the space group (whose general element is $\{E|\tau\}$) and either its right or left cosets is isomorphic with the point group denoted by the crystal class (the symbol for which forms the basis of the Schönflies symbol for the space group). In this factor group, the *translation group* fills the role of the unit element. This is possible since the operations of T, operating on each other, just reproduce themselves. Because of this possibility, we may use the representations $\mathbf{k}$ and of the factor group *separately* to write symmetry coordinates in which the vibrational secular determinant of the crystal will be factorable. The entire procedure we shall use in deriving selection rules for vibrational transitions in crystals rests upon this factorization. The knowledge that the point group is isomorphous with the factor group is of great importance to the spectroscopist. It is for this reason that the Schönflies system is of greater currency in our field.

4-3 THE BRILLOUIN ZONE

In the first section of the present chapter it was shown how to construct coordinates which have the symmetry of the irreducible representations of the translation group. When expressed in terms of these coordinates, the dynamical matrices F and G are factored with respect to the $\mathbf{k}$, or wave vector. Thus in a crystal with n atoms per (primitive) unit cell, N^3 unit cells, or $3nN^3$ degrees of freedom, the secular determinant has in principle been factored into N^3 blocks, each $3n \times 3n$, and each corresponding to one of the allowed values of the vector $\mathbf{k}$.

In the previous section we have discussed the manner in which the translational symmetry combines with point symmetry and leads to 230 possible space groups. We shall now investigate how the point symmetry may be used to produce further factoring of the dynamical matrices within each block. But to do this, it is convenient first to examine more closely the specification of the allowed range of **k**. It will prove feasible to state this restriction in such a way that the symmetry of **k** under the point group operations will be clearer.

In the case of the one-dimensional lattice we simply chose the interval $-N/2 \leq k \leq +N/2$ because symmetry coordinates defined by (3-2-5) with k, $k \pm N$, $k \pm 2N$, etc., are identical and because symmetry coordinates defined by $+k$ and $-k$ are degenerate. In three dimensions these two statements are readily generalized as follows:

1. Symmetry coordinates defined by $\mathbf{k}$ and by $\mathbf{k} + \bar{\mathbf{k}}$, where $\bar{\mathbf{k}}$ is a vector of the reciprocal lattice, are identical;
2. Symmetry coordinates defined by $+\mathbf{k}$ and by $-\mathbf{k}$ are degenerate.

In other words, (1) informs us that symmetry coordinates have the three-dimensional periodicity of the reciprocal lattice, and (2) makes it clear that we need only solve our problem say for positive k. For these reasons it is advantageous to choose a primitive unit cell in **k**-space, and, while there is no unique way of doing this, it is certainly desirable to choose this cell in such a way that the point symmetry is in evidence. This can be achieved by following this rule: starting from a point, call it $\mathbf{k} = 0$, in the reciprocal lattice, one constructs vectors to neighboring points of the reciprocal lattice, and bounds the unit cell by planes which are perpendicular bisectors of the vectors to the neighboring points. Such a cell is called the Brillouin zone.[11]

It is easiest to visualize this construction when the crystal lattice is primitive and belongs to the cubic, tetragonal, or orthorhombic systems because the reciprocal lattice will again be primitive cubic, tetragonal, or orthorhombic (in the latter two cases the relative lengths of the sides of the cell will of course be inverted). Moreover, the prescription that the Brillouin zone be bounded by planes which are perpendicular bisectors of reciprocal lattice vectors, leads to zones which are in each case *rectangular parallelopipeds*.

Even in the case of the cubic system, the situation is more complicated if the crystal lattice is body-centered (I) or face-centered (F). However, these two Bravais lattices are reciprocal to one another, so that the Brillouin zone for a body-centered cubic crystal must be constructed in a face-centered reciprocal space and vice versa. Details of the reciprocal relations are developed in Appendix VIII with the aid of matrix techniques.

In Fig. 4-3 these Brillouin zones are exhibited. In these figures, points of special interest corresponding to special k values are labelled following a widely used convention introduced by Bouckaert, Smoluchowski, and Wigner (hereinafter BSW).[12] (See also Appendix IX.)

The key to our understanding of the further analysis of the dynamical problem is the knowledge of how our symmetrized $s_i(\mathbf{k})$ coordinates transform with respect

11. L. Brillouin, *J. Phys. Radium*, **1**, 377 (1930).
12. L. P. Bouckaert, R. Smoluchowski, and E. P. Wigner, *Phys. Rev.*, **50**, 58 (1936).

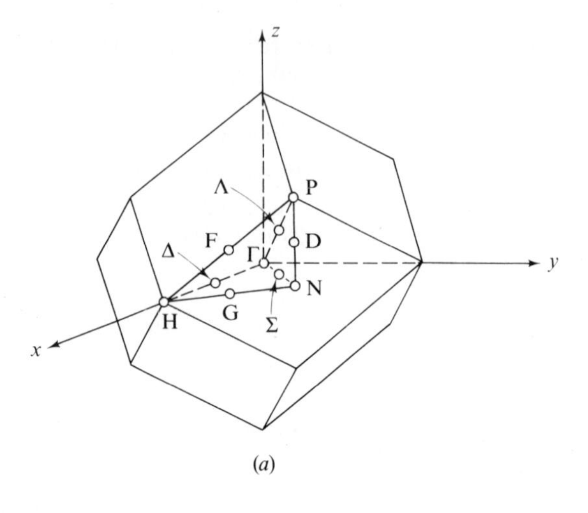

(*a*)

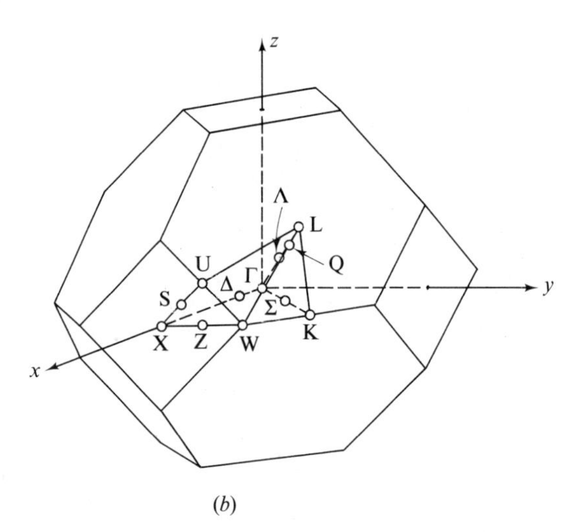

(*b*)

Figure 4-3 (*a*) Brillouin zone for body-centered cubic Bravais lattice, with symmetry notations of Bouckaert, Smoluchowski, and Wigner. (*b*) Brillouin zone for the face-centered Bravais lattice. (After J. C. Slater, *Quantum Theory of Molecules and Solids,* Vol. 2, McGraw-Hill, New York, 1965)

to the point symmetry operations, or more precisely, with respect to the factor group of the full space group. Anticipating a result which will be established below (Section 4-5(a)), one finds that the transformation of the $s_i(\mathbf{k})$ can be classified according to the effect of the rotational operations upon the vector $\mathbf{k}$ (or $\boldsymbol{\kappa}$) in reciprocal space. For each $\mathbf{k}$ in the Brillouin zone, there exists a group, herein designated as $H(\mathbf{k})$, which leaves $\mathbf{k}$ invariant or sends $\mathbf{k}$ into itself plus a lattice vector $\bar{\mathbf{k}}$ of reciprocal space. In the interior of the Brillouin zone, $H(\mathbf{k})$ is always a subgroup of the point group G; of course it may be G itself. $H(\mathbf{k})$ admits operations which send $\mathbf{k} \to \mathbf{k} + \bar{\mathbf{k}}$ only for certain points on the zone boundary; for some zone boundary points in non-symmorphic groups it often happens that $H(\mathbf{k})$ is not a point group. In section 4-5 we shall develop a simple criterion to test whether $H(\mathbf{k})$ is or is not isomorphic with a point group; if it is *not,* the reader must be referred to more extensive tabulations of the space group irreducible representatives.

Perhaps the easiest way to study these transformation properties is to start by investigating the situation at $\mathbf{k} = 0$ and then use correlation methods to classify the symmetry species of $s_i(\mathbf{k})$ for $\mathbf{k} \neq 0$. For example, consider the points Γ, Λ, and L in Fig. 4-3(*b*) which shows the Brillouin zone for the face-centered cubic lattice. The point Γ corresponding to $\mathbf{k} = 0$ has the full symmetry of the unit cell which is O_h. Proceeding along the line from Γ to L, one sees that the point Λ is left invariant by the operations of a subgroup of O_h—in this case C_{3v}. Thus all modes for which $k_1 = k_2 = k_3 = k$ with $0 < k < a^{-1}$ can be classified according to the species (irreducible representations) of C_{3v}, and moreover can be precisely correlated with the appropriate species of O_h in the limit as $\mathbf{k} \to 0$. We shall see in more detail below that for the NaCl type of crystal that at $\mathbf{k} = 0$ these modes, six in number, belong to the species F_{1u} of O_h, i.e., the vibrational representation is $2F_{1u}$. Then Appendix IV shows that when the point symmetry of $\mathbf{k}$ is lowered from O_h to C_{3v}, the species correlation is $2F_{1u} \to 2A_1 + 2E$.

On the zone boundary at the point L the symmetry of $\mathbf{k}$ is higher, since operations which transform the point L into a translationally equivalent point in the reciprocal lattice must also be counted in the group of $\mathbf{k}$. In the present case, the only such equivalent point is the one obtained, say, by inversion (recall that the zone faces are halfway between reciprocal lattice points). The addition of i to C_{3v} implies that the subgroup of O_h is now D_{3d} for the point L.

4-4 SYMMETRY SPECIES AT $\kappa = 0$

As indicated in the last section, the point Γ in the Brillouin zone, corresponding to $\kappa = 0$, has the full symmetry of the unit cell, i.e., that of the factor group of the space group. The matter was illustrated for the face-centered cubic lattice, but it has general validity. Since at $\kappa = 0$ $\chi_{\mathbf{t}}^{(0)} = 1$ for all $\mathbf{t}$, $\Gamma_{\mathbf{t}}^{(0)} = \Gamma_1$. In other words, at Γ the transformations which factorize the dynamical matrices are those whose coefficients come from the representation of the factor group of the space group alone.

As examples we shall consider the diamond, rock salt, fluorite, and tungsten lattices. Diamond belongs to the space group $O_h^7 = Fd3m$. The factor group is O_h, hence the symmetry coordinates $s_i(\mathbf{0})$ must form bases for the irreducible representations of the group O_h. The unit cell which is illustrated in Fig. 4-4 has a population of eight atoms. As is well known, however, we may take another view of this cell. Taking the origin of the cell axes as indicated in the figure and focussing our attention only on those atoms which are shared by other unit cells, we see that these four atoms define a face-centered cubic lattice with side a. Figure 4-4 shows that the four interior atoms define an identical face-centered cubic lattice, displaced from the first by the vector $(a/4)(1, 1, 1)$. Because of this, the diamond lattice is one with a basis and can be viewed as two *interpenetrating* face-centered cubic lattices. Furthermore, we may also take another view of the

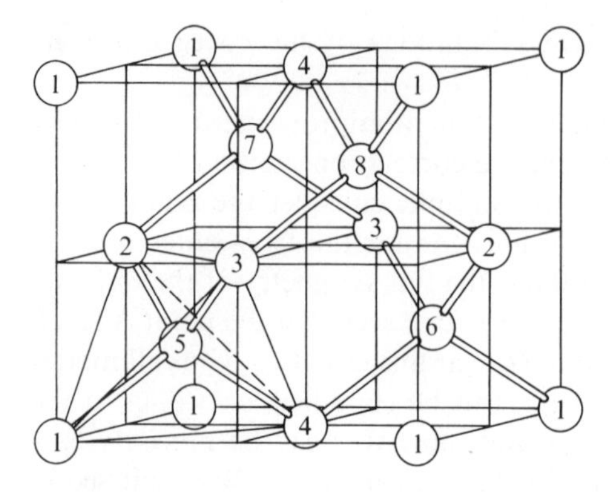

Figure 4-4 The diamond lattice. (After J. C. Slater, *Quantum Theory of Molecules and Solids*, Vol. 2, McGraw-Hill, New York, 1965)

structure. Namely, we can form a new unit cell, the sides of which are formed by the "bonds" connecting atoms numbered 2, 3, and 4 to the origin atom (1) and to each other. This new cell fulfills all the requirements of a lattice type or Bravais lattice. It is in fact the one denoted Γ_c^f in Table 4-1.

The translational symmetry of Γ_c^f may not be obvious in Fig. 4-4. If the (111) direction is oriented vertically, as in Fig. 4-5, we see that the carbon tetrahedra are related by the basis vectors of Γ_c^f. Furthermore, the structure is a close-packed array of tetrahedra, half with carbon atoms at their centers and half empty. The basis vectors of Γ_c^f take each type of tetrahedron (filled or empty) only into others of the same type. Also, the filled tetrahedra account for all the atoms. Hence, we need only count them in, say, computing the population of a unit cell.

In that each (numbered) atom at a corner of the new cell is shared by three other unit cells (neighboring unit cells still share faces; three faces of the new unit cell have a common vertex at any numbered atom) the population of the new unit cell is $n = 2$ atoms, counting the interior atoms.

The process by means of which we have gone from the original, eight-atomic unit cell to the new, diatomic cell, must always be used. If we attempt it again in this example we find that there is no further redefinition of the Bravais lattice

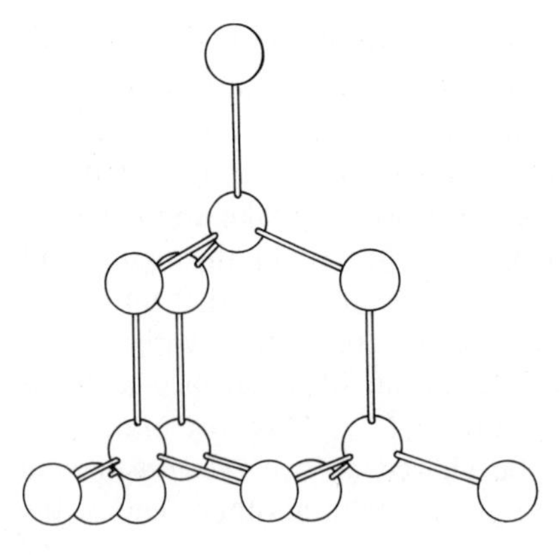

Figure 4-5 Perspective view of atomic arrangement in diamond, with 111 axis vertical. (After J. C. Slater, *Quantum Theory of Molecules and Solids*, Vol. 2, McGraw-Hill, New York, 1965)

appropriate to diamond which will yield a unit cell of smaller volume. The least populous unit cell from which the lattice can be generated will be called the *primitive unit cell.* A primitive unit cell should not be confused with a primitive lattice as defined in Section 4-2.

As we have seen in Section 4-3, at each value of $\mathbf{k}$ including $\mathbf{k} = 0$, we expect $3n$ degrees of freedom. Hence, in the case of diamond there will be six $\mathbf{k} = 0$ modes of motion. Three of these must correspond to the translational motions of the crystal as a whole, analogous to the situation encountered at $k = 0$ in the case of the linear monatomic lattice (Section 3-2). Therefore we shall eventually seek a description of three vibrational degrees of freedom.

We proceed in essentially the same way as we did in analyzing the motions of a molecule. This is particularly pertinent at $\mathbf{k} = 0$ since we may then utilize the complete translational symmetry of the crystal, i.e., $\Gamma(0)$ of T. We examine a model of the primitive unit cell and for each operation of the factor group we find the characters of the representation with the displacement coordinates of the atoms of that cell as a basis. The restriction of the analysis to one unit cell is valid since we have taken translational symmetry into account. Extending the procedures of Appendix II to crystals, this is done as follows.

1. List the classes of the factor group.
2. Apply an operation from each class of the factor group to the primitive unit cell (pictured as a molecule). In non-symmorphic space groups some factor group operations will have non-primitive translations associated with them, as we shall see below.
3. Compute the quantity $(\pm 1 + 2\cos\theta_R)$ for each such operation (R). Note that θ_R depends only on the rotational part of the operation.
4. Compute the quantity u_R as follows:
   (a) If R leaves an atom *unmoved* or moves it to a translationally equivalent point in another unit cell, then the contribution to u_R is *unity* for that atom.
   (b) Otherwise the atom contributes *zero* to u_R. It is especially important to include the translational part of the operation in carrying out R in this test, should such exist.
5. Now compute $\chi_m(R) = u_R(\pm 1 + 2\cos\theta_R)$.

We apply the rules given above, using Fig. 4-4. The character table of the factor group is given in Table 4-3. The symmetry elements of the factor group O_h are illustrated in Fig. 4-6.

It is important to take note of the fact that those operations of the factor group O_h which define the subgroup T_d are pure point operations; that is, they have the form $\{R\,|\,\mathbf{0}\}$. When T_d is adjoined by the inversion, we obtain O_h. The inversion operation of this space group, however, is of the form

$$\mathbf{i} = \{i\,|\,\tau_i\} \tag{4-4-1}$$

where $\tau_i = \frac{1}{4}(1, 1, 1)$, that is, it includes a non-primitive translation. Hence, all new operations derived by means of $O_h = T_d \times C_i$, with $\mathbf{i}$ as defined in (4-4-1), will also have the same translation, $(a/4)(1, 1, 1)$, associated with them. These

Table 4-3

O_h^7	$\{E\|0\}$	$8\{C_3\|0\}$	$3\{C_2\|0\}$	$6\{C_4\|\tau_i\}$	$6\{C_2'\|\tau_i\}$	$\{i\|\tau_i\}$	$8\{S_6\|\tau_i\}$	$3\{\sigma_h\|\tau_i\}$	$6\{S_4\|0\}$	$6\{\sigma_d\|0\}$	
A_{1g}	1	1	1	1	1	1	1	1	1	1	α
A_{2g}	1	1	1	-1	-1	1	1	1	-1	-1	
E_g	2	-1	2	0	0	2	-1	2	0	0	α
F_{1g}	3	0	-1	1	-1	3	0	-1	1	-1	R
F_{2g}	3	0	-1	-1	1	3	0	-1	-1	1	α
A_{1u}	1	1	1	1	1	-1	-1	-1	-1	-1	
A_{2u}	1	1	1	-1	-1	-1	-1	-1	1	1	
E_u	2	-1	2	0	0	-2	1	-2	0	0	
F_{1u}	3	0	-1	1	-1	-3	0	1	-1	1	T
F_{2u}	3	0	-1	-1	1	-3	0	1	1	-1	
u_R	2	2	2	0	0	0	0	0	2	2	
θ_R	0	$2\pi/3$	π	$\pi/4$	π	π	$\pi/3$	0	$\pi/2$	0	
$\pm 1 + 2\cos\theta_R$	3	0	-1	1	-1	-3	1	$+1$	-1	$+1$	
$\chi_m(R)$	6	0	-2	0	0	0	0	0	-2	2	

$\tau_i \equiv (\frac{1}{4}, \frac{1}{4}, \frac{1}{4})$

include **i**, $\mathbf{S}_6$, $\boldsymbol{\sigma}_h$, $\mathbf{C}_4$ and $\mathbf{C}_2'$. That these operations have this translational character must be borne in mind when computing u_R; however, θ_R is still determined solely by the "rotational" part of each space group operation.

Reduction of Γ_m among the irreducible representations of the factor group O_h is carried out in the usual fashion, using the decomposition formula (2-2-27). In the case of diamond, at $\mathbf{k} = 0$, we find

$$\Gamma_m = F_{2g} + F_{1u} \tag{4-4-2}$$

At this point we observe that Γ_m is the representation with all of the displacement coordinates of the primitive unit cell as a basis. Since $n = 2$, $\chi_m(E) = 6$. Just as in the case of free molecules, however, three of these degrees of freedom are translational. Clearly the symmetrized linear combinations of displacement coordinates which correspond to the translation of the crystal as a whole must belong to the representations $\Gamma_T = F_{1u}$ of O_h (Appendix I). Furthermore, of all the $6N^3$ degrees of freedom of diamond only three are translational; hence, we have completely determined their symmetry. Their representation under T can only be $\Gamma(0)$. The situation is exactly analogous to that first encountered in Section 3-2, except that there are now three S_0 modes, triply degenerate with one another. Indeed, similar dynamics will be shown later to yield $\lambda_0 = 0$, as in the linear example. Thus in diamond, the F_{1u} mode is the $\mathbf{k} = 0$ limit of the acoustic branches.

Although a crystal as a whole may be subjected to translational motions which preserve its full translational symmetry, our insistence upon this invariance

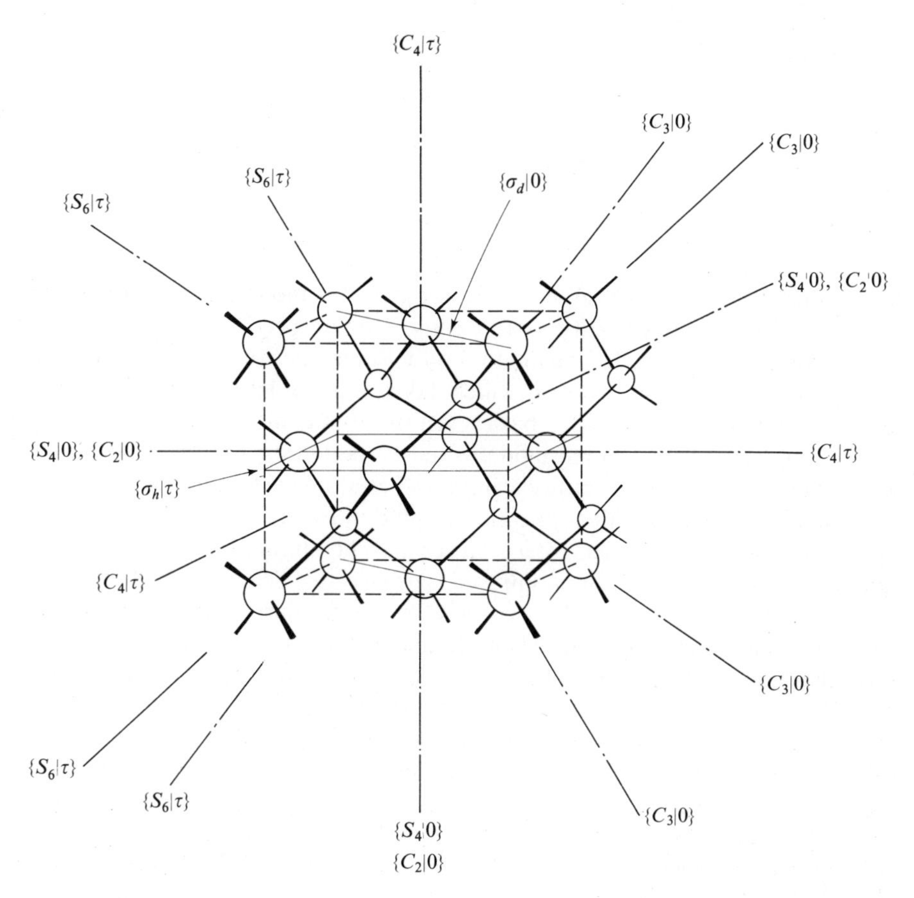

Figure 4-6 Symmetry elements of the factor group O_h. (After J. Waser, K. N. Trueblood, and C. M. Knobler, *Chem One*, McGraw-Hill, New York, 1976, page 91)

condition precludes crystal rotations as proper degrees of freedom contained in the enumeration $\chi_m(E) = 6$. Hence the remaining three degrees of freedom at $\mathbf{k} = 0$ must all be vibrational. From (4-4-2) we see that these are also triply degenerate. Furthermore, inspection of Table 4-3 instructs us that the $\mathbf{k} = 0$ limit of the optic branch of diamond will be Raman-, but not infrared-active.

Had we evinced no interest in the translational modes, we could have proceeded to subtract another $(\pm 1 + 2\cos\theta_R)$ from $\chi_m(R)$ and thus obtain $\chi_v(R)$ directly. In this case, reduction of Γ_v of course yields

$$\Gamma_v = F_{2g} \tag{4-4-3}$$

We may use any of several basis sets and the Wigner Projection Operator (2-2-30) to generate symmetry coordinates for the F_{2g} modes. If the basis set is composed of cartesian displacement coordinates at each atom (see Fig. 4-4), the three (degenerate) symmetry coordinates have the form

$$S_x^{F_{2g}} = \sum_\tau (x_{1\tau} - x_{5\tau})$$
$$S_y^{F_{2g}} = \sum_\tau (y_{1\tau} - y_{5\tau}) \qquad (4\text{-}4\text{-}4)$$
$$S_z^{F_{2g}} = \sum_\tau (z_{1\tau} - z_{5\tau})$$

where τ identifies the unit cell. We thereby see that this mode consists of the two sublattices of diamond vibrating out-of-phase with respect to one another, as expected for an optic branch. The mode is only Raman active, of course, because of the presence of the operation **i**, which takes one sublattice into the other. Were diamond not a lattice with a basis, even this (optic) mode would be missing and only the trivial translational mode would be permitted. On the other hand, if we could replace one of the lattices with another atom (as an example, conceive of the cubic form of ZnS,[13] the optic mode would indeed be infrared active.

We now see that an elementary lattice has no infrared spectrum due to fundamentals, i.e., no one-phonon infrared spectrum (see Section 1-2). This statement is quite generally true. As we shall see in Section 4-7(b), two-phonon combination bands are both infrared and Raman active, in *any* crystal. It is for this reason that diamond does exhibit some weak infrared absorption in the 2–6 μm region.

There is another, extraordinarily powerful way to arrive at the symmetry species of the $\mathbf{k} = 0$ modes of a crystal, which is of particular value when the crystal lattice of interest is one with a basis. This is the way of correlation, previously discussed [Section 2-5(c)] in connection with the effects of isotopic substitution in molecules with attendant change of symmetry.[14] In the diamond crystal, we have already noted that the primitive unit cell is occupied by two atoms which are translationally inequivalent—that is, there is no vector **t** (Eq. (4-1-1)) which takes one into the other. Hence there are two "kinds" of carbon atoms in diamond, related properly by operations of the type defined in Eq. (4-4-1). *Each* of these atoms, however, is tetrahedrally bonded to those of the other kind; hence the *local* symmetry for each is that of the regular tetrahedron, T_d. Whereas the unit cell is invariant only under the operations of the factor group O_h, with some operations (e.g., $\{i|\tau_i\}$) being both rotational and translational in nature, each inequivalent atom is invariant under the purely rotational operations of the point group T_d. The operations of T_d are insufficient to describe the complete symmetry of the crystal because of the exclusion of operations like $\{i|\tau_i\}$ or, indeed, of the pure translations. However, the operations of T_d do describe the symmetry of the local potential field—due to the remainder of the crystal—in which one or the other of each inequivalent carbon atom vibrates.

Each of the 230 space groups has a finite number of positions of such symmetry,

13. R. W. G. Wyckoff, *Crystal Structures*, Interscience, New York, 1963, Volume I, page 108.
14. The correlation technique was first applied to the vibrational analysis of molecular crystals by D. F. Hornig, *J. Chem. Phys.*, **16**, 1063 (1948).

excepting of course the trivial positions of symmetry C_1, which abound in each unit cell. These positions are called *sites*.[15] Their enumeration rests on the fact that each factor group is generally a point group of sufficient order to afford numerous subgroups, each defined by its own rotation axes and reflection planes, but which necessarily have certain relationships to those of the full factor group—hence, to the symmetry axes and planes of the unit cell. The number of possible sites of a given point group, called *site groups*, which are admitted by a particular factor group, is limited only to those which are subgroups of the factor group. On this basis, the factor group C_{2h} could have site groups C_{2h}, C_2, C_s, and C_i. The distribution of sites differs, however, with the space group for each of the factor groups. For example, the symmorphic space group $C_{2h}^1(P2/m)$ has representatives of all the site groups just listed. On the other hand, in $C_{2h}^2(P2_1/m)$, the reflection plane σ_h maintains its purity, but the C_2 and i operations acquire translatory nature. The C_2 operation thus becomes a screw rotation with a pitch of 1/2. In going from C_{2h}^1 to C_{2h}^2, therefore, the C_2 and (hence) C_{2h} sites are lost whereas just one of the C_s sites is retained.

The precise number of each variety of site group for each space group is thus a complicated function of which kind of axes and planes are present in each particular case. Site distributions in all of the 230 space groups have been analytically determined by Zachariasen.[16] They have been individually tabulated for each case.[17] In Appendix VII the positions are identified.

Inspection of Table VII-11 demonstrates that T_d is indeed a site of the space group O_h^7. In fact, there are two distinct sets of T_d sites. Now if we refer to Table 4-3, we note that there are five classes of operations of O_h for which u_R is finite and five for which $u_R = 0$. The first set are the operations of T_d and the second are those of the coset $T_d \cdot i$. In applying the operations of T_d to one of the two carbon atoms it is left unmoved while the operations $T_d \cdot i$ take one T_d site into another. In general there will be $n = g/h$ of a given set of sites, where g is the order of the factor group and h is the order of the site group. This statement can be proven by use of the remarkable correlation theorem (2-5-5) together with the orthogonality condition (2-2-24).[18] Thus in diamond there are two of each distinct set of T_d sites, since $2 = 48/24$. Either set may be used to define the diamond lattice, as each may be obtained from the other by a shift of the origin of $(a/4)(1, 1, 1)$, the "inter-site vector."

The utility of the site group is as follows. The representation with the displacement coordinates of the site occupant (in this case, one C atom) as a basis can be reduced among the representations of the site group. One simply looks for the representation of a translational vector at the site, and finds it to be

$$\Gamma_m = F_2 \text{ of } T_d \tag{4-4-5}$$

15. R. S. Halford, *J. Chem. Phys.*, **14**, 8 (1946).
16. W. H. Zachariasen, *Theory of X-ray Diffraction in Crystals*, Wiley, New York, 1945.
17. *International Tables for Crystallography*, Volume I, N. F. M. Henry and K. Lonsdale, Eds., Kynoch Press, Birmingham, 1952.
18. H. Winston and R. S. Halford, *J. Chem. Phys.*, **17**, 607 (1949).

Since $\Gamma_T = F_2$ in T_d and since $u_R = 1$ for all R of T_d, each atom contributes only to the translational motion of its sublattice. Using either the correlation theorem (2-5-5) or the correlation tables (Appendix IV) we can correlate the irreducible representations of S (the site group) with those of F (the factor group)—i.e., find the number of times Γ^η of S appears in Γ^γ of F. Alternatively, the result can be given by a *correlation diagram* composed of the correlating Γ^η's and Γ^γ's. In such tables the number of lines radiating from a particular Γ^η of S is given by $\sum_\gamma d_\eta/d_\gamma$, which may also be proven by use of the orthogonality conditions.

In diamond, the correlation diagram is

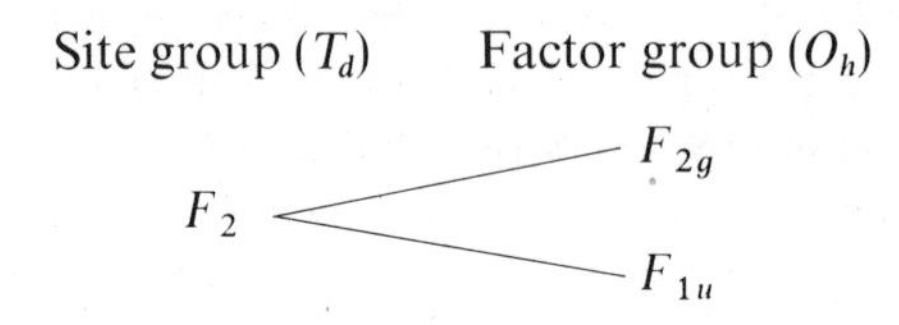

and we thereby see that the identical information as was obtained by analysis of the factor group is more quickly obtained using the site group. We recall that the correlation indicated in the diagram demands that the correct crystal coordinates (in this case, at $\mathbf{k} = 0$) can be in principle obtained by a perturbation calculation in which the zeroth-order coordinates describe only the translational motion of each sublattice (see Section 1-3). In this case, then, the perturbation which mixes the two sets and yields one optic mode and one mode corresponding to translation of the crystal as a whole can hardly be viewed as a small perturbation. In molecular crystals, however, we shall see that very often analogous perturbations suggested by correlation diagrams are small and are hence amenable to low-order perturbation treatments.

Having treated the diamond lattice at some length, we proceed quickly to our other two examples, rock salt and fluorite.

The sodium chloride structure is face-centered cubic (fcc), space group $O_h^5(Fm3m)$. It is in fact composed of *two* fcc lattices, one of Na^+ ions and the other of Cl^- ions, displaced from one another by $(a/2)\boldsymbol{l}$, where $\boldsymbol{l} = \mathbf{i}, \mathbf{j}$, or $\mathbf{k}$, the latter being unit vectors along the sides of the cube. *All* ions are at sites O_h, thus both sets of O_h sites (a and b in the Wyckoff notation, see Table VII-11) are utilized. Thus S and F are the same; consequently, O_h^5 must be symmorphic—there are no space group operations $\{R \,|\, \boldsymbol{\tau}_R\}$ in which $\boldsymbol{\tau}_R$ is a fractional translation.

The dynamically primitive unit cell is a rhombohedron, the primitive translations of which are related to those of the cube by

$$\mathbf{a}_1 = \frac{a}{2}(\mathbf{j} + \mathbf{k})$$

$$\mathbf{a}_2 = \frac{a}{2}(\mathbf{i} + \mathbf{k}) \qquad (4\text{-}4\text{-}6)$$

$$\mathbf{a}_3 = \frac{a}{2}(\mathbf{i} + \mathbf{j})$$

The non-primitive NaCl lattice, constructed in an orthogonal basis, has ions of one kind at all cube corners and face centers, and counter ions at all edge and body centers. Thus $n = 4$. In the rhombohedral basis, the members of the first set of ions are positioned only at the cell corners while the counter ions are all at the cell centers ($n = 1$). Hence, the rhombohedron has one-fourth the volume of the cube. This may also be seen from $V = \mathbf{a}_3 \cdot (\mathbf{a}_1 \times \mathbf{a}_2) = a^3/4$.

Since there is but one ion of each kind in the dynamically primitive cell, the total number of degrees of freedom at each $\mathbf{k}$ is six, i.e., three per ion. Were we to adopt a site group analysis, these would necessarily each form a basis for the F_{1u} representation of $S = O_h$, since taken individually, each set can only be translational and $\Gamma_T = F_{1u}$ of O_h. While in the factor group analysis the results are formally the same, in that all six degrees of freedom belong to F_{1u}, clearly three of these form the triply degenerate and trivial translations of the crystal as a whole, while the other three comprise the $\mathbf{k} = 0$ limit of the optic branch, which to the extent that we have discussed it so far (see Section 4-7) is also triply degenerate. As in the diamond example, this mode consists of one set of atoms (here, ions) vibrating against the other set. In contrast to the situation in diamond, the mode in rock salt is infrared- but not Raman-active. Since both factor groups contain the inversion operation, the contrast may at first perplex; recall, however, that in O_h^7, $\mathbf{i} = \{i|\tau_i\}$, whereas in O_h^5, $\mathbf{i} = \{i|\mathbf{0}\}$. Another manifestation of this difference is contained in the distinction between the sites T_d and O_h: only in rock salt does each nucleus sit on a center of symmetry—the lattice distortion removes that center, hence the mode is infrared-active and not Raman-active.

The correlation diagram for rock salt is

$O_h(S)$		$O_h(F)$
$F_{1u}(Na^+)$	⟍	
		$2F_{1u}$
$F_{1u}(Cl^-)$	⟋	

which indicates again the substantial perturbation that must occur in order to convert six (local) translational modes into three (crystal) vibrational and three (crystal) translational modes.

Our last example is of CaF_2 which also belongs to the O_h^5 space group; however, the site distribution differs from that in NaCl. The cations are still at the same sites (O_h) as before, but the anions are at the T_d positions (see Table VII-11), located at $a(\frac{1}{4}, \frac{1}{4}, \frac{1}{4})$, $a(\frac{3}{4}, \frac{3}{4}, \frac{1}{4})$, $a(\frac{3}{4}, \frac{1}{4}, \frac{3}{4})$, and $a(\frac{1}{4}, \frac{3}{4}, \frac{3}{4})$ in the cubic basis. In the fcc rhombohedral basis, there is again one formula unit per cell. Each fluoride ion sits at the center of a tetrahedron of Ca^{++} ions, one of which is at a cube corner and the remaining three are at centers of cube faces. Hence, the T_d site symmetry.

Site group analysis is more powerful in this example. The correlation table is

Ca^{+2} (O_h site)	Unit cell group (O_h)	$2F^-$ (T_d sites)
	F_{1u}	
F_{1u}		$2F_2$
	F_{2g}	

from which we conclude that the $\mathbf{k} = \mathbf{0}$ modes of the crystal number nine. Three of these are again translational (F_{1u}). There are, however, two optic modes, one of which is infrared- and the other of which is Raman-active. The perturbation implied in the correlation table is that the in-phase (u) motions of the fluoride ions mix with the calcium ion motions to yield the two F_{1u} modes, but that the out-of-phase (g) fluoride ion motions are alone responsible for the F_{2g} lattice vibration. Thus the Raman-active mode should not be affected by calcium isotope substitution. Analogous situations occur in a number of crystals (e.g., $\nu_3(B_{1g})$ of LiOH ($D_{4h}^7 = P4nmm$)) but there have been no experimental tests of this prediction.

Having illustrated the forming of symmetry species at $\mathbf{k} = 0$ with several common fcc lattices, it is of some interest, particularly for use in Section 4-6, to quickly repeat the process for a body-centered cubic (bcc) elementary lattice, such as that of α–W, space group O_h^9. The dynamically primitive unit cell is again rhombohedral, and its primitive translations are related to those of the cube by

$$\mathbf{a}_1 = \frac{a}{2}(-\mathbf{i} + \mathbf{j} + \mathbf{k})$$

$$\mathbf{a}_2 = \frac{a}{2}(\mathbf{i} - \mathbf{j} + \mathbf{k}) \qquad (4\text{-}4\text{-}7)$$

$$\mathbf{a}_3 = \frac{a}{2}(\mathbf{i} + \mathbf{j} - \mathbf{k})$$

This cell has $n = 1$, and thus has half the volume of the non-primitive cell with an orthogonal basis. Since $n = 1$, there are but three degrees of freedom at each $\mathbf{k}$. Each atom is at a corner of a rhombohedral cell, an O_h site. In this case, correlation adds no information. The three modes at $\mathbf{k} = 0$ clearly belong to F_{1u} of O_h since Γ_T also belongs to that representation.

In summary, we see the power of the way of correlation, in finding the species of lattice vibrations at $\mathbf{k} = 0$. In the next section we shall continue to admire the same method, as we move across the Brillouin zone and in Chapter 6 we shall see that the technique is equally valuable in treating the internal vibrations of molecular crystals.

We close this section with a final remark concerning the distribution of sites for all of the space groups, which will be of particular value in the next section. Inspection of Tables VII-1 through VII-11 reveals that symmorphic groups always have at least one site whose point group S is identical with the factor group F. This follows from the fact that in a symmorphic group, no operation of F may contain non-primitive translations. Now if a molecule is situated on a site such that $S = F$, the orthogonality conditions require that there can be only one such molecule per unit cell. Hence *all* operations of F either leave the molecule fixed or shift it to an equivalent position in another unit cell. More importantly, when the space group is symmorphic, all operations of F can be written in the form $\{R|\mathbf{0}\}$, whence it is appreciated that they do only what the associated operations of S do; hence, $S = F$.

4-5 SYMMETRY SPECIES AT $\kappa \neq 0$

A complete theory of lattice vibrations in three dimensions requires the treatment of modes with all $\boldsymbol{\kappa}$ values occurring in the Brillouin zone, even though the modes with $\boldsymbol{\kappa} \approx 0$ are the only ones which may appear as fundamentals in infrared and Raman spectra. Combination modes, which include the so-called two- (or multi-) phonon processes, in general will involve all $\boldsymbol{\kappa}$ values.

We have seen in the previous sections that the symmetry species of modes at $\boldsymbol{\kappa} = 0$ can be obtained by investigating the transformation of the cartesian coordinates of the atoms in a single unit cell; whenever a factor group operation carried an atom into a translationally equivalent atom in another unit cell the effect of the operation was considered to be the identity. The analytical foundation of this rule will now be developed in order to extend the discussion to $\boldsymbol{\kappa} \neq 0$. The cartesian (or any other local) coordinates, x_i, may be translationally symmetrized in the form

$$s_i(\boldsymbol{\kappa}) = N^{-3/2} \sum_{\boldsymbol{\tau}} e^{-i\boldsymbol{\kappa}\cdot\boldsymbol{\tau}} x_i(\boldsymbol{\tau}) \tag{4-5-1}$$

In the limit $\boldsymbol{\kappa} = 0$, the effect of the factor group operations upon $s_i(\mathbf{0})$ is identical with the results stated in Section 4-4, since the $s_i(\mathbf{0})$ are invariant to any lattice translation.

(a) Transformation of the Translationally Symmetrized Coordinates

We shall now investigate the transformation of the $s_i(\boldsymbol{\kappa})$ by the operations of the unit cell group. In Section 4-3 a subgroup $H(\boldsymbol{\kappa})$ was defined such that $R\boldsymbol{\kappa} = \boldsymbol{\kappa} + \bar{\boldsymbol{\kappa}}_R$ where $\bar{\boldsymbol{\kappa}}_R$ is a lattice translation in reciprocal space. In the interior of the Brillouin zone, $\bar{\boldsymbol{\kappa}}_R = 0$, and $H(\boldsymbol{\kappa})$ consists only of those operations which leave the vector $\boldsymbol{\kappa}$ invariant, but at certain points on the boundary $H(\boldsymbol{\kappa})$ may include operations which send $\boldsymbol{\kappa}$ into $\boldsymbol{\kappa} + \bar{\boldsymbol{\kappa}}$. If R belongs to G (the unit cell group) but not to $H(\boldsymbol{\kappa})$, $R\boldsymbol{\kappa} = \boldsymbol{\kappa}'$. The set of all distinct $\boldsymbol{\kappa}'$ is called the *star* of $\boldsymbol{\kappa}$.

In order that we may discuss non-symmorphic and symmorphic cases together, it is necessary to specify the local coordinates with some care; in particular, we need to label local coordinates with indices i, $\boldsymbol{\tau}$, and $\boldsymbol{\tau}_\sigma$. These have the following meaning: $\boldsymbol{\tau}$ identifies the unit cell; $\boldsymbol{\tau}_\sigma$, a fractional vector, labels one of a set of equivalent sites; and i identifies a coordinate at a site. The index σ takes on g/s distinct values, g being the order of the factor group and s the order of some site group. In the symmorphic case there exist one or more sites whose order $s = g$, so in such a case, the choice of such a maximal site as an origin allows the single $\boldsymbol{\tau}_\sigma$ value to be set equal to zero. The choice of a maximal site in the non-symmorphic case, for example $T_d(a)$ in the diamond lattice $O_h^7 = Fd3m$, or in calcite, either the site $a = \frac{1}{4}\frac{1}{4}\frac{1}{4}$ (point symmetry D_3 occupied by the carbon of $CO_3^=$) or the site $b = 000$ (point symmetry C_{3i}, occupied by Ca^{+2}), would yield the smallest set of $\boldsymbol{\tau}_\sigma$ with $\sigma = 1, 2$. This follows since in each of these examples $g/s = 2$. In the case of diamond, $\boldsymbol{\tau}_\sigma = 0$ and $\frac{1}{4}\frac{1}{4}\frac{1}{4}$, whereas in calcite, the choice either of an anion or cation origin leads to $\tau_\sigma = 0$ and $\frac{1}{2}\frac{1}{2}\frac{1}{2}$. In the case of calcite if we were studying the transformation properties of the six carbon oxygen bonds, we could label the three bonds of the carbonate at $(\frac{1}{4}\frac{1}{4}\frac{1}{4})$ r_1, r_2, r_3 and the three bonds of the carbonate at $(\frac{3}{4}\frac{3}{4}\frac{3}{4})$ as r_4, r_5, r_6. Then choosing $\frac{1}{4}\frac{1}{4}\frac{1}{4}$ as an origin we would have

$$\tau_1 = 0$$

and

$$\tau_2 = \tfrac{1}{2}\tfrac{1}{2}\tfrac{1}{2}$$

The transformation law for our local coordinate is

$$\{R \mid \tau_R\} r_i(\tau + \tau_\sigma) = \sum_j R_{ij} r_j(R\tau + R\tau_\sigma + \tau_R) \tag{4-5-2}$$

As shown in Appendix VI an important consequence of the fact that the pure translations $\{E \mid \tau\}$ (τ = integer) constitute an invariant subgroup of the space group is the simple result that $R\tau$ is necessarily also an integer vector of the lattice. Moreover, it is also true that

$$R(\tau + \tau_\sigma) + \tau_R = \tau' + \tau_{\sigma'} \tag{4-5-3}$$

where τ' is an integer (lattice) vector and $\tau_{\sigma'}$ is some fractional vector.

Now combining (4-5-2) with (4-5-3) we can write

$$\{R \mid \tau_R\} r_i(\tau + \tau_\sigma) = \sum_j R_{ij} r_j[\tau' + \tau_{\sigma'}] \tag{4-5-4}$$

We now define the translationally symmetrized coordinates by

$$s_{i\sigma}(\boldsymbol{\kappa}) = N^{-3/2} \sum_\tau e^{-i\boldsymbol{\kappa}\cdot(\tau + \tau_\sigma)} r_i(\tau + \tau_\sigma) \tag{4-5-5}$$

Note that a constant phase factor of the form $e^{-i\boldsymbol{\kappa}\cdot\tau_\sigma}$ appears in this definition, which is a generalization of our earlier (4-5-1). The law of transformation becomes

$$\{R \mid \tau_R\} s_{i\sigma}(\boldsymbol{\kappa}) = N^{-3/2} \sum_\tau e^{-i\boldsymbol{\kappa}\cdot(\tau + \tau_\sigma)} \sum_j R_{ij} r_j(\tau' + \tau_{\sigma'}) \tag{4-5-6}$$

In order to reduce the right-hand side of the last equation to a function of translationally symmetrized coordinates, one needs to replace the summation over τ by a summation over τ' where clearly the τ' are integer lattice vectors which range over the entire lattice when i and R are fixed and τ is allowed to range over the entire lattice. We accomplish this desired end by transforming the phase factor as follows. From (4-5-3)

$$\tau + \tau_\sigma = R^{-1}(\tau' + \tau_{\sigma'} - \tau_R)$$

so it follows that

$$e^{-i\boldsymbol{\kappa}\cdot(\tau + \tau_\sigma)} = e^{-i\boldsymbol{\kappa}\cdot R^{-1}(\tau' + \tau_{\sigma'} - \tau_R)} \tag{4-5-7}$$

Now regard the operator R^{-1} in the exponent of (4-5-7) as an operation upon

$\boldsymbol{\kappa}$ instead of the $\boldsymbol{\tau}$ vectors. We limit the rotations R to members of $H(\kappa)$, so that

$$R\boldsymbol{\kappa} = \boldsymbol{\kappa} + \bar{\boldsymbol{\kappa}}_R$$

which is equivalent to

$$\boldsymbol{\kappa}R^{-1} = \boldsymbol{\kappa} + \bar{\boldsymbol{\kappa}}_R$$

since in the exponent the transpose of the κ vector is implied ($\boldsymbol{\kappa} \cdot \boldsymbol{\tau}$ signifies a scalar product). Thus the exponent in (4-5-7) becomes

$$\begin{aligned} &-i\boldsymbol{\kappa}R^{-1} \cdot (\boldsymbol{\tau}' + \boldsymbol{\tau}_{\sigma'} - \boldsymbol{\tau}_R) \\ &= -i(\boldsymbol{\kappa} + \bar{\boldsymbol{\kappa}}_R) \cdot (\boldsymbol{\tau}' + \boldsymbol{\tau}_{\sigma'} - \boldsymbol{\tau}_R) \\ &= -i[\boldsymbol{\kappa} \cdot (\boldsymbol{\tau}' + \boldsymbol{\tau}_{\sigma'}) + \bar{\boldsymbol{\kappa}}_R \cdot \boldsymbol{\tau}' - \boldsymbol{\kappa} \cdot \boldsymbol{\tau}_R + \bar{\boldsymbol{\kappa}}_R \cdot (\boldsymbol{\tau}_{\sigma'} - \boldsymbol{\tau}_R)] \end{aligned}$$

Since $e^{-i\bar{\boldsymbol{\kappa}}_R \cdot \boldsymbol{\tau}'} = 1$ because $\bar{\boldsymbol{\kappa}}_R$ and $\boldsymbol{\tau}'$ are integer vectors of their respective lattices, it is now possible to write (4-5-6) in the form

$$\begin{aligned} \{R \mid \boldsymbol{\tau}_R\} s_{i\sigma}(\boldsymbol{\kappa}) &= N^{-1/2} e^{i\boldsymbol{\kappa} \cdot \boldsymbol{\tau}_R} e^{-i\bar{\boldsymbol{\kappa}}_R \cdot (\boldsymbol{\tau}_{\sigma'} - \boldsymbol{\tau}_R)} \sum_{\boldsymbol{\tau}'} e^{-i\boldsymbol{\kappa} \cdot (\boldsymbol{\tau}' + \boldsymbol{\tau}_{\sigma'})} \sum_j R_{ij} r_j(\boldsymbol{\tau}' + \boldsymbol{\tau}_{\sigma'}) \\ &= e^{i\boldsymbol{\kappa} \cdot \boldsymbol{\tau}_R} e^{-i\bar{\boldsymbol{\kappa}}_R \cdot (\boldsymbol{\tau}_{\sigma'} - \boldsymbol{\tau}_R)} \sum_j R_{ij} s_{j\sigma'}(\boldsymbol{\kappa}) \end{aligned} \tag{4-5-8}$$

This central result is the main tool which we require for further symmetrization of the basis functions under the point symmetry operations throughout the Brillouin zone. Some of its important implications are the following:

1. At $\boldsymbol{\kappa} = \mathbf{0}$, the $s_{i\sigma}(\mathbf{0})$ provide a basis for a representation of the point group G (we have already used this idea in Section 4-4). For when $\boldsymbol{\kappa} = \bar{\boldsymbol{\kappa}}_R = 0$, the phase factor disappears from (4-5-8) and the representation can be reduced simply by studying the R_{ij} in precisely the manner employed in molecular cases in Chapter 2.
2. Had we not limited R to members of $H(\boldsymbol{\kappa})$ we would have found for $\boldsymbol{\kappa} \neq 0$ that operations of G but not of $H(\kappa)$ send $s(\boldsymbol{\kappa})$ into $s(\boldsymbol{\kappa}')$ where $\boldsymbol{\kappa}' \neq \boldsymbol{\kappa} + \bar{\boldsymbol{\kappa}}_R$. Since the dynamical equations are degenerate for distinct $\boldsymbol{\kappa}'$ related by $\boldsymbol{\kappa}R^{-1} = \boldsymbol{\kappa}'$ it is therefore only necessary to solve the problem for a single $\boldsymbol{\kappa}$ and hence permissible to restrict the point symmetry analysis to $R \subset H(\boldsymbol{\kappa})$. Physically this amounts to the equivalence of the various propagation directions which lie in the star of $\boldsymbol{\kappa}$.
3. For *all* points of the Brillouin zone in *symmorphic* groups the translationally symmetrized coordinates $s_i(\boldsymbol{\kappa})$ behave like basis functions of point groups $H(\boldsymbol{\kappa})$. This follows because in (4-5-8), *no* fractional vectors, either for transformations $\boldsymbol{\tau}_R$ or for sites, $\boldsymbol{\tau}_\sigma$, are required, and the phase factor preceding R_{ij} is unity.
4. If $\boldsymbol{\kappa}$ is in the interior of the Brillouin zone, or even if it is at a low symmetry point on the zone boundary, so that $\bar{\boldsymbol{\kappa}}_R = 0$ for all R in $H(\boldsymbol{\kappa})$, (4-5-8) may be simplified, i.e., the only phase factor is $e^{i\boldsymbol{\kappa} \cdot \boldsymbol{\tau}_R}$, the remaining expression, $e^{-i\bar{\boldsymbol{\kappa}}_R \cdot (\boldsymbol{\tau}_{\sigma'} - \boldsymbol{\tau}_R)}$ being simply unity. The $s_{i\sigma}$ will therefore transform just like coordinates under a point group $H(\boldsymbol{\kappa})$ except for the factor $e^{i\boldsymbol{\kappa} \cdot \boldsymbol{\tau}_R}$. But this factor only means that the representation of the group $H(\boldsymbol{\kappa})$ for which $s_{i\sigma}(\kappa)$ are basis functions differs from the representation of the corresponding point group by the factor $e^{i\boldsymbol{\kappa} \cdot \boldsymbol{\tau}_R}$ because the numbers $e^{i\boldsymbol{\kappa} \cdot \boldsymbol{\tau}_R}$ themselves constitute a representa-

tion: to see this we must consider the result of two transformations which we shall call R and S. The group property demands that

$$\{S|\tau_S\}\{R|\tau_R\} = \{T|\tau_T\}$$

with

$$SR = T$$

and

$$S\tau_R + \tau_S = \tau_T \tag{4-5-9}$$

If we apply $\{S|\tau_S\}\{R|\tau_R\}$ to $s_{i\sigma}(\boldsymbol{\kappa})$ we shall obtain a phase factor of the form $e^{i\boldsymbol{\kappa}\cdot(\tau_R+\tau_S)}$; but if these phase factors are to constitute a representation by themselves, we must have

$$e^{i\boldsymbol{\kappa}\cdot(\tau_R+\tau_S)} = e^{i\boldsymbol{\kappa}\cdot\tau_T} = e^{i\boldsymbol{\kappa}\cdot(S\tau_R+\tau_S)} \tag{4-5-10}$$

Now if S operates upon $\boldsymbol{\kappa}$ instead of $\boldsymbol{\tau}_R$, we have

$$e^{i\boldsymbol{\kappa}\cdot(S\tau_R)} = e^{i(\boldsymbol{\kappa}+\bar{\boldsymbol{\kappa}}_{S^{-1}})\cdot\tau_R} \tag{4-5-11}$$

i.e., we obtain an extra factor of the form $e^{i\bar{\boldsymbol{\kappa}}_{S^{-1}}\cdot\tau_R}$. But we are here considering only those points of the Brillouin zone for which all $\bar{\boldsymbol{\kappa}}$ vanish, so the extra factor is unity, and our assertion that the factors $e^{i\boldsymbol{\kappa}\cdot\tau_R}$ constitute a representation is validated. It is important to note that the basis coordinates must contain the phase factor $e^{-i\boldsymbol{\kappa}\cdot\tau_\sigma}$ in the non-symmorphic cases.

5. Even at high symmetry zone boundary points where $\bar{\boldsymbol{\kappa}}_R \neq 0$ for some $R \subset H(\boldsymbol{\kappa})$, it may still happen that the $s_{i\sigma}(\boldsymbol{\kappa})$ afford the basis of a representation which is simply related to a point group representation. The condition for this to be true is that

$$e^{i\bar{\boldsymbol{\kappa}}_S\cdot\tau_R} = 1 \tag{4-5-12}$$

for all S and R in $H(\boldsymbol{\kappa})$. A special case in which the condition (4-5-12) is satisfied occurs if $H(\boldsymbol{\kappa})$ is contained in a site group of G because then for all R in $H(\boldsymbol{\kappa})$ $\boldsymbol{\tau}_R = 0$. First, we mention some examples and subsequently give the proof of this assertion.

For the diamond lattice O_h^7, important zone boundary points include $X(D_{4h})$ and $L(D_{3d})$. The point symmetry of X is not contained in any site of O_h^7 and detailed examination shows that the criterion (4-5-12) is *not* obeyed. The point L has a symmetry D_{3d} which is a site (consult Appendix VII) so that (4-5-12) is satisfied since the choice of D_{3d}, rather than T_d, as an origin, permits all $\boldsymbol{\tau}_R$ to be put equal to zero. This test must naturally be applied with caution in cases where the sites in direct space involve a point group with the same name as $H(\boldsymbol{\kappa})$ but do not necessarily involve the same elements. In the space group D_{6h}^4 whose Brillouin zone is depicted in Fig. IX-3* the point

* Figure IX-3 applies to D_{6h}^4 as well as to C_6^6, although generally with groups $H(\mathbf{k})$ of higher order. The *location* of the BSW special points is nevertheless the same.

H has D_{3h} symmetry. Despite the fact that no less than three sites, namely b, c, d have D_{3h} symmetry, different C_2 operations are involved in H from those in b, c, d, so H is not in fact identical with any D_{3h} site.

The point K in this same Brillouin zone provides another example of the $e^{i\bar{\kappa}_S \cdot \tau_R} = 1$ criterion. If the site b in D_{6h}^4 is chosen as an origin, one finds from the *International Tables of X-Ray Crystallography* that the only non-vanishing $\tau_R = (0\,0\,\frac{1}{2})$. But at $H(\kappa) = \mathrm{K}$, the only non-vanishing $\bar{\kappa}_S$ will be of the form $(2\pi, 2\pi, 0)$, so that indeed $e^{i\bar{\kappa}_S \cdot \tau_R} = 1$.

In order to establish the validity of the criterion (4-5-12), we must again examine the result of two successive operations, $\{R \mid \tau_R\}$ and $\{S \mid \tau_S\}$ upon $s_{i\sigma}(\kappa)$. If we have $\sigma \xrightarrow{R} \sigma' \xrightarrow{S} \sigma''$, then the phase factor obtained in two steps should equal that obtained for the single step $\sigma \xrightarrow{T} \sigma''$. From (4-5-8) we see that this requirement is expressed by:

$$e^{i\kappa \cdot (\tau_R + \tau_S)}\, e^{-i\bar{\kappa}_R \cdot (\tau_{\sigma'} - \tau_R)}\, e^{-i\bar{\kappa}_S \cdot (\tau_{\sigma''} - \tau_S)} = e^{i\kappa \cdot \tau_T}\, e^{-i\bar{\kappa}_T \cdot (\tau_{\sigma''} - \tau_T)} \tag{4-5-13)}$$

We have already seen that

$$e^{i\kappa \cdot (\tau_R + \tau_S)} = e^{i\kappa \cdot \tau_T}$$

if

$$e^{i\bar{\kappa}_{S^{-1}} \cdot \tau_R} = 1$$

which is assured by the condition (4-5-12). Thus it remains to examine the remaining factors in (4-5-13). Subject to the condition (4-5-13) these factors are:

$$e^{-i\bar{\kappa}_T \cdot (\tau_{\sigma''} - \tau_T)} = e^{-i\bar{\kappa}_T \cdot \tau_{\sigma''}} \tag{4-5-14A}$$

and

$$e^{-i\bar{\kappa}_R \cdot (\tau_{\sigma'} - \tau_R)}\, e^{-i\bar{\kappa}_S \cdot (\tau_{\sigma''} - \tau_S)} = e^{-i(\bar{\kappa}_R \cdot \tau_{\sigma'} + \bar{\kappa}_S \cdot \tau_{\sigma''})} \tag{4-5-14B}$$

Since $T = SR$ or $T^{-1} = R^{-1}S^{-1}$, we have

$$\begin{aligned} \bar{\kappa}_T &= \kappa(T^{-1} - E) = \kappa(R^{-1}S^{-1} - E) \\ &= (\bar{\kappa}_R + \kappa)S^{-1} - \kappa = \bar{\kappa}_R S^{-1} + \bar{\kappa}_S \end{aligned} \tag{4-5-15}$$

If we substitute (4-5-15) for $\bar{\kappa}_T$ in (4-5-14A), we see that equality of (4-5-14A) and (4-5-14B) requires that

$$e^{-i\bar{\kappa}_R S^{-1} \tau_{\sigma''}} = e^{-i\bar{\kappa}_R \cdot \tau_{\sigma'}} \tag{4-5-16}$$

But by an extension of (4-5-3),

$$S(\tau' + \tau_{\sigma'}) + \tau_S = \tau'' + \tau_{\sigma''}$$

or

$$S^{-1}\tau_{\sigma''} = \tau' - S^{-1}\tau'' + \tau_{\sigma'} + S^{-1}\tau_S \tag{4-5-17}$$

If we substitute (4-5-17) on the left-hand side of (4-5-16), the integer vectors τ'

and $\boldsymbol{\tau}''$ drop out since $e^{i\bar{\kappa}\cdot\boldsymbol{\tau}} = 1$ for any integer $\boldsymbol{\tau}$. Also, $S^{-1}\boldsymbol{\tau}_S = -\boldsymbol{\tau}_{S^{-1}}$, so that

$$e^{-i\bar{\kappa}_R\cdot S^{-1}\boldsymbol{\tau}_{\sigma''}} = e^{-i\bar{\kappa}_R\cdot\boldsymbol{\tau}_{\sigma'}}\,e^{+i\bar{\kappa}_R\cdot\boldsymbol{\tau}_{S^{-1}}}$$

$$= e^{-i\bar{\kappa}_R\cdot\boldsymbol{\tau}_{\sigma'}}$$

because the factor $e^{i\bar{\kappa}_R\cdot\boldsymbol{\tau}_{S^{-1}}} = 1$ from the condition (4-5-12). We have now demonstrated that (4-5-14A) and (4-5-14B) are equal subject to the condition (4-5-12), and that therefore the representations of $H(\boldsymbol{\kappa})$ are the same as those of the point group isomorphic with $H(\boldsymbol{\kappa})$, *provided suitable phase factors are included.* Representations of this kind are called *multiplier representations.*

6. There remain cases for high symmetry points on the zone boundary (not all $\bar{\boldsymbol{\kappa}}_R \neq 0$) in non-symmorphic space groups for which the condition (4-5-12) is not satisfied. In such cases, the appropriate representations cannot be obtained from those of the point groups, and it is necessary to extend the theory by basing the representations upon irreducible representations of other finite groups which are not isomorphic with the 32 point groups, which define the possible site symmetries of the space groups. A detailed and lucid account of the necessary representation theory is given by Bradley and Cracknell[19] together with all the necessary character tables or the reader may be referred to tables of all the necessary irreducible representations published by Miller and Love.[20] The tables as given by Miller and Love occupy some 410 pages. To include all of these is beyond the scope of the present work, but we give in Appendix X a few of the more important examples.

Summarizing this lengthy discussion, it has been demonstrated that the representations based upon the coordinates $s_{i\sigma}(\boldsymbol{\kappa})$ are simply related to point group representations as follows:

1. Everywhere in the Brillouin zone for symmorphic space groups
2. Everywhere in the interior and also on the zone boundary if $\bar{\boldsymbol{\kappa}}_R = 0$ all R in $H(\boldsymbol{\kappa})$
3. Even if some $\bar{\boldsymbol{\kappa}}_R \neq 0$ provided $e^{i\bar{\kappa}_R\cdot\boldsymbol{\tau}_S} = 1$ for all R and S in $H(\boldsymbol{\kappa})$.

In any of the above circumstances, the ultimate construction of symmetry coordinates and factoring of the dynamical matrices F and G with respect to the point symmetry operations proceeds from the results at $\boldsymbol{\kappa} = 0$ simply by correlation. At $\boldsymbol{\kappa} = 0$, the appropriate symmetry coordinates are

$$S_{\gamma\alpha}(0) = \sum_{i\sigma} U_{\gamma\alpha,i\sigma} s_{i\sigma}(0) \tag{4-5-18}$$

in which the coefficients $U_{\gamma\alpha,i\sigma}$ can be found with the aid of the Wigner Projection Operator summed over the group G neglecting any fractional translations. The indices γ and α refer to a particular irreducible representation (γ) of G and a

19. C. J. Bradley and A. P. Cracknell, *The Mathematical Theory of Symmetry in Solids,* Clarendon Press, Oxford, 1972.
20. S. C. Miller and W. F. Love, *Tables of Irreducible Representations of Space Groups and Co-Representations of Magnetic Space Groups,* Pruett Press, Boulder, 1967.

particular component (α) of a degenerate representation. Provided any degenerate symmetry coordinates in G are properly oriented, they remain symmetry coordinates for any subgroup of G, so that (4-5-18) is equally applicable to all values of $\boldsymbol{\kappa}$ which are specified in the three categories above, so that

$$S_{\gamma\alpha}(\boldsymbol{\kappa}) = \sum_{i\sigma} U_{\gamma\alpha,i\sigma} s_{i\sigma}(\boldsymbol{\kappa}) \tag{4-5-19}$$

Correlation tables may be used to find the names of the symmetry species in $H(\boldsymbol{\kappa})$ which correlate with species γ of G.

The dynamical matrices $F(\boldsymbol{\kappa})$ and $G(\boldsymbol{\kappa})$ are of course factored at $\boldsymbol{\kappa} = 0$ by applying the transformation

$$\begin{aligned} F(0) &= UF(0)U^{\dagger} \\ G(0) &= UG(0)U^{\dagger} \end{aligned} \tag{4-5-20}$$

The discussion given above shows that the same transformation also factors the dynamical matrices for most other values of $\boldsymbol{\kappa}$ throughout the Brillouin zone, i.e.,

$$\begin{aligned} F(\boldsymbol{\kappa}) &= UF(\boldsymbol{\kappa})U^{\dagger} \\ G(\boldsymbol{\kappa}) &= UG(\boldsymbol{\kappa})U^{\dagger} \end{aligned} \tag{4-5-21}$$

(b) The Symmetrization of Coordinates under $H(\kappa)$

In Section 4-4 and in Section 4-5(a) we have seen how to classify the transformation properties of the normal modes at any point in the Brillouin zone. The actual construction of symmetry coordinates involves the use of the Wigner Projection Operator for the operations of the factor group applied to the translationally symmetrized coordinates $s_i(\boldsymbol{\kappa})$, introduced in (4-1-7) and generalized in (4-5-5). In general one applies the WPO in the form

$$S_{\eta}(\boldsymbol{\kappa}) = N^{-3/2} \sum_{R \subset H(\kappa)} \chi_R^{(\eta)*}\{R \mid \tau_R\} s_1(\boldsymbol{\kappa}) \tag{4-5-22}$$

in order to produce a coordinate of species η under the group $H(\boldsymbol{\kappa})$. However, if one wishes to use correlation along some selected direction in the Brillouin zone, it may be expedient to choose an "oriented" initial linear combination of certain $s_i(\boldsymbol{\kappa})$ in place of $s_1(\boldsymbol{\kappa})$ in (4-5-22), in order to take advantage of the correlation of species at $\boldsymbol{\kappa} = 0$ and at finite $\boldsymbol{\kappa}$. No examples of this process will be given because the introduction of such coordinates is simply a step in the factoring of the dynamic matrices which will be illustrated in Section 4-6.

(c) The Body-Centered Lattice

In Section 4-4 we saw that for this monatomic lattice, the coordinates for the single atom give rise to a representation of species F_{1u} at $\boldsymbol{\kappa} = 0$ and that their frequencies are zero at this point. The possible $H(\boldsymbol{\kappa})$ are found by recalling from

Section 4-3 that the lattice reciprocal to body-centered O_h^9 is O_h^5, whose special points are listed in Table VII-11.* Consider first a $\boldsymbol{\kappa}$ with three equal components, i.e., one which lies along one of the body diagonals which are threefold axes (see Fig. 4-3). In the interior of the zone, the symmetry of $\boldsymbol{\kappa}$ is C_{3v} (point f in Table VII-11), but at the boundary it becomes T_d (point c). The BSW nomenclature for this sequence of points in Fig. 4-3 is $\Gamma \rightarrow \Lambda \rightarrow P$. The correlation between certain of the species of O_h and some of its subgroups is given in Table 4-4. Complete correlation tables for all crystallographic point groups are given in Appendix IV. Inspection of Table 4-4 shows that for the sequence of points in the Brillouin zone just described, $O_h \rightarrow C_{3v} \rightarrow T_d$, the vibrational species are $F_{1u} \rightarrow A_1 + E \rightarrow F_2$. A similar discussion for a mode propagating along a cartesian direction (fourfold axis) corresponds to the sequence $H(\boldsymbol{\kappa}) = O_h \rightarrow C_{4v} \rightarrow O_h$ ($\Gamma \rightarrow \Delta \rightarrow H$ in Fig. 4-3) so the species correlation is $F_{1u} \rightarrow A_1 + E \rightarrow F_{1u}$. Still another interesting sequence consists of modes propagating in the (110) direction as expressed in cartesian components in **k** space. These are the points $\Gamma \rightarrow \Sigma \rightarrow \mathrm{N}$ in Fig. 4-3 or point groups $O_h \rightarrow C_{2v} \rightarrow D_{2h}$ so that correlation indicates the symmetry species $F_{1u} \rightarrow A_1 + B_1 + B_2 \rightarrow B_{1u} + B_{2u} + B_{3u}$. Note that for these three distinct directions, namely (111), (100), and (110) in cartesian coordinates, one predicts in the first two cases that the modes which are triply degenerate at the origin will split into two frequencies in the interior, one non-degenerate (A_1), the other doubly degenerate (E), and ultimately coalesce into a triply degenerate frequency at the boundary. In the third case, the triple degeneracy at the origin is completely lifted, i.e., there is a splitting into three distinct frequencies in the interior, and this splitting persists at the zone boundary.

(d) The Rocksalt, Diamond, and Fluorite Lattices

As shown in Section 4-4, the modes at $\boldsymbol{\kappa} = 0$ are of species $2F_{1u}$ in rocksalt and $F_{1u} + F_{2g}$ in diamond. Also, the modes of CaF_2 are of species $2F_{1u} + F_{2g}$ at $\boldsymbol{\kappa} = 0$. This is easily shown by noting that calcium ions occupy O_h sites and F^- ions occupy T_d sites, the former yielding F_{1u} and the latter $F_{1u} + F_{2g}$ when one notes the correlation of F_2 and T_d with the species of O_h which is formally identical with the case of diamond. In Table 4-5 we summarize the symmetry species for these cases as obtained by correlation at several interesting points in the Brillouin zone (see Fig. 4-3). For all three direct lattices, the reciprocal lattice

Table 4-4 Correlation of certain species of O_h with those of some of its subgroups

O_h	T_d	D_{4h}	D_{3d}	C_{4v}	C_{3v}	D_{2h}	C_{2v}
F_{1u}	F_2	$A_{2u} + E_u$	$A_{2u} + E_u$	$A_1 + E$	$A_1 + E$	$B_{1u} + B_{2u} + B_{3u}$	$A_1 + B_1 + B_2$
F_{2g}	F_2	$B_{2g} + E_g$	$A_{1g} + E_g$	$A_2 + E$	$A_1 + E$	$B_{1g} + B_{2g} + B_{3g}$	$A_2 + B_1 + B_2$

* See also Appendix VIII.

Table 4-5 Symmetry of vibrations in rocksalt, diamond, and fluorite

	Point coordinates in the Brillouin zone				
k	000	$00k$	$00\frac{1}{a}$	kkk	$\frac{1}{a}\frac{1}{a}\frac{1}{a}$
κ	000	$\kappa\kappa 0$	$\pi\pi 0$	$\kappa\kappa\kappa$	$\pi\pi\pi$
BSW	Γ	Δ	X	Λ	L
Wyckoff	a	e	b	f	c
Point symmetry	O_h	C_{4v}	D_{4h}	C_{3v}	D_{3d}
Rocksalt	$2F_{1u}$	$2(A_1 + E)$	$2(A_{2u} + E_u)$	$2(A_1 + E)$	$2(A_{2u} + E_u)$
Diamond	$F_{2g} + F_{1u}$	$A_1 + B_2 + 2E$	$B_{2g}^* + E_g^* + A_{2u}^* + E_u^*$	$2(A_1 + E)$	$A_{1g} + E_g + A_{2u} + E_u$
Fluorite	$F_{2g} + 2F_{1u}$	$2A_1 + B_2 + 3E$	$B_{2g} + E_g + 2A_{2u} + 2E_u$	$3(A_1 + E)$	$A_{1g} + E_g + 2A_{2u} + 2E_u$

* Since this is a non-symmorphic space group and the criterion given by (4-5-12) is *not* satisfied, different representations are encountered here; B_{2g} is actually degenerate with A_{2u}; see Appendix X.

is O_h^9.* Summarizing what has been developed thus far, we can confidently claim that the symmetry species for any lattice and at any point in the interior of the Brillouin zone and at many of the boundary points can readily be identified with the aid of the site group and the correlation tables. Indeed, there is no need to restrict this method of analysis to the crystals composed of monatomic units which are the subject of this chapter. However, it is physically more instructive to treat crystals containing molecular units in a different fashion as will be shown in Chapter 6.

It is very important, however, for a correct understanding of the foregoing remarks to understand certain limitations in the application of the correlation methods to the assignment of the symmetry species along dispersion curves. These problems are well illustrated by Fig. 4-7(a, b) which depict the phonon frequencies for potassium bromide (a) and diamond (b) in the principal symmetry directions, $\mathbf{k} = (k\,0\,0)$ which in the face-centered lattice correspond to $\boldsymbol{\kappa} = (\kappa\,\kappa\,0)$, and $\mathbf{k} = (k\,k\,k)$ equivalent to $\boldsymbol{\kappa} = (\kappa\,0\,0)$. The first surprise is that even at $\mathbf{k} = (0\,0\,0)$ in KBr the optic mode does not exhibit the formally predicted triple degeneracy but is instead split into a doubly degenerate TO = *T*ransverse *O*ptic branch and a nondegenerate LO = *L*ongitudinal *O*ptic branch. This phenomenon, characteristic of all infrared active modes, will be discussed at greater length in Section 4-9 and Chapter 5. Aside from this deviation, the formal symmetry theory is easily applied as k increases towards the respective zone boundary points. The longitudinal modes, both optic and acoustic (LA) are of species A_1, in the interior, becoming A_{2u} at each zone boundary point here illustrated, and similarly, the transverse modes are E modes in the interior and E_u modes on both the D_{4h} and D_{3d} boundary points.

* See Appendix VIII.

In the diamond lattice, the situation is more ambiguous. Consider first the direction of three-fold symmetry: $O_h \rightarrow C_{3v} \rightarrow D_{3d}$. In the interior, the situation is similar to the case for rocksalt, though note that the optic mode, species F_{2g} (infrared inactive) is not split into L and T branches at the origin. Moreover, when **k** reaches the D_{3d} boundary point, it is *not* obvious how the lower and upper A_1 branches, which clearly must be designated as LA and LO respectively, are to be assigned to the D_{3d} species A_{1g} and A_{2u}, and correspondingly how TA and TO are to be assigned to E_g and E_u. This is a consequence of the fact that everywhere in the interior there is mixing of g and u, and when the boundary point D_{3d} is reached, a branch cannot "remember" the parity it had at the zone origin. Of course, if one knew the proper potential function, calculation would settle the matter. In practice the study of the two-phonon combination spectrum in diamond and related crystals has been employed to determine the correct assignment which appears to correspond with the first column of alternatives in Fig. 4-7(*b*).

To conclude this section, we remark that at $\mathbf{k} = (\pi 0 0)$, i.e., D_{4h}, in diamond a further complication has arisen, namely the $LO(B_2)$ and $LA(A_1)$ branches become degenerate at the zone boundary, despite their formal correlation with the distinct, respective species, B_{2g} and A_{2u}. Moreover, the modes at this point have neither pure g nor pure u parity. This is one of the examples mentioned in section 4-5a in connection with the criterion $e^{i\bar{\kappa}_S \cdot \tau_R} = 1$ which fails in this instance, requiring that the representations be found with the aid of groups which are not necessarily isomorphic with point groups. This is the reason for the asterisk and footnote in Table 4-5. The appropriate representation theory for these cases is

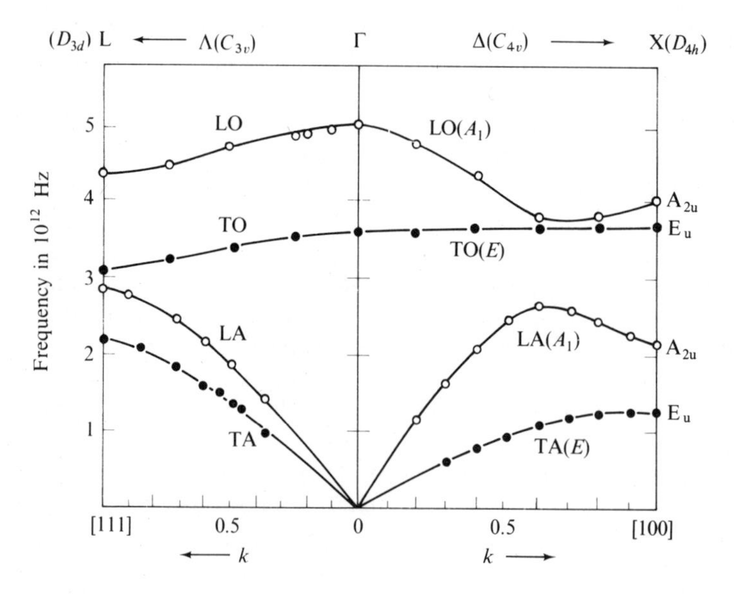

Figure 4-7(*a*) Dispersion curves for KBr at 90° K. k is in units of π/a. (After A. D. B. Woods, B. N. Brockhouse, R. A. Cowley, and W. Cochran, *Phys. Rev.*, **131**, 1025 (1963))

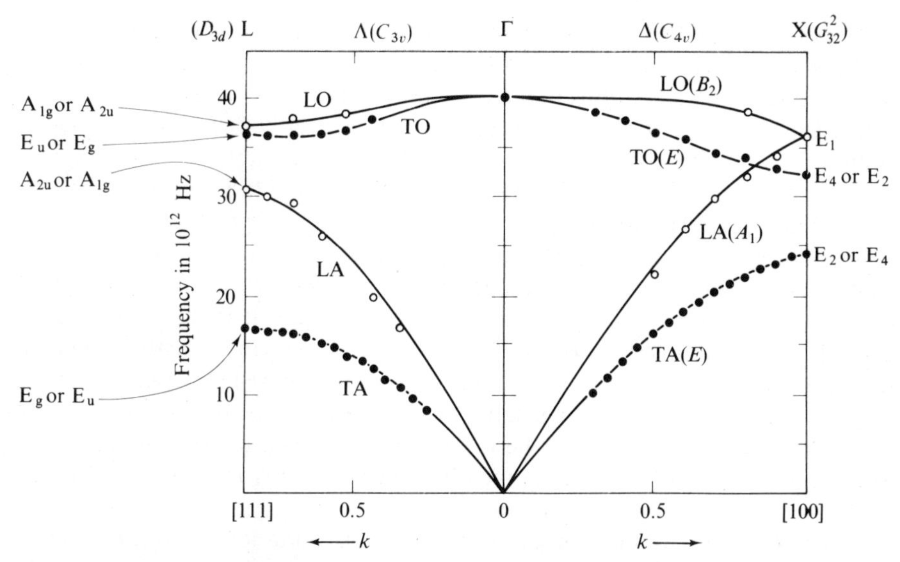

Figure 4-7(*b*) Dispersion curves for diamond. k is in units of π/a. (After J. L. Warren, R. G. Wenzel, and J. L. Yarnell, *Inelastic Scattering of Neutrons,* International Atomic Energy Agency, Vienna, 1965)

summarized in Appendix X, which shows, for example, that at the X point in diamond, the *irreducible* representations must be sought amongst those of a group of order 32 and that the physically relevant representations are in fact all doubly degenerate.

4-6 THE SECULAR DETERMINANT AS A FUNCTION OF κ

In this section, we shall extend the procedures of Chapter 3 to three-dimensional crystals. In Section 4-4 we developed the symmetry species at $k = 0$ of several types of cubic crystals composed of monatomic units, either atoms or ions, with the exception of the simple cubic (sc) structure O_h^1 ($Pm3m$). In all cases the dynamically primitive unit cell is a rhombohedron. Rhombohedral unit cells are so frequently encountered that it would be valuable to derive the secular determinant as a function of $\boldsymbol{\kappa}$ for the general rhombohedral structure with $n = 1$.* We shall study the feasibility of this goal in the cases of the sc, bcc, and fcc lattices, and comment on the extension of the treatment to non-cubic lattices.

* The only difference between the fcc, bcc, and sc rhombs is in the scalar products of the basis vectors. In each case it is convenient to choose the origin at a corner exhibiting threefold point symmetry. When this is done, the three scalar products $\mathbf{a}_i \cdot \mathbf{a}_j$, $i, j = 1, 2, 3$ are all equal. In the bcc rhomb $(\mathbf{a}_i \cdot \mathbf{a}_j)a^{-2} = -1/3$, so at the origin the angle between any two lattice vectors is tetrahedral. For fcc, the lattice vectors make 60° angles with each other at the origin. In sc, of course, this angle is 90°. In each case the $3n = 3$ displacement coordinates are vectors with origin as defined above and respectively parallel to the three lattice vectors.

The problem we are undertaking at this point is one of the classic exercises of solid state physics. Most of the early treatments have been based upon the search for plane-wave solutions of the equations of motion of the origin particle, either in the Newtonian or the Lagrangian form, and usually in vector-component notation.[21–24] The equations of motion include some assumption of the Hooke's law constants and each wave is described by a vector amplitude, a frequency and a propagation (wave) vector **k**.

More recently a matrix formalism has been adopted but, as with the earlier approaches, the displacement coordinates used are of the cartesian variety.[25,26,27] This has the advantage that the G matrix is very simple: it is the inverse mass matrix. Even in a lattice with two different kind of atoms, the G matrix element which connects any pair of atoms i and j is simply $(M_i M_j)^{-1/2}$. On the other hand, the analog of the force constant matrix, with elements usually denoted $\Phi_{\alpha\beta}(l\kappa; l'\kappa')$ in these treatments (where α denotes the α-cartesian component of the displacement of the κth atom in the lth unit cell), clearly requires much more detailed consideration in its evaluation.

It is of the nature of the internal coordinate representation, in contrast to the cartesian, to reverse the relative difficulty of evaluation of these two matrices. It has obviously seemed preferable in the past to throw the burden of the effort upon F instead of G. In molecular dynamics the reversal of this effort yields a net simplification of the formalism. We proceed similarly in this approach to lattice dynamics.

A similar simplification is realizable because in crystals as in molecules, displacement coordinates, even when translationally symmetrized at $k \neq 0$ (i.e., delocalized displacement coordinates) group themselves into equivalent sets. The set of (3) displacement coordinates in the basis of the dynamically primitive cell is necessarily kinematically complete. Further, none of the displacement coordinates needs to be an angle change. But as in the case of numerous molecular dynamic problems, a coordinate set which is kinematically complete is frequently not simultaneously symmetrically complete.*

21. M. Blackman, *Proc. Roy. Soc.*, **A148,** 384 (1935).
22. P. C. Fine, *Phys. Rev.*, **56,** 355 (1939).
23. E. Montroll, *J. Chem. Phys.*, **11,** 481 (1943).
24. R. B. Leighton, *Rev. Mod. Phys.*, **20,** 165 (1948).
25. G. Leibfried and W. Ludwig, *Solid State Physics,* F. Seitz and D. Turnbull, Eds., Academic Press, New York, 1961, Vol. 12, page 276.
26. A. A. Maradudin, E. W. Montroll, G. H. Weiss, and I. P. Ipatova, *Theory of Lattice Dynamics in the Harmonic Approximation,* Academic Press, New York, 1971. (Supplement 3 to *Solid State Physics,* H. Ehrenreich, F. Seitz, and D. Turnbull, Eds.)
27. J. L. Warren, *Rev. Mod. Phys.*, **40,** 38 (1968).
28. E. Wigner and F. Seitz, *Phys. Rev.*, **43,** 804 (1933).
29. J. C. Slater, *Quantum Theory of Molecules and Solids,* McGraw-Hill, New York, 1965, Vol. 2, pages 16–17.

* A coordinate basis is said to be kinematically complete when the rank of its kinetic energy matrix is equal to the number of degrees of freedom. A well-known example is that illustrated by the bending coordinates of methane (WDC, Sections 6-2 and 6-8). The analogous situation in solid state physics was first recognized by Wigner and Seitz.[28] The significance of their contribution in the present connection is thoroughly discussed by Slater.[29]

Symmetrical completeness in lattice dynamics is achieved by use of (4-5-6): If necessary, the set of three lattice displacement coordinates in the basis of the dynamically primitive cell is augmented by others, such that the new set is invariant under the operations of $H(\kappa)$. Sometimes it is also necessary to replace one or two members of the primitive set with others such that the prescribed invariance is satisfied. Thus kinematic completeness is still assured. The new, symmetrically complete cell defined by the new set will vary in its definition, depending on the direction of κ. The selection of the extra vectors must be done with considerable finesse and is based upon the concept of *orientation* (see WDC, Section 6-4). Obviously, the extra coordinates, when mixed with those of the dynamically primitive cell via the operations $\{R|\tau_R\}$ of $H(\kappa)$, will give rise to redundant coordinates as numerous as the coordinates added.

$H(\kappa)$ is always of highest order at Γ, when it equals G.* At that point the symmetrically complete cell is the so-called Wigner–Seitz cell.[30] Its boundary planes are the perpendicular bisectors of each "bond" connecting an atom with others in its first coordination shell. Accordingly, we might expect the number of redundant coordinates to always be three less than the coordination number. This point of view, however, has not properly accounted for the unique translational symmetry of $\kappa = 0$, which enables us to further subdivide each Wigner-Seitz cell into that defined by *half* the number of bonds leading from an atom to those in its first coordination sphere.† The omission of these extra coordinates can be done in a large number of ways, testifying to the extensive symmetry (O_h) at Γ; the manner in which this is done depends on the direction of κ which is to be investigated as we allow $|\kappa|$ to become finite. Further discussion of Γ is therefore postponed. We recognize, however, that at Γ, the redundant coordinates will always number three less than *half* the coordination number (3 in fcc, 1 in bcc, 0 in sc).

At $\kappa \neq 0$ there are circumstances when the redundancies can actually be further reduced. These circumstances occur whenever κ is a "special" point or site of the Brillouin Zone (BZ), either on a boundary plane or in the interior. In all cases, the first step in augmenting the primitive set of coordinates is to add sufficient new equivalent coordinates (see WDC, Section 6-4) such that the set can serve as a basis for a representation of $H(\kappa)$. In some cases, it may be necessary to add other coordinates, but in these cases these will form *another* set of equivalent coordinates $r_{j'}$. In all cases, the maximal number is that dictated at Γ. Thus one may apply the operations of $H(\kappa)$ to each member of the maximal set with the result that it will be resolved into *several* sets of equivalent coordinates, $(r_i, r_j, \ldots)$, $(r_{i'}, r_{j'}, \ldots)$, etc. Some of these sets may be related to others by the

30. E. Wigner and F. Seitz, *Phys. Rev.*, **43,** 804 (1933).

* This statement is only valid in *symmorphic crystals*, and even then there may be other points in the Brillouin zone, notably those at some of its boundaries, where $H(\kappa)$ has an order equal to that at Γ. In non-symmorphic crystals there may exist some $H(\kappa)$ whose order is formally greater than that at Γ. See Appendix X.

† This is a result of the fact that in these lattices (which lack bases) each internal coordinate is shared by *two* atoms.

inversion operation; we need not concern ourselves with these since they arise in the shared nature of each coordinate.*

Notwithstanding the fact that we may choose to omit members of the primitive set, each internal coordinate is still described in the "τ-system" (see Section 4-1). That is, each coordinate will be indexed $r_i(\tau)$, where τ is the lattice vector of the unit cell with which the internal coordinate is associated. Furthermore, it must be appreciated that each s_i is in fact a *delocalized* coordinate since it is implicit that each has already been translationally symmetrized, using (4-1-7).

If we temporarily assign six bond-stretching coordinates $r_1, r_2, \ldots, r_6$ to the Wigner–Seitz cell with $\mathbf{i} \cdot r_1 = r_4$, $\mathbf{i} \cdot r_2 = r_5$, $\mathbf{i} \cdot r_3 = r_6$, and then evaluate the $s_i(\boldsymbol{\kappa})$ we find

$$s_1(\boldsymbol{\kappa}) = s_4(\boldsymbol{\kappa}) \cdot e^{-i\kappa_1}$$

$$s_2(\boldsymbol{\kappa}) = s_5(\boldsymbol{\kappa}) \cdot e^{-i\kappa_2}$$

$$s_3(\boldsymbol{\kappa}) = s_6(\boldsymbol{\kappa}) \cdot e^{-i\kappa_3}$$

which shows that indeed only three coordinates are independent.

We now see that the coordinates, $\mathbf{i} \cdot r_i$, $i = 1, 2, 3$ are assigned to another atom, i.e., unit cell. Hence $H(\boldsymbol{\kappa})$ can never be C_i or another group containing C_i as a subgroup. This statement is valid throughout the interior of the Brillouin zone, but not at its origin or at the centers of certain boundary planes, namely, those which are separated from other, identical planes by $2\bar{\boldsymbol{\kappa}}$, where $\bar{\boldsymbol{\kappa}}$ is a vector of the reciprocal lattice. All operations $\boldsymbol{S}_{2n}$ ($\mathbf{i} \equiv \mathbf{S}_2$) take a direction of space into its negative. Γ can admit of $\mathbf{i}$ since at Γ, $\boldsymbol{\kappa} = 0$ and $\mathbf{i} \cdot s_i(0) = s_i(0)$, where $s_i(\boldsymbol{\kappa})$ was defined in (4-1-7). At $\boldsymbol{\kappa} = 0$, $\mathbf{i} \cdot s_i = s_{i+3}$, which is *translationally equivalent* to s_i. Hence $\mathbf{i}$ can be in G, but not in $H(\boldsymbol{\kappa})$. We shall discuss the special points on the surface of the BZ toward the end of this section.

We have already remarked about the internal coordinate representation with respect to the ease of construction of the F matrix. In general we begin with the assumption that F is exceedingly simple: we take it to be the constant matrix $f_0 E$, of dimension appropriate to the lattice (6×6 in fcc, 4×4 in bcc, 3×3 in sc). From the considerations of Section 4-1, we have the analog of (4-1-8):

$$G_{ij}(\boldsymbol{\kappa}) = N^{-1} \sum_{\Delta\tau} G_{ij}(\Delta\tau)\, e^{i\boldsymbol{\kappa} \cdot \Delta\tau} \tag{4-6-1}$$

Since all coordinates are of the bond-stretching variety, there can be only two kinds of elements G_{ij}, namely, G_{rr}^2 and G_{rr}^1 (see WDC, Appendix VI). To make use of (4-6-1) we must evaluate $\Delta\tau$. For each *diagonal* element of the G matrix, a given coordinate is coupled with itself (G_{rr}^2) plus those ahead and behind it (G_{rr}^1). Hence each of the components of $\Delta\tau$ can only differ by 0 or ± 1.

Because of this

$$\begin{aligned} G_{ii}(\boldsymbol{\kappa}) &= 2\mu - \mu(e^{i\kappa_i} + e^{i\kappa_i}) \\ &= 2\mu(1 - \cos \kappa_i) \end{aligned} \tag{4-6-2}$$

* This is again a result of the fact that each internal coordinate is shared by *two* atoms.

Owing to the relationship (4-1-3) (see footnote, page 103), the scalar product in a cartesian basis

$$\mathbf{k} \cdot \mathbf{t} = \mathbf{k}'\mathbf{t} = \boldsymbol{\kappa}'\mathbf{B}'\mathbf{A}\boldsymbol{\tau} = \boldsymbol{\kappa}'\mathbf{A}^{-1}\mathbf{A}\boldsymbol{\tau} = \boldsymbol{\kappa}'\boldsymbol{\tau} = \boldsymbol{\kappa} \cdot \boldsymbol{\tau} \tag{4-6-3}$$

is preserved upon transformation to a rhombohedral basis. Hence, no other components of $\boldsymbol{\kappa}$ enter into (4-6-2). (The reader should realize, however, that only in sc is $\boldsymbol{\kappa}_i$ parallel to $\mathbf{r}_i$.)

Contributing to each *off-diagonal* element of $G(\boldsymbol{\kappa})$ of an elementary lattice there are only four terms, since at each terminus of the coordinate $r_i(000)$ there are two representatives of the coordinate r_j which have an atom in common with r_i. If $r_i(000)$ extends from (000) to (ijk), the four terms are of the form

$$r_i(000)[r_j(000) + r_j(\overline{ijk}) + r_j(lmn) + r_j(l-i, m-j, n-k)]$$

where the notation indicates that the term $r_i(\tau)r_j(\tau')$ refers to $G_{ij}(\boldsymbol{\tau} - \boldsymbol{\tau}')$. Furthermore, because of the trigonometry of the rhombohedral lattice, the scalar products necessary in evaluating G_{rr}^1 at any bond terminus are supplementary. Thus each off-diagonal element of $G(\boldsymbol{\kappa})$ is given by

$$G_{ij}(\boldsymbol{\kappa}) = \mu \cos \phi_{ij}[1 - e^{-i\kappa_j} - e^{i\kappa_i} + e^{i(\kappa_i - \kappa_j)}] \tag{4-6-4}$$

where $\mathbf{r}_i(000) \cdot \mathbf{r}_j(000) = r^2 \cos \phi_{ij}$. Note that (4-6-2) is a special case of (4-6-4), valid for $\kappa_i = \kappa_j$ and $\phi_{ij} = 0$.

Should a member of one equivalent set $r_{j'}$ be perpendicular to $\boldsymbol{\kappa}$, all elements $G_{ij'}(\boldsymbol{\kappa})$ and $G_{j'j'}(\boldsymbol{\kappa})$ will vanish. This statement follows from the definition of equivalent coordinates and from inspection of (4-6-4) and (4-6-2). In this circumstance, the set $r_{j'}$ is obviously redundant, as are all linear (symmetrized) combinations $\sum_{j'} a_{i'j'} r_{j'}$. This conclusion is a general one and is valid for any primitive Bravais lattice. Its consequences in monatomic lattices will be examined below.

Since F is real and the secular determinant must have real roots, G is Hermitian:

$$G_{ij}(\boldsymbol{\kappa}) = G_{ji}^*(\boldsymbol{\kappa}) \tag{4-6-5}$$

The G matrix elements assume a simplified form for the particular $\boldsymbol{\kappa}$ which is parallel to an axis about which a rotation takes each member of the set r_i into a different member, such as the (111) direction in any rhombohedral lattice, provided the set r_i consists of nearest-neighbor bonds. This simplified form is

$$G_{ij}(\boldsymbol{\kappa}) = 2\mu \cos \phi_{ij}(1 - \cos \kappa) \tag{4-6-6}$$

where κ is one of the equal components of $\boldsymbol{\kappa}$ and $G_{ii}(\boldsymbol{\kappa})$ is obtained from the last equation when $\phi = 0$. Thus the G matrix is of the simple form

$$G(\boldsymbol{\kappa}) = 2\mu(1 - \cos \kappa) \begin{pmatrix} 1 & \cos \phi_{12} & \cos \phi_{13} \cdots \\ & 1 & \cos \phi_{23} \cdots \\ \text{(Symmetric)} & & 1. \\ & & & \ddots \end{pmatrix} \tag{4-6-7}$$

The value of ϕ, however, depends on the particular lattice under consideration. As a consequence of this form of the G matrix and the assumed constancy of F, the dispersion curve along $\boldsymbol{\kappa}$ will have only the simple periodicity dictated by the factor $(1 - \cos \kappa)$.

At $\boldsymbol{\kappa} = 0$ we notice that $G(\mathbf{0}) = 0$, as a consequence of the κ-dependent factor. In those cases where there still remain redundant columns (and rows) in G (i.e., when the G matrix has a dimension greater than three), we are unable to distinguish between the acoustic modes and the redundant coordinates at Γ, since the detailed information which distinguishes these modes and which is contained in the non-κ-dependent part of G is then no longer of any value.

Frequently, however, members of more than one equivalent set $r_{j'}$ are not orthogonal to the direction of $\boldsymbol{\kappa}$. In these cases, $G(\boldsymbol{\kappa})$ is necessarily of higher dimension than three. Then (4-6-4) applies and the extra columns and rows have other κ-dependence than that simply given by $(1 - \cos \kappa)$. The result is that some of the acoustic branches have a more complicated periodicity than that dictated by $(1 - \cos \kappa)$ and, as a further consequence, degeneracy lost at $\boldsymbol{\kappa} \neq 0$ is sometimes recovered at the Brillouin Zone Boundary (BZB) along $\boldsymbol{\kappa}$.

We proceed to demonstrate these generalizations using examples of $\boldsymbol{\kappa}$ along specific directions in sc, bcc, and fcc.

(a) The Simple Cubic (sc) Lattice

In sc these general procedures assume deceptively simple forms. The Wigner–Seitz cell is identical with its BZ (see Fig. IX-2). The rank of $G(\boldsymbol{\kappa})$ depends upon $\boldsymbol{\kappa}$, and for some $\boldsymbol{\kappa}$ will be less than three, as we shall see. For example, along (001), where $H(\boldsymbol{\kappa}) = C_{4v}$, there is only one set of equivalent coordinates whose solitary member is not orthogonal to the direction of $\boldsymbol{\kappa}$. We therefore directly conclude that

$$\omega_{A_1}^2(\Delta) = 2\mu f_0(1 - \cos \kappa) \tag{4-6-8}$$

and

$$\omega_E^2(\Delta) = 0 \tag{4-6-9}$$

which states that a three-dimensional lattice of the sc structure, with no angle-preserving forces could exhibit shear waves with zero restoring force. In other words, the unit cell was not kinematically complete and we must introduce at least one other set of equivalent coordinates. These can be either changes in the angles between the "bonds" connecting nearest-neighbor atoms or the stretchings of bonds between non-nearest neighbors. Obviously there is no benefit if the new set is in a plane perpendicular to (001). However, we can select another set of stretching coordinates which extend from any atom to its next-nearest neighbor in the plane *above* it along (001).* The new set, r_i', is another example of a set, the

* An equally good (and equivalent) set would be the four angle changes between $r(001)$ and, respectively, $r(100)$, $r(\bar{1}00)$, $r(010)$, and $r(0\bar{1}0)$.

members of which are taken into each other by a rotation about an axis parallel to $\boldsymbol{\kappa}$. As with the set r_i, we assign the subset $\mathbf{i}r_i'$ to other unit cells. The remaining bonds in this second coordination sphere are equatorial. They are therefore orthogonal to $\boldsymbol{\kappa}$ and are not needed for the discussion. The geometry of the two sets (the first is lonely) clearly illustrates $H(\boldsymbol{\kappa})$. The G matrix is then

$$G(\Delta) = \mu(1 - \cos\kappa)\begin{array}{c} \begin{matrix} r_1' & r_2' & r_3' & r_4' & r_3 \end{matrix} \\ \begin{pmatrix} 2 & 1 & 0 & 1 & \sqrt{2} \\ & 2 & 1 & 0 & \sqrt{2} \\ & & 2 & 1 & \sqrt{2} \\ & \text{(Symmetric)} & & 2 & \sqrt{2} \\ & & & & 2 \end{pmatrix} \end{array} \tag{4-6-10}$$

The new set of four members transforms as $A_1 + B_1 + E$ of C_{4v}, of which both A_1 and B_1 must be redundant, in view of (4-6-8). The lonely set belongs to A_1, hence the redundancy is mixed with a genuine acoustic mode. Equation (4-6-10) can be diagonalized by a matrix U constructed in the usual manner, by collecting the coefficients in (4-5-19), to wit

$$U = \frac{1}{2}\begin{pmatrix} 1 & 1 & 1 & 1 & 0 \\ 1 & -1 & 1 & -1 & 0 \\ \sqrt{2} & 0 & -\sqrt{2} & 0 & 0 \\ 0 & \sqrt{2} & 0 & -\sqrt{2} & 0 \\ 0 & 0 & 0 & 0 & 2 \end{pmatrix} \tag{4-6-11}$$

which diagonalizes G to

$$\mathsf{G}(\Delta) = UG(\Delta)U^{\dagger} = 2\mu(1 - \cos\kappa)\begin{pmatrix} 2 & \sqrt{2} & 0 & 0 & 0 \\ \sqrt{2} & 1 & 0 & 0 & 0 \\ 0 & 0 & 2 & 0 & 0 \\ 0 & 0 & 0 & 2 & 0 \\ 0 & 0 & 0 & 0 & 0 \end{pmatrix} \tag{4-6-12}$$

The redundancy of the B_1 coordinate is immediately revealed. However, another redundancy lies in the 2×2 (A_1) block; hence the determinant of that block must vanish. Expansion of the 2×2 secular determinant is equivalent to diagonalizing it to

$$\hat{\mathsf{G}}_{A_1}(\Delta) = 2\mu(1 - \cos\kappa)\begin{pmatrix} 3 & 0 \\ 0 & 0 \end{pmatrix} \tag{4-6-13}$$

which shows that one root of $(\mathsf{FG})_{A_1}$ must vanish.

The introduction of the second set of equivalent coordinates r_i' necessitates revision of the F matrix. With the same ordering of the r_i and r_i' as was used in composing (4-6-10), and with the introduction of all possible (quadratic) cross terms in the potential energy, we obtain

$$F(\Delta) = \begin{pmatrix} f & f' & f'' & f' & f''' \\ & f & f' & f'' & f''' \\ & & f & f' & f''' \\ & \text{(Symmetric)} & & f & f''' \\ & & & & f_0 \end{pmatrix} \tag{4-6-14}$$

Symmetrization of (4-6-14) with (4-6-11) gives

$$\mathsf{F}_{A_1}(\Delta) = \begin{pmatrix} f + 2f' & 2f''' \\ 2f''' & f_0 \end{pmatrix} \tag{4-6-15}$$

$$\mathsf{F}_{B_1}(\Delta) = f - 2f' + f'' \tag{4-6-16}$$

and

$$\mathsf{F}_E(\Delta) = \begin{pmatrix} f - f'' & 0 \\ 0 & f - f'' \end{pmatrix} \tag{4-6-17}$$

Solution of the $2 \times 2\ (\mathsf{FG})_{A_1}$ yields

$$\omega_{A_1}^2 = 2\mu F(1 - \cos\kappa) \tag{4-6-18}$$

where

$$F = f_0 + 4\sqrt{2}f''' + 2(f + 2f') \tag{4-6-19}$$

From the E factor of the secular determinant one finds

$$\omega_E^2(\Delta) = 4\mu(f - f'')(1 - \cos\kappa) \tag{4-6-20}$$

in contrast to (4-6-9). We note the persistent lack of dependence of $\omega_E^2(\Delta)$ upon f_0.

In monatomic lattices, all modes are of course acoustic in character. Since in the A_1 mode the resultant motion is parallel to **k** and in the E mode it is transverse to **k**, these modes are often referred to as *LA* (longitudinal acoustic) and *TA* (transverse acoustic), respectively.

Other directions in the BZ of sc will demonstrate few additional principles; it is more instructive to investigate the more common structures of elementary solids, fcc and bcc.

(b) The Face-Centered Cubic (fcc) Lattice

We first consider $\boldsymbol{\kappa}$ parallel to (111)—i.e., along Λ. The specification of the direction of **k** is usually stated in a cartesian basis (in this case, (111)). Now, in a matrix notation, the transformation between **t** and τ in fcc is stated

$$\mathbf{t} = \frac{a}{2} \cdot \begin{pmatrix} 0 & 1 & 1 \\ 1 & 0 & 1 \\ 1 & 1 & 0 \end{pmatrix} \tau \tag{4-6-21}$$

while in bcc

$$\mathbf{t} = \frac{a}{2} \cdot \begin{pmatrix} -1 & 1 & 1 \\ 1 & -1 & 1 \\ 1 & 1 & -1 \end{pmatrix} \tau \tag{4-6-22}$$

Because of the reciprocal nature of the two lattices discussed in Section 4-3, the transformation matrix in (4-6-21) takes $\mathbf{k}$ into $\boldsymbol{\kappa}$. Hence, if $\mathbf{k}$ parallels (111), so does $\boldsymbol{\kappa}$. Here $H(\boldsymbol{\kappa}) = C_{3v}$.

The set r_i is disposed along (011), (101), and (110) of the cube. Applying (4-6-22) these are respectively the (100), (010), and (001) directions in the rhombohedral basis. Hence (4-6-7) applies with all $\cos \phi_{ij} = \frac{1}{2}$. Hence

$$G(\Lambda) = \mu(1 - \cos \kappa) \cdot \begin{pmatrix} 2 & 1 & 1 \\ 1 & 2 & 1 \\ 1 & 1 & 2 \end{pmatrix} \tag{4-6-23}$$

$G(\Lambda)$ can be diagonalized with a matrix constructed in the usual manner by collecting the coefficients in (4-5-19), i.e.,

$$U = \frac{1}{\sqrt{6}} \begin{pmatrix} \sqrt{2} & \sqrt{2} & \sqrt{2} \\ 2 & -1 & -1 \\ 0 & \sqrt{3} & -\sqrt{3} \end{pmatrix} \tag{4-6-24}$$

If $F = f_0 E$, the roots of the secular determinant are found to be

$$\omega_{A_1}^2(\Lambda) = 4\mu f_0(1 - \cos \kappa) \tag{4-6-25}$$

and

$$\omega_E^2(\Lambda) = \mu f_0(1 - \cos \kappa) \tag{4-6-26}$$

Note that the LA mode is always higher than the TA.

At the BZB along Λ, the appropriate symmetrically complete unit cell is that portion of the Wigner–Seitz cell which is a trigonal bipyramid. At the boundary, $H(\boldsymbol{\kappa}) = D_{3d}$ admits of $\mathbf{i}$ since $\mathbf{i} \cdot \boldsymbol{\kappa} = -\boldsymbol{\kappa} = \boldsymbol{\kappa} + \bar{\boldsymbol{\kappa}}$. The solutions of the secular determinant are still degenerate with respect to $\pm\boldsymbol{\kappa}$, but $\pm\boldsymbol{\kappa}$ mean the same thing. Hence we can continue to use the C_{3v} pyramid of r_i as a unit cell of sufficient symmetrical completeness. Accordingly, we may find the roots at the boundary along Λ (L in the BSW notation) by simply allowing $\kappa = \kappa_1 = \kappa_2 = \kappa_3 = \pi$ whence

$$\omega_{A_{1u}}^2(L) = 8\mu f_0 \tag{4-6-27}$$

$$\omega_{E_u}^2(L) = 2\mu f_0 \tag{4-6-28}$$

Equations (4-6-25) to (4-6-28) are in complete agreement with the results obtained in those calculations based upon a cartesian displacement representation.[31,32] Note that $LA > TA$ throughout the zone.

Our next example is of **k** along (001) in fcc, $H(\Delta) = C_{4v}$. Using (4-6-21), we find $\boldsymbol{\kappa} = \kappa(110)$ in the rhombohedral basis. The set r_i has four members disposed respectively along (101), ($\bar{1}$01), (011), and (0$\bar{1}$1) of the cube. Using (4-6-22) these directions are respectively the same as (010), (10$\bar{1}$), (100), and (01$\bar{1}$) in the rhombohedral basis, the consequence of which is that all κ_i, $\kappa_j = \kappa$ in (4-6-4). As the set is an equivalent one, (4-6-7) applies with $\cos\phi_{i,i+1} = \frac{1}{2}$ and $\cos\phi_{i,i+2} = 0$, where $r_{i+1} = C_4 r_i$ and $r_{i+2} = C_2 r_i$. Thus

$$G(\Delta) = \mu(1 - \cos\kappa)\begin{pmatrix} 2 & 1 & 0 & 1 \\ & 2 & 1 & 0 \\ \multicolumn{2}{c}{\text{(Symmetric)}} & 2 & 1 \\ & & & 2 \end{pmatrix} \tag{4-6-29}$$

Since the set r_i has four members, the secular determinant will have one zero root, corresponding to the redundant coordinate. The species of this mode (as well as of the acoustic modes) at Δ must be found in order to diagonalize (4-6-29). One procedure is as follows.

The reducible representation with the coordinate set consisting of the four r_i can be decomposed in C_{4v} to $A_1 + B_1 + E$. The cartesian representation at Γ of the three degrees of freedom gives F_{1u} of O_h. Correlation from O_h to C_{4v} gives rise to A_1 and E species, from which we conclude that B_1 is the redundancy. The transformation matrix is then

$$U = \frac{1}{2}\begin{pmatrix} 1 & 1 & 1 & 1 \\ 1 & -1 & 1 & -1 \\ \sqrt{2} & 0 & -\sqrt{2} & 0 \\ 0 & \sqrt{2} & 0 & -\sqrt{2} \end{pmatrix} \tag{4-6-30}$$

Again we let $F = f_0 E$ and the roots of the secular determinant are

$$\omega_{A_1}^2(\Delta) = 4\mu f_0(1 - \cos\kappa) \tag{4-6-31}$$

$$\omega_{B_1}^2(\Delta) = 0 \tag{4-6-32}$$

$$\omega_E^2(\Delta) = 2\mu f_0(1 - \cos\kappa) \tag{4-6-33}$$

As expected, it is the B_1 mode which is redundant.

At the BZB along Δ, $H(X)$ is D_{4h}. The symmetrically complete cell is a tetragonal bipyramid. For the same reasons as were discussed with regard to the L point, we can continue to use the C_{4v} pyramid of the r_i at Δ, even when X is

31. J. de Launay, *Solid State Physics*, F. Seitz and D. Turnbull, Eds., Academic Press, New York, 1956, Vol. 2, page 220.
32. G. Leibfried and W. Ludwig, *Solid State Physics*, F. Seitz and D. Turnbull, Eds., Academic Press, New York, 1961, Vol. 12, page 276.

reached. The results are found from (4-6-31) and (4-6-33) by setting $\kappa = \pi$:

$$\omega^2_{A_{1g}}(X) = 8\mu f_0 \tag{4-6-34}$$

$$\omega^2_{E_g}(X) = 4\mu f_0 \tag{4-6-35}$$

Again, $LA > TA$, throughout the zone along Δ.

(c) The Body-Centered Cubic (bcc) Lattice

We shall discuss only two directions of $\boldsymbol{\kappa}$ in the bcc lattice. The first will be along, say, the (100) axis of the cube ((1̄11) in the rhombohedral basis). $H(\boldsymbol{\kappa})$ is again C_{4v}; the direction is called Δ in the BSW notation. There is only one set of equivalent coordinates r_i and it is composed of (111), (111̄), (11̄1̄), and (1̄11) in the cartesian basis or of (111), (001), (1̄00), and (010) in the rhombohedral basis. Equation (4-6-7) becomes

$$G(\Delta) = 2\mu(1 - \cos\kappa)\cdot\begin{pmatrix} 1 & \frac{1}{3} & -\frac{1}{3} & \frac{1}{3} \\ & 1 & \frac{1}{3} & -\frac{1}{3} \\ & & 1 & \frac{1}{3} \\ \text{(Symmetric)} & & & 1 \end{pmatrix} \tag{4-6-36}$$

The four coordinates belong to $A_1 + B_1 + E$ of C_{4v}, so that the redundancy lies in B_1. The transformation matrix is

$$U = \frac{1}{2}\begin{pmatrix} 1 & 1 & 1 & 1 \\ 1 & -1 & 1 & -1 \\ 1 & -1 & -1 & 1 \\ 1 & 1 & -1 & -1 \end{pmatrix} \tag{4-6-37}$$

With only a diagonal F matrix, the allowed solutions are

$$\omega^2_{A_1}(\Delta) = \tfrac{8}{3}\mu f_0(1 - \cos\kappa) \tag{4-6-38}$$

$$\omega^2_{B_1}(\Delta) = 0$$

$$\omega^2_{E}(\Delta) = \tfrac{8}{3}\mu f_0(1 - \cos\kappa) \tag{4-6-39}$$

The result belies the expectation of one A_1 and two E (finite) roots. Rather, it states that Δ has at least tetrahedral symmetry! The trouble is in the assumption of a diagonal force field, i.e., one of cubic symmetry. We shall therefore modify F to include interaction constants, with the result that the resulting GF matrix will be seen to factor in the expected way.

Since there is only one set of equivalent coordinates r_i, $F(\Delta)$ will manifest the same symmetry among its off-diagonal elements as did $G(\kappa)$ (see (4-6-7)). We may derive $F(\Delta)$ by replacing $\cos\phi_{ij}$ with the appropriate interaction constant. Thus

$$\cos\phi_{ii} = 1 \rightarrow f'$$

$$\cos\phi_{ij} = \tfrac{1}{3} \rightarrow f''$$

and

$$\cos \phi_{ij+1} = -\tfrac{1}{3} \rightarrow f'''$$

which give

$$F(\Delta) = \begin{pmatrix} f_0 + 2f' \cos \kappa & 2(f'' + f''' \cos \kappa) & 2(f''' + f'' \cos \kappa) & 2(f'' + f''' \cos \kappa) \\ & f_0 + 2f' \cos \kappa & 2(f'' + f''' \cos \kappa) & 2(f''' + f'' \cos \kappa) \\ & & f_0 + 2f' \cos \kappa & 2(f'' + f''' \cos \kappa) \\ \text{(Symmetric)} & & & f_0 + 2f' \cos \kappa \end{pmatrix}$$

(4-6-40)

which is diagonalized by (4-6-37) to give

$$\mathsf{F}(\Delta) = \begin{pmatrix} f_0 + 4f'' + 2f''' + 2(f' + f'' + 2f''') \cos \kappa & & & \\ & f_0 - 4f'' + 2f''' + 2(f' + f'' - 2f''') \cos \kappa & & \\ & & f_0 - 2f''' + 2(f' - f'') \cos \kappa & \\ & & & f_0 - 2f''' + 2(f' - f'') \cos \kappa \end{pmatrix} \quad (4\text{-}6\text{-}41)$$

Substituting $\mathsf{F}_{11}(\Delta)$ and $\mathsf{F}_{33}(\Delta)$ for f_0 in (4-6-38) and (4-6-39) respectively we find the refined solutions

$$\omega_{A_1}^2(\Delta) = \tfrac{8}{3}\mu[f_0 + 4f'' + 2f''' + 2(f' + f'' + 2f''') \cos \kappa](1 - \cos \kappa) \quad (4\text{-}6\text{-}42)$$

and

$$\omega_E^2(\Delta) = \tfrac{8}{3}\mu[f_0 - 2f''' + 2(f' - f'') \cos \kappa](1 - \cos \kappa) \quad (4\text{-}6\text{-}43)$$

Thus the introduction of interaction force constants lifts the degeneracy indicated when a diagonal force field was used. Note that f' plays no role in this splitting. Examination of (4-6-40) shows it to be a diagonal F matrix element; hence, it only *shifts* the roots at each value of κ.

Although the diagonal force field led to the inappropriate solutions (4-6-38) and (4-6-39), we may still profit from another aspect of their comparison with the refined solutions (4-6-42) and (4-6-43). The first solutions are simply periodic while the second pair of roots depend, in part, on $\cos^2 \kappa$ (i.e., on $\cos 2\kappa$). Note that this dependence arises with the introduction of interaction force constants. There is a parallel situation in the cartesian representation in which it may be shown[33] that along a direction such as (100) of bcc

$$\omega_{A_1}^2(\Delta) = \mu \sum_n \Phi_n(1 - \cos n\kappa)$$

where Φ_n is the nth nearest neighbor central force constant.* Thus the presence of higher-order periodicity in a dispersion curve betrays the importance of interaction or non-first-nearest neighbor force constants.

33. B. N. Brockhouse, in *Phonons and Phonon Interactions*, T. A. Bak, Ed., W. A. Benjamin, Inc., New York, 1964, page 264.

* This kind of dependence in linear lattices was discussed in Section 3-2 (see (3-2-12)).

The intersection of Δ at the BZB is the point H, one of O_h symmetry. Examination of the shape of the entire BZ shows that for all $+\boldsymbol{\kappa}$ solutions along Δ (*or* for all $-\boldsymbol{\kappa}$ solutions, which join with those of $+\boldsymbol{\kappa}$ at H), a tetragonal pyramid suffices as a symmetrically complete unit cell. Hence we may let $\kappa \to \pi$ in (4-6-42) and (4-6-43) and obtain

$$\omega_{F_{1u}}^2(H) = \tfrac{16}{3}\mu[f_0 - 2(f' - f'' + f''')] \tag{4-6-44}$$

which shows that the threefold degeneracy *is* valid at H, even with consideration of interaction constants. The *form* of the two modes may be found by using (4-6-37) and remembering the phase change (of π) at each successive atom (unit cell). Figure 4-8 illustrates the result. In (*a*) the planes perpendicular to (100) appear to be alternately compression and expansion waves moving parallel to (100), while in (*b*) the same planes seem to be shear waves moving perpendicular to (100). Another view of (*b*), however, is that it is just (*a*) rotated by 90°; hence, the modes must be degenerate. Clearly they are *u*-type modes, as required by correlation Δ with H.

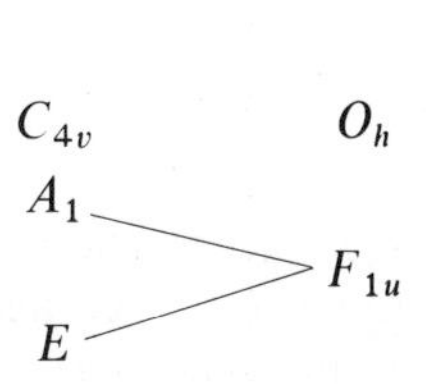

Without making any special assumptions about the relative magnitudes of f'' and f''', it is impossible to say which branch is higher, even near $\kappa = 0$. However, they must join at $\kappa = \pi$. Hence, one or both must depart from the simply periodic functional dependence on κ given by the factor $(1 - \cos\kappa)$. This is seen in (4-6-42) and (4-6-43). If by experiment we could determine the two dispersion curves, we could determine the four parameters $f_0 + 4f'' + 2f'''$, $f' + f'' + 2f'''$, $f_0 - 2f'''$ and $f' - f''$ and the four force constants would then be known.

The conclusions we have reached concerning the roots of the secular determinant when $\mathbf{k}$ is along the (100) direction of the cube are identical with those obtained in the calculations based upon a cartesian representation, but offer more

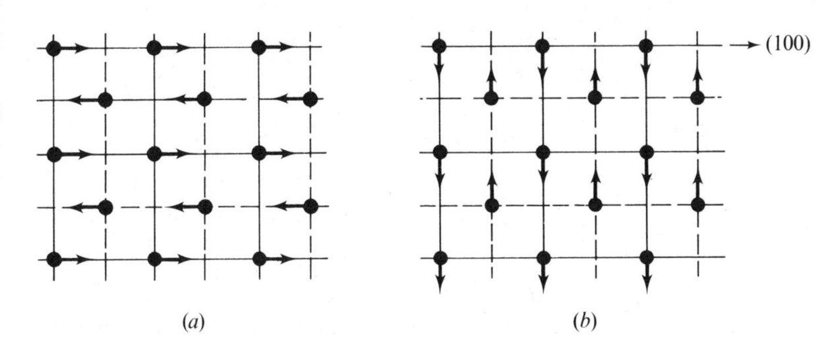

Figure 4-8 The form of the LA and TA modes of a bcc monatomic lattice at $\boldsymbol{\kappa} = (\bar{\pi}\pi\pi)$: (*a*) The longitudinal mode; (*b*) The transverse mode

detail, since that model uses only a two-parameter potential energy function, based upon a nearest-neighbor and next-to-nearest neighbor central force constants.

We conclude our discussion of bcc by considering the (111) direction of the cube ((111) in the κ-system as well). In the BSW notation, this direction is denoted Λ and its boundary on the surface of the BZ is P. $H(\Lambda) = C_{3v}$ and $H(P) = T_d$. Along Λ, there are two sets of equivalent coordinates, one consisting of the three cartesian-based vectors $r_1 = (\bar{1}11)$, $r_2 = (1\bar{1}1)$, and $r_3 = (11\bar{1})$, which form a trigonal pyramid and another, lonely set consisting of $r_4 = (\overline{111})$ by itself. Neither set is orthogonal to $\boldsymbol{\kappa}$. Hence $G(\boldsymbol{\kappa})$ is 4×4 and is found to be (taking $\kappa_1 = \kappa_2 = \kappa_3 = \kappa$; $\kappa_4 = -3\kappa$)

$$G(\Lambda) = 2\mu \begin{pmatrix} 1 - \cos\kappa & -\frac{1}{3}(1 - \cos\kappa) & -\frac{1}{3}(1 - \cos\kappa) & \frac{1}{3}(\cos\kappa - \cos 2\kappa)\, e^{2i\kappa} \\ & 1 - \cos\kappa & -\frac{1}{3}(1 - \cos\kappa) & \frac{1}{3}(\cos\kappa - \cos 2\kappa)\, e^{2i\kappa} \\ & & 1 - \cos\kappa & \frac{1}{3}(\cos\kappa - \cos 2\kappa)\, e^{2i\kappa} \\ & \text{(H.S.)} & & 1 - \cos 3\kappa \end{pmatrix} \tag{4-6-45}$$

The four-component basis $(r_1 r_2 r_3 r_4)$ transforms as $2A_1 + E$ of C_{3v} and the redundancy condition clearly belongs to A_1. A diagonalizing matrix U, constructed from the coefficients of (4-5-6), will factor $G(\Lambda)$ into one 2×2 and two 1×1 sub-matrices, the 2×2 corresponding to the redundancy and the genuine, longitudinal acoustical mode, which are mixed. This situation parallels that encountered in section 4-6(a) (($\kappa\kappa\kappa$) of sc). Using

$$U = 12^{-1/2} \begin{pmatrix} \sqrt{3} & \sqrt{3} & \sqrt{3} & \sqrt{3} \\ -1 & -1 & -1 & 3 \\ -2\sqrt{2} & \sqrt{2} & \sqrt{2} & 0 \\ 0 & \sqrt{6} & -\sqrt{6} & 0 \end{pmatrix} \tag{4-6-46}$$

we obtain

$$G(\Lambda) = \frac{\mu}{6} \left(\begin{array}{c:c:c:c} 3(A + 2B\cos C + D) & \sqrt{3}[-A + B(3\, e^{iC} - e^{-iC}) + 3D] & & \\ & A - 6B\cos C + 9D & & \\ & & 16A & \\ \text{(H.S.)} & & & 16A \end{array}\right) \tag{4-6-47}$$

in which $A = (1 - \cos\kappa)$, $B = (\cos\kappa - \cos 2\kappa)$, $C = 2\kappa$ and $D = 1 - \cos 3\kappa$. Since the redundancy lies in the 2×2 block, the determinant of that block must vanish; indeed, a little algebra will show that it does. Hence a unitary transformation exists which will diagonalize the A_1 block. The transformation matrix can be found by the use of first-order perturbation theory for degenerate states. In carrying out this procedure, since the off-diagonal element of the 2×2 matrix to be transformed is complex, a complex phase factor will be introduced into the statement of the *perturbed* coordinates, i.e.,

$$\begin{aligned} \mathscr{S}_1 &= aS_1 + e^{i\varphi}\, bS_2 \\ \mathscr{S}_2 &= -e^{-i\varphi}\, bS_1 + aS_2 \end{aligned} \tag{4-6-48}$$

where

$$\begin{aligned} S_1 &= r_1 + r_2 + r_3 + r_4 \\ S_2 &= 3r_4 - (r_1 + r_2 + r_3) \end{aligned} \tag{4-6-49}$$

$$a, b = \left(\frac{\Delta \pm \delta}{2\Delta}\right)^{1/2}, \text{ taken positive,} \tag{4-6-50}$$

$$\delta = \mathsf{G}_{11} - \mathsf{G}_{22}, \tag{4-6-51}$$

and

$$\Delta^2 = \delta^2 + 4|\mathsf{G}_{12}|^2 \tag{4-6-52}$$

The phase φ can be determined, within a choice of sign, by the requirement that the off-diagonal elements of $V\mathsf{G}V^\dagger$ must vanish, where V, the matrix of the transformation, is that of the coefficients in (4-6-48). The diagonal G matrix elements, associated with $\mathscr{S}_1$ and $\mathscr{S}_2$, are

$$\tilde{G}_{1,2} = \tfrac{1}{2}\{\mathsf{G}_{11} + \mathsf{G}_{22} \pm \Delta\} \tag{4-6-53}$$

However, when G_{12} is complex and its dependence on κ is non-trivial, the calculation of Δ (4-6-52) is tedious. As in the analogous problem in molecular dynamics (see WDC, Section 6-8), one can *arbitrarily* set *either* S_1 *or* S_2 equal to zero; in other words, *either* row (and associated column) of the 2×2 block of $\mathsf{G}(\Lambda)$ can be set equal to zero. Were $F(\Lambda)$ not diagonal, the resulting solutions of the secular determinant would contain more parameters (f', f'', etc.) than observables, and associated with the arbitrary omission of a particular row and column of G there would be an implied conventional relationship among the parameters. When F is diagonal, this problem does not arise.

We also seek continuity in the roots at the BZB, which imposes another requirement upon the selection of the row and column of G which are to be omitted—i.e., these should be the row and column which are to be associated with the redundant coordinate at P. These may not be the row and column which are associated with the redundant coordinate at Γ. Since, however, $G(\Gamma)$ identically vanishes, we have no way of making that association.

At P, where $H(\boldsymbol{\kappa}) = T_d$ it is clear that (4-5-19) with coefficients from the first row of (4-6-46) is the redundant coordinate; hence we omit the *first* row and column of (4-6-47). With a diagonal and constant F matrix, the genuine roots are

$$\omega^2_{A_1}(\Lambda) = \frac{\mu f_0}{6}\,[(1 - \cos\kappa) - 6(\cos\kappa - \cos 2\kappa)\cos 2\kappa + 9(1 - \cos 3\kappa)] \tag{4-6-54}$$

and

$$\omega^2_E(\Lambda) = \tfrac{8}{3}\mu f_0(1 - \cos\kappa) \tag{4-6-55}$$

In the vicinity of $\kappa = 0, \omega_E^2(\Lambda) - \omega_{A_1}^2(\Lambda)$ is dominated by the term $9(1 - \cos 3\kappa)$, which is the negative. Hence to begin with, $LA > TA$, as in both fcc examples. However, in some similarity to the last example (Δ in bcc), the terms of the extra periodicity in $\omega_{A_1}^2(\Delta)$ serve to make the two roots approach each other at the BZB. That they do join can be seen by evaluating (4-6-47) at P. The result is

$$\mathsf{G}(P) = \frac{16\mu}{3} \begin{pmatrix} 0 & 0 & 0 & 0 \\ 0 & 1 & 0 & 0 \\ 0 & 0 & 1 & 0 \\ 0 & 0 & 0 & 1 \end{pmatrix} \tag{4-6-56}$$

which also demonstrates the choice of the redundant row in (4-6-47). The threefold degenerate root is

$$\omega_{F_2}^2(P) = \frac{16\mu f_0}{3} \tag{4-6-57}$$

in essential agreement with that obtained in the cartesian representation, called there $16\mu\alpha_1/3$, where α_1 is the nearest-neighbor central force constant.[34] The numerical correspondence is no accident: α_1 and f_0 are the same parameters.

There remains only one principal direction in the interior of the bcc BZ, namely $\Sigma(C_{2v})$ (see Fig. 4-3(a)). The simplicity of $G(\kappa)$ in a direction such as this has already been indicated. Along the same direction of fcc, the same simplicity is not encountered but there is a close analog with Λ of bcc. Symmetrizations along Σ in both lattices are left as exercises for the reader.

Earlier in this section we postponed certain general remarks concerning the inclusion of **i** in $H(\boldsymbol{\kappa})$ at the BZB. We now recall to the reader the discussion of the Δ direction in bcc and its intersection with the surface of the BZ at H. Along Δ, $H(\boldsymbol{\kappa}) = C_{3v}$; at H it is O_h. We know that at Γ, the acoustic modes must belong to F_{1u} of O_h. Upon leaving the origin, they lose their *ungerade* character, but this is restored at H. Correlation tables (Appendix IV) will sometimes be ambiguous in these cases; their careful use is required. For example, in this case the tables read

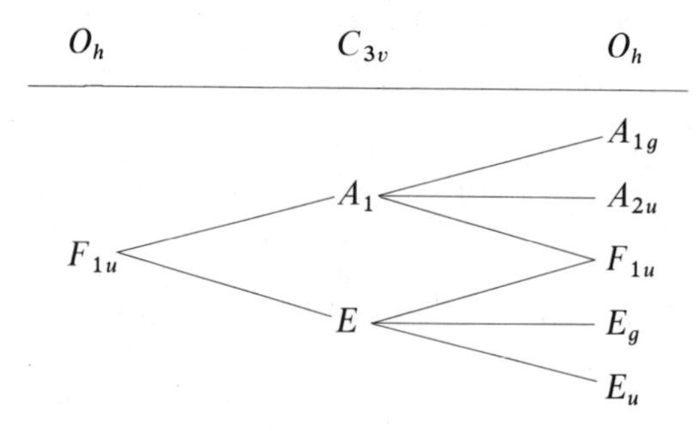

34. J. de Launay, *Solid State Physics*, F. Seitz and D. Turnbull, Eds., Academic Press, New York, 1956, Vol. 2, page 220.

In this use of correlation, however, we do not seek multiplicity; rather, conservation is the rule. In any case, if the reader is in doubt, he may construct and inspect an illustration of the mode at the BZB (as we did in Fig. 4-8).

(d) A Lattice with a Basis; CsCl

Throughout this section we have illuminated a number of the intricacies which characterize the construction of the secular determinant as a function of $\boldsymbol{\kappa}$ by a series of examples, all of which have dealt with crystals with one atom/cell, so that only acoustic modes were present. We now give an example of a lattice with a basis, CsCl, so that one should expect optic branches as well.

CsCl belongs to space group O_h^1 ($Pm3m$). The structure is best visualized as one consisting of two interpenetrating sc lattices, with one set of ions at the corner of the cube and the other at the body center. As such, the structure strongly resembles the elementary bcc, except that the centered ion is different from that at the cube corner. Since there are two atoms/cell, $f = 6$ at each $\boldsymbol{\kappa}$. At Γ, the symmetrically complete cell is precisely the octahedral XY_8 molecule and

$$\Gamma_{r_i} = A_{1g} + A_{2u} + F_{2g} + F_{1u} \tag{4-6-58}$$

Clearly the optic branches belong to F_{1u}; however, since all other modes (including the redundant ones) are non-genuine, we cannot be any more specific. Along Δ (001), $H(\boldsymbol{\kappa}) = C_{4v}$ and

$$\Gamma_{r_i} = 2(A_1 + B_2 + E) \tag{4-6-59}$$

in which we can say with certainty that both the optic and the acoustic branches belong to $A_1 + E$, while the redundancies transform as B_2. Furthermore, one of these must have the same form as the A_{2u} mode at Γ, whereby we have been just a little more specific about the form of a non-genuine zone origin mode.

Orientation (along Δ) requires that we choose *two* sets of equivalent coordinates s_i and s_j, which together form the tetragonal bipyramidal XY_8 "molecule" illustrated in Fig. 4-9 on page 150.* This and (4-6-59) determine the U matrix; it is

$$U = \frac{1}{2}\begin{pmatrix} 1 & 1 & 1 & 1 & 0 & 0 & 0 & 0 \\ 0 & 0 & 0 & 0 & 1 & 1 & 1 & 1 \\ 1 & -1 & 1 & -1 & 0 & 0 & 0 & 0 \\ 0 & 0 & 0 & 0 & 1 & -1 & 1 & -1 \\ 1 & 1 & -1 & -1 & 0 & 0 & 0 & 0 \\ 0 & 0 & 0 & 0 & 1 & 1 & -1 & -1 \\ 1 & -1 & -1 & 1 & 0 & 0 & 0 & 0 \\ 0 & 0 & 0 & 0 & 1 & -1 & -1 & 1 \end{pmatrix} \tag{4-6-60}$$

* It is essential to recall that the s_i and s_j are *delocalized* coordinates. That is, they have already been translationally symmetrized according to (4-1-7). Hence the XY_8 figure is repeated along (001) but with each translation the figure is multiplied by the phase factor, $e^{i\kappa_3 t_3}$. This is what is meant by the statement that s_i and s_j form Fig. 4-9.

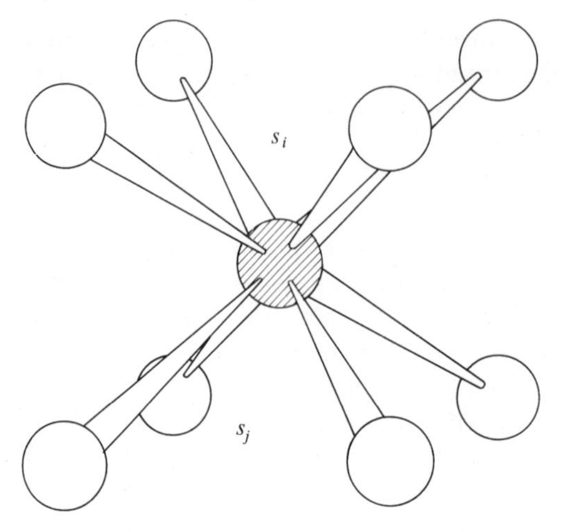

Figure 4-9 "Orientation" of translationally symmetrized coordinates along $\Delta(001)$ in CsCl.

Using U we can directly write the form of the submatrices of $G(\Delta)$, factored under $H(\kappa) = C_{4v}$.

$$\mathsf{G}_{A_1}(\Delta) = \begin{pmatrix} G_{11} + G_{12} + G_{13} + G_{14} & G_{15} + G_{25} + G_{35} + G_{45} \\ (\text{H.S.}) & G_{55} + G_{56} + G_{57} + G_{58} \end{pmatrix} \tag{4-6-61}$$

$$\mathsf{G}_{B_2}(\Delta) = \begin{pmatrix} G_{11} - G_{12} + G_{13} - G_{14} & G_{15} - G_{25} + G_{35} - G_{45} \\ (\text{H.S.}) & G_{55} - G_{56} + G_{57} - G_{58} \end{pmatrix} \tag{4-6-62}$$

$$\mathsf{G}_{E}(\Delta) = \begin{pmatrix} G_{11} + G_{12} - G_{13} - G_{14} & G_{15} + G_{25} - G_{35} - G_{45} \\ (\text{H.S.}) & G_{55} + G_{56} - G_{57} - G_{58} \end{pmatrix} \tag{4-6-63}$$

Calculation of the individual G_{ii} and G_{ij} is straightforward but we note that the effect of the inversion operator upon a bond stretching coordinate yields another coordinate in the same cell rather than the same coordinate in a different cell, as it does in the elementary lattices. Hence any off-diagonal element G_{ij} has only *two* contributions—one from the origin cell plus one from a neighboring cell. Similarly, a diagonal element G_{ii} has only one term, of the form $G_{rr}^2 = \mu_a + \mu_b$, where a and b are the two unlike atoms or ions. Along Δ (001), $\kappa_1 = \kappa_2 = 0$; $\kappa_3 = \kappa$. The several G_{ij} are:

$$G_{12} = G_{14} = G_{56} = G_{58} = -G_{13} = -G_{57} = \tfrac{1}{3}(\mu_a + \mu_b) \tag{4-6-64}$$

and

$$G_{25} = G_{45} = -G_{15} = \tfrac{1}{3}G_{35} = -\tfrac{1}{3}(\mu_a + \mu_b\, e^{i\kappa}) \tag{4-6-65}$$

Substitution of (4-6-64) and (4-6-65) in (4-6-61) through (4-6-63) yields

$$\mathsf{G}_{A_1}(\Delta) = \begin{pmatrix} G_{11} + G_{14} & G_{15} + G_{35} \\ \text{H.S.} & G_{11} + G_{14} \end{pmatrix} = \frac{4}{3}\begin{pmatrix} \mu_a + \mu_b & -(\mu_a + \mu_b\, e^{i\kappa}) \\ \text{H.S.} & \mu_a + \mu_b \end{pmatrix} \tag{4-6-66}$$

$$\mathsf{G}_{B_2}(\Delta) = 0 \tag{4-6-67}$$

$$\mathsf{G}_E(\Delta) = \begin{pmatrix} G_{11} - G_{13} & G_{15} - G_{35} \\ \text{H.S.} & G_{11} - G_{13} \end{pmatrix} = \frac{4}{3}\begin{pmatrix} \mu_a + \mu_b & \mu_a + \mu_b\, e^{i\kappa} \\ \text{H.S.} & \mu_a + \mu_b \end{pmatrix} \tag{4-6-68}$$

We note that the redundancies were correctly identified. However, the use of a diagonal force field, $f_0 E$, will not correctly describe $H(\boldsymbol{\kappa})$. We therefore add interaction constants in a parallel fashion to the G matrix elements (4-6-64) and (4-6-65), as we did in the discussion of the Δ point of bcc:

$$\begin{aligned} F_{12} &= F_{14} = F_{56} = F_{58} = f \\ F_{13} &= F_{57} = f' \\ F_{15} &= f\, e^{i\kappa} \\ F_{25} &= F_{45} = f'\, e^{i\kappa} \\ F_{35} &= f''\, e^{i\kappa} \end{aligned} \tag{4-6-69}$$

Construction of the F matrix and symmetrization with (4-6-60) gives the submatrices

$$\mathsf{F}_{A_1}(\Delta) = \begin{pmatrix} f_0 + 2f + f' & (f + 2f' + f'')\, e^{i\kappa} \\ \text{H.S.} & f_0 + 2f + f' \end{pmatrix} \tag{4-6-70}$$

and

$$\mathsf{F}_E(\Delta) = \begin{pmatrix} f_0 - f' & (f - f'')\, e^{i\kappa} \\ & f_0 - f' \end{pmatrix} \tag{4-6-71}$$

The equation pairs (4-6-66), (4-6-70) and (4-6-68), (4-6-71) can be used to form the symmetrized dynamical matrices (GF) for each representation, A_1 and E. The roots of each of these, however, can only be expressed in the form of lengthy quadratic formulas. It is easily determined, however, that at $\boldsymbol{\kappa} = 0\, |GF| = 0$, for each representation. The reason for this is that one root in each case vanishes; these are the acoustic modes at $\boldsymbol{\kappa} = 0$. Hence the traces of each dynamical matrix at $\boldsymbol{\kappa} = 0$ must equal the finite roots, that is, the squared frequencies of the optic branches at Γ. Furthermore, since the optic branches belong to F_{1u} of O_h at Γ, the two traces at $\kappa = 0$ must be equal. They are indeed

$$\omega^2(\Gamma) = \text{trace}\,(GF) = \tfrac{8}{3}(\mu_a + \mu_b)(f_0 + f - f' - f'') \tag{4-6-72}$$

The dispersion curves along Δ can be found by making an approximate separation of the high and low frequency roots, in a manner similar to that used in the related one-dimensional problem, the diatomic lattice without symmetry (see Section 3-3(c) and compare (4-6-66) and (4-6-68) with (3-3-7)). As in that problem,

$$G^{-1} = \frac{1}{|\mathbf{G}|}\begin{pmatrix} G_0 & -G_{12} \\ -G_{12}^* & G_0 \end{pmatrix} \tag{4-6-73}$$

where $|\mathbf{G}| = \frac{32}{9}\mu_a\mu_b(1 - \cos\kappa)$. We proceed to solve $|\mathbf{F} - \mathbf{G}^{-1}\lambda| = 0$, dropping out the first row and column of (4-6-70), (4-6-71), and (4-6-73). We obtain

$$\omega_h^2(A_1) \cong \tfrac{4}{3}\mu(f_0 + 2f + f') \tag{4-6-74}$$

$$\omega_h^2(E) \cong \tfrac{4}{3}\mu(f_0 - f') \tag{4-6-75}$$

$$\omega_l^2(A_1) \cong \frac{8}{3}\frac{(f_0 + 2f + f')}{M}(1 - \cos\kappa) \tag{4-6-76}$$

$$\omega_l^2(E) \cong \frac{8}{3}\frac{(f_0 - f')}{M}(1 - \cos\kappa) \tag{4-6-77}$$

where $\mu = \mu_a + \mu_b$ and $M = m_a + m_b$. In all of these we have indicated that the solutions are approximate, which may be seen by comparison of the first two with (4-6-72). Making use of the latter, the dispersion curves of Fig. 4-10 are obtained.

Note that to obtain an LO–TO splitting at $\kappa \neq 0$, it is necessary to include the off-diagonal interaction force constants f and f'. Without them, the crystal vibrations are indeed isotropic.

The results of Sections 4-6(a), (b), (c), and (d) are summarized in Table 4-6.

4-7 SELECTION RULES

Having developed a method of symmetry classification of normal modes valid anywhere in the BZ, we now turn to the question of the selection rules imposed by the factor group operations, recalling that in earlier chapters we excluded the possibility of the appearance of acoustic modes in the optical spectra and further

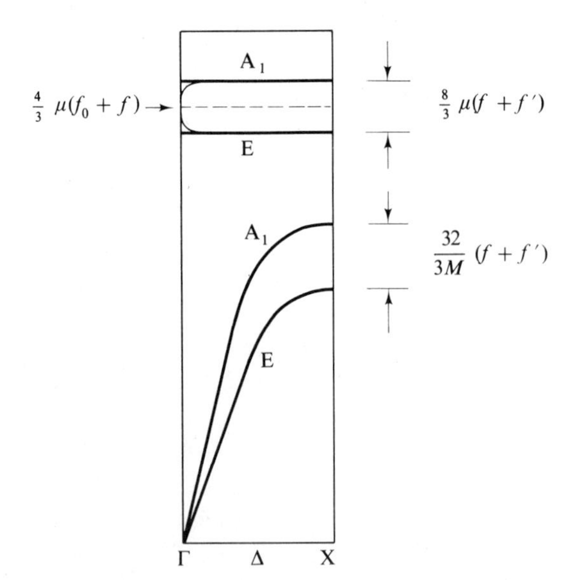

Figure 4-10 Dispersion curves of CsCl along Δ.

Table 4-6

Point coordinates in the BZ of the lattices

	sc (O_h^1, $Pm3m$)			fcc (O_h^5, $Fm3m$)				
k	000	$00k$	$00\frac{1}{a}$	000	$00k$	$00\frac{1}{a}$	kkk	$\frac{1}{a}\frac{1}{a}\frac{1}{a}$
$\boldsymbol{\kappa}$	000	00κ	00π	000	$\kappa\kappa 0$	$\pi\pi 0$	$\kappa\kappa\kappa$	$\pi\pi\pi$
BSW	Γ	Δ	X	Γ	Δ	X	Λ	L
Wyckoff	a	e	d	a	e	b	f	c
$H(\boldsymbol{\kappa})$	O_h	C_{4v}	D_{4h}	O_h	C_{4v}	D_{4h}	C_{3v}	D_{3d}
LA	0	$\sim 6\mu f_0(1 - \cos\kappa)$	$\sim 12\mu f_0$	0	$4\mu f_0(1 - \cos\kappa)$	$8\mu f_0$	$4\mu f_0(1 - \cos\kappa)$	$8\mu f_0$
TA	0	$\sim 4\mu f(1 - \cos\kappa)$	$\sim 8\mu f$	0	$2\mu f_0(1 - \cos\kappa)$	$4\mu f_0$	$\mu f_0(1 - \cos\kappa)$	$2\mu f_0$

	bcc (O_h^9, $Im3m$)					CsCl (O_h^1, $Pm3m$)		
k	000	$k00$	$\frac{1}{a}00$	kkk	$\frac{1}{a}\frac{1}{a}\frac{1}{a}$	000	$00k$	$00\frac{1}{2}$
$\boldsymbol{\kappa}$	000	$\bar{\kappa}\kappa\kappa$	$\bar{\pi}\pi\pi$	$\kappa\kappa\kappa$	$\pi\pi\pi$	000	00κ	00π
BSW	Γ	Δ	H	Λ	P	Γ	Δ	X
Wyckoff	a	e	b	f	c	a	e	d
$H(\boldsymbol{\kappa})$	O_h	C_{4v}	O_h	C_{3v}	T_d	O_h	C_{4v}	D_{4h}
LA	0	(4-6-42)	(4-6-44)	(4-6-54)	$\frac{16\mu f_0}{3}$	0	(4-6-70)	(4-6-76)
TA	0	(4-6-43)	(4-6-44)	$\frac{8}{3}\mu f_0(1 - \cos\kappa)$	$\frac{16\mu f_0}{3}$	0	(4-6-71)	(4-6-77)
LO						$\frac{16\mu}{3}(f_0 + f - f' - f'')$	(4-6-70)	(4-6-74)
TO							(4-6-71)	(4-6-75)

that we showed that only transitions in which $\Delta\kappa \approx 0$ were permitted as fundamentals. Such transitions may in crystals more properly be designated as one-phonon processes; multiphonon processes will be discussed below.

(a) One-Phonon Transitions

With the additional restriction that $\Delta\kappa \approx 0$, the rules for crystals are identical with those for molecules, i.e., infrared absorption or emission is permitted if the vibration has the species of some component of the dipole moment, and Raman activity is permitted if the vibration has the species of some component of the polarizability. Thus the character tables (Appendix I), which indicate the species of T_x, T_y, T_z show the allowed species for absorption, since μ_x, μ_y, μ_z transform exactly like the corresponding components of a translation, and further designate the Raman active modes by giving the species of α, the polarizability. It is understood, of course, that the point group in question is a unit cell group.

The crystals treated in Sections 4-4, 4-5, and 4-6 will now be discussed as examples. The elementary body-centered lattice, having only acoustic modes, yields no infrared or Raman spectra via one-phonon processes. The rocksalt, diamond, and fluorite lattices all have O_h as their factor group. In O_h (see Appendix I) the infrared active species is triply degenerate F_{1u}; Raman activity is allowed in A_{1g}, E_g, and F_{2g}. Since the species of optical modes in rocksalt, diamond, and fluorite were F_{1u}, F_{2g}, and $F_{1u} + F_{2g}$ respectively, one concludes that rocksalt will exhibit a single infrared fundamental, but no Raman fundamental, that the opposite will be the case for diamond, i.e., one Raman and no infrared fundamental, and finally that fluorite will exhibit two distinct frequencies, one in the infrared and one in the Raman.

(b) Multiphonon Transitions

When the vibrational quantum number of a single mode changes by more than unity, or the numbers for two or more modes change simultaneously, there exist new phenomena which are not strictly analogous to the situations encountered with single molecules. Consider first the case in which the quantum number associated with a single normal coordinate changes by two, which would simply be called an overtone in the molecular case. In the crystal this can occur in the following distinct ways, assuming that the initial state is the ground state.

1. the final state is $v_k(\mathbf{0}) = 2$ for the kth branch of the optical spectrum at $\boldsymbol{\kappa} = \mathbf{0}$ in that branch, all other modes remaining in their ground states;
2. the final state is $v_k(\boldsymbol{\kappa}) = 1$ and $v_k(-\boldsymbol{\kappa}) = 1$ for the kth branch; in this case the net change from the ground state is again $\Delta\kappa = 0$.

Case (2) is a simple example of a two-phonon process. Since all $\boldsymbol{\kappa}$ values are allowed, the two-phonon intensity appears as a *band* rather than a single line. In the absence of mechanical anharmonicity, the transition intensity for case (1) is incorporated in the two-phonon band, but if there is appreciable anharmonicity, the case (1) transition may appear as a sharp line separated from the two-phonon band.

The selection rules for transitions of type (1), and indeed those for any combination bands in which the final state involves a set of finite $v_k(\mathbf{0})$, $v_l(\mathbf{0})$, etc., may be found exactly by the methods discussed in Sections 2-4(b) and (c), but those for the multiphonon processes require further discussion. If the final state is simply $v_k(\boldsymbol{\kappa}) = v_k(-\boldsymbol{\kappa}) = 1$, one inquires whether the transition moment between the final state and the ground state is non-vanishing. This is equivalent to asking whether the product $Q_k(\boldsymbol{\kappa})Q_k(-\boldsymbol{\kappa})$ has the same species as some component of the electric dipole moment or the polarizability. At first it may seem unclear which point group one uses in this analysis, but since $H(\boldsymbol{\kappa}) = H(-\boldsymbol{\kappa})$ the species of $Q_k(\boldsymbol{\kappa})$ is identical with the species of $Q_k(-\boldsymbol{\kappa})$ under $H(\boldsymbol{\kappa})$.

Thus in general the product $Q_k(\boldsymbol{\kappa}) \cdot Q_k(-\boldsymbol{\kappa})$ will transform, *with respect to operations R of* $H(\boldsymbol{\kappa})$, according to the rule

$$\begin{aligned}\{R\,|\,\tau\}Q_k(\boldsymbol{\kappa})Q_k(-\boldsymbol{\kappa}) &= e^{i\boldsymbol{\kappa}\cdot\tau}\,\chi_R^{(\eta)}\,e^{-\boldsymbol{\kappa}\cdot\tau}\,\chi_R^{(\eta)}Q_k(\boldsymbol{\kappa})Q_k(-\boldsymbol{\kappa})\\ &= (\chi_R^{(\eta)})^2 Q_k(\boldsymbol{\kappa})Q_k(-\boldsymbol{\kappa})\end{aligned}$$

Any R not in $H(\boldsymbol{\kappa})$ will change the values of $\boldsymbol{\kappa}$ and $-\boldsymbol{\kappa}$ and thus contribute zero to the character of the transformation, since $Q_k(\boldsymbol{\kappa})Q_k(-\boldsymbol{\kappa})$ will be sent into $Q_k(\boldsymbol{\kappa}')Q_k(-\boldsymbol{\kappa}')$. Since the product $Q_k(\boldsymbol{\kappa})Q_k(-\boldsymbol{\kappa})$ is translationally totally symmetric, we must seek the species of this product at the origin of the Brillouin zone. This is simply a matter of (1) reducing the character $(\chi_R^{(\eta)})^2$ in H; and (2) correlating the resultant species in H with those of G. Obviously if η is a non-degenerate species, $(\chi_R^{(\eta)})^2 = 1$ is the totally symmetric species under H.

There is an important exception to the above result which arises if the space group contains the inversion $\mathbf{i}$.[35] In such a case the product $Q_k(\boldsymbol{\kappa})Q_k(-\boldsymbol{\kappa})$ *is sent into itself* by $\mathbf{i}$ since $\mathbf{i}Q_k(\boldsymbol{\kappa}) \rightarrow Q_k(-\boldsymbol{\kappa})$ and $\mathbf{i}Q_k(-\boldsymbol{\kappa}) \rightarrow Q_k(\boldsymbol{\kappa})$, even though $\mathbf{i}$ does not belong to $H(\boldsymbol{\kappa})$ for any point in the interior of the BZ except the origin. This implies that the product in question transforms like certain g species of the point group obtained by combining H with $\mathbf{i}$. Again, if η is a non-degenerate species, $(\chi_R^{(\eta)})^2 = 1$, so the species of H is totally symmetric. If $\boldsymbol{\kappa}$ is a general point, i.e., one of no symmetry in the Brillouin zone, $Q_k(\boldsymbol{\kappa})Q_k(-\boldsymbol{\kappa})$ transforms like A_g under C_i. Correlation may ultimately be used to find the symmetry modes at $\boldsymbol{\kappa} = 0$. Since A_g in C_i correlates with *all* g species of $G = H(\mathbf{0})$, one perceives the basis of the quite general selection rules for general points in any crystal with a center of inversion: all such two-phonon overtone processes at general points are Raman allowed and infrared forbidden, since α occurs amongst the g species and μ amongst the u species. Moreover, the same rule clearly applies at interior points with non-trivial symmetry for a non-degenerate mode, since the correlation process must necessarily include the totally symmetric species at $\boldsymbol{\kappa} = 0$ and exclude all u species.

35. R. Loudon, *Phys. Rev.*, **137A,** 1784 (1965).

The selection rules for two-phonon combination spectra require the analysis of the symmetry of $Q_k(\boldsymbol{\kappa})Q_l(-\boldsymbol{\kappa})$. Here the presence of the inversion operation does not affect the situation in the interior of the Brillouin zone, and the analysis of the symmetry species is carried out by identifying the species η and η' of $Q_k(\kappa)$ and $Q_l(-\kappa)$, reducing the product $\chi_R^{(\eta)}\chi_R^{(\eta')}$ for R in $H(\boldsymbol{\kappa})$ and correlating with $\boldsymbol{\kappa} = 0$.

Although the presence of a center of symmetry has thus been seen to have a significant effect in prohibiting one type of overtone—the two-phonon case—in the infrared spectrum, on the whole the selection rules for overtones and combinations are really quite permissive.

This point was made in 1949 by Winston and Halford[36] who emphasized that the overwhelming majority of modes in the BZ correspond to $H(\boldsymbol{\kappa})$ of trivial (C_1) or very low symmetry, where essentially no selection rules are operative. It must, however, be remembered that the spectroscopist sees intensities, and that therefore even though many combination modes arising from points in the interior of the BZ are allowed, their cumulative effect on the spectrum may be weak if the transition frequencies are distributed. In other words, it is important to consider the density of states, a topic which will be discussed in detail in the next section and Section 6-6, as well as the electrical and mechanical coupling responsible for the breakdown of the harmonic oscillator and linear dipole model $(\mu \sim Q_k)$ which permits only fundamentals.

(c) Density of States and Critical Points

We have obtained, in Sections 3-2, 3-3, and 4-6, the phonon energies of a variety of crystals as a function of wave vector, and have also discussed selection rules for both their single and multiphonon spectra. Insofar as fundamental transitions are concerned, we have observed that the dispersion of $\omega(\mathbf{k})$ is of slight concern to us, inasmuch as such transitions involve $\mathbf{k} \cong 0$. On the other hand, multiphonon transitions involve all possible values of $\mathbf{k}$. Since along any one dispersion curve $\mathbf{k}$ may assume N different values, the maxima in multiphonon spectra must represent energies, the occurrence of which are more probable than those of others. That there are such energies is illustrated by a reconsideration of the one-dimensional, monatomic crystal.

From (3-2-10) we have

$$\omega(k) = \omega_M \sin \frac{\pi k}{N} \tag{4-7-1}$$

where $\omega_M = (F_0 \mu)^{1/2}$, and from which we find

$$\frac{d\omega_k}{dk} = \frac{\pi \omega_M}{N} \cos \frac{\pi k}{N} \tag{4-7-2}$$

36. H. Winston and R. S. Halford, *J. Chem. Phys.*, **17**, 607 (1949).

which we recognize as the number of states per unit interval of **k**, $g(\mathbf{k})$. Since we spectroscopically detect states in frequency space, we are more interested in the density of states per unit interval of frequency,

$$g(\omega) = \frac{dk}{d\omega_k} = \frac{N}{\pi}(\omega_M^2 - \omega_k^2)^{-1/2} \tag{4-7-3}$$

which is displayed in Fig. 4-11. Thus the density of states per unit frequency interval, $g(\omega)$, becomes singular at the BZB, $\omega = \omega_M$, and we should expect multiphonon spectra in these crystals to manifest transitions primarily involving such states.

In Section 3-3 we also obtained the dispersion relation for one-dimensional, diatomic crystals (Eq. (3-3-8)). For simplicity, when the atomic masses are equal ($\mu_a = \mu_b = \mu$), we have

$$\omega^2(k) = 2\mu F_0(1 \pm \cos \pi k/N) \tag{4-7-4}$$

and the acoustical $(-)$ and optical $(+)$ branches meet at the BZB $(k = N/2)$. Then considering the two branches separately, it is easily found that

$$g_-(\omega) = \frac{dk}{d\omega_-} = \frac{2N}{\pi}[2\omega_M^2 - \omega_-^2(k)]^{-1/2} \tag{4-7-5}$$

and

$$g_+(\omega) = \frac{dk}{d\omega_+} = \frac{2N}{\pi}[2\omega_M^2 - \omega_+^2(k)]^{-1/2} \tag{4-7-6}$$

where $\omega_M^2 = 2\mu F_0$ for both branches. The total density-of-states, $g(\omega)$, is found by adding (4-7-5) and (4-7-6), the result of which resembles Fig. 4-11 when $\mu_a = \mu_b$.

The case for $\mu_a \neq \mu_b$ is illustrated in Fig. 4-12. There are now several maxima in $g(\omega)$. Besides that reached on the optical branch at $k = 0$, there are others on

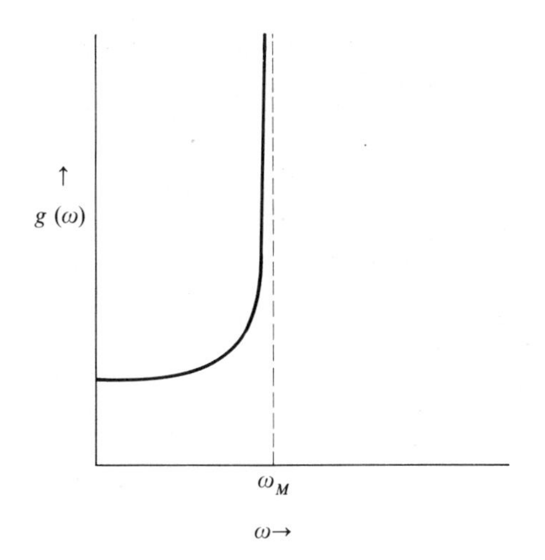

Figure 4-11 Frequency distribution function of a one-dimensional, monatomic crystal.

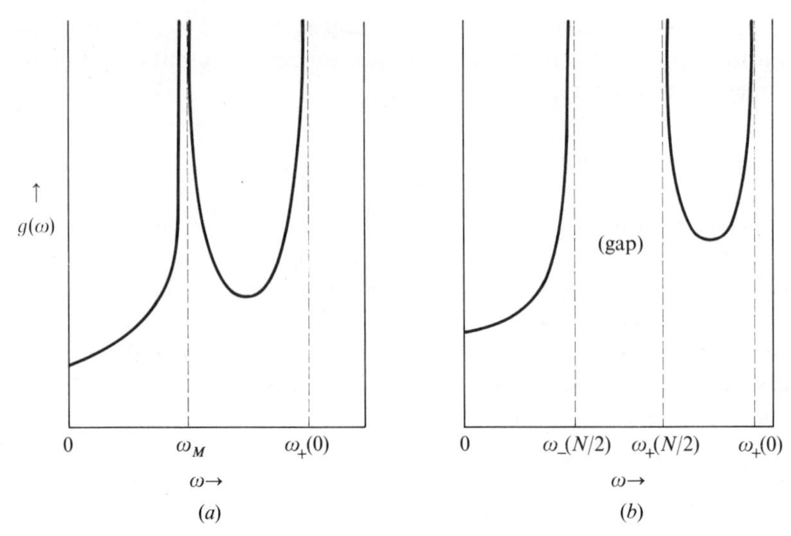

Figure 4-12 Frequency distribution functions of one-dimensional, diatomic crystals: (a) $\mu_a \cong \mu_b$; (b) $\mu_a \neq \mu_b$

both branches at $k = N/2$ (Fig. 4-12). When $\mu_a \neq \mu_b$, there is a gap between the acoustic and optical branches at the BZB proportional to the difference in the square roots of the reduced masses (see Fig. 3-8), and this gap is evident in $g(\omega)$ for this system. As $\mu_a \rightarrow \mu_b$, the gap narrows and $g(\omega)$ becomes continuous at $k = N/2$.

We may also consider the *joint density* of the "summation states" $\omega_+(k) + \omega_-(k)$, in which an acoustic and an optical phonon are simultaneously created. For this kind of state,

$$g(\omega) = \frac{dk}{d\omega} = \frac{2N}{\pi}\,[\omega_M^2(s) - \omega^2(s)]^{-1/2}$$

where $\omega(s)$ is the frequency of the summation state. The joint density of states, like that of the negative (acoustic) branch, has a singularity only at the BZB (ω_M).

Singularities or discontinuities in $g_j(\omega)$ for the several branches j of one-dimensional crystals also occur as $\omega \rightarrow \omega_+(0)$ or $\omega \rightarrow \omega_M$. These discontinuities have long been recognized as *critical points* (c.p.), and the behavior of the density of states functions in their vicinity is seen to depend on $(\omega_M^2 - \omega^2)^{-1/2}$. It has been shown that, in one dimension, with nearest-neighbor forces only, such behavior is quite general.[37]

The density of states functions of three-dimensional crystals have not been obtained in closed form. However, for each branch j we can indicate its relationship to the dispersion relation $\omega_j(\mathbf{k})$.

In one dimension, a dispersion relation is a curve ω_j *vs.* k. In three dimensions

37. G. H. Weiss, *Bull. Res. Council Israel*, **7F**, 165 (1958).

each branch is a surface which joins ω_j *vs.* k curves along particular directions. Figure 4-13 illustrates some possible surfaces for a two-dimensional square lattice. Because of the curvature of this surface, it is not generally a surface of constant energy. We can, however, define surfaces of constant energy $\omega(\mathbf{k})$ = constant in this same space (**k**-space), and a few are illustrated in Fig. 4-14.

Since the density function is defined as the number of states per unit frequency interval, it is normalized by

$$\sum_{j=1}^{3r} \int_0^{\omega_M} g_j(\omega)\, d\omega = 3rN$$

where r = the number of atoms per unit cell and N = the number of unit cells.

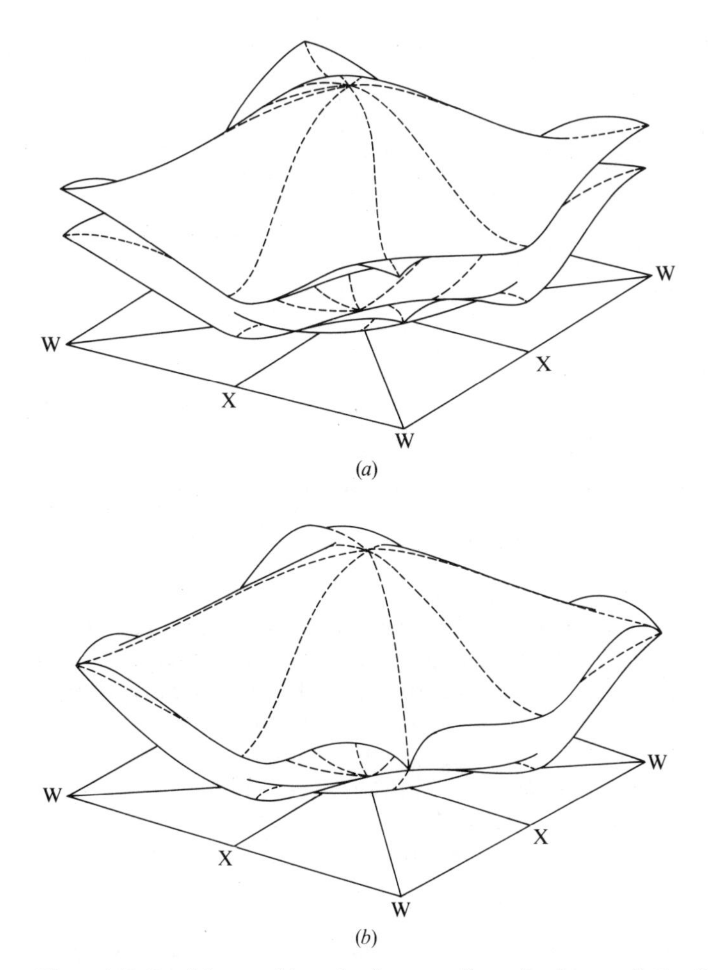

Figure 4-13 Possible ω_j *vs.* k branches for a two-dimensional square lattice. In (*a*) the surfaces overlap, the second branch being lower at X than the first branch at W. In (*b*) there is a degeneracy at W. (After W. A. Harrison, *Solid State Theory*, McGraw-Hill, New York, 1970, pages 62–63)

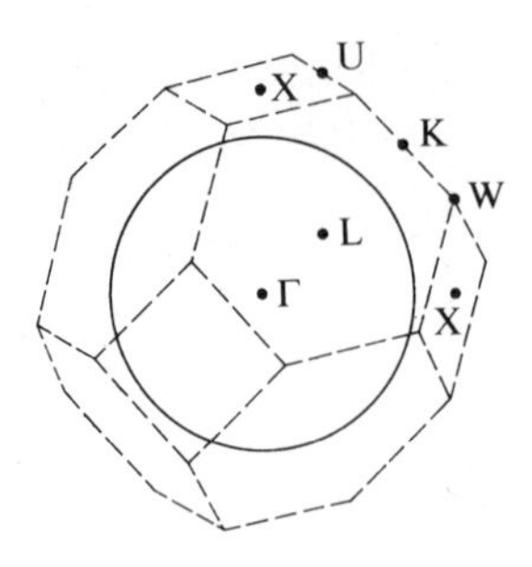

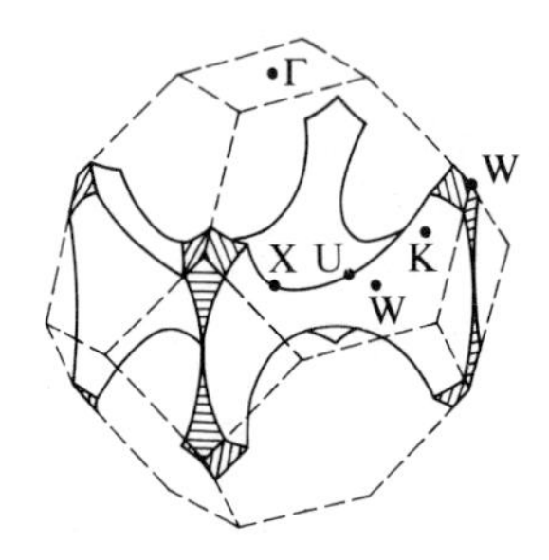

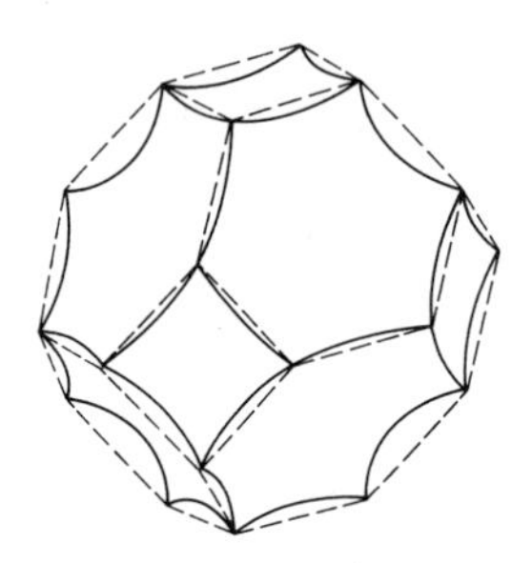

Figure 4-14 Some surfaces of constant energy in a face-centered cubic lattice. (After W. A. Harrison, *Solid State Theory*, McGraw-Hill, New York, 1970, page 118)

Although we can draw dispersion curves and surfaces through the points $\omega_j(\mathbf{k})$ in $\mathbf{k}$-space, in fact, $3rN$ is such a large number of points that the states $\omega_j(\mathbf{k})$ can be regarded as being uniformly distributed in $\mathbf{k}$-space. Because of this, the number of states in the frequency interval ω_j to $\omega_j + d\omega_j$ is

$$g_j(\omega)\, d\omega = \frac{NV}{8\pi^3} \int d_3\mathbf{k}$$

where the integration over the volume element of $\mathbf{k}$-space $d_3\mathbf{k}$ extends from a surface of constant energy ω_j to another at $\omega_j + d\omega_j$, and V is the volume of one unit cell.

If we call the infinitesimal element on the surface of constant energy $\omega_j\, d\mathbf{S}$, then $\nabla_\mathbf{k}\omega_j(\mathbf{k})$ is its normal and

$$d\omega = \nabla_\mathbf{k}\omega_j(\mathbf{k}) \cdot d\mathbf{k}$$

or

$$d\mathbf{k} = \frac{d\mathbf{S}}{|\nabla_{\mathbf{k}}\omega_j(\mathbf{k})|}d\omega$$

Thus we have

$$g_j(\omega)\,d\omega = \iiint \frac{d\mathbf{S}}{|\nabla_{\mathbf{k}}\omega_j(\mathbf{k})|}d\omega$$

and the density of states function is therefore

$$g_j(\omega) = \iint \frac{d\mathbf{S}}{|\nabla_{\mathbf{k}}\omega_j(\mathbf{k})|} \tag{4-7-7}$$

and the integration is carried out over the constant energy surface ω_j.

In a three-dimensional crystal a critical point is a point where every component of $\nabla_{\mathbf{k}}\omega_j(\mathbf{k})$ is either zero or changes sign discontinuously. In the first case, a c.p. at $\omega_j(\mathbf{k})$ is said to be *analytic*. When one or more components of $\nabla_{\mathbf{k}}\omega_j(\mathbf{k})$ change sign discontinuously while the remaining components vanish, the c.p. is called *singular*. We adopt Phillips' notation[38] for c.p.'s in which $P_i(n)$ denotes a c.p. for which the index $i = 0, 1, 2, 3$ is the number of principal directions in $\mathbf{k}$-space along which ω decreases from its value ω_0 at the c.p. and $n = 0, 1, 2, 3$ is the number of discontinuous components of $\nabla_{\mathbf{k}}\omega_j(\mathbf{k})$ at the c.p. If the c.p. is analytic, $n = 0$ and the c.p. is generally denoted with the shortened symbol P_i.

In the immediate vicinity of an analytic c.p., the frequency can be expanded in a Taylor's series about ω_0, the frequency at the c.p. Since by definition of an analytic c.p., $\nabla_{\mathbf{k}}\omega_j(\mathbf{k}) = 0$, the linear terms are absent, and the variation in the frequency is approximately parabolic:

$$\omega(\mathbf{k}) = \omega_0 + \sum_{i=1}^{3} a_i(k_i - k_i^\circ)^2 + \cdots \tag{4-7-8}$$

where k_i is a component of $\mathbf{k}$ along one of the principal directions of the BZ, and $\mathbf{k}_0$ is the wave vector at the c.p. The index i of the c.p. is the number of negative second derivatives a_i of ω in the neighborhood of k_0. There are four possibilities, each of which may be obtained by substitution of (4-7-8) in (4-7-7).[39]

1. All of the a_i are negative. In the vicinity of the c.p., the surface $\omega_j(\mathbf{k})$ is approximated by a sphere in $\mathbf{k}$-space centered at $\mathbf{k}_0$ with radius proportional to $(\omega_0 - \omega)^{1/2}$. Since the surface element of a sphere is proportional to the square of its radius,

$$g_j(\omega) \sim \frac{\omega_0 - \omega}{d\omega/dk}$$

and because $dk/d\omega \sim (\omega_0 - \omega)^{1/2}$ for $\omega < \omega_0$,

$$g_j(\omega) \sim (\omega_0 - \omega)^{1/2}$$

38. J. C. Phillips, *Phys. Rev.*, **104,** 1263 (1956).
39. L. Van Hove, *Phys. Rev.* **89,** 1189 (1953).

For $\omega > \omega_0$ there is no contribution to $g_j(\omega)$ from the jth branch in the vicinity of the c.p. Other branches, however, may contribute to the frequency distribution $g(\omega) = \sum_j g_j(\omega)$, but not discontinuously. These contributions, to a first approximation, consist of two parts, one of which is constant and another depends linearly on $\omega - \omega_0$. Neither part produces a discontinuity in $g_j(\omega)$. In this case, then, the frequency distribution function $g(\omega)$ falls parabolically as $\omega \to \omega_0$, at which point it discontinuously changes slope, as illustrated in Fig. 4-15(*a*). In the region $\omega > \omega_0$, $g(\omega)$ contains only the constant and linear parts. In the Phillips notation, a c.p. of this type is denoted by P_3, as there are three independent directions in **k**-space along which ω_j decreases from its value ω_0 at the c.p. Thus P_3 represents a *maximum* in the ω_j surface.

2. If all of the a_i are positive, all of the considerations of the first case are repeated, except that the surface $\omega_j(\mathbf{k})$ is spherical about **k** with radius $\sim(\omega - \omega_0)^{1/2}$ and $g_j(\omega) \sim (\omega - \omega_0)^{1/2}$ for $\omega > \omega_0$. For $\omega < \omega_0$, $g(\omega)$ has only constant and linear contributions. In this case there are no directions in **k**-space along which ω_j decreases from its value ω_0 at the c.p. This kind of singularity is denoted P_0, and it represents a *minimum* in the ω_j surface. The form of the frequency distribution functions is illustrated in Fig. 4-15(*b*).
3. A third possibility is that two of the a_i are positive and one is negative. The dispersion surface is then *saddle-shaped.* ω_0 is a local maximum along one direction, a minimum along another, orthogonal to the first. Substitution of (4-7-8) in (4-7-7) shows that for $\omega < \omega_0$, the frequency distribution function $g(\omega)$ resembles that for the first case (P_3), except that the part that depends on $(\omega_0 - \omega)^{1/2}$ is of opposite sign to that of the constant and linear portions. ω decreases from ω_0 only along one direction in **k**-space. We thus obtain Fig. 4-15(*c*) for $g(\omega)$ near the saddle point, P_1.
4. The only remaining possibility is for two of the a_i to be negative and one to be positive. The ω_j surface is again saddle-like, indeed, it resembles an inverted P_1. This surface is also related to P_0, in that the part that $g_j(\omega) \sim (\omega - \omega_0)^{1/2}$ in the region $\omega > \omega_0$, but this part of the frequency distribution function $g(\omega)$ is again of opposite sign to the constant and linear portions. Thus the ω_j surface has some of the properties of a minimum, and ω decreases from ω_0 along two directions in **k**-space. Figure 4-15(*d*) illustrates $g(\omega)$ near the c.p. P_2.

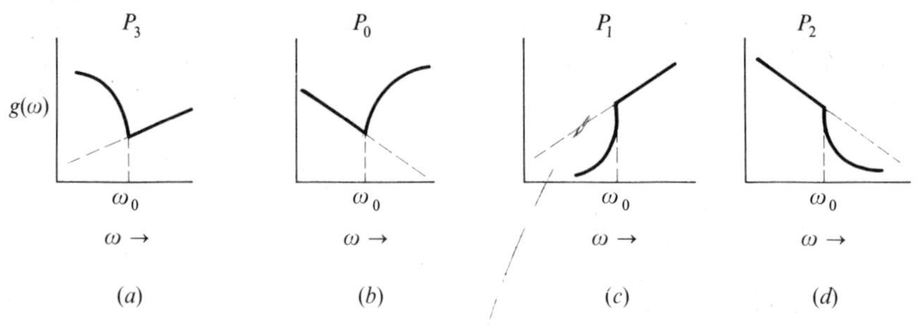

Figure 4-15 Frequency distribution functions in the neighborhood of analytical critical points, P_j: (a) P_3; (b) P_0; (c) P_1; (d) P_2

The c.p.s discussed above are analytic; they occur only when *all* components of $\nabla_{\mathbf{k}}\omega_j(\mathbf{k})$ vanish. There remain the cases when one or more of the components of $\nabla_{\mathbf{k}}\omega_j(\mathbf{k})$ changes sign discontinuously while the remainder vanish, that is, the so-called *singular* c.p.s. All of these arise as a result of contacts between different phonon branches, either accidental or as a result of symmetry, for example at some special points of the BZ.

Phillips has extended Van Hove's analysis of $\omega(k)$ surfaces by continuing the expansion begun in (4-7-8).* Defining $\xi = \mathbf{k} - \mathbf{k}_0$, the eigenvalues $\omega_j^2(\mathbf{k})$ of the dynamic matrix $\mathbf{FG}$ are, to second degree in ξ, those of the matrix

$$W_{ij}(\xi) = \omega_{j0}^2\delta_{ij} + \langle i|\xi\cdot\nabla_{\mathbf{k}}\omega^2|j\rangle + \tfrac{1}{2}\langle i|\xi\xi:\nabla_{\mathbf{k}}\nabla_{\mathbf{k}}\omega^2|j\rangle + \sum_m \frac{\langle i|\xi\cdot\nabla_{\mathbf{k}}\omega^2|m\rangle\langle m|\xi\cdot\nabla_{\mathbf{k}}\omega^2|j\rangle}{\omega_{i0}^2 - \omega_{m0}^2} \tag{4-7-9}$$

We have previously noted that for an analytic c.p., the linear terms in this expansion vanish. By writing the expansion in this way, we see that a more general statement of this condition is

$$\langle i|\xi\cdot\nabla_{\mathbf{k}}\omega^2|j\rangle = 0 \tag{4-7-10}$$

Because the states $|i\rangle$, $|j\rangle$, etc., may serve as bases for representations of the group of the $\mathbf{k}$-vector $H(\mathbf{k})$, Eq. (4-7-10) makes it possible to identify those c.p. which are the consequence of symmetry alone. The integral (4-7-10) is the basis of a triple direct product of representations, the decomposition of which must contain the totally symmetric irreducible representation. Since we seek those cases where (4-7-10) *vanishes,* the condition for a c.p. becomes

$$\Gamma_i \times \Gamma_j \not\supset \Gamma_{\mathbf{v}} \tag{4-7-11}$$

where $\Gamma_{\mathbf{v}}$ is a representation of a vector, since $\Gamma_{\nabla_k} = \Gamma_{\mathbf{v}}$.† Thus, to detect a c.p., the irreducible representations of $H(\mathbf{k})$ are tested using (4-7-11). When it is satisfied, the branch to which $|i\rangle$ belongs will have a c.p. at $\mathbf{k}$. Phillips refers to this kind of c.p. as *ordinary*.‡

If Γ_i is non-degenerate, since the linear term in the expansion (4-7-9) must vanish, (4-7-10) can be satisfied only if *all* of the gradients $\nabla_{\mathbf{k}}\omega^2 = 0$, i.e., the c.p. is analytic. If Γ_i is degenerate, however, (4-7-11) will identify not only analytic c.p.,

* Reference 38. Since $g(\omega) = 2G(\omega^2)$, this expansion can be carried out either in terms of ω_j, as in (4-7-8), or ω_j^2, as in (4-7-9). The reader should note that the minimum in $\omega_j(\mathbf{k})$ for the acoustic branches as $\mathbf{k} \to 0$ is not a c.p., although it is an absolute minimum. The Taylor expansion (4-7-8) is not valid in the neighborhood of $\mathbf{k} = 0$; hence, it may not be used to demonstrate this, one way or another. However, Van Hove has shown that $\lim_{\mathbf{k}\to 0}|\text{grad}\,\omega(\mathbf{k})| \neq 0$. It is for this reason that the minimum in the acoustic branches at $\mathbf{k} \to 0$ does not yield a singularity in $g(\omega)$. Instead, in three dimensions, $g(\omega) \sim \omega^2$ near $\mathbf{k} = 0$, as illustrated in Figs. 4-11 and 4-12.

† The reader is referred to Section 6-6(e) for an example of the application of (4-7-11).

‡ This procedure for identifying all ordinary c.p. required by symmetry was first developed for electronic energy bands by R. H. Parmenter, *Phys. Rev.*, **100,** 573 (1955).

but also those points for which only *some* of the components of the gradient $\xi \cdot \nabla_{\mathbf{k}}$ vanish. In these cases zero gradients in $\omega^2(k)$ are predicted only in particular planes, or even along only particular lines, of the BZ. Examples of such cases are given in Section 6-6(e).

In addition to ordinary c.p.'s (those required by symmetry), there may be additional ones due to accidental degeneracy. This kind of c.p. was first discussed by Herring,[40] and he has qualitatively discussed the effects of symmetry with respect to "contacts between inequivalent manifolds" (accidental degeneracy). These effects have also been discussed by Cracknell,[41] and a variety of possibilities in the vicinity of symmetry points of the BZ have been sketched by Parmenter.[42] In these cases, we may envisage two non-degenerate branches approaching one another as $\mathbf{k} \to \mathbf{k}_0$ so as to become degenerate at $\mathbf{k}_0$, as illustrated in Fig. 4-16.

In terms of the expansion (4-7-9), we see that the second-order perturbation term

$$\sum_m \frac{\langle i|\xi \cdot \nabla_{\mathbf{k}}\omega^2|m\rangle\langle m|\xi \cdot \nabla_{\mathbf{k}}\omega^2|j\rangle}{\omega_{i0}^2 - \omega_{m0}^2}$$

is now involved. $\omega^2(\mathbf{k})$ can now be found only by solving a secular equation, the solution of which yields the dashed curves in Fig. 4-16. For obvious reasons, Phillips refers to this kind of c.p. as *fluted*.[38]

We now have an ambiguity in labeling branches in the vicinity of a c.p. Referring to Fig. 4-16, we see that $\mathbf{k}_0$ terminates in a symmetry plane of the BZ.[41] If the branches are labelled (1)–(4) and (2)–(3), the degeneracy is lost as soon as $\mathbf{k} \neq \mathbf{k}_0$. The $\omega^2(k)$ surfaces are then discontinuous. We therefore adopt what is called *ordered* labeling, in which the ith branch at $\mathbf{k}$ is the one of frequency $\omega_i(\mathbf{k})$, where $i < j$ implies $\omega_i(k) < \omega_j(k)$.

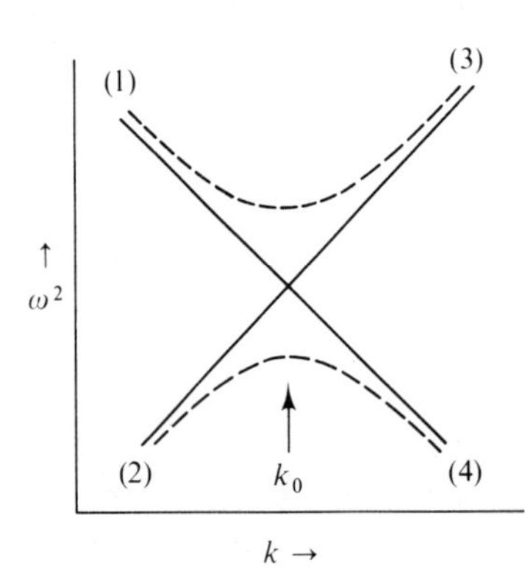

Figure 4-16 A "fluted" critical point. (After J. C. Phillips, *Phys. Rev.*, **104**, 1263 (1955))

40. C. Herring, *Phys. Rev.*, **52**, 361 (1937). Degeneracies due to the additional symmetry of time reversal (see Appendix X) are also discussed in this article.
41. A. P. Cracknell, *Adv. Phys.*, **23**, 673 (1974).
42. R. H. Parmenter, *Phys. Rev.*, **100**, 573 (1955).

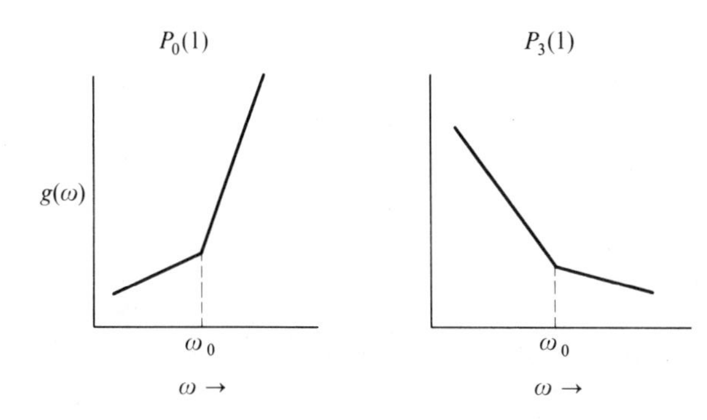

Figure 4-17 Frequency distribution functions due to singular critical points

In terms of Fig. 4-16, (2)–(4) < (1)–(3). Instead of discontinuous $\omega^2(\mathbf{k})$ surfaces, we now must admit of points where one or more components of the gradient change sign discontinuously while the remaining components vanish. Referring to the definition given earlier, such c.p.'s are called *singular*.

In three dimensional crystals, ordinary c.p.'s cannot produce infinities in $g(\omega)$. There are, however, infinities in $g'(\omega)$ in these cases (see Fig. 4-15). There are also discontinuities in the slope of $g(\omega)$ at the singular c.p.'s, but no infinities in $g'(\omega)$. In the case of $P_0(1)$ and $P_3(1)$, the contribution of such branches to $g(\omega)$ is linear in $(\omega - \omega_0)$, as illustrated in Fig. 4-17.

A classical example of a singular c.p. is encountered at the X point in diamond (the BZB along (100)). As $\mathbf{k}$ approaches X along S, the sign of $(\omega - \omega_0)$ is the opposite of that as $\mathbf{k}$ approaches X along Z (see Fig. 4-3(*b*)). Thus, as $\mathbf{k}$ circles about X on the square face of the BZ, there are four regions where $\omega - \omega_0 > 0$ and four where $\omega - \omega_0 < 0$. Each of these regions is denoted a *sector*, and the number of positive (P) and negative (N) sectors which are encountered as $\mathbf{k}$ encircles $\mathbf{k}_0$ is called the sector number (P, N). Study of particular kinds of secular equations for different degrees of degeneracy shows that the shape of the $\omega(\mathbf{k})$ surface in the vicinity of a singular c.p. resembles that of either the saddle-type analytic c.p.'s; otherwise the $\omega(\mathbf{k})$ surfaces can be fluted, in which case they resemble analytic c.p.s. that are either maxima or minima. In each case, the contribution to $g(\omega)$ is similar to that of the analytic c.p. resembled by the singular c.p.

Van Hove[39] has shown that when there are only analytic c.p., the number N_i of P_i in a given phonon branch is given by the Betti numbers.[43] In this case the N_i satisfy the following relationships:

$$
\begin{aligned}
N_0 &= 1 \\
N_1 - N_0 &= 2 \\
N_2 - N_1 + N_0 &= 1 \\
N_3 - N_2 + N_1 - N_0 &= 0
\end{aligned}
\tag{4-7-12}
$$

43. M. Morse, *Fundamental Topology and Abstract Variational Theory, Mémorial des Sciences Mathématiques*, Gauthier-Villars, Paris, 1938, *Fasc. 92*. The Betti number is the maximum number of closed j-dimensional surfaces (tori) on a manifold which cannot be transformed into one another or into a point by continuous deformation on the manifold.

equivalent to $N_0 = N_3 = 1$ and $N_1 = N_2 = 3$. Thus we know how many c.p.'s $P_i(0)$ will be encountered in each branch. If there are singular c.p.'s, however, each $P_i(n)$ will be encountered n times as often as each $P_i(0)$. The first three relationships in (4-7-12) are then changed into inequalities in which the numbers on the right-hand sides become minimum values. Thus, the *minimal set* of c.p.'s can be derived. For further details the reader is referred to Phillips' detailed analysis.[38]

In any critical point analysis, care must be exercised to allow for "compensation." For example, a c.p. of type P_3 in one branch and one of type P_2 in another may cause a cancellation of the singularities in $g(\omega)$. Similarly, P_1 and P_0 may compensate for one another.

Finally, it must also be appreciated that accidental degeneracies may produce extra maxima (and, similarly, minima) in phonon dispersion curves (called "kinks" by Phillips) which are consequences of particular non-nearest neighbor (or off-diagonal) force constants (see Section 3-2).

Using (4-7-11) and calculated dispersion curves for various metals and semimetals, a few critical point analyses have been carried out, such as those by Phillips in the case of aluminum[38] and by Johnson and Loudon for diamond, silicon, and germanium.[44] There has been no complete critical point analysis of a molecular crystal, although a partial analysis has been carried out by Neto, Oehler, and Hexter,[45] and is discussed in Section 6-6(e).

Using direct techniques (i.e., counting occupied cells indexed by ω and k in the BZ), nomographs which represent densities of state functions can be prepared. Provided the one-phonon states (and, hence, dispersion curves) are sufficiently well-separated, the nomographs so produced will represent the one-phonon density of state functions $g_j(\omega)$ and not the frequency distribution functions $g(\omega)$. The labor of this procedure can be considerably shortened by the computer sampling techniques developed by Gilat and Raubenheimer,[46] which in effect carry out the integration indicated in (4-7-7).

As discussed in Section 4-7(b), many infrared and Raman spectra of crystals are the result of higher-order multiphonon processes for which wave-vector selection rules are much less rigorous than they are for fundamentals. For example, two-phonon processes, such as overtone and binary combination bands have the selection rule

$$\mathbf{k}_1 + \mathbf{k}_2 = 0 \tag{4-7-13}$$

as a result of which over many directions in **k**-space, both branches are entirely active. In effect, however, the two-phonon spectrum is a representation of the two-phonon, or joint density of states function, that is,

$$I_{f\leftarrow i} \sim \left[\frac{\partial^2 \mu}{\partial Q(\mathbf{k})\,\partial Q(\mathbf{k}')}\right]^2 |\langle i|\,Q(\mathbf{k})Q(\mathbf{k}')\,|f\rangle|^2 g[\omega_i(\mathbf{k}) + \omega_j(\mathbf{k}')]\ \delta(\mathbf{k} + \mathbf{k}')$$

44. F. A. Johnson and R. Loudon, *Proc. Roy. Soc.* (*London*), **A281**, 274 (1964).
45. N. Neto, O. Oehler, and R. M. Hexter, *J. Chem. Phys.*, **58**, 5661 (1973).
46. G. Gilat and L. J. Raubenheimer, *Phys. Rev.*, **144**, 320 (1966).

where because of (4-7-13) we have the selection rule $\mathbf{k}' = -\mathbf{k}$. Thus a knowledge of the c.p.'s of the two-phonon dispersion surfaces can be most useful in sketching the joint density-of-states function $g[\omega_i(\mathbf{k}) + \omega_j(\mathbf{k}')]$.

By simple addition of the one-phonon dispersion surfaces $\omega_j(\mathbf{k})$ we obtain the two-phonon dispersion surfaces $\omega_i(\mathbf{k}) + \omega_j(\mathbf{k})$. In analogy to (4-7-7), we define the joint density of states function by

$$g_{ij}(\omega) = \iint \frac{d\mathbf{S}_{ij}}{\nabla_{\mathbf{k}}[\omega_i(\mathbf{k}) + \omega_j(-\mathbf{k})]} \tag{4-7-14}$$

The analytical c.p.'s of $\mathbf{S}_{ij}$ are similarly the points where all components of

$$\nabla_{\mathbf{k}}[\omega_i(\mathbf{k}) + \omega_j(-\mathbf{k})] = 0 \tag{4-7-15}$$

Clearly (4-7-15) is satisfied if

$$\nabla_{\mathbf{k}}\omega_i(\mathbf{k}) = \nabla_{\mathbf{k}}\omega_j(\mathbf{k}) = 0 \tag{4-7-16}$$

so that wherever there is a c.p. in both one-phonon dispersion surfaces, there should also be a c.p. in the two-phonon surface. However (4-7-15) shows that, in addition, the two-phonon dispersion surface may exhibit a c.p. where

$$\nabla_{\mathbf{k}}\omega_i(\mathbf{k}) = \pm\nabla_{\mathbf{k}}\omega_j(\mathbf{k}), \tag{4-7-17}$$

the sign choice referring respectively to subtractive or additive combinations (see section 4-7(b)).

4-8 DIRECTIONAL AND POLARIZATION PROPERTIES IN SINGLE CRYSTALS

The basic point to be made in this section is that the symmetry species of the several allowed infrared modes, and correspondingly the Raman modes, may be distinguished by the use of polarized radiation. However, the simple cubic examples described in Section 4-7 fail to illustrate in full detail the information which can be extracted from experiments with single crystals of known orientation. Many such crystals involve molecular units and a number will be discussed in Chapter 6. Here we shall mention a single example chosen to illustrate this point. A number of interesting crystals have a structure similar to *rutile* (TiO_2) such as MgF_2, etc.

The space group is D_{4h}^{14} with two formula units per unit cell. Thus two cations are accommodated at D_{2h} sites, and four anions on C_{2v} sites, points a and f in Table VII-6 respectively. The derivation of the species of the modes at $k = 0$ by correlation requires a little care, since Table VII-6 alone does not indicate which specific subgroups of D_{4h} are involved; a study of the correlation table of D_{4h} with its subgroups (Table IV-1) reveals that *two* correlations with D_{2h} are possible, depending upon the choice of the C_2' or C_2'' axes within D_{4h} and similarly, no less than four distinct correlations with C_{2v} arise depending upon the choice of the twofold axis and planes for C_{2v}. In the structure under consideration we are free to adopt either C_2' or C_2'' in D_{2h} and elect the former. Having done this, we must note that the anions on site f which is described in ITXRC[47] as $\pm[u, u, 0; u + 1/2, 1/2 - u, u]$ must necessarily lie on the C_2'' axis so that a consistent correlation is that given in Table 4-7. The 18 degrees of freedom for the "bimolecular" unit cell are thus:

$$A_{1g} + A_{2g} + B_{1g} + B_{2g} + E_g + 2A_{2u} + B_{1u} + B_{2u} + 4E_u$$

At $k = 0$, these modes include the zero frequency acoustic modes of species $A_{2u} + E_u$ under D_{4h}.

Absorption (or reflection) experiments may be conducted using plane polarized light. Let us imagine first that the crystal sample is cut perpendicular to the fourfold axis, so that the polarization ($\mathscr{E}$ vector) of the incident radiation is necessarily in the XY plane of the crystal. In this situation the applied field can only couple with crystal modes whose $\boldsymbol{\mu}$ vector also lies in this plane. Now since μ_X and μ_Y are of species E_u under D_{4h}, the conclusion is that only the three optical E_u modes may be active in such an experiment. Further, because of the degeneracy, no changes are to be expected if the crystal (or the polarization of $\mathscr{E}$) is rotated about the Z-axis, whose direction coincides with C_4.

Another possible crystal sample may be prepared by cutting a slab parallel to the unique axis; it does not matter whether the second axis contained in the slab is X or Y or any intermediate axis in the XY plane. However for this "parallel cut" sample, interesting polarization effects will be observed when plane polarized incident radiation is employed. If the $\mathscr{E}$ vector is parallel to the Z-axis

Table 4-7 Correlation of site and unit cell symmetries in rutile, TiO_2

$D_{2h}(2Ti^{4+})$	D_{4h} (unit cell)	$C_{2v}(4\,O^{2-})$
	A_{1g}	$A_1(Z)$
	B_{2g}	$A_1(Z)$
$(X)B_{3u}$, $(Y)B_{2u}$	E_u	$A_1(Z)$, $B_1(X)$
	A_{2g}	$B_1(X)$
	B_{1g}	$B_1(X)$
	E_g	$B_2(Y)$
$(Z)B_{1u}$	A_{2u}	$B_2(Y)$
	B_{1u}	$B_2(Y)$
$(Z)B_{1u}$	B_{2u}	

47. *International Tables of X-Ray Crystallography*, N. F. M. Henry and K. Lonsdale, Eds., Kynoch Press, Birmingham, 1952, Vol. I.

of the crystal, the optical A_{2u} mode will be active in absorption since its $\boldsymbol{\mu}$ vector lies in the Z direction, but the three E_u modes will not absorb at all. Conversely, if the $\mathscr{E}$ vector is polarized in a direction perpendicular to Z, only the E_u modes will absorb. For intermediate directions all five modes may absorb. This is an example of infrared *dichroism*, the term having been introduced in reference to analogous effects in the visible absorption spectra of crystals.

In the Raman effect one has available an even more detailed scheme for the assignment of modes. The character table (Appendix I) for D_{4h} exhibits the following non-vanishing components of the polarizability derivative:

Table 4-8

A_{1g}:	$\alpha'_{XX} = \alpha'_{YY} = a;\ \alpha'_{ZZ} = b$
B_{1g}:	$\alpha'_{XX} = -\alpha'_{YY} = c$
B_{2g}:	$\alpha'_{XY} = d$
E_g:	$\alpha'_{XZ} = \alpha'_{YZ} = e$

Now one is prepared to examine the possible results of Raman experiments using plane polarized incident light and, in order to obtain the maximum information, also employing an *analyzer*, i.e., selecting only a single polarized component of the scattered light. To describe a particular experiment of this type, we follow the convention of Damen, Porto, and Tell[48] and use for the case of right angle scattering the simple notation $X(ZZ)Y$ to indicate an experiment in which incident light propagating in the X direction and polarized in the Z direction is scattered into the Y direction with Z polarization. Similarly, $Z(XY)X$ signifies incident radiation propagating in the Z direction, X polarized, which is scattered into the X direction with Y polarization, etc.* Following this convention, one then finds from Table 4-8 the following activities:

$X(ZZ)Y$: only the A_{1g} modes with intensities proportional to b^2
$X(ZX)Y$: only the E_g modes with intensities proportional to e^2
$Z(YY)X$: A_{1g} modes proportional to a^2
and B_{1g} modes proportional to c^2
$Z(XY)X$: only the B_{2g} modes proportional to d^2

While other directional and polarized experiments can be described, they will not yield any information in addition to that already obtained. For the crystal under consideration it is thus evident that:

(i) the single A_{1g} mode will appear by itself in $X(ZZ)Y$
(ii) the single E_g mode will appear by itself in $X(ZX)Y$

48. T. C. Damen, S. P. S. Porto, and B. Tell, *Phys. Rev.*, **142**, 570 (1966).
* The mnemonic, *i*(*pa*)*s*, is helpful: *i* = incident, *p* = polarizer, *a* = analyzer, *s* = scattered.

(iii) one A_{1g} and one B_{1g} mode will appear in $Z(YY)X$
(iv) the single B_{2g} mode will appear in $Z(XY)X$

Just as in molecular examples there are a number of modes which are inactive as fundamental transitions (from the ground state): in the present case these are the $A_{2g} + B_{1u} + B_{2u}$ modes.

4-9 LONGITUDINAL AND TRANSVERSE MODES

In Section 4-7(a) it was concluded that rocksalt, space group O_h^5 ($Fm3m$), will exhibit a single infrared active fundamental, of species F_{1u}. It was noted even earlier (Section 4-5) that at $\mathbf{k} = 0$ the optic mode of rocksalt does not, in fact, manifest the predicted triple degeneracy. Instead, it is split into a doubly degenerate TO mode and a non-degenerate LO mode. This phenomenon is characteristic of all states, transitions to which are dipole-allowed. It is a consequence of the long-range nature of the dipole–dipole interaction, which may couple local and resonant transition moments throughout a crystal. We shall examine this *microscopic* view in Section 5-7.

Because of its fundamental nature, it is important to first discuss the LO–TO splitting from a *macroscopic* point of view. For the most part, this chapter has been devoted to monatomic crystals. Most of these have been cubic. Interestingly, the LO–TO splitting represents an anisotropy which may be encountered even in a cubic crystal, although the splitting cannot be encountered in a monatomic crystal. As shown in Section 4-6(d), a lattice must be one with a basis in order for an optic mode to be observable. We therefore choose to discuss the phenomenon here.

We shall first discuss the optic modes of an infinite, cubic ionic crystal (like CsCl or NaCl) in the *absence* of an externally applied electromagnetic field. Although the system is in vibrational motion, we do not inquire as to how it was acquired; later we shall examine the consequence of interaction with an exciting field. We follow the development first introduced by Born and Huang.[49]

The lattice motion is described by a single coordinate which is defined as the (mass-weighted) displacement $\mathbf{r}$ of neighboring ions from their equilibrium separation. As a result of the motion of the ions, an electric field $\mathscr{E}$ is produced. The equation of motion of the ions is then

$$\ddot{\mathbf{r}} = b_{11}\mathbf{r} + b_{12}\mathscr{E} \tag{4-9-1}$$

that is, there is a restoring force proportional both to the local displacement (Hooke's Law) as well as to the long-range electric field produced in the infinite lattice by the implied concerted ($\mathbf{k} = 0$) motion.

The static ionic lattice may also be thought of as a lattice of dipoles having

49. M. Born and K. Huang, *Dynamical Theory of Crystal Lattices*, Oxford University Press, 1954, pages 82–86.

a net dielectric polarization $\mathscr{P}$ induced by and proportional to the field $\mathscr{E}$. $\mathscr{P}$ is defined as the average dipole moment per unit volume, where each cation-anion pair may be identified as a dipole. The vibrating lattice, however, modifies the polarization according to

$$\mathscr{P} = b_{21}\mathbf{r} + b_{22}\mathscr{E} \tag{4-9-2}$$

These two equations are *coupled* and they must be solved simultaneously.*

For each of the vector quantities $\mathscr{E}$, $\mathscr{P}$, and $\mathbf{r}$, we seek plane-wave solutions. For simplicity we set $\mathbf{k} = 0$. Thus

$$\begin{aligned} \mathscr{E} &= \mathscr{E}_0\, e^{i\omega t} \\ \mathscr{P} &= \mathscr{P}_0\, e^{i\omega t} \\ \mathbf{r} &= \mathbf{r}_0\, e^{i\omega t} \end{aligned} \tag{4-9-3}$$

Taking the necessary derivatives we obtain

$$\begin{aligned} -\omega^2\mathbf{r} &= b_{11}\mathbf{r} + b_{12}\mathscr{E} \\ \mathscr{P} &= b_{21}\mathbf{r} + b_{22}\mathscr{E} \end{aligned} \tag{4-9-4}$$

Elimination of $\mathbf{r}$ gives

$$\mathscr{P} = \left\{ b_{22} - \frac{b_{12}b_{21}}{b_{11} + \omega^2} \right\} \mathscr{E} \tag{4-9-5}$$

In a system containing bound charges such as ions, the electric displacement $\mathscr{D}$ differs from the electric field,

$$\mathscr{D} = \mathscr{E} + 4\pi\mathscr{P} \tag{4-9-6}$$

But in material systems, it is also true that $\mathscr{D}$ and $\mathscr{E}$ are linearly related by the dielectric tensor, ε

$$\mathscr{D} = \varepsilon\mathscr{E} \tag{4-9-7}$$

Elimination of $\mathscr{D}$ and $\mathscr{P}$ between (4-9-5), (4-9-6), and (4-9-7) results in the dispersion relation for the dielectric constant

$$\varepsilon = 1 + 4\pi \left\{ b_{22} - \frac{b_{12}b_{21}}{b_{11} + \omega^2} \right\} \tag{4-9-8}$$

This relation is usually written

$$\varepsilon = \varepsilon_\infty + \frac{\varepsilon_0 - \varepsilon_\infty}{1 - (\omega/\omega_0)^2} \tag{4-9-9}$$

* Neither (4-9-1) nor (4-9-2) includes a damping term, which must certainly be included in a more exact discussion, in order to account for the finite lifetime of any excited state. Such terms are included in the more detailed discussion of Section 5-7. By restricting the discussion here to the restoring forces, the system is non-dissipative, but it is intended only to reveal the origin of the *LO–TO* splitting.

ω_0 is seen to be the frequency (usually in the infrared) where the dielectric constant has a singularity. ε_0 is the value of ε at $\omega \ll \omega_0$; ε_∞ is the value of ε for $\omega \gg \omega_0$.

It can be shown* that the part of the restoring force which arises from the long-range electric field is identical to the dependence of the polarizability upon the local displacement vector; that is,

$$b_{12} = b_{21} \tag{4-9-10}$$

Comparing (4-9-8) and (4-9-9) the following identifications can be made:

$$\begin{aligned} b_{11} &= -\omega_0^2 \\ b_{12} = b_{21} &= \left(\frac{\varepsilon_0 - \varepsilon_\infty}{4\pi}\right)^{1/2} \omega_0 \\ b_{22} &= \frac{\varepsilon_\infty - 1}{4\pi} \end{aligned} \tag{4-9-11}$$

We now proceed to determine the phonon frequency $\omega(\mathbf{k} = 0)$ as a function of ω_0, ε_0, and ε_∞.

In the absence of conduction currents and a magnetic field, $\mathscr{E}$ is irrotational, $\nabla \times \mathscr{E} = 0$. This also means that there is no electromagnetic field *in the crystal.* At the end of this section we consider the consequences of removing this assumption. Since $\mathscr{P}$ is by definition irrotational,

$$\nabla \times (\mathscr{E} + 4\pi\mathscr{P}) = 0 \tag{4-9-12}$$

If the divergence *and* the curl of a field vanish, the field itself is zero. Hence,

$$\mathscr{E} = -4\pi\mathscr{P} \tag{4-9-13}$$

Making use of (4-9-2) we find

$$\nabla \cdot \mathscr{E} = -\frac{4\pi b_{21}}{1 + 4\pi b_{22}} \nabla \cdot \mathbf{r} \tag{4-9-14}$$

Now $\mathbf{r}$, too, is a vector which may have an irrotational part, $\mathbf{r}_l$, and a solenoidal part, $\mathbf{r}_t$ (for *l*ongitudinal and *t*ransverse, respectively) defined by

$$\begin{aligned} \nabla \times \mathbf{r}_l &= 0 \qquad \text{(irrotational)} \\ \nabla \cdot \mathbf{r}_t &= 0 \qquad \text{(solenoidal)} \end{aligned} \tag{4-9-15}$$

From (4-9-15) we see that only the irrotational part of $\mathbf{r}$ may appear in (4-9-14), that is

$$\nabla \cdot \mathscr{E} = -\frac{4\pi b_{21}}{1 + 4\pi b_{22}} \nabla \cdot \mathbf{r}_l \tag{4-19-16}$$

* Reference 49, Appendix V.

The unique solution of which is simply

$$\mathscr{E} = -\frac{4\pi b_{21}}{1 + 4\pi b_{22}}\mathbf{r}_l \tag{4-19-17}$$

Since $\mathscr{E}$ is irrotational, substitution of (4-9-17) into (4-9-1) gives

$$\ddot{\mathbf{r}}_t + \ddot{\mathbf{r}}_l = \left\{b_{11} - \frac{4\pi b_{12}b_{21}}{1 + 4\pi b_{22}}\right\}\mathbf{r}_l + b_{11}\mathbf{r}_t \tag{4-9-18}$$

Separating the solenoidal and irrotational parts, with the aid of (4-9-11) we have

$$\ddot{\mathbf{r}}_t = b_{11}\mathbf{r}_t = -\omega_0^2\mathbf{r}_t \tag{4-9-19}$$

and

$$\begin{aligned}\ddot{\mathbf{r}}_l &= \left\{b_{11} - \frac{4\pi b_{12}b_{21}}{1 + 4\pi b_{22}}\right\}\mathbf{r}_l \\ &= -\left(\frac{\varepsilon_0}{\varepsilon_\infty}\right)\omega_0^2\mathbf{r}_l\end{aligned} \tag{4-9-20}$$

The solutions to each of these equations of motion can be written in the plane waves:

$$\mathbf{r}_t = \mathbf{r}_t^0\, e^{i\omega_{TO}t} \qquad \text{and} \qquad \mathbf{r}_l = \mathbf{r}_l^0\, e^{i\omega_{LO}t} \tag{4-9-21}$$

Note that we have again taken $\mathbf{k} = 0$.

According to (4-9-19) and (4-9-20) these are respectively the transverse and longitudinal solutions, with

$$\omega_{TO} = \omega_0 \tag{4-9-22}$$

and

$$\omega_{LO} = \left(\frac{\varepsilon_0}{\varepsilon_\infty}\right)^{1/2}\omega_0 \tag{4-9-23}$$

Forming their ratio, we have

$$\left(\frac{\omega_{LO}}{\omega_{TO}}\right)^2 = \frac{\varepsilon_0}{\varepsilon_\infty} \tag{4-9-24}$$

which is known as the Lyddane, Sachs, Teller (LST) relation, first introduced by those authors[50] to estimate the LO–TO splitting. The relationship has been generalized to all degrees of freedom of a vibrating crystal by Cochran and Cowley.[51] Since $\varepsilon_0 > \varepsilon_\infty$ (see Section 5-3), $\omega_{LO} > \omega_{TO}$, in general.

The crucial distinction of ω_{LO} and ω_{TO} is seen to derive from the irrotational nature of $\mathscr{E}$. In other words, for a transverse mode (see Fig. 3-2) the local

50. R. H. Lyddane, R. G. Sachs, and E. Teller, *Phys. Rev.*, **59,** 673 (1941).
51. W. Cochran and R. A. Cowley, *J. Phys. Chem. Solids*, **23,** 447 (1962).

macroscopic field vanishes. Thus, the propagation direction is an axis of symmetry, but its very presence spoils the cubic symmetry. In this case, examination of (4-9-1) gives (4-9-19) directly and convinces us that $\omega_{TO} = \omega_0$. The tighter binding provided by $\mathscr{E}$ necessarily makes $\omega_{LO} > \omega_{TO}$.

Much has been written about the *LO–TO* splitting to qualify its appearance in cubic crystals as a function of the wave vector **k**, as well as certain physical aspects of the crystal sample, such as its size and shape, particularly in relation to the wavelength of an exciting electromagnetic field.[52–58] These are valid considerations with respect to the *computation* of the actual splitting for specific solids under particular conditions of excitation, such as wavelength, polarization of the electric field, sample size, and crystallinity, and we shall consider them in detail in Section 5-7. For our present purposes we have not had to take these matters into consideration. The anisotropy described by the *LO–TO* splitting is not predicted from symmetry considerations—the optic mode of CsCl belongs to F_{1u} of O_h^1 $(Pm3m)$ at $\mathbf{k} = (000)$. But this conclusion is based upon a consideration of the crystal at equilibrium—not under the artificial anisotropic displacement **r**. The fundamental anisotropy is that of the long-range field $\mathscr{E}$. Since no magnetic field is present, $\mathscr{E}$ can only be irrotational. The *LO–TO* splitting follows.

In calculating the actual splittings, we shall use a microscopic approach (see Section 5-7). According to (4-9-13), we can base this on a specific recipe for either $\mathscr{E}$ or $\mathscr{P}$. Traditionally $\mathscr{P}$ is calculated, and this must be done with great care, special attention being given to the various matters overlooked here, such as wavelength, polarization of the electric field, etc.

Before proceeding to a re-determination of ω_{LO} and ω_{TO} in the presence of electromagnetic radiation, we wish to emphasize that, in an absorption experiment at normal incidence, under no circumstances can ω_{LO} be directly observed, whatever its value or the origin of its difference from ω_{TO}. The reason for this is simply that the electric vector of the external radiation field is a transverse mode of the vacuum, and it can only interact with a transverse mode of the material system. *This* anisotropy can be appreciated by consideration of the interaction of a transverse electric field with a transverse phonon, such that **K** (photon) · **k** (phonon) = 0. We conclude that in cubic crystals, the directions called transverse and longitudinal must be operationally defined.

In the preceding pages we have presented the simple phenomenological basis of the *LO–TO* splitting following Huang.[59] In doing so, we effectively made use of only two of Maxwell's equations, in the form

$$\nabla \cdot (\mathscr{E} + 4\pi\mathscr{P}) = 0 \tag{4-9-25}$$

52. R. H. Lyddane and K. F. Herzfeld, *Phys. Rev.*, **54,** 846 (1938).
53. E. W. Kellerman, *Phil. Trans. Roy. Soc.* (*London*), **238,** 63 (1940).
54. W. R. Heller and A. Marcus, *Phys. Rev.*, **84,** 809 (1951).
55. M. H. Cohen and F. Keffer, *Phys. Rev.*, **99,** 1128 (1955).
56. A. D. B. Woods, W. Cochran, and B. N. Brockhouse, *Phys. Rev.*, **119,** 980 (1960).
57. H. B. Rosenstock, *Phys. Rev.*, **121,** 416 (1961).
58. T. H. K. Barron, *Phys. Rev.*, **123,** 1995 (1961).
59. K. Huang, *Proc. Roy. Soc.*, **A203,** 178 (1950).

and

$$\nabla \times \mathscr{E} = 0 \tag{4-9-26}$$

Thus, we implicitly assumed that $\dot{\mathscr{H}} = \dot{\mathscr{E}} = \dot{\mathscr{P}} = 0$, at least within time intervals which are long compared to vibrational periods. This assumption also results in another: $\mathscr{H} = 0$, where $\mathscr{H}$ is the external magnetic field. Similarly, there was the implicit assumption that **r**, $\mathscr{P}$, and $\mathscr{E}$ varied little over distances comparable with **a**, the lattice constant. Thus in (4-9-3) there is no wave-vector dependence of these quantities. We now proceed to recapitulate Huang's more detailed analysis[60] where these assumptions are not made.

Substitution of the plane wave solutions

$$\mathbf{r} = \mathbf{r}_0\, e^{i(\mathbf{k}\cdot\mathbf{x} - \omega t)} \tag{4-9-27}$$

$$\mathscr{P} = \mathscr{P}_0\, e^{i(\mathbf{k}\cdot\mathbf{x} - \omega t)} \tag{4-9-28}$$

$$\mathscr{E} = \mathscr{E}_0\, e^{i(\mathbf{k}\cdot\mathbf{x} - \omega t)} \tag{4-9-29}$$

$$\mathscr{H} = \mathscr{H}_0\, e^{i(\mathbf{k}\cdot\mathbf{x} - \omega t)} \tag{4-9-30}$$

and the equations of motion (4-9-4) in Maxwell's equations

$$\nabla\cdot(\mathscr{E} + 4\pi\mathscr{P}) = 0 \tag{4-9-31}$$

$$\nabla\cdot\mathscr{H} = 0 \tag{4-9-32}$$

$$\nabla \times \mathscr{E} = -\frac{1}{c}\dot{\mathscr{H}} \tag{4-9-33}$$

$$\nabla \times \mathscr{H} = -\frac{1}{c}(\dot{\mathscr{E}} + 4\pi\dot{\mathscr{P}}) \tag{4-9-34}$$

gives

$$\begin{aligned} -\omega^2\mathbf{r} &= b_{11}\mathbf{r} + b_{12}\mathscr{E} \\ \mathscr{P} &= b_{21}\mathbf{r} + b_{22}\mathscr{E} \end{aligned} \tag{4-9-4}$$

$$\mathbf{k}\cdot(\mathscr{E} + 4\pi\mathscr{P}) = 0 \tag{4-9-35}$$

$$\mathbf{k}\cdot\mathscr{H} = 0 \tag{4-9-36}$$

$$\mathbf{k} \times \mathscr{E} = \frac{\omega}{c}\mathscr{H} \tag{4-9-37}$$

$$\mathbf{k} \times \mathscr{H} = -\frac{\omega}{c}(\mathscr{E} + 4\pi\mathscr{P}) \tag{4-9-38}$$

In deducing (4-9-37) and (4-9-38), use is made on the LHS of the right-hand rule, in order to correctly find the sign of the RHS.

60. K. Huang, *Proc. Roy. Soc. (London)*, **A208,** 352 (1951).

Elimination of **r** from (4-9-4) yields (4-9-5), as before; substitution in (4-9-35) produces

$$(\mathbf{k} \cdot \mathscr{E})\left\{1 + 4\pi b_{22} - \frac{4\pi b_{12} b_{21}}{b_{11} + \omega^2}\right\} = 0 \tag{4-9-39}$$

There are two ways to achieve (4-9-39). Either,

$$\left\{1 + 4\pi b_{22} - \frac{4\pi b_{12} b_{21}}{b_{11} + \omega^2}\right\} = 0 \tag{4-9-40}$$

or

$$\mathbf{k} \cdot \mathscr{E} = 0 \tag{4-9-41}$$

Consider the first case. Because of (4-9-7), (4-9-40) requires

$$\mathscr{E} + 4\pi\mathscr{P} = 0 \tag{4-9-42}$$

Consequently (4-9-38) reduces to

$$\mathbf{k} \times \mathscr{H} = 0 \tag{4-9-43}$$

which means that either $\mathscr{H} = 0$ or $\mathscr{H} \parallel \mathbf{k}$. But (4-9-36) states that either $\mathscr{H} = 0$ or $\mathscr{H} \perp \mathbf{k}$. The only joint possibility, which must be true in this case, is

$$\mathscr{H} = 0 \tag{4-9-44}$$

This means that (4-9-37) becomes

$$\mathbf{k} \times \mathscr{E} = 0 \tag{4-9-45}$$

Since $\mathscr{E} \neq 0$, $\mathscr{E} \parallel \mathbf{k}$. Evidently $\mathscr{E}$, **k**, $\mathscr{P}$, and **r** are all parallel in this case, and we have the longitudinal mode. From (4-9-40),

$$\omega_l^2 = -b_{11} + \frac{4\pi b_{12} b_{21}}{1 + 4\pi b_{22}} \tag{4-9-46}$$

which can be shown to be the same as (4-9-23) by using (4-9-11).

We now consider the second way to achieve (4-9-39), $\mathbf{k} \cdot \mathscr{E} = 0$. This is the *TO* case, $\mathbf{k} \perp \mathscr{E}$, and $\mathscr{H} \neq 0$. Hence, **k**, $\mathscr{E}$, and $\mathscr{H}$ form the usual right-hand system of orthogonal vectors. Equation (4-9-37) becomes

$$k\mathscr{E} = \frac{\omega}{c}\mathscr{H} \tag{4-9-47}$$

(4-9-36) is satisfied, and (4-9-38) becomes

$$k\mathscr{H} = \frac{\omega}{c}(\mathscr{E} + 4\pi\mathscr{P}) \tag{4-9-48}$$

Combining (4-9-6)–(4-9-8) and (4-9-47)–(4-9-48), we obtain

$$\varepsilon = \left(\frac{ck}{\omega_t}\right)^2 = 1 + 4\pi b_{22} - \frac{4\pi b_{12} b_{21}}{b_{11} + \omega_t^2} \tag{4-9-49}$$

Comparison with (4-9-9) yields the identities (4-9-11), whereby we may re-write (4-9-49) as

$$\left(\frac{ck}{\omega_t}\right)^2 = \varepsilon_\infty + \frac{(\varepsilon_0 - \varepsilon_\infty)}{1 - (\omega_t/\omega_0)^2} \tag{4-9-50}$$

the asymptotes of which are the approximate TO dispersion curve (4-9-22) and $\omega = ck$, the photon dispersion curve. In the vicinity of the crossing of the two curves, the more exact expression (4-9-50) applies. The inclusion of $\mathscr{H}$ and its derivatives has served to mix the vector field and the displacements. Since $\nabla \times \mathscr{E} \neq 0$, $\mathscr{E}$ is not irrotational and the substitution which gave rise to (4-9-18) and its subsequent factoring is not exact. We see again the importance of the magnetic field in mixing the mechanical ($\mathbf{r}$) and optical ($\mathscr{E}$) waves. This effect, known as retardation, is discussed from a microscopic point of view in Section 5-7.

Mixing of the field and the displacements creates a new particle—the *polariton*, which is neither pure photon nor pure phonon, but some of each. Recalling that in an absorption experiment only ω_{TO} can be detected, we now enquire into how ω_{TO} depends on the experimental conditions. In order to do so, let us consider first another experiment, namely one in which the phonon, or more accurately, the polariton, is detected in a scattering experiment.

For this purpose we consider a Raman scattering experiment, in which the incident photon has momentum $\mathbf{k}_i = \omega_L n_i \mathbf{s}_i / c$, where ω_L is the laser frequency in a cubic crystal of index n. The scattered photon then has momentum $\mathbf{k}_s = \omega_s n_s \mathbf{s}_s / c$. Conservation of momentum requires that

$$\mathbf{k}_i = \mathbf{k}_s + \mathbf{k}_p \tag{4-9-51}$$

where $\mathbf{k}_p$ is the momentum of the polariton. The scattering geometry is illustrated in Fig. 4-18.

Equation (4-9-51) can also be written as

$$k_p = (k_i^2 + k_s^2 - 2k_i k_s \cos\theta)^{1/2} \tag{4-9-52}$$

where $\theta = \mathbf{s}_i \cdot \mathbf{s}_s$, the scattering angle. Since $k_i = n_i\omega_i/c$ and $k_s = n_s\omega_s/c$ at small scattering angles k_p is

$$\begin{aligned} k_p &= [(n_i\omega_i/c - n_s\omega_s/c)^2 + k_i k_s \theta^2]^{1/2} \\ &\cong \left\{\omega^2(\mathbf{k})\left[\frac{dk(\omega)}{d\omega}\right]^2_{\omega=\omega_i} + k_i k_s \theta^2\right\}^{1/2} \end{aligned} \tag{4-9-53}$$

where

$$\frac{dk(\omega)}{d\omega} = \frac{1}{c}\left(n + \frac{dn}{d\omega}\right) \tag{4-9-54}$$

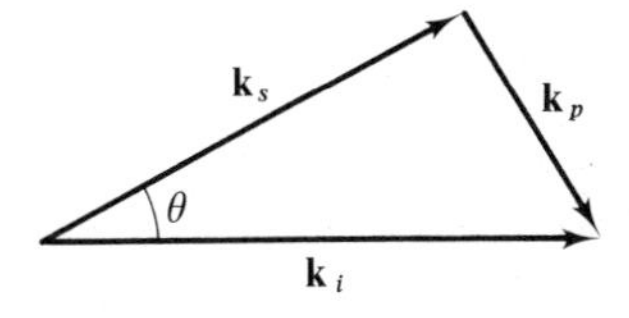

Figure 4-18 Scattering geometry in a Raman experiment. $\mathbf{k}_i$ is the wave vector of the incident photon, $\mathbf{k}_s$ is the wave vector of the scattered photon, and $\mathbf{k}_p$ is the wave vector of the polariton

Detection of a polariton in an experiment at a particular scattering angle θ requires simultaneous satisfaction of (4-9-50) and (4-9-53). Results for a cubic crystal (GaP) were first obtained by Henry and Hopfield,[61] who showed that significant deviations of ω_{TO} from ω_0 were only found for $\theta < 2°$. At larger scattering angles, for all practical purposes the polariton may be considered a pure TO phonon.

Frequency shifts of Raman scattering as a function of angle of observation from the forward direction have also been observed in ZnO,[62] α-quartz,[63] $LiIO_3$,[64] and $LiNbO_3$.[65] The angular dependence of these Raman shifts, interpreted using (4-9-52), can be used to obtain the polariton dispersion curves. It should be noted that the region in k-space in which significant photon–phonon interaction takes place is generally in the vicinity of $10^3 < |k| < 10^4$ cm^{-1}. Recalling that the BZB corresponds to $|k| \sim \pi/a$ or $|k| \sim 10^8$ cm^{-1}, this means that a phonon is perturbed by a photon in what should be considered as the $k \approx 0$ region. This is sensible, since the unperturbed photon dispersion curve hardly departs from the $k \approx 0$ region.

An absorption experiment carried out at normal incidence may be viewed as a scattering experiment at zero scattering angle. Such an experiment would therefore be in the polariton regime. On the other hand, since absorption experiments are generally carried out with poorly collimated light, the particles observed are, on the average, pure phonons.

61. C. H. Henry and J. J. Hopfield, *Phys. Rev. Letters,* **15,** 964 (1965).
62. S. P. S. Porto, B. Tell, and T. C. Damen, *Phys. Rev. Letters,* **16,** 450 (1966).
63. J. P. Scott, L. E. Cheesman, and S. P. S. Porto, *Phys. Rev.,* **162,** 834 (1967).
64. R. Claus, *Z. Naturf.,* **25a,** 306 (1970).
65. R. Claus and H. W. Schrötter, *Second International Conference on Light Scattering in Solids,* Paris, 1971.

CHAPTER

FIVE

OPTICAL PROPERTIES OF SINGLE CRYSTALS

5-1 THE DIELECTRIC CONSTANT

The behavior of a crystal subjected to externally applied electromagnetic radiation can be described with the aid of Maxwell's equations and the dielectric tensor $\boldsymbol{\varepsilon}$, which relates the $\mathscr{E}$ vector to the $\mathscr{D}$ (displacement) vector according to

$$\mathscr{D}_\sigma = \sum_{\sigma'} \varepsilon_{\sigma\sigma'}\mathscr{E}_{\sigma'} \qquad \sigma, \sigma' = X, Y, Z \tag{5-1-1}$$

In crystals with orthorhombic or higher symmetry, the principal axes of the dielectric tensor are simply related to the structure as described in Table 5-1, so that one need only consider

$$\varepsilon_\sigma = \varepsilon_\sigma(\infty) + \sum_j S_j^{(\sigma)}\nu_j^2/(\nu_j^2 - \nu^2 - i\gamma_j\nu) \tag{5-1-2}$$

The microscopic justification of (5-1-2) will be given below in section 5-8, where it will be shown how $S_j^{(\sigma)}$, the *transition strength*, and ν_j may be related to molecular properties. It is more difficult to relate the *damping constant* γ_j to such properties, and we shall content ourselves with the inclusion of this term as a phenomenological parameter.

The dielectric constant is, of course, dimensionless, and according to the present definition, $S_j^{(\sigma)}$ is also dimensionless. The frequency ν may be chosen in units of sec^{-1}, but in optical spectroscopy it is more convenient to express it in cm^{-1}, which we shall do throughout this chapter. The quantity $\varepsilon_\sigma(\infty)$ represents the high frequency dielectric constant, i.e., the part due to electronic transitions, and is justifiably treated as a constant term at the frequencies (wave numbers) of atomic vibrations.

Table 5-1 Relation of principal axes of the dielectric tensor to crystal axes

System	Orientation	Remarks
Cubic	Any	$\varepsilon_X = \varepsilon_Y = \varepsilon_Z$
Hexagonal	Z parallel to C_6 or C_3; X, Y any directions in plane perpendicular to Z	$\varepsilon_X = \varepsilon_Y$
Rhombohedral (Trigonal)	Z parallel to C_3; X, Y any directions in plane perpendicular to Z	$\varepsilon_X = \varepsilon_Y$
Tetragonal	Z parallel to C_4 or S_4; X, Y any directions in plane perpendicular to Z	$\varepsilon_X = \varepsilon_Y$
Orthorhombic	X, Y, Z parallel to the three C_2 axes in D_2 or D_{2h}; Z parallel to C_2 and X and Y perpendicular to σ_v planes in C_{2v}	
Monoclinic	Z parallel to C_2 or perpendicular to σ; X and Y not fixed by symmetry, and frequency dependent	X and Y different for real and imaginary parts of ε
Triclinic	No axes fixed by symmetry and all three axes may be frequency dependent	Real and imaginary parts of ε have different principal axes

The quantity $S_j^{(\sigma)}$ will subsequently be shown to be proportional, *inter alia,* to $(\partial \mu_\sigma / \partial Q_j)^2$ where ν_j is the frequency of a fundamental transition, and Q_j is the normal coordinate associated with a phonon mode. The summation over j ought to be extended over all transition frequencies ν_j which are infrared active, including overtones and combinations, but for simplicity we will confine our attention to the active fundamentals.

Although the dielectric tensor completely specifies the optical properties of the crystal, it is necessary to prescribe carefully the direction and polarization of wave propagation in the crystal and to match this specification with the boundary conditions and the directions of propagation and polarization external to the crystal before the response of the crystal can be fully determined. Although interesting results arise when the crystal boundary plane is varied in a general manner, most of the remainder of this chapter will assume that the boundary face is a principal plane of the dielectric tensor. However, before we analyze the boundary conditions, we must relate the dielectric constant to the properties of a plane wave propagating in the crystal. This is done in the next section.

5-2 PLANE WAVE PROPAGATION IN ANISOTROPIC CRYSTALS

With the aid of Maxwell's equations, it is possible to show that[1]

$$\mathscr{D} = n^2[\mathscr{E} - \mathbf{s}(\mathbf{s} \cdot \mathscr{E})] \tag{5-2-1}$$

when one assumes that the field vectors have the plane wave form proportional to $e^{i(\mathbf{k}\cdot\mathbf{r} - \omega t)}$ where

$$\mathbf{k} = 2\pi \nu n \mathbf{s} \tag{5-2-2}$$

n is the refractive index, and $\mathbf{s}$ is a unit vector in the propagation direction. Then in view of (5-1-1) we must have

$$\sum_{\sigma'} [(n^2 - \varepsilon_\sigma)\delta_{\sigma\sigma'} - n^2 s_\sigma s_{\sigma'}]\mathscr{E}_{\sigma'} = 0 \tag{5-2-3}$$

where σ and σ' are taken in principal axis directions. Thus the existence of the $\mathscr{E}$ field in the crystal, for an assumed propagation direction $\mathbf{s}$ requires that the square of the refractive index, n^2, satisfies the characteristic equation

$$|(n^2 - \varepsilon_\sigma)\delta_{\sigma\sigma'} - n^2 s_\sigma s_{\sigma'}| = 0 \tag{5-2-4}$$

Suppose the wave to be propagating in the z direction: then (5-2-4) becomes

$$\begin{vmatrix} n^2 - \varepsilon_X & 0 & 0 \\ 0 & n^2 - \varepsilon_Y & 0 \\ 0 & 0 & -\varepsilon_Z \end{vmatrix} = 0 \tag{5-2-5}$$

This has *three* solutions, either (i) $n^2 = \varepsilon_X$, or (ii) $n^2 = \varepsilon_Y$, or (iii) $\varepsilon_Z = 0$. The first two solutions admit $\mathscr{E}$ vectors of the forms: (i) $\mathscr{E}_X \neq 0$, $\mathscr{E}_Y = \mathscr{E}_Z = 0$, or (ii) $\mathscr{E}_Y \neq 0$, $\mathscr{E}_X = \mathscr{E}_Z = 0$, provided $\varepsilon_X \neq \varepsilon_Y$. If $\varepsilon_X = \varepsilon_Y$, we have a crystal of at least uniaxial symmetry, and the first two solutions admit any combination of $\mathscr{E}_X \neq 0$, and $\mathscr{E}_Y \neq 0$. These first two solutions are *transverse* waves, the $\mathscr{E}$ vector being perpendicular to the propagation vector. The third solution represents a purely *longitudinal* wave, since (5-2-3) requires that $\mathscr{E}_X = \mathscr{E}_Y = 0$, if ε_X and ε_Y are different from $\varepsilon_Z = 0$.

If instead of propagating parallel with a principal axis, the propagation vector of the wave lies in a principal plane, say YZ, the characteristic equation (5-2-4) reduces to

$$\begin{vmatrix} n^2 - \varepsilon_X & 0 & 0 \\ 0 & n^2(1 - s_Y^2) - \varepsilon_Y & -n^2 s_Y s_Z \\ 0 & -n^2 s_Y s_Z & n^2(1 - s_Z^2) - \varepsilon_Z \end{vmatrix} = 0 \tag{5-2-6}$$

One solution is obviously $n^2 = \varepsilon_X$ and in general $\mathscr{E} = (\mathscr{E}_X, 0, 0)$. This wave will be called the "ordinary" wave.

1. See M. V. Klein, *Optics*, Wiley, New York, 1970, page 597.

The other solution is obtained by choosing n^2 so that the subdeterminant

$$\begin{vmatrix} n^2(1 - s_Y^2) - \varepsilon_Y & -n^2 s_Y s_Z \\ -n^2 s_Y s_Z & n^2(1 - s_Z^2) - \varepsilon_Z \end{vmatrix}$$

vanishes. The equation for n^2 may be simplified since $s_Y^2 + s_Z^2 = 1$, and the result is

$$\frac{1}{n^2} = \frac{s_Z^2}{\varepsilon_Y} + \frac{s_Y^2}{\varepsilon_Z} \tag{5-2-7}$$

The wave which arises in this way is called the "extraordinary" wave and it is clear that the refractive index, unlike that of the ordinary wave, depends upon the propagation direction, assuming $\varepsilon_Y \neq \varepsilon_Z$.

5-3 SEPARATION OF THE DIELECTRIC CONSTANT AND REFRACTIVE INDEX INTO REAL AND IMAGINARY PARTS

All of the preceding discussion has contemplated the existence of a complex dielectric constant and a correspondingly complex refractive index. The description of the wave propagation in a crystal requires in general that, given

$$\varepsilon = \varepsilon' + i\varepsilon'' \tag{5-3-1}$$

one works with

$$\begin{aligned} n &= n' + in'' \\ n^2 &= \varepsilon = (n')^2 - (n'')^2 + 2in'n'' \end{aligned} \tag{5-3-2}$$

Thus

$$\varepsilon' = (n')^2 - (n'')^2 \tag{5-3-3}$$

and

$$\varepsilon'' = 2n'n'' \tag{5-3-4}$$

The frequency dependence follows from (5-1-2); if there are several resonant modes, the analysis of ε into its real and imaginary parts is rather cumbersome, but if one first puts

$$\varepsilon = \varepsilon(\infty) + \sum_j S_j \nu_j^2(\nu_j^2 - \nu^2 + i\gamma_j\nu)/[(\nu_j^2 - \nu^2)^2 + \gamma_j^2\nu^2] \tag{5-3-5}$$

and then approximates the contribution of the terms for $j' \neq j$ in the vicinity of the frequency ν_j by using

$$\begin{aligned} \varepsilon \cong \varepsilon(\infty) &+ S_j \nu_j^2(\nu_j^2 - \nu^2 + i\gamma_j\nu)/[(\nu_j^2 - \nu^2)^2 + \gamma_j^2\nu^2] \\ &+ \sum_{j' \neq j} S_{j'} \nu_{j'}^2(\nu_{j'}^2 - \nu_j^2 + i\gamma_{j'}\nu_j)/[(\nu_{j'}^2 - \nu_j^2)^2 + \gamma_{j'}^2\nu_j^2] \end{aligned} \tag{5-3-6}$$

it is possible to express the real and imaginary parts, in the vicinity of ν_j, as

$$\varepsilon' = \varepsilon(\infty) + \bar{\varepsilon}'(\nu_j) + \frac{S_j\nu_j^2(\nu_j^2 - \nu^2)}{(\nu_j^2 - \nu^2)^2 + \gamma_j^2\nu^2} \tag{5-3-7}$$

$$\varepsilon'' = \bar{\varepsilon}''(\nu_j) + \frac{S_j\nu_j^2\gamma_j\nu}{(\nu_j^2 - \nu^2)^2 + \gamma_j^2\nu^2} \tag{5-3-8}$$

where $\bar{\varepsilon}'(\nu_j)$ and $\bar{\varepsilon}''(\nu_j)$ are the (approximately) constant real and imaginary parts arising from the sum over $j' \neq j$ in (5-3-6).

In many cases, one can utilize the *well separated mode* approximation which is defined as follows:

$$\text{(i)}\ S_{j'}\nu_{j'}^2 \ll |\nu_{j'}^2 - \nu_j^2|$$

and

$$\text{(ii)}\ \gamma_{j'}\nu_j \ll |\nu_{j'}^2 - \nu_j^2| \tag{5-3-9}$$

When (5-3-9) is satisfied, the approximate expressions given in (5-3-7) and (5-3-8) reduce to

$$\begin{aligned} \varepsilon' &= \varepsilon(\infty) + S_j\nu_j^2(\nu_j^2 - \nu^2)/[(\nu_j^2 - \nu^2)^2 + \gamma_j^2\nu^2] \\ \varepsilon'' &= S_j\nu_j^2\gamma_j\nu/[(\nu_j^2 - \nu^2)^2 + \gamma_j^2\nu^2] \end{aligned} \tag{5-3-10}$$

which simply mean that only the single near-resonant term need be considered so far as the imaginary part of ε is concerned and that the real part similarly contains a single term together with the high frequency contribution. Of these approximations (5-3-9), the first is usually less well justified, and it may be better to use a slightly different approximation for $\bar{\varepsilon}'(\nu_j)$, obtained by noting that

$$\begin{aligned} S_{j'}\nu_{j'}^2/(\nu_{j'}^2 - \nu_j^2) &\cong 0 \qquad \text{if } \nu_j > \nu_{j'} \\ &\cong S_{j'} \qquad \text{if } \nu_j < \nu_{j'} \end{aligned}$$

so that

$$\bar{\varepsilon}'(\nu_j) = \sum_{\{j' : (\nu_j < \nu_{j'})\}} S_{j'} \tag{5-3-11}$$

5-4 REFLECTIVITY AT A SINGLE CRYSTAL SURFACE

Although we have described the relation between the parameters of an electromagnetic wave propagating in a crystal and the dielectric tensor regarded as a phenomenological function of frequency, we have not yet described how the crystal interacts with external radiation. The initial discussion will be devoted to the case of reflection from a single, principal plane surface, the crystal being assumed to be sufficiently thick so that no energy is returned from internal reflection at the back surface.

(a) Normal Incidence

We consider first the case of normal incidence upon the XY plane of the crystal, the external radiation being X polarized, i.e., $\mathscr{E} = (\mathscr{E}_X, 0, 0)$. From the discussion in Section 5-2 one sees that $n^2 = \varepsilon_X$. Now a further consequence of Maxwell's equations is the requirement that the transverse component of $\mathscr{E}$, which is the only component which exists in this experiment, be continuous across the boundary and that similarly the derivative of the transverse component of $\mathscr{E}$ with respect to Z be continuous. These two requirements can only be satisfied if the external field is split into an incident part i and a reflected part r; the transmitted part in the medium is designated with subscript t.

$$\mathscr{E}_X = \mathscr{E}^\circ_{Xi}\, e^{ik_z Z} + \mathscr{E}^\circ_{Xr}\, e^{-ik_z Z}$$

which leads to the continuity equations (in the boundary $Z = 0$):

$$\mathscr{E}^\circ_{Xi} + \mathscr{E}^\circ_{Xr} = \mathscr{E}^\circ_{Xt}$$

and

$$\mathscr{E}^\circ_{Xi} - \mathscr{E}^\circ_{Xr} = n\mathscr{E}^\circ_{Xt}$$

From the above, it is easily found that

$$\frac{\mathscr{E}^\circ_{Xr}}{\mathscr{E}^\circ_{Xi}} = \frac{1 - n_X}{1 + n_X} \tag{5-4-1}$$

and the reflectivity R defined as the ratio of the intensities is then

$$R = \frac{I_r}{I_i} = \left|\frac{\mathscr{E}^\circ_{Xr}}{\mathscr{E}^\circ_{Xi}}\right|^2 = \left|\frac{1 - n_X}{1 + n_X}\right|^2 \tag{5-4-2}$$

Were the incident radiation Y polarized, one would replace n_X in (5-4-2) by n_Y.

Since by (5-2-5), the refractive index is the square root of the appropriate principal value of the dielectric tensor, which values are in general complex, the reflectivity will depend upon the parameters of the dielectric constant in a rather involved manner. For the purpose of a qualitative understanding, it is useful to consider the phenomenon for the case of a vanishing damping constant, $\gamma_j \to 0$. The frequency dependence of a component of the dielectric constant may then be illustrated as in Fig. 5-1. One sees that for a frequency region just above ν_j the real part of the dielectric constant is negative, so that the refractive index in such regions is pure imaginary. Figure 5-2 shows n' and n'' versus frequency for the same parameters assumed in Fig. 5-1. For the case of $\gamma = 0$, Eq. (5-4-2) predicts that the reflection will be total ($R = 1$) in such a region. This conclusion will, of course, be somewhat modified if a finite damping constant is included; this is shown in Fig. 5-3, in which $|\varepsilon|$ and R are plotted on the same frequency scale. Note that the longitudinal frequency as judged by the upper edge of the reflection band closely matches the minimum in $|\varepsilon|$.

It should be obvious from the foregoing that by the choice of an appropriate crystal face together with polarization of the incident beam, it is possible to

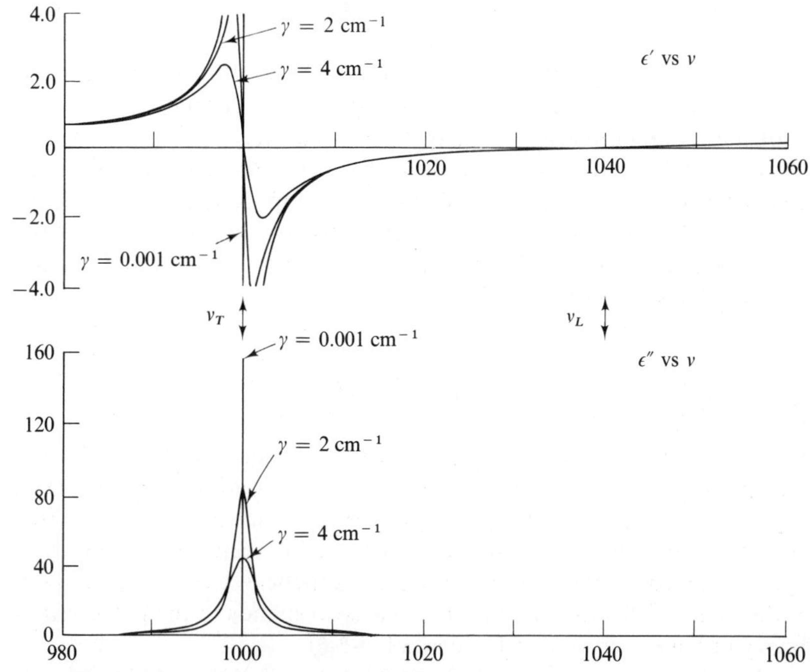

Figure 5-1 Real (ε') and imaginary (ε'') parts of dielectric constant versus frequency ν in cm^{-1}, showing the effect of varying damping constant, $\gamma = 0.001 \approx 0$, 2, and 4 cm^{-1}. Fixed parameters: $\nu_T = 1000$, $S = 0.18$ ($\nu_L = 1040$), $\varepsilon(\infty) = 2.24$

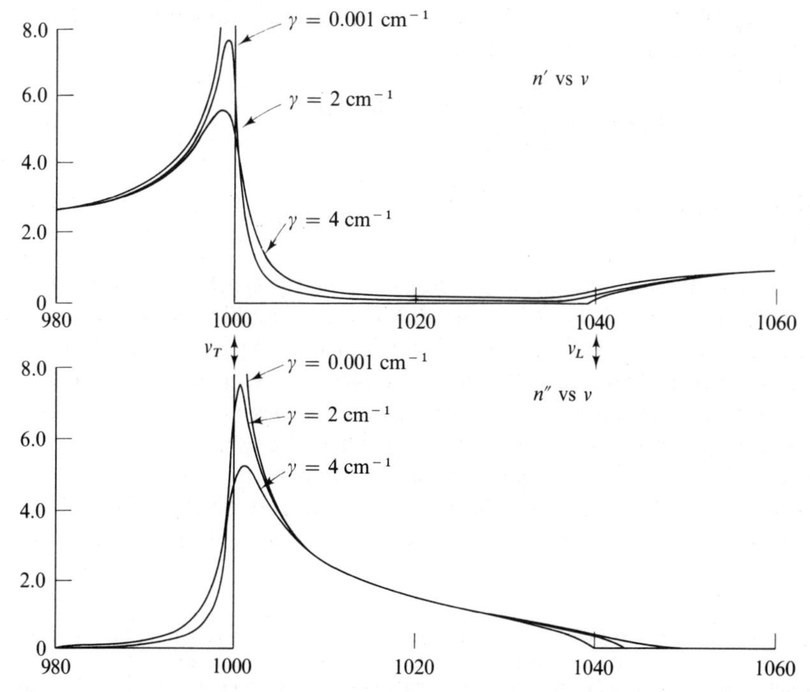

Figure 5-2 Refractive index (n') and absorption index (n'') versus frequency ν in cm^{-1}. Fixed parameters: $\nu_T = 1000$, $S = 0.18$ ($\nu_L = 1040$), $\varepsilon(\infty) = 2.24$

study separately ε_X, ε_Y, and ε_Z. For each direction, the reflection spectrum taken at normal incidence will locate more or less accurately, depending upon the magnitude of the damping constant, the resonant frequency ν_j, together with a neighboring frequency ν_{jL} at which the dielectric constant vanishes. We have already seen (Section 5-2) that the vanishing of a component of the dielectric constant corresponds with a longitudinal wave in the crystal. The microscopic theory of such a mode (as distinguished from the transverse mode, which has the frequency ν_j) is discussed in Section 5-7.

(b) Oblique Incidence

When radiation is incident upon a principal face of a crystal at a finite angle, the reflectivity equations, as would be expected, become more complicated. Moreover, it is necessary to distinguish between two polarization cases which are often called TE (the transverse electric vector, i.e. the external $\mathscr{E}$ vector is perpendicular to the plane of incidence) and TM (transverse magnetic vector, which implies that $\mathscr{E}$ is in the plane of incidence).* The TE case corresponds to the propagation of the ordinary ray in the crystal in the sense of (5-2-6).

We use the numerical subscripts 1, 2, 3 to identify the principal directions of the dielectric tensor (any permutation of X, Y, Z in crystals of orthorhombic or higher symmetry). Then let the crystal face be the 12 plane and the plane of incidence the 23 plane (see Fig. 5-4). The appropriate generalizations[2] of the normal incidence reflectivity equation (5-4-2) are

$$R(TE) = \left| \frac{\cos\theta_i - (\varepsilon_1 - \sin^2\theta_i)^{1/2}}{\cos\theta_i + (\varepsilon_1 - \sin^2\theta_i)^{1/2}} \right|^2 \tag{5-4-3}$$

$$R(TM) = \left| \frac{\varepsilon_2^{1/2}\cos\theta_i - (1 - \sin^2\theta_i/\varepsilon_3)^{1/2}}{\varepsilon_2^{1/2}\cos\theta_i + (1 - \sin^2\theta_i/\varepsilon_3)^{1/2}} \right|^2 \tag{5-4-4}$$

In the TE case, the incident electric vector has a component only in direction 1, so only ε_1 is involved. In the TM case, the incident electric vector has components in both the 2 and 3 directions, of which, according to our convention, 3 is normal to the crystal face. A particularly interesting case arises if ε_3 contains a resonant term in a certain frequency region, but ε_2 does not. In such a case, strong reflection will occur in the frequency region where the term $(1 - \sin^2\theta_i/\varepsilon_3)^{1/2}$ is pure imaginary, or nearly so. Neglect of the damping term makes ε_3 real; the term $1 - \sin^2\theta_i/\varepsilon_3$ will be negative where ε_3 is small and positive, i.e., where $0 < \varepsilon_3 < \sin^2\theta_i$. The frequency region in which this occurs is just above the

* These two cases are also often distinguished as s and p (or σ and π) where s, from the German, means "senkrecht," i.e., the electric vector is perpendicular to the plane of incidence.

2. L. P. Mosteller and F. Wooten, *J. Opt. Soc. Amer.*, **58,** 511 (1968); J. C. Decius, R. E. Frech, and P. Brüesch, *J. Chem. Phys.*, **58,** 4056 (1973).

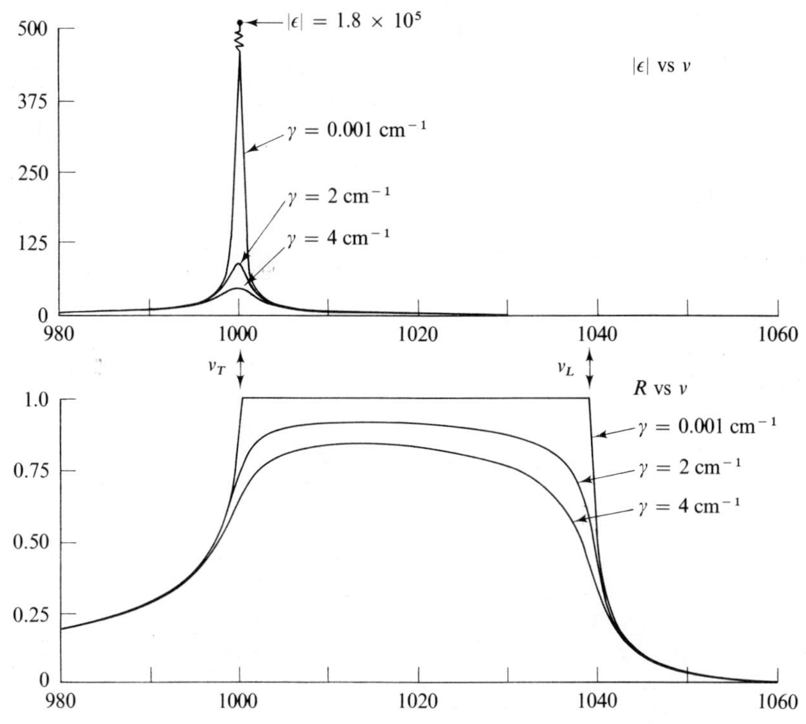

Figure 5-3 Absolute value of dielectric constant, $|\varepsilon|$, and reflectivity R versus frequency ν in cm^{-1}, showing effect of varying damping constant (γ). Fixed parameters: $\nu_T = 1000$, $S = 0.18$ ($\nu_L = 1040$), $\varepsilon(\infty) = 2.24$

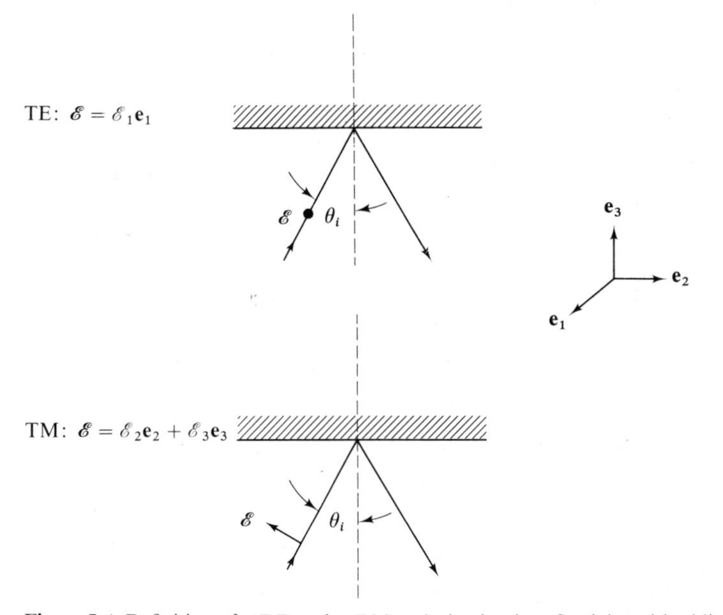

Figure 5-4 Definition of $s(TE)$ and $p(TM)$ polarization in reflectivity with oblique incidence

longitudinal frequency, hence it is expected that strong reflection will occur beginning at ν_{jL} and extending to higher frequencies to an extent determined by the magnitude of θ_i.

Observations of this type are useful for a number of reasons. First, there are numerous crystals whose growth habit makes it difficult to prepare specimens of useful area except for faces normal to a certain axis. Second, such an off-axis experiment (as compared with normal incidence) is helpful in establishing a precise value for ν_{jL}. Third, for purely experimental reasons, it may be difficult to limit the observation to purely normal incidence.

(c) Overlapping Modes

When several nearby resonances of the same polarization occur, as is often the case in crystals with several molecular units per unit cell, the analysis of the resultant reflectance, even at normal incidence, may prove to be more difficult. If the damping constants are sufficiently small, it will be possible to resolve the regions associated with two or more resonances, but the zeros in the dielectric constant between modes will evidently no longer depend simply upon the transverse frequency of a single mode ν_j and upon its strength S_j. In fact, for finite damping constants, the dielectric constant only has zeros for complex frequencies. Some examples of reflectance at normal incidence for two resonances, varying damping constants, and varying strengths are shown in Fig. 5-5. Note particularly the asymmetry for the case of equal strengths which clearly arises from the partial cancellation of the large negative value of the dielectric constant just above the first resonance due to the second (higher frequency) resonance.

(d) Reflectivity from an Arbitrary Face

If the crystal face is no longer a principal plane of the dielectric tensor, the situation may become inconveniently complicated. An interesting discussion is possible, however, for uniaxial crystals.[3] If *normal* incidence is observed on a face whose normal subtends an angle θ with the unique (Z) axis, then the propagation direction lies in a principal plane which we call YZ, and Eq. (5-2-6) applies. If the radiation is polarized in the X direction, then $n^2 = \varepsilon_X$ and the reflectivity is governed by the normal incidence equation (5-4-2). However, if the radiation is polarized in the YZ plane, we have according to Eq. (5-2-7), putting $s_Z = \cos\theta$ and $s_Y = \sin\theta$, and inverting

$$n_\theta^2 = \varepsilon_\theta = \frac{\varepsilon_Y \varepsilon_Z}{\varepsilon_Y \sin^2\theta + \varepsilon_Z \cos^2\theta} \qquad (5\text{-}4\text{-}5)$$

3. L. Couture, J. P. Mathieu, J. A. A. Ketelaar, W. Vedder, and J. Fahrenfort, *J. Chem. Phys.*, **20**, 1492 (1952).

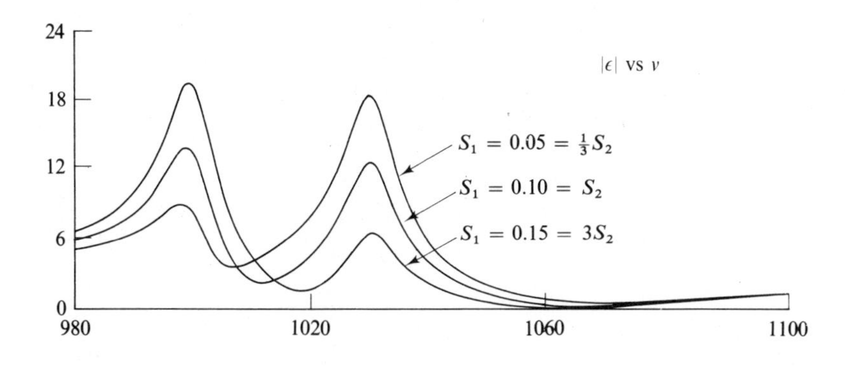

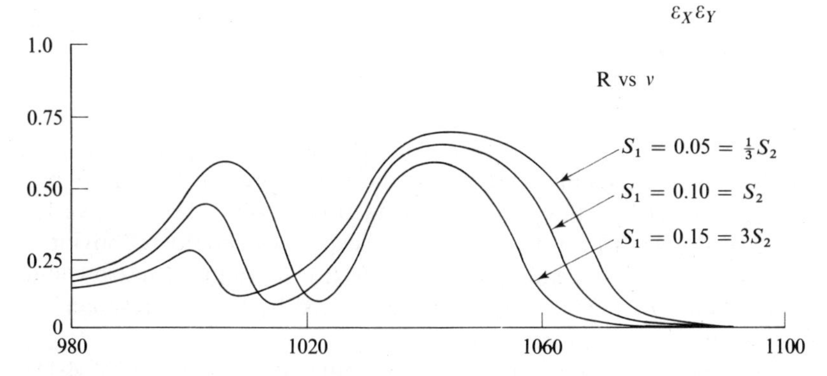

Figure 5-5 Absolute value of dielectric constant, $|\varepsilon|$, and reflectivity R versus frequency ν in cm^{-1}. Parameters: $\nu_{T1} = 1000$, $\nu_{T2} = 1030$; damping constant for both modes = 8 cm^{-1}; $\varepsilon(\infty) = 2.24$. Strengths varied as shown

Considering only a single resonance and neglecting the damping constant, one may write

$$\varepsilon = \varepsilon_\infty + \frac{S}{\omega_T^2 - \omega^2} = \varepsilon_\infty \frac{\omega_L^2 - \omega^2}{\omega_T^2 - \omega^2} \tag{5-4-6}$$

where ω_T is the (transverse) resonant frequency and ω_L is the longitudinal frequency defined by

$$\varepsilon(\omega_L) = 0 = \varepsilon_\infty + \frac{S}{\omega_T^2 - \omega_L^2} \tag{5-4-7}$$

According to (5-4-6), ε has a pole at $\omega = \omega_T$ and a zero at $\omega = \omega_L$. We now analyze the frequency dependence of the angle dependent dielectric constant ε_θ. Suppose first the ε_Y has a resonance in a certain frequency region but ε_Z is constant.

Then one finds by combining (5-4-6) and (5-4-5) that

$$n_\theta^2 = \frac{\varepsilon_{Y\infty}\varepsilon_Z(\omega_L^2 - \omega^2)}{\varepsilon_{Y\infty}(\omega_L^2 - \omega^2)\sin^2\theta + \varepsilon_Z(\omega_T^2 - \omega^2)\cos^2\theta} \tag{5-4-8}$$

By defining

$$\omega_\theta^2 = \frac{\omega_L^2\varepsilon_{Y\infty}\sin^2\theta + \omega_T^2\varepsilon_Z\cos^2\theta}{\varepsilon_{Y\infty}\sin^2\theta + \varepsilon_Z\cos^2\theta} \tag{5-4-9}$$

and

$$\varepsilon_{\theta\infty} = \frac{\varepsilon_{Y\infty}\varepsilon_Z}{\varepsilon_{Y\infty}\sin^2\theta + \varepsilon_Z\cos^2\theta} \tag{5-4-10}$$

(5-4-8) can be written in the form

$$n_\theta^2 = \varepsilon_\theta = \varepsilon_{\theta\infty}\frac{\omega_L^2 - \omega^2}{\omega_\theta^2 - \omega^2} \tag{5-4-11}$$

which shows that $\omega = \omega_L$ is still a zero of the dielectric constant for any θ, but that ω_T has been replaced by the expression defined in (5-4-9). Clearly ω_θ varies from ω_T to ω_L as θ goes from 0 to 90°. The consequences should be obvious: the intense reflection region, extending from ω_θ to ω_L, will gradually narrow and disappear in the limit at $\theta = 90°$, being then determined by the non-resonant ε_Z component of the dielectric tensor.

If, on the contrary, ε_Y is constant and ε_Z resonant, Eqs. (5-4-10) and (5-4-11) are replaced by

$$\varepsilon_{\theta\infty} = \frac{\varepsilon_Y\varepsilon_{Z\infty}}{\varepsilon_Y\sin^2\theta + \varepsilon_{Z\infty}\cos^2\theta} \tag{5-4-12}$$

$$\omega_\theta^2 = \frac{\omega_T^2\varepsilon_Y\sin^2\theta + \omega_L^2\varepsilon_{Z\infty}\cos^2\theta}{\varepsilon_Y\sin^2\theta + \varepsilon_{Z\infty}\cos^2\theta} \tag{5-4-13}$$

from which it is clear that ω_θ ranges from ω_T to ω_L as θ goes from 90° to 0.

The dependence of the transverse mode resonant frequency upon θ has effects not only upon reflection but also transmission spectra.

5-5 TRANSMISSION THROUGH A SLAB

The expression for the transmission by an absorbing dielectric slab is derived in several textbooks on optics.[4] If the crystal sample has a thickness l and the media both in front of and behind the sample have a refractive index $n = 1$, the transmission equation is

4. M. Born and E. Wolf, *Principles of Optics*, third revised edition, Pergamon, Oxford, 1965.

$$T = \frac{nn^*}{(n')^2}(1 - R)^2 e^{-\alpha l}[1 + R^2 e^{-\alpha l} - 2R e^{-\alpha l} \cos(2\phi + \delta)]^{-1} \qquad (5\text{-}5\text{-}1)$$

in which R is the reflectivity at a single crystal surface and

$$\alpha = 4\pi\nu n'' \qquad (5\text{-}5\text{-}2)$$

$$\delta = 4\pi\nu n' l \qquad (5\text{-}5\text{-}3)$$

$$\tan\phi = 2n''/(nn^* - 1) \qquad (5\text{-}5\text{-}4)$$

This result contains interference effects and depends upon the coherence of the radiation over the sample length. A simpler expression is available if the sample is "rough," i.e., if the reflection–transmission processes at the front and back surfaces of the sample add incoherently. In this case, the interference terms disappear and

$$T = (1 - R)^2 e^{-\alpha l}/(1 - R^2 e^{-2\alpha l}) \qquad (5\text{-}5\text{-}5)$$

However, (5-5-5) should not be used in the limit of thin films, i.e., those for which l is small compared with a wavelength. The limiting form of the appropriate equation (5-5-1) can be shown to be

$$T \simeq 1 - 4\pi n' n'' \nu l = 1 - 2\pi\varepsilon'' \nu l \qquad (5\text{-}5\text{-}6)$$

Equation (5-5-6) shows that for thin slabs, the minimum of the transmission occurs at the frequency for which $\nu\varepsilon''$ is a maximum, which with the aid of (5-3-10) is readily shown to be $\nu = \nu_j$.

It should thus be noted that T never depends simply upon the absorption coefficient α as it does with gases, for which $n' \simeq 1$ and $n'' \ll 1$.

It is possible to obtain information about the transition strengths S independent of the damping term γ by taking the limit as $l \to 0$ of a suitable integral over one transition provided it does not overlap others. For quite different reasons, gas-phase spectroscopists have used

$$B_j = \frac{-1}{l}\int_j \ln T \, d\nu \qquad (5\text{-}5\text{-}7)$$

where the $\int_j$ means that the frequency integration is extended over the range of the jth transition. Then

$$A_j = \lim_{l \to 0} B_j$$

is defined as the "integrated" absorbance. In the present context one sees that

$$A_j = \lim_{l \to 0} -\frac{1}{l}\int \ln(1 - 2\pi\varepsilon''\nu l)\, d\nu = 2\pi \int_j \varepsilon''\nu \, d\nu \qquad (5\text{-}5\text{-}8)$$

When ε'' is defined by (5-3-10), the integral in (5-5-8) becomes*

$$A_j = \pi^2 S_j \nu_j^2 \qquad (5\text{-}5\text{-}9)$$

* This simple result is based on the approximation in which $\gamma_j \ll \nu_j$.

In Section 5-8 below it will be shown how such integrals as A_j obtained from experiment can be used to obtain the molecular parameter, $\partial\mu/\partial Q_j$. It must be kept in mind that in this chapter ν always signifies a radiation "frequency" in units of cm^{-1}.

For thick slabs, (5-5-1) predicts a more complicated transmission spectrum, from which it may be difficult to identify the characteristic transverse frequency. For such samples, the region of low transmission spreads out unsymmetrically on both sides of ν_T and interference fringes may develop, particularly on the low frequency side of the band. This is illustrated in Fig. 5-6 which compares transmission spectra for $l = 0.04$, 0.4, 4, and 40 μm with the reflection spectrum of a thick crystal. The parameters chosen to describe the dielectric constant, which are typical of those encountered for moderately strong infrared bands, are $\varepsilon(\infty) = 2.0$, $\nu_T = 1000\ cm^{-1}$, $S = 0.1$, and $\gamma = 4\ cm^{-1}$.

Although the index σ on $S_j^{(\sigma)}$ has been omitted in this section, just as in 5-4 it is clearly possible to identify separately those modes which have non-vanishing $S_j^{(X)}$, $S_j^{(Y)}$, and $S_j^{(Z)}$. Although we give no details here for oblique incidence, it has been shown[5] that thin film transmission through isotropic samples at finite θ_i with TM polarization yields transmission minima at both $\nu = \nu_j = \nu_{jT}$ and $\nu = \nu_{jL}$.

Other optical techniques are available for the study of single crystals. An important example is attenuated total reflectance (ATR).[6]

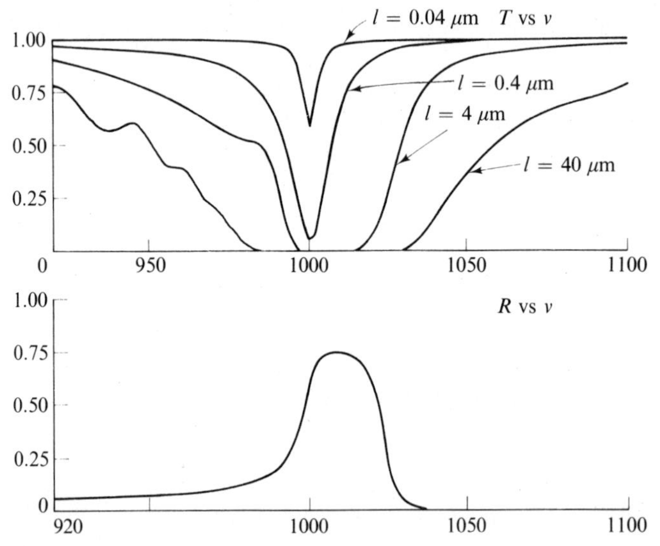

Figure 5-6 Comparison of transmission spectra for 0.04, 0.4, 4, and 40 μm thick crystals with the reflection spectrum of a thick crystal. Fixed parameters: $\nu_T = 1000\ cm^{-1}$, $S = 0.1$, $\gamma = 4\ cm^{-1}$; $\varepsilon(\infty) = 2.0$

5. D. W. Berreman, *Phys. Rev.*, **130,** 2193 (1963).
6. N. J. Harrick, *Internal Reflection Spectroscopy,* Interscience, New York, 1967, pages 41, 69, 227 and 293. See also J. L. Bass, Ph.D. Dissertation, University of Minnesota, 1965 and R. E. Frech, Ph.D. Dissertation, University of Minnesota, 1968.

Although the theory just outlined is adequate for the case of an unsupported thin crystal, it does not take into account the effect of a substrate window which must be employed for studies of condensible gases or thin frozen liquid films. In such cases an appropriate modification of the transmission function is available. If one designates the real, approximately constant refractive index of the substrate window as n_w, for thin samples one finds

$$T = [4n_w/(1 + n_w)^2]\left[1 - \frac{2}{1 + n_w}\cdot 2\pi\varepsilon''\nu l\right]$$

or

$$\ln T = \ln [4n_w/(1 + n_w)^2] + \ln\left(1 - \frac{2}{1 + n_w}2\pi\varepsilon''\nu l\right) \simeq c - \frac{2}{1 + n_w}2\pi\varepsilon''\nu l$$

The first term presumably would be removed by double beam measurements and adjustments of the baseline, so that the absorbance integral would differ from Eq. (5-5-8) only by inclusion of the multiplicative factor $2/(1 + n_w)$. This analysis has been described,[7] but the authors felt it necessary to express the results in terms of the integral

$$\int \alpha\, d\nu = \int 2\pi n''\nu\, d\nu$$

whereas the viewpoint adopted here is that the integral involving as integrand $\varepsilon''\nu = 2n'n''\nu$ is of more fundamental significance in condensed phase studies.

5-6 POLARIZATION OF RAMAN SCATTERING

The observation of single crystals with known orientation makes possible a more detailed assignment of the infrared active modes than is possible with polycrystalline samples as described in previous sections. The advantages of single crystal work are even more pronounced in connection with the Raman effect, since a precise analysis of the emission originating from any of the six components of the polarizability derivative tensor may in many cases define the symmetry species of the mode.

(a) Right-Angle Scattering

Raman radiation may be observed for any angle of scattering, and near-forward scattering yields information not available from other experiments (see Section 4-9) but right-angle scattering is, perhaps, most commonly employed, and most of the remainder of this section will be devoted to it. We have described previously in

7. S. Maeda, G. Thyagaragan, and P. N. Schatz, *J. Chem. Phys.*, **39**, 3474 (1963).

Section 4-8, a convenient convention which defines the propagation direction of the incident radiation, the polarization of the incident and of the scattered radiation, and the propagation direction of the scattered radiation by a four letter symbol.

The theoretically expected intensities may be deduced in the following way from the forms of the polarizability derivative tensors given with the character tables of Appendix I.

(b) Example of a T_d Unit Cell

Consider a unit cell with T symmetry. According to Appendix I, the polarizability derivatives when written in matrix form are:

$$A_1: \frac{\partial \boldsymbol{\alpha}}{\partial Q} = \begin{pmatrix} a & 0 & 0 \\ 0 & a & 0 \\ 0 & 0 & a \end{pmatrix} \tag{5-6-1}$$

$$E: \frac{\partial \boldsymbol{\alpha}}{\partial Q_a} = \begin{pmatrix} b & 0 & 0 \\ 0 & b & 0 \\ 0 & 0 & -2b \end{pmatrix}; \qquad \frac{\partial \boldsymbol{\alpha}}{\partial Q_b} = 3^{1/2} \begin{pmatrix} b & 0 & 0 \\ 0 & -b & 0 \\ 0 & 0 & 0 \end{pmatrix} \tag{5-6-2}$$

$$F_2: \frac{\partial \boldsymbol{\alpha}}{\partial Q_X} = \begin{pmatrix} 0 & 0 & 0 \\ 0 & 0 & c \\ 0 & c & 0 \end{pmatrix}; \qquad \frac{\partial \boldsymbol{\alpha}}{\partial Q_Y} = \begin{pmatrix} 0 & 0 & c \\ 0 & 0 & 0 \\ c & 0 & 0 \end{pmatrix}; \qquad \frac{\partial \boldsymbol{\alpha}}{\partial Q_Z} = \begin{pmatrix} 0 & c & 0 \\ c & 0 & 0 \\ 0 & 0 & 0 \end{pmatrix} \tag{5-6-3}$$

In the four experiments described as $Z(XX)Y$, $Z(YX)Y$, $Z(XZ)Y$, and $Z(YZ)Y$, one therefore sees that a fundamental mode of A_1 or E symmetry may appear only in $Z(XX)Y$, but that modes of F_2 symmetry may appear in $Z(YX)Y$, $Z(XZ)Y$, and $Z(YZ)Y$, though not in $Z(XX)Y$. Modes of A_2 or F_1 symmetry are, of course, completely inactive. The intensities are proportional to the squares of the $\partial\boldsymbol{\alpha}/\partial Q$ matrix elements, and although the several diagonal components of $\partial\boldsymbol{\alpha}/\partial Q$ for the two degenerate E modes are different, the intensities are the same for XX, YY, ZZ polarization since

$$I_{XX} \simeq \left(\frac{\partial \alpha_{XX}}{\partial Q_a}\right)^2 + \left(\frac{\partial \alpha_{XX}}{\partial Q_b}\right)^2 = b^2 + 3b^2$$

$$I_{YY} \sim \left(\frac{\partial \alpha_{YY}}{\partial Q_a}\right)^2 + \left(\frac{\partial \alpha_{YY}}{\partial Q_b}\right)^2 = b^2 + 3b^2$$

$$I_{ZZ} \sim \left(\frac{\partial \alpha_{ZZ}}{\partial Q_a}\right)^2 + \left(\frac{\partial \alpha_{ZZ}}{\partial Q_b}\right)^2 = 4b^2$$

The *frequencies* of the two degenerate modes Q_a and Q_b are equal, but they are independent, so their intensities add incoherently. Note that this polarization behavior is indistinguishable from that of an A_1 mode. It is premised, however, upon orientation of the cartesian axes as cube (i.e., two-fold symmetry) axes. If some other orientation of the crystal is chosen, the polarizability derivative with respect to an A_1 mode will be unchanged, since $\partial(\boldsymbol{\alpha})/\partial Q_j$, where Q_j is an A_1 type coordinate, is a constant matrix, which is invariant to transformation by the rotational matrix **R** which expresses the changed orientation of the crystal:

$$\mathbf{R}\frac{\partial(\boldsymbol{\alpha})}{\partial Q_j}\tilde{\mathbf{R}} = \frac{\partial(\boldsymbol{\alpha})}{\partial Q_j} \tag{5-6-4}$$

but this will not be the case in general for E modes, i.e.,

$$\mathbf{R}\frac{\partial(\boldsymbol{\alpha})}{\partial Q_{ja}}\tilde{\mathbf{R}} \neq \frac{\partial(\boldsymbol{\alpha})}{\partial Q_{ja}}$$

and for certain **R**, off-diagonal components will appear.

A simple and useful way of discriminating between the A and E modes uses a rotation of the crystal by 45° around, say, the Z-axis. Calling the directions 110 and $\bar{1}10$ respectively $\bar{X}$ and $\bar{Y}$, in the new coordinate system with basis $\bar{X}\bar{Y}Z$ the polarizability derivative tensors become

$$\frac{\partial\boldsymbol{\alpha}}{\partial Q_a} = \begin{pmatrix} b & 0 & 0 \\ 0 & b & 0 \\ 0 & 0 & -2b \end{pmatrix} \quad \text{and} \quad \frac{\partial\boldsymbol{\alpha}}{\partial Q_b} = 3^{1/2}\begin{pmatrix} 0 & b & 0 \\ b & 0 & 0 \\ 0 & 0 & 0 \end{pmatrix}$$

thus making it possible to discriminate between A and E modes, since the latter are now allowed in $Z(\bar{X}\bar{Y})X$, etc. For other examples see Reference 8.

The form of the polarizability derivative tensors for various symmetries is summarized in an article by Loudon.[9]

(c) Transverse and Longitudinal Modes in the Raman Spectrum[10]

In crystals without an inversion center, some modes can be simultaneously active in both infrared and Raman spectra, for example, F_2 modes in T_d. We have previously seen that *all* infrared active modes have different frequencies for the longitudinal and transverse relations between the polarization (dipole) and propagation vectors, even in the limit of $|\mathbf{k}| \to 0$. The question now arises as to how these modes, whose splitting is proportional to $(\partial\mu/\partial Q)^2$, will manifest themselves in the Raman spectrum. Let us first assume, incorrectly, that transverse and longitudinal Raman modes have equal intrinsic intensity, so that it is merely a matter of working out their components for a given polarization and scattering configuration. To do this, we shall clearly have to define the direction of $\mathbf{k}_p$, the propagation vector of the (phonon) mode. If $\mathbf{k}_i$ and $\mathbf{k}_s$ are respectively the propagation vectors for the incident and scattered (Raman) radiation, the

8. L. Couture and J. P. Mathieu, *Ann. Phys.*, Ser. 12, **3**, 52 (1948).
9. R. Loudon, *Adv. Phys.*, **13**, 423 (1964); **14**, 62 (1965).
10. J. P. Mathieu, L. Couture-Mathieu, and H. Poulet, *J. Phys Radium*, **16**, 781 (1955).

momentum conservation condition

$$\mathbf{k}_i = \mathbf{k}_p + \mathbf{k}_s \tag{5-6-5}$$

must hold. Energy conservation requires of course that

$$\omega_i = \omega_s \pm \omega_p \tag{5-6-6}$$

but since $\omega_i \gg \omega_p$, the radiation wavelengths λ_i and λ_s are approximately equal, so that $|\mathbf{k}_i|$ will be approximately equal to $|\mathbf{k}_s|$, and $\mathbf{k}_p$, which necessarily lies in the scattering plane, will have approximately equal components along the directions of $\mathbf{k}_i$ and $-\mathbf{k}_s$.

In order to gauge the intensities of the T and L modes, it is now necessary to express the derivatives of the polarizability tensor in terms of normal coordinates which are oriented transverse and longitudinal to the propagation direction, $\mathbf{k}_p$, rather than along the cartesian axes X, Y, Z of the unit cell. For the present case of right-angle scattering, $\mathbf{k}_p$ makes an angle of approximately 45° with the incident radiation beam and this defines the orientation of the longitudinal mode. Taking $\mathbf{k}_i$ parallel with Z and $\mathbf{k}_s$ parallel with Y, we see that the equations of transformation are

$$\begin{aligned} Q_{T1} &= Q_X \\ Q_{T2} &= \frac{1}{\sqrt{2}}(Q_Y + Q_Z) \\ Q_L &= \frac{1}{\sqrt{2}}(-Q_Y + Q_Z) \end{aligned} \tag{5-6-7}$$

whose inverse is

$$\begin{aligned} Q_X &= Q_{T1} \\ Q_Y &= \frac{1}{\sqrt{2}}(Q_{T2} - Q_L) \\ Q_Z &= \frac{1}{\sqrt{2}}(Q_{T2} + Q_L) \end{aligned} \tag{5-6-8}$$

so that by the customary laws of partial differentiation and from (5-6-3),

$$\frac{\partial \boldsymbol{\alpha}}{\partial Q_{T1}} = \frac{\partial \boldsymbol{\alpha}}{\partial Q_X} = \begin{pmatrix} 0 & 0 & 0 \\ 0 & 0 & c \\ 0 & c & 0 \end{pmatrix} \tag{5-6-9}$$

$$\frac{\partial \boldsymbol{\alpha}}{\partial Q_{T2}} = \frac{1}{\sqrt{2}}\left(\frac{\partial \boldsymbol{\alpha}}{\partial Q_y} + \frac{\partial \boldsymbol{\alpha}}{\partial Q_Z}\right) = \frac{1}{\sqrt{2}}\begin{pmatrix} 0 & c & c \\ c & 0 & 0 \\ c & 0 & 0 \end{pmatrix} \tag{5-6-10}$$

and

$$\frac{\partial \boldsymbol{\alpha}}{\partial Q_L} = \frac{1}{\sqrt{2}}\left(-\frac{\partial \boldsymbol{\alpha}}{\partial Q_Y} + \frac{\partial \boldsymbol{\alpha}}{dQ_Z}\right) = \frac{1}{\sqrt{2}}\begin{pmatrix} 0 & c & -c \\ c & 0 & 0 \\ -c & 0 & 0 \end{pmatrix} \tag{5-6-11}$$

From the form of these polarizability derivative matrices, it is now possible to make the following predictions with regard to the relative T and L intensities. In an experiment of the type $Z(YZ)Y$, *only* the transverse component should appear, with an intensity proportional to c^2. In contrast, for either $Z(YX)Y$ or $Z(XZ)Y$, *both* transverse and longitudinal components should appear with intensities proportional to $\frac{1}{2}c^2$; it should be remembered that the transition strength is proportional to $(\partial\alpha/\partial Q)^2$.

Although these considerations properly account for the purely directional characteristics of the transverse and longitudinal modes, they fail to describe the quantitative relative intensities of the T and L peaks. Moreover, it is observed that the intensity ratio is different for different modes of the same symmetry species in the same polarization.

This was first explained by Poulet[11] as a consequence of the fact that what we have previously termed the polarizability of the unit cell depends not only upon the normal coordinates but upon the electric field arising from the vibration (not to be confused with the externally applied electric field of the incident radiation).

For the transverse mode, such a vibrationally induced electric field vanishes, but for the longitudinal mode,

$$\mathscr{E} = -4\pi\mathscr{P} = \frac{-4\pi}{v\varepsilon_\infty}\left(\frac{\partial \boldsymbol{\mu}}{\partial Q}\right)Q \tag{5-6-12}$$

where $\mathscr{E}$, $\mathscr{P}$, and $\partial\boldsymbol{\mu}/\partial Q$ all are parallel to $\mathbf{k}_p$. The proportionality constant between $\boldsymbol{\alpha}$ and $\mathscr{E}$ is related to the macroscopic second order susceptibility,[12] and in the molecular context has been termed the *hyperpolarizability*.[13] It is a third rank tensor, the selection rules for which have been tabulated in the literature.[14] Thus, the total dependence of $\boldsymbol{\alpha}$ upon Q_j may be written as:

$$\alpha_{\sigma\sigma'} = \left(\frac{\partial\alpha_{\sigma\sigma'}}{\partial Q_j}\right)Q_j + \left(\frac{\partial\alpha_{\sigma\sigma'}}{\partial\mathscr{E}_\tau}\right)\mathscr{E}_\tau = \left[\frac{\partial\alpha_{\sigma\sigma'}}{\partial Q_j} + \frac{\partial\alpha_{\sigma\sigma'}}{\partial\mathscr{E}_\tau}\frac{\partial\mathscr{E}_\tau}{\partial Q_j}\right]Q_j \tag{5-6-13}$$

(summation of τ over the cartesian indices is understood). Putting

$$\frac{\partial\alpha_{\sigma\sigma'}}{\partial\mathscr{E}_\tau} = \beta^\tau_{\sigma\sigma'} \tag{5-6-14}$$

11. H. Poulet, *Ann. Phys.*, **10,** 908 (1955).
12. F. Zernike and J. E. Midwinter, *Applied Nonlinear Optics*, Wiley-Interscience, New York, 1972.
13. C. A. Coulson, A. Maccoll, and L. E. Sutton, *Trans. Faraday Soc.*, **48,** 106 (1952).
14. S. J. Cyvin, J. E. Rauch, and J. C. Decius, *J. Chem. Phys.*, **43,** 4083 (1965).

and using

$$\frac{\partial \mathscr{E}_\tau}{\partial Q_j} = \frac{-4\pi}{v\varepsilon_\infty} \frac{\partial \mu_\tau}{\partial Q_j} \tag{5-6-15}$$

we have

$$\alpha_{\sigma\sigma'} = \left[\frac{\partial \alpha_{\sigma\sigma'}}{\partial Q_j} - \frac{4\pi}{v} \frac{\beta^\tau_{\sigma\sigma'}}{\varepsilon_\infty} \frac{\partial \mu_\tau}{\partial Q_j} \right] Q_j \tag{5-6-16}$$

We emphasize again that the second term above drops out for the transverse mode. Thus the intrinsic intensity ratio for the two modes in a $\sigma\sigma'$ polarized experiment is

$$\frac{I_L(\sigma\sigma')}{I_T(\sigma\sigma')} = \left[1 - \frac{4\pi}{v} \frac{\beta^\tau_{\sigma\sigma'}}{\varepsilon_\infty} \frac{\partial \mu_\tau}{\partial Q_j} \bigg/ \frac{\partial \alpha_{\sigma\sigma'}}{\partial Q_j} \right]^2 \tag{5-6-17}$$

whose magnitude depends upon the relative infrared and Raman strengths and upon $\boldsymbol{\beta}$. In the T_d example only elements of $\boldsymbol{\beta}$ of the form $\beta^Z_{XY} = \beta^X_{YZ} = \beta^Y_{ZX}$ are non-vanishing; $I_L(\sigma\sigma')$ can be either greater or less than $I_T(\sigma\sigma')$. Since it is probably true that $\partial\mu/\partial Q$ is easier to measure than $\partial\alpha/\partial Q$ (from the absolute Raman intensity) Eq. (5-6-17) affords an interesting alternative method of determining $\partial\alpha/\partial Q$ given $\beta^\tau_{\sigma\sigma'}$.

(d) Errors in Relative Intensities for Polarized Raman Spectra

Despite the theoretical simplicity of the expression for the relative intensities of Raman lines in various polarizations, experimental determinations are fraught with error. Well-aligned crystal specimens are obviously necessary. The theoretical expressions assume a single scattering angle (90° in all the previous discussion) so the incident beam should not be convergent, nor should the scattered beam be divergent. With a laser source, the first condition is quite easily satisfied, but unless the f number of the collection optics at the entrance to the monochromator is kept large, with concomitant reduction of the energy, the second condition is not really satisfied. More subtle errors originating in the birefringence of anisotropic crystals have been discussed.[15] If the crystal is optically active (the unit cell group has no improper rotations or is C_s, C_{2v}, S_4 or D_{2d}) the plane of the incident radiation will not be conserved as it penetrates the crystal. Finally, we mention *twinning*, which invalidates the assumption of a uniquely oriented specimen and which may be insidious for certain cubic crystals, since it is not easily detected by ordinary optical examination. For example, in a cubic crystal like CaF_2 it cannot be detected.

15. S. P. S. Porto, J. A. Giordmaine, and T. C. Damen, *Phys. Rev.*, **147**, 608 (1951)

(e) Raman Scattering at Other Angles

Although most Raman experiments are conducted using 90° scattering, it is occasionally useful to employ other arrangements. A back-scattering experiment with the beam direction along the X-axis would give rise to $X(YY)\bar{X}$, $X(YZ)\bar{X}$, $X(ZY)\bar{X}$, and $X(ZZ)\bar{X}$ as possible polarizations. Since (5-6-5) and (5-6-6) still apply, but $\mathbf{k}_s$ is along the negative X-axis, the magnitude of

$$|\mathbf{k}_p| = |\mathbf{k}_i| + |\mathbf{k}_s| \cong 2|\mathbf{k}_i| \qquad (5\text{-}6\text{-}18)$$

which is still very small compared with the Brillouin zone boundary although large compared with the wave vector of radiation whose frequency equals that of the phonon; the frequency of the phonon, observed as $\omega_p = \omega_i - \omega_s$ (Stokes) will still correspond with a point close to the origin of the Brillouin zone.

For near forward scattering, a quite different situation can arise, if the mode in question is simultaneously infrared and Raman active. For a cubic crystal with only a single infrared active fundamental, the formula for the dielectric constant, neglecting damping, is

$$\varepsilon = \varepsilon_\infty + S\omega_T^2/(\omega_T^2 - \omega^2)$$

where ω_T is the transverse mode frequency. It has been tacitly assumed that the phonon dispersion curve is constant, i.e., $\omega_p = \omega_T$, in the vicinity of $\mathbf{k} = \mathbf{0}$. But in the medium, $\lambda_p = \lambda_0/n$ or $k_p = k_0 n$ where λ_0 is the vacuum wavelength of radiation and $n^2 = \varepsilon$. Therefore

$$k_p^2 = k_0^2 n^2 = \frac{\omega^2}{c^2}\varepsilon = \frac{\omega_p^2}{c^2}[\varepsilon_\infty + S\omega_T^2/(\omega_T^2 - \omega_p^2)] \qquad (5\text{-}6\text{-}19)$$

is the correct dispersion curve for an infrared active transverse phonon. A more detailed development of this sort of relation was first given by Huang,[16] as we have reviewed in Section 4-9.

A plot of (5-6-19) is shown in Fig. 5-12 (page 213) where one sees that there are two branches. The lower branch approaches zero with asymptotic slope $c/\varepsilon_\infty^{1/2}$ and is bounded by ω_T at $k_p \to \infty$. The upper branch approaches ω_L at $k_p = 0$ and has an asymptotic slope of $c/\varepsilon_0^{1/2}$ at large k_p. The essential point is that for small forward scattering angles, it is possible to observe points on the lower branch, i.e., to note a decrease of the Raman frequency below ω_T. These are the polaritons discussed in Section 4-9. In principle, they do not appear to provide any information not available from the infrared spectrum, but observations at several small angles certainly produces a more direct analysis of the frequency dependence of the dielectric constant than is possible from the rather indirect analysis of infrared reflectivity.

In the multi-mode crystals of greater interest in the present work, the polariton situation may be summarized as follows. Suppose there are two modes; then there will be two "forbidden" frequency bands $\omega_{T1} < \omega < \omega_{L1}$ and $\omega_{T2} < \omega < \omega_{L2}$

16. K. Huang, *Proc. Roy. Soc. Lond.*, **A208**, 352 (1951).

where $\omega_{T1} < \omega_{L1} < \omega_{T2} < \omega_{L2}$. The lowest and highest branches, $\omega < \omega_{T1}$ and $\omega > \omega_{L2}$ will be similar to the branches of the single mode system which we have just discussed. In the middle branch, ω will tend to ω_{L1} as $k_p \to 0$ and to ω_{T2} as $k_p \to \infty$. The extension to more modes is obvious.

In concluding this section we remark that the comparison of observed infrared and Raman frequencies of a mode allowed in both spectra poses a problem if the two spectra can be observed with very high precision. Even at large angles, the polariton theory shows that $\omega_p < \omega_T$ everywhere in the lower branch. By assuming that the refractive index for the incident and scattered radiation is n, using $k_p^2 = k_i^2 + k_s^2$ for 90° scattering and making a suitable expansion assuming that $\omega_p = \omega_T - \Delta$ where Δ is very small, one finds

$$\Delta = \frac{S\omega_T^3}{4n^2\omega_i^2} \tag{5-6-20}$$

Now although S can be of the order of unity or larger, this usually happens only for external modes in ionic crystals for which the transverse mode frequency is a few hundred cm^{-1} compared with incident frequencies of the order of 2×10^4 cm^{-1}. Calculations of Δ using (5-6-20) using representative examples show that Δ is always less than 1 cm^{-1}, and usually, even for fairly intense infrared modes, of the order of 10^{-2}–10^{-1} cm^{-1} and thus insignificant for right-angle scattering.

5-7 MICROSCOPIC THEORY OF TRANSVERSE AND LONGITUDINAL MODES

The fundamental basis of TO–LO splittings was discussed in Section 4-9 for monatomic crystals. In the present section we wish to examine the same phenomenon, but for a molecular crystal, so that the individual vibrational degrees of freedom include either "internal" or "external" (lattice) modes. We focus our attention on the internal modes. The Hamiltonian for such a system may be divided into an uncoupled, zero order term, and a coupling term. The zero order term has the form:

$$H^{(0)} = \sum_{i=1}^{N} H_i^{(0)} \tag{5-7-1}$$

where N is the number of molecules in the crystal, and

$$H_i^{(0)} = \tfrac{1}{2}\sum (p_k^2 + \omega_k^2 q_k^2) \tag{5-7-2}$$

as in (2-3-1) and (2-3-2), except that the molecular momenta and coordinates are now written in lower-case letters.

For simplicity, we consider a lattice without a basis, and further restrict our immediate discussion to a single non-degenerate degree of freedom. If μ_i and μ_j are the dipole moment operators for molecules i and j, then the term in the coupling Hamiltonian for this pair of molecules is

$$H_{ij} = -\boldsymbol{\mu}_i^\dagger \boldsymbol{D}_{ij} \boldsymbol{\mu}_j \tag{5-7-3}$$

where

$$\boldsymbol{D}_{ij} = \frac{3\mathbf{rr}}{r^5} - \frac{E}{r^3} \tag{5-7-4}$$

is the field propagation tensor, $\mathbf{r}$ is the vector between molecules i and j, and r is its magnitude. The total coupling Hamiltonian H' is the sum of H_{ij} over all pairs. According to the vibrational exciton description of the crystal, developed further in Section 6-4, the crystal ground state is described by putting all the molecules in their ground states, and an excited state is described by taking a linear combination of all the states in which a single molecule is excited, the coefficient of the term in which molecule j is excited being $e^{i\mathbf{k}\cdot\mathbf{r}_j}$, where $\mathbf{k}$ is the propagation vector for the exciton. If the molecular dipole moment operator is now expanded in its normal coordinate

$$\boldsymbol{\mu}_i = \boldsymbol{\mu}_0 + \left(\frac{\partial \boldsymbol{\mu}}{\partial q}\right) q_i$$

it is found that the coupling terms H_{ij} make no contribution to the energy of the ground state, since $\langle 0 | q_i | 0 \rangle = 0$, but that when H_{ij} is summed over all pairs, the energy of the first excited state is modified by the term

$$\Delta(\mathbf{k}) = -\sum (p^\mu)^* D^{\mu\nu}(\mathbf{k}) p^\nu$$

where μ, ν refer to the components, in a crystal-fixed cartesian frame, of the molecular transition dipole moment vector

$$\mathbf{p} = \left(\frac{\partial \boldsymbol{\mu}}{\partial q}\right) \langle 0 | q | 1 \rangle \tag{2-4-13}$$

The field propagation tensor of (5-7-4) has been replaced by its Fourier component,

$$D^{\mu\nu}(\mathbf{k}) = \sum_r{}' \left[\frac{3r^\mu r^\nu}{r^5} - \frac{\delta^{\mu\nu}}{r^3} \right] e^{i\mathbf{k}\cdot\mathbf{r}} \tag{5-7-5}$$

The sum is over all molecular pairs and the phase factor $e^{i\mathbf{k}\cdot\mathbf{r}}$ comes from the vibrational exciton. The prime on the summation sign denotes exclusion of the origin cell from the summation, which is taken over all other unit cells in the crystal. Because of the plane-wave term, the summation is called a dipole-wave sum.

If the molecules are polarizable, (5-7-3), which may be written in tensor form as

$$H' = -\mathbf{p}_i^* \cdot \boldsymbol{D} \cdot \mathbf{p}_j \tag{5-7-6}$$

is modified to (see Section 5-8)

$$H' = -\mathbf{p}_i^* \cdot \boldsymbol{D}(E - \alpha \boldsymbol{D})^{-1} \cdot \mathbf{p}_j \tag{5-7-7}$$

where the **p** and D are defined as before. $\boldsymbol{\alpha}$ and E are respectively 3×3 polarizability (time-independent) and unit matrices.

We can relate the polarizability contribution to the dielectric constant at optical frequencies (ε_∞). In the case of NaCl, its neglect results in a 60 percent error in the difference of the longitudinal and transverse limiting normal mode frequencies, $\omega_L^2 - \omega_T^2$. In the case of molecular crystals, the error will be somewhat smaller. Accordingly, we shall first discuss the effect of (5-7-6), for a crystal of non-polarizable molecules.

The effect of H' is primarily to shift the allowed states under $H^{(0)}$. It is easiest to discuss these shifts in terms of the spectroscopic observable, the excitation energy of the crystal mode which results from the coupling of the individual $\mathbf{p}_i$. We therefore will be calculating matrix elements of H' for the ground state of the crystal and this excited state. The zero of the energy scale will be taken as the analogous matrix element of $H^{(0)}$.

From the form of the field propagation tensor we see that it is a function only of lattice structure; all "molecular" information is contained in the **p**'s. It is for this reason that the summation indicated in (5-7-5) can be evaluated without reference to a particular molecular system. The field propagation tensor was first evaluated by Cohen and Keffer,[17] using the Ewald–Kornfeld summation procedure.* For the (effectively) infinite crystal, in the limit of small **k**,

$$\lim_{\mathbf{k}\to 0} D^{\mu\nu}(\mathbf{k}) = d^{\mu\nu} - \frac{4\pi}{a^3}\frac{k^\mu k^\nu}{k^2} \tag{5-7-8}$$

where $d^{\mu\nu}$ depends on crystal structure but not **k**, and the second term depends on the *direction* of **k** but not on its magnitude or the crystal structure. To take advantage of certain direct evaluations of (5-7-8), following Fox and Hexter[18] we rewrite it as

$$\lim_{\mathbf{k}\to 0} D^{\mu\nu}(\mathbf{k}) = R^{\mu\nu} + \frac{4\pi}{3a^3}\delta^{\mu\nu} - \frac{4\pi}{a^3}\frac{k^\mu k^\nu}{k^2} \tag{5-7-9}$$

The advantage of this formulation is that the $R^{\mu\nu}$ have been tabulated for a number of sites of cubic sublattices at various points in the BZ. The significance of the several terms in (5-7-9) is discussed below.

Before accepting (5-7-9), we must inquire whether the limit $\mathbf{k} \to 0$ is the regime appropriate to our interest. The question has to do with the magnitude of **k** relative to its possible range. Also, since **k** measures the wave vector of the dipole propagation wave in the crystal medium, we are in fact inquiring into the size of the wavelength of that excitation relative to the size of the medium itself.

It has long been realized that real crystals are imperfect, and are composed of small regions in which the three-dimensional, microscopic order is perfect,

17. M. Cohen and F. Keffer, *Phys. Rev.*, **99**, 1128 (1955).
18. D. Fox and R. M. Hexter, *J. Chem. Phys.*, **41**, 1125 (1964).

* See Appendix XI.

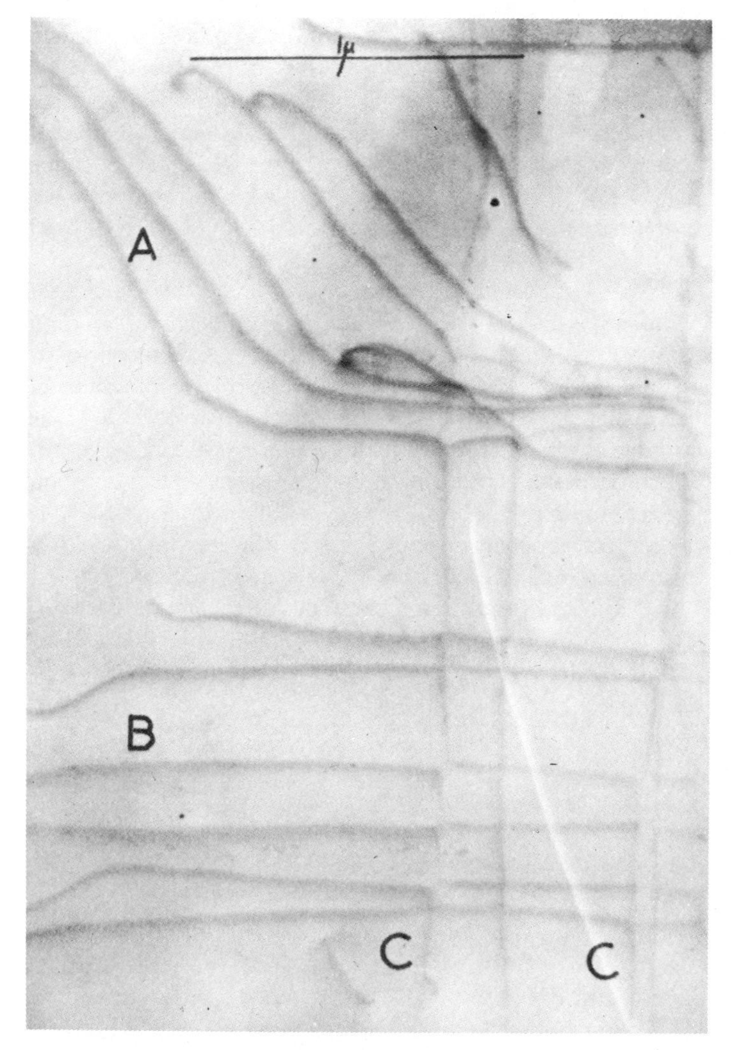

Figure 5-7 Dislocation interactions in MgO. (From J. Washburn, *et al.*, *Phil. Mag.* **5**, 8, Plate 139 (1960)). Note scale.

sometimes called domains. Figure 5-7 illustrates an actual crystal at the microscopic level. In the early history of X-ray diffraction, crystal perfection was gauged by a comparison of the coefficients of reflection at various wavelengths with those calculated by Darwin, based upon his theory of X-ray reflection by perfect crystals.[19] A more precise measure of crystal perfection has been developed by Hirsch and Ramachandran[20] in which the variation of the integrated reflection

19. C. G. Darwin, *Phil. Mag.*, **27**, 315 (1914); **27**, 675 (1914).
20. P. B. Hirsch and G. N. Ramachandran, *Acta Cryst.*, **3**, 187 (1950).

of polarized X-rays with the Bragg angle by real crystals is compared with the exact relationships expected for perfect crystals. A "degree of perfection" is then deduced from the comparison. As a result, it has become recognized that absolute parallelism of crystallographic plans persists, in general, for not more than, say 10,000 repeat distances; in other words, for distances of not more than $\sim 1\ \mu m$.

In recent years a variety of new topographical methods for the measurement of crystal imperfection have been developed, such as the Borrmann effect, X-ray interferometry, and the observation of Pendellösung.[21] Most imperfections arise due to the development of dislocation lines, the occurrence of which is measured by their "density." The most perfect crystals produced to date (e.g., diamond, Si, Ge) have dislocation lines present with line densities of 10^3 cm^{-2}. This corresponds to an average domain diameter of $\sim 25 \mu m$. Maintenance of phase can thus be absolute only over such distances.

There is little information about domain size in other crystals. It is known that sample preparation and treatment (e.g., thermal shock) affect perfection. [See, for example, the experiments on NaCl by Bragg and coworkers[22] and on calcite, etc., by Sakisaka.[23] An excellent review of this subject has been given by Gunier.[24]] The general situation for molecular crystals has been summarized by Peiser[25] as follows: "More commonly, 'single crystals' are composed of tiny 'mosaics'—crystallites of linear dimension of around 10^{-4}–10^{-6} cm within which the lattice order is perfect. Adjacent crystallites are inclined to each other by angles up to about one degree."

In an infrared experiment $|\mathbf{k}|$ is determined by the selection rule

$$\mathbf{k}\,(\text{photon}) = \sum_i \mathbf{k}_i\,(\text{phonon}) \tag{5-7-10}$$

Thus the *total* momentum of the *system* of photons and phonons is conserved. We have previously discussed (Section 4-6) the dispersion of phonons in crystals. Photons are also dispersed, even in free space, according to

$$\omega_{\text{photon}} = ck_{\text{photon}} \tag{5-7-11}$$

where c is the velocity of light. Equation (5-7-11) is illustrated in Fig. 5-8(a). In crystals, the photon dispersion curve, illustrated in Fig. 5-8(b), has the diminished slope c/n where n is the crystal's refractive index. Resonant absorption occurs at the intersection with the phonon dispersion curve.*

21. B. W. Batterman and H. Cole, *Rev. Mod. Phys.*, **36,** 682 (1964).
22. W. L. Bragg, R. W. James and C. H. Bosanquist, *Phil. Mag.* **42,** 1 (1921).
23. Y. Sakisaka, *Proc. Phys.-Math. Soc. Japan,* **12** [3], 189 (1930); **13** [3], 211 (1930).
24. A. Gunier, *X-Ray Diffraction in Crystals, Imperfect Crystals and Amorphous Bodies,* W. H. Freeman & Co., San Francisco, 1963, Sections 4.6 and 4.10.
25. H. S. Peiser, *Formation and Trapping of Free Radicals,* A. M. Bass and H. P. Broida, Eds., Academic Press, New York, 1960. Chapter 9.

* A still more complete crystal Hamiltonian than that which includes (5-7-3) also includes terms which directly couple the external electromagnetic field to the phonons (i.e., terms such as $\beta \cdot \mathscr{E}$ and $\mathscr{E}'\alpha \cdot \mathscr{E}$, where α is the polarizability. These terms can mix transverse photon and phonon states, particularly in the vicinity of the intersection of their dispersion curves. We defer discussion of the importance of this mixing until later in this section.

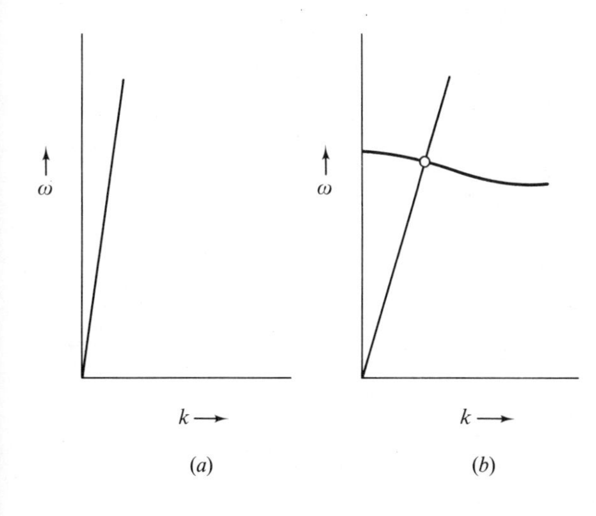

Figure 5-8 (*a*) Photon dispersion curve in vacuum; (*b*) photon and phonon dispersion curves in a dielectric

Strictly speaking, **k** (photon) is always finite, since all photons have finite energy. In the crystal medium, even without photon–phonon mixing, if a phonon state has finite energy and occupancy, $\mathbf{k} \neq 0$. At the BZB, the limiting modes have $|\mathbf{k}| \sim a^{-1}$. Thus $k_{\text{BZB}} \sim 10^8\ \text{cm}^{-1}$. At the intersection of the photon and phonon dispersion curves, k in an infrared process is determined approximately by

$$\omega_{\text{phonon}} = \omega_{\text{photon}} = ck \sim 2\pi \times 10^{13}\ \text{sec}^{-1} \tag{5-7-12}$$

Hence, $k \sim 10^3\ \text{cm}^{-1}$. Therefore, resonant phonons do indeed have $\mathbf{k} \to 0$, compared to $\mathbf{k}_{\text{BZB}}$.

Dipole excitation waves are therefore long. Just as with the interaction of a dipole field with a free atom or a molecule, the phase change across a domain is negligible. Hence, we may set $\mathbf{k} = 0$ in (5-7-5) and the summation becomes similar to that encountered in the classical theory of the dielectric constant (Section 5-8).[26] We construct a sphere, centered at the origin cell, of large enough radius so that the material outside the sphere can be treated as a uniformly polarized continuum. The summation is explicitly carried out in the interior of the sphere, but it is replaced by an integral in the region extending from the surface of the sphere to the crystal surface. As is shown in a number of standard texts,[27] a uniformly polarized continuum may be replaced by its equivalent surface polarization charges, here that of the inner sphere* and that of the exterior surface of the sample, to which it is equivalent. The result is

$$D^{\mu\nu} = R^{\mu\nu} + \frac{4\pi}{3a^3}\,\delta^{\mu\nu} + N^{\mu\nu} \tag{5-7-13}$$

26. H. A. Lorentz, *The Theory of Electrons and Its Application to the Phenomenon of Light and Radiant Heat*, Dover Publications, Inc., New York, 1952, second edition, pages 137–39; 305–8.
27. See, for example, M. V. Klein, *Optics*, John Wiley & Sons, Inc., New York, 1970.

* The "Lorentz sphere."

The tensor $\boldsymbol{N}$ represents the surface polarization charge; it depends on the shape of the crystal surface, but not on the crystal structure. Defining

$$S^{\mu\nu} \equiv N^{\mu\nu} + \frac{4\pi}{3a^3}\delta^{\mu\nu} \tag{5-7-14}$$

we may write

$$\boldsymbol{D} = \boldsymbol{R} + \boldsymbol{S} \tag{5-7-15}$$

in which $\boldsymbol{R}$ depends on crystal structure but not on sample shape, while $\boldsymbol{S}$ depends on a crystal's shape but not on its structure. It is, of course, still an open question whether the domain boundaries, with dimensions of the order of 10^{-6} m, are to be identified with the crystal surface and hence govern the determination of the tensor $\boldsymbol{S}$. In what follows we shall discuss the implications which such an assumption has for the splittings of modes in microcrystals.

Table 5-2(a) The eight basic arrays (After J. M. Luttinger and L. Tisza, *Phys. Rev.*, **70**, 954 (1956))

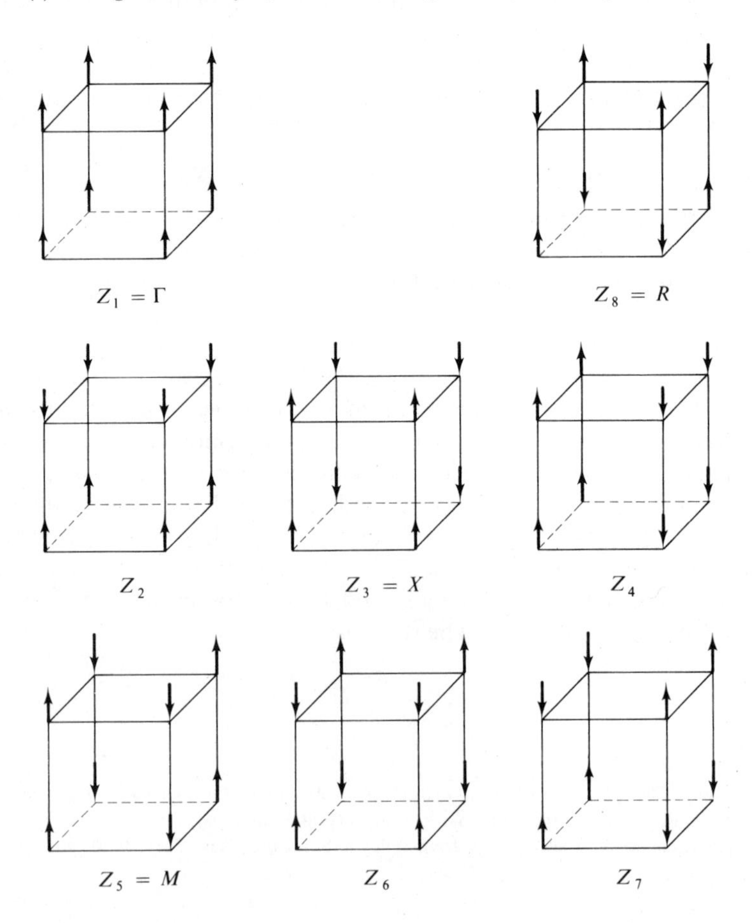

Table 5-2(b) ***R* tensors**

Basic array	Origin at: Lattice point	Body center	Face center
Z_1	$\begin{pmatrix} 0 & 0 & 0 \\ 0 & 0 & 0 \\ 0 & 0 & 0 \end{pmatrix}$	$\begin{pmatrix} 0 & 0 & 0 \\ 0 & 0 & 0 \\ 0 & 0 & 0 \end{pmatrix}$	$h_1\begin{pmatrix} 1 & 0 & 0 \\ 0 & 1 & 0 \\ 0 & 0 & -2 \end{pmatrix}$
Z_2	$f_2\begin{pmatrix} 0 & 0 & 0 \\ 0 & 0 & 0 \\ 0 & 0 & 1 \end{pmatrix}$	$\begin{pmatrix} 0 & 0 & 0 \\ 0 & 0 & 0 \\ 0 & 0 & 0 \end{pmatrix}$	$h_2\begin{pmatrix} 0 & 0 & 0 \\ 0 & 0 & 0 \\ 0 & 0 & -2 \end{pmatrix}$
Z_3	$f_3\begin{pmatrix} 0 & 0 & 0 \\ 0 & 0 & 0 \\ 0 & 0 & 1 \end{pmatrix}$	$\begin{pmatrix} 0 & 0 & 0 \\ 0 & 0 & 0 \\ 0 & 0 & 0 \end{pmatrix}$	$h_2\begin{pmatrix} 1 & 0 & 0 \\ 0 & 0 & 0 \\ 0 & 0 & 0 \end{pmatrix}$
Z_4	$f_4\begin{pmatrix} 0 & 0 & 0 \\ 0 & 0 & 0 \\ 0 & 0 & 1 \end{pmatrix}$	$\begin{pmatrix} 0 & 0 & 0 \\ 0 & 0 & 0 \\ 0 & 0 & 0 \end{pmatrix}$	$h_2\begin{pmatrix} 0 & 0 & 0 \\ 0 & 1 & 0 \\ 0 & 0 & 0 \end{pmatrix}$
Z_5	$f_5\begin{pmatrix} 0 & 0 & 0 \\ 0 & 0 & 0 \\ 0 & 0 & 1 \end{pmatrix}$	$\begin{pmatrix} 0 & 0 & 0 \\ 0 & 0 & 0 \\ 0 & 0 & 0 \end{pmatrix}$	$\begin{pmatrix} 0 & 0 & 0 \\ 0 & 0 & 0 \\ 0 & 0 & 0 \end{pmatrix}$
Z_6	$f_6\begin{pmatrix} 0 & 0 & 0 \\ 0 & 0 & 0 \\ 0 & 0 & 1 \end{pmatrix}$	$g\begin{pmatrix} 0 & 0 & 0 \\ 0 & 0 & 1 \\ 0 & 1 & 0 \end{pmatrix}$	$h_3\begin{pmatrix} 0 & 0 & 0 \\ 1 & 0 & 0 \\ 0 & 0 & 0 \end{pmatrix}$
Z_7	$f_7\begin{pmatrix} 0 & 0 & 0 \\ 0 & 0 & 0 \\ 0 & 0 & 1 \end{pmatrix}$	$g\begin{pmatrix} 0 & 0 & 1 \\ 0 & 0 & 0 \\ 1 & 0 & 0 \end{pmatrix}$	$h_3\begin{pmatrix} 0 & 1 & 0 \\ 0 & 0 & 0 \\ 0 & 0 & 0 \end{pmatrix}$
Z_8	$\begin{pmatrix} 0 & 0 & 0 \\ 0 & 0 & 0 \\ 0 & 0 & 0 \end{pmatrix}$	$\begin{pmatrix} 0 & 0 & 0 \\ 0 & 0 & 0 \\ 0 & 0 & 0 \end{pmatrix}$	$h_4\begin{pmatrix} 0 & 1 & 0 \\ 1 & 0 & 0 \\ 0 & 0 & 0 \end{pmatrix}$

R tensors for a number of sites and special points of cubic crystals have been calculated by Luttinger and Tisza,[28] and are tabulated in Table 5-2.

R-tensors for some non-cubic crystals have been evaluated by Mueller[29] using the Ewald-Kornfeld[30] technique, and by deWette[31] using the planewise summation method. deWette and Schacher have generalized this method to any crystal[32] and Dickmann[33] has developed a Fortran IV program based upon it.[34]

28. J. M. Luttinger and L. Tisza, *Phys. Rev.*, **70,** 954 (1946); **72,** 257 (1947).
29. H. Mueller, *Phys. Rev.*, **47,** 947 (1935).
30. See Appendix XI.
31. F. W. deWette, *Phys. Rev.*, **123,** 103 (1961).
32. F. W. deWette and G. E. Schacher, *Phys. Rev.*, **137,** A78 (1965).
33. D. B. Dickmann, Thesis, U.S. Naval Postgraduate School, 1966.
34. Control Data Corporation CO-OP Program No. Z1-NPGS-LATSUM.

Table 5-2(c) Values of coefficients in Table 5-2(b)

$$f_2 = -\tfrac{1}{2}[S_z(0\tfrac{1}{2}\tfrac{1}{2}) - S_z(\tfrac{1}{2}00)] = -9.687$$
$$f_3 = f_4 = \tfrac{1}{4}[S_z(0\tfrac{1}{2}\tfrac{1}{2}) - S_z(\tfrac{1}{2}00)] = -f_2/2$$
$$f_5 = -\tfrac{1}{2}[S_z(0\tfrac{1}{2}\tfrac{1}{2}) + S_z(\tfrac{1}{2}00)] = 5.351$$
$$f_6 = f_7 = \tfrac{1}{4}[S_z(0\tfrac{1}{2}\tfrac{1}{2}) + S_z(\tfrac{1}{2}00)] = f_5/2$$
$$g = S_y(\tfrac{1}{4}\tfrac{1}{4}\tfrac{1}{4}) = 10.620$$
$$h_1 = S_z(0\tfrac{1}{2}\tfrac{1}{2}) = 4.334$$
$$h_2 = S_z(0\tfrac{1}{4}\tfrac{1}{4}) - S_z(0\tfrac{1}{2}\tfrac{1}{2}) = 7.995$$
$$h_3 = \tfrac{1}{2}[S_y(0\tfrac{1}{4}\tfrac{1}{4}) + S_y(\tfrac{1}{2}\tfrac{1}{4}\tfrac{1}{4})] = 17.060$$
$$h_4 = \tfrac{1}{2}[S_y(0\tfrac{1}{4}\tfrac{1}{4}) - S_y(\tfrac{1}{2}\tfrac{1}{4}\tfrac{1}{4})] = 14.461$$

where

$$S_z(\mathbf{r}) = \sum_{l_1=-\infty}^{\infty} \sum_{l_2} \sum_{l_3} \frac{2(l_3 - z)^2 - (l_1 - x)^2 - (l_2 - y)^2}{[(l_1 - x)^2 + (l_2 - y)^2 + (l_3 - z)^2]^{5/2}}$$

$$S_y(\mathbf{r}) = \sum\sum\sum \frac{3(l_2 - y)(l_3 - z)}{[(l_1 - x)^2 + (l_2 - y)^2 + (l_3 - z)^2]^{5/2}}$$

$$S_x(\mathbf{r}) = \sum\sum\sum \frac{3(l_1 - x)(l_3 - z)}{[(l_1 - x)^2 + (l_2 - y)^2 + (l_3 - z)^2]^{5/2}}$$

x, y, and z are cartesian components of $\mathbf{r}$ and the l_i index the cube corners.

Since the planewise summation method directly calculates $\boldsymbol{D}$ for a slab-shaped crystal, a correction must be added to yield a result appropriate to a spherically-shaped sample. The shape tensor $\boldsymbol{S}$ can then be used in much the same manner as with cubic crystals; the only difference is to replace the factor a^{-3} with the unit cell volume, v. While $\mathbf{S}$ depends on the position of the origin cell with respect to the surface of the sample, for ellipsoidal samples it may be shown that this dependence vanishes. For ellipsoids of revolution

$$\boldsymbol{S} = -\frac{1}{2}\frac{g}{a^3}\begin{pmatrix} 1 & & \\ & 1 & \\ & & -2 \end{pmatrix} \tag{5-7-16}$$

where the coordinate axes coincide with the principal axes of the sample ellipsoid; g is a parameter which depends on the axial ratio of the ellipsoid, as illustrated in Fig. 5-9.

From (5-7-16) we see that, even for a cubic crystal, provided the domain shape is not itself spherical, there will be two distinct states as a result of H'. In the simple cubic crystal the high symmetry dictates that, in the absence of H', the crystal state arising from the degree of freedom of interest will be threefold degenerate, as in Fig. 5-10. In the presence of H', due to (5-7-16), the degeneracy is partially removed. Thus we see again $\omega_L > \omega_T$. This is the microscopic basis of the splitting

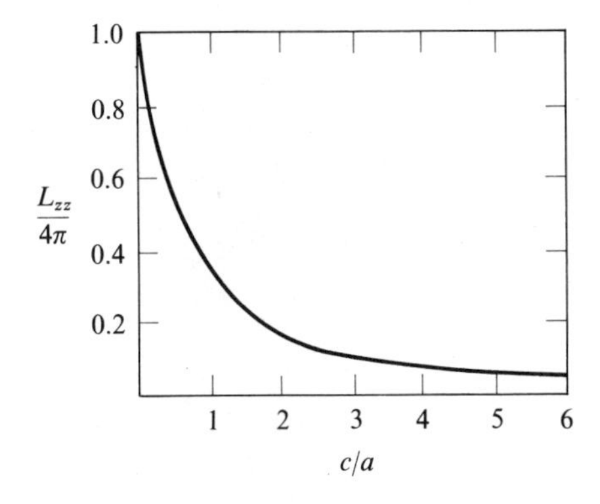

Figure 5-9 Depolarization factor in the z direction versus the axial ratio c/a. (*See* J. A. Osborn, *Phys. Rev.*, **67,** 351 (1945).) The parameter $g/4\pi$ is found by subtracting $L_{zz}/4\pi$ from 1/3

of TO and LO modes. In terms of the microscopic parameter, $(\partial\mu/\partial Q)^2$, the splitting can be shown to be given by

$$\omega_L - \omega_T \cong \frac{ga^{-3}}{\omega}\left(\frac{\partial\mu}{\partial Q}\right)^2 \tag{5-7-17}$$

where ω is the frequency of the transition in the absence of the splitting—say, for the same transition when the molecule is isolated in a suitable matrix (see Chapter 7).

It is to be noted that, in the circumstance of spherically-shaped, sc crystals, $\boldsymbol{R} = \boldsymbol{S} = \boldsymbol{D} = 0$, and $\omega_L = \omega_T$. It is of course difficult to produce such crystals experimentally. Martin has examined the infrared absorption of KCl crystals with square cross-sections.[35] These crystals had approximate diameters of 5–10 μ and thicknesses of 2 μ; absorption at ω_{TO} and at a higher frequency was always observed. The so-called Berreman conditions[5] ($\theta_i \neq 0$, TM polarization; see Section 5-5) obtained; hence, these observations are not surprising. It is interesting,

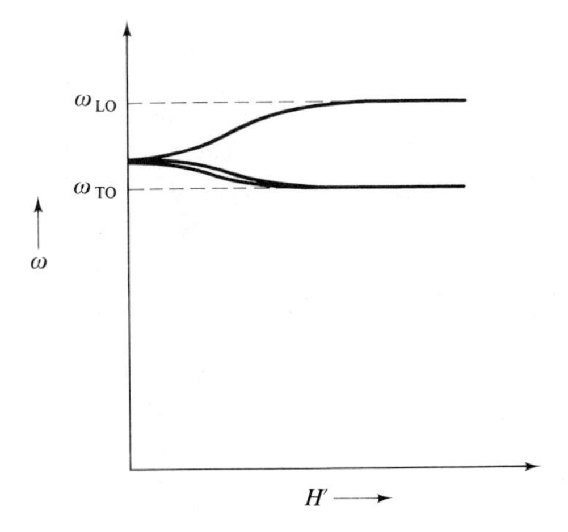

Figure 5-10 The optical mode of a cubic crystal as a function of the dipole coupling H' (5-7-6)

35. T. P. Martin, *Phys. Rev.*, **177,** 1349 (1969).

however, that on going from crystals in which the diameter was twice the thickness to those in which the diameter was approximately four times the thickness, the relative intensity of the entire region $\omega > \omega_T$ decreased. When the axial ratio is more "slab-like," ω_T becomes more prominent, which agrees with the predictions of Fox and Hexter.[18]

Inclusion of polarizability involves the use of (5-7-7) instead of (5-7-6). We may write the field propagation tensor as

$$\mathbf{D}_{t,t'} = r_{tt'}^{-3}(3\mathbf{r}_t\mathbf{r}_{t'}/r_{tt'}^2 - \mathbf{E}) \tag{5-7-18}$$

where $r_{tt'} = |\mathbf{r}_t - \mathbf{r}_{t'}|$ is the distance between the cells t and t'. For $\mathbf{k} = 0$, Eq. (5-7-5) becomes

$$\mathbf{D} = \sum_{t'} \mathbf{D}_{t,t'} \tag{5-7-19}$$

Inspection of (5-7-7) shows that we need to evaluate another sum,

$$\sum_{t'} \mathbf{D}_{t,t'}(\mathbf{E} - \alpha\mathbf{D}_{t,t'})^{-1}$$

however, it can be easily shown[36] that

$$\sum_{t'} (\mathbf{E} - \alpha\mathbf{D}_{t,t'})^{-1} = (\mathbf{E} - \alpha\mathbf{D})^{-1} \tag{5-7-20}$$

Hence, (5-7-7) becomes

$$\begin{aligned} H'_{tt'} &= -\mathbf{p}_t^*\mathbf{D}(\mathbf{E} - \alpha\mathbf{D})^{-1}\mathbf{p}_{t'} \\ &= -\mathbf{p}_t^*\mathbf{D}\mathbf{B}\mathbf{p}_{t'} \end{aligned} \tag{5-7-21}$$

where

$$\mathbf{B} \equiv (\mathbf{E} - \alpha\mathbf{D})^{-1} \tag{5-7-22}$$

In the next section the tensor **B** will be shown to be related to the dielectric constant at optical frequencies.

The results of this section will now be reviewed in the light of the fact that in the vicinity of the crossing of the photon and phonon dispersion curves [see Fig. 5-8(*b*)], the normal modes of the crystal are not accurately phonon modes. As must occur at and near the crossing, the photon mixes with the phonon, and the propagating mode is a hybrid particle, first called a "polariton" by J. J. Hopfield.[37] In the absence of radiation, phonons can indeed interact by way of (5-7-6). From the point of view of second-order perturbation theory, radiative transitions occur only when there is a manifold of intermediate states to which the initial and final states are coupled.[38] It is for this reason that, contrary to

36. J. C. Decius, *J. Chem. Phys.*, **49,** 1387 (1968).
37. J. J. Hopfield, *Phys. Rev.*, **112,** 1555 (1958).
38. B. Kursunoglu, *Modern Quantum Theory*, W. H. Freeman and Company, San Francisco, 1962, Chapter XIV.

the usual view, radiative transitions do not really occur with the phonons. Instead, energy is constantly being exchanged between the photons and the phonons. This exchange of virtual photons gives rise to a *retarded* dipole interaction

$$I_{n,m} = \sum [\{-R^{-1}(1 - \hat{\mathbf{R}}\hat{\mathbf{R}})(\omega/c)^2 + R^{-3}(1 - 3\hat{\mathbf{R}}\hat{\mathbf{R}})(1 - i\omega R/c)\} \exp(i\omega R/c)]\, e^{i\mathbf{k}\cdot\mathbf{R}} \tag{5-7-23}$$

where $\mathbf{R}$ is the vector separation of the cells r_n and r_m. $\hat{\mathbf{R}}$ is a unit vector in the direction of $\mathbf{R}$. Equation (5-7-23) replaces (5-7-5), but reduces to it in the absence of radiation. Equation (5-7-23) was first obtained in this form by Mahan,[39] who recognized that the coupled photon–phonon problem is equivalent to the classical interaction of the electric and polarization fields within a crystal and solved the classical problem directly.

When they are coupled, the dispersion of both the phonons and the photons is drastically altered, as was first shown by Huang from a macroscopic point of view.[16] In the absence of photon–phonon mixing, the dispersion of a photon is given by $\omega = ck$, or in a medium of refractive index $n = \varepsilon^{1/2}$,

$$\omega = \frac{c}{\sqrt{\varepsilon}} k \tag{5-7-24}$$

As derived in the next section, for a single undamped vibrational mode, (5-1-2) assumes the form

$$\varepsilon(\omega) = \varepsilon(\infty) + \frac{S\omega_T^2}{\omega_T^2 - \omega^2} \tag{5-7-25}$$

where S is a transition strength parameter. This function, as we noted in Section 5-4, has a pole at $\omega = \omega_T$ and a zero at $\omega = \omega_L$; it is illustrated in Fig. 5-11. The family of curves (5-7-24) is also illustrated in Fig. 5-11, and there is one such equilateral hyperbola for each value of k; as k increases, a new hyperbola is encountered, further displaced from the origin. These curves represent the dispersion of pure *photons,* in the crystal medium.

The dispersion curve of the polariton is found by plotting the two-valued intersections of each member of the photon family of curves with the phonon dispersion curve (5-7-25). The result is Fig. 5-12. Note that the *upper* branch of Fig. 5-12 derives from the *lower* branch of Fig. 5-11, and vice versa.

The dispersion of the polariton, shown by solid curves, is similar to quantum mechanical resonance, as originally pointed out by Fano.[40] By a calculation of the Poynting vector, Huang was able to calculate the percentage phonon emerging in each polariton branch as a function of k. Not surprisingly, each branch is

39. G. D. Mahan, *J. Chem. Phys.*, **43,** 1569 (1965).
40. U. Fano, *Phys. Rev.*, **103,** 1202 (1956).

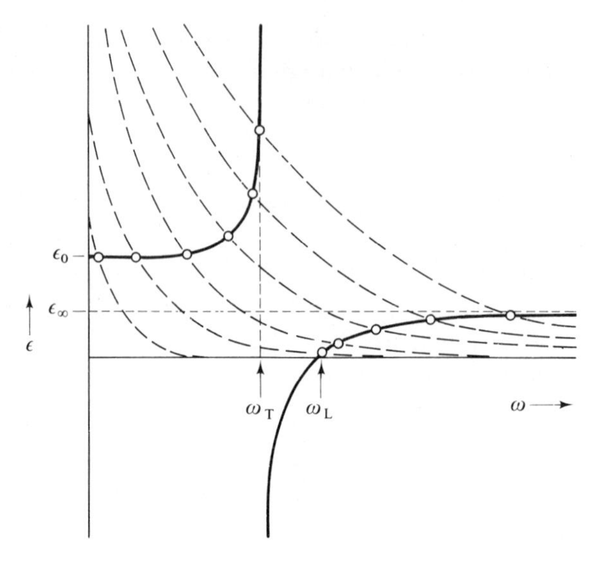

Figure 5-11 Dielectric functions of photons and phonons. Solid lines: the function $\varepsilon(\omega)$ from (5-7-25). Dashed lines: the functions $\varepsilon(\omega)$ from (5-7-24)

50 percent phonon at the crossing point. Even so, numerous authors have shown that retardation is of little importance in problems of this kind.

For example, Stephen has shown that provided $\lambda > R$, where λ is the wavelength of the transition and R is the separation of the interacting atoms, the energy shift due to the retarded interaction is exactly the (non-retarded) dipole–dipole interaction.[41] Similarly, Mahan proved that retardation has a negligible influence on the van der Waals binding energy, which is equivalent to the phonon zero point energy, as originally conjectured by Hopfield. The result is related to the conclusion that the retarded van der Waals interaction between two molecules calculated by Casimir and Polder[42] to be $\sim l^{-7}$ is adequately given by l^{-6} at small separations. Since in the crystal there are so many more molecular pairs (energy-weighted) in the near zone than in the far, (5-7-5) suffices instead of (5-7-23). Finally, Philpott and Lee,[43] by actual and exact calculation of two strong *electronic* transitions (large $\mathbf{p}_i$) in crystalline anthracene, have shown that the corrections to (5-7-5) for retardation are minute.

Throughout this section we have attempted to demonstrate the microscopic basis of *TO* and *LO* modes using the simplest of models—the sc crystal without a basis. One of the great beauties of molecular crystals is the complexity of their unit cells, both with respect to symmetry and population. The methods of this section have been further developed for such crystals by the authors in two ways, one of which stresses high symmetry and population (Z) but with neglect of polarizability,[18] and the other the examination of the importance of the polarizability, particularly in anisotropic crystals.[36]

41. M. J. Stephen, *J. Chem. Phys*, **40,** 669 (1964).
42. H. B. G. Casimir and D. Polder, *Phys. Rev.*, **73,** 360 (1948).
43. M. R. Philpott and J. W. Lee, *J. Chem. Phys.*, **58,** 595 (1973).

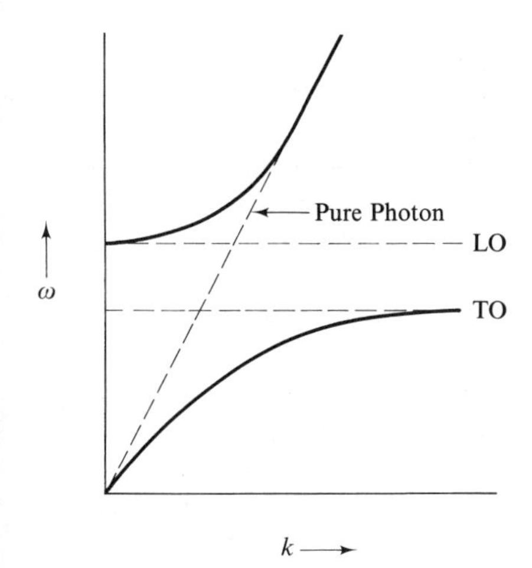

Figure 5-12 Polariton dispersion curves

The extension to crystals with more than one molecule per unit cell is straightforward. Each lattice sum (5-7-5) describes an interaction between two sublattices. The interaction energies (matrix elements of R) at $k = 0$ are found by standard matrix diagonalization procedures, based upon the symmetry of the factor groups. The procedures of Section 6-6 of Wilson, Decius, and Cross may be used. Calculation of the relative intensities of multiplet components is easily accomplished using the usual methods of perturbation theory of degenerate states. The effects of the shape tensor (5-7-14) may be found by finding the matrix elements of $\boldsymbol{D}$ (5-7-15) instead of $\boldsymbol{R}$ (Table 5-2).

Inclusion of the polarizability is also straightforward, provided the components of α are known. The polarizabilities may in fact be determined from the refractive indices using the dipole coupling tensor $\boldsymbol{D}$ (see the following section). Even when they are not, however, certain symmetry-based predictions about the L–T splitting in anisotropic lattices with bases can be made.

We close this section with a discussion of LO–TO splitting of an "infinite" cubic crystal from the point of view of the microscopic theory presented in this section. Equation (5-7-9) applies to such crystals provided, of course, that $\mathbf{k} \cong 0$. If we select a *particular* direction for the propagation vector, say z, the field propagation tensor then has the special form

$$\lim_{\mathbf{k}\to 0} D^{\mu\nu}(\mathbf{k}) = R^{\mu\nu} + \frac{4\pi}{3a^3}\begin{pmatrix}1 & 0 & 0\\ 0 & 1 & 0\\ 0 & 0 & 1\end{pmatrix} - \frac{4\pi}{a^3}\begin{pmatrix}0 & 0 & 0\\ 0 & 0 & 0\\ 0 & 0 & 1\end{pmatrix}$$
$$= R^{\mu\nu} + \frac{4\pi}{3a^3}\begin{pmatrix}1 & 0 & 0\\ 0 & 1 & 0\\ 0 & 0 & -2\end{pmatrix} \tag{5-7-26}$$

Comparison with (5-7-15) and (5-7-16) shows that from this point of view "infinite" crystals have LO–TO splittings quite like those of slab-shaped microcrystals, in which the symmetry axis of the slab (the slab normal) is parallel to the propagation vector. Indeed, we now see that even when the $\boldsymbol{R}$ tensor vanishes, such as at a lattice point, the propagation vector, now a symmetry axis, *spoils* the cubic symmetry, and therefore $\omega_{LO} \neq \omega_{TO}$. In terms of the interacting dipoles, the convergence of the summation of all of the interactions is determined by the wavelength. Thus, for an infinite crystal, the wavelength is much smaller than the domain size; hence, the validity of the planewise summation method.

There remains the question of phase memory across domain boundaries, in this case boundaries perpendicular to the propagation direction. In microscopic crystals, where the wavelength is greater than the domain size, the phase change across the entire sample is small, so that effectively $\mathbf{k} \cong 0$. Although we have earlier recognized, along with Peiser, that "single crystals" are really mosaics, we must conclude that there is little or no phase change across domain boundaries, for if there were, we should have to interpret observed L–T splittings as the resultant of those of all the domains which make up the crystal. Aside from the broadening which could then result from scattering at domain boundaries, rotation of the crystal about an axis perpendicular to the propagation direction would result in an exchange of the roles of ω_T and ω_L. Experimentally $\omega_L > \omega_T$ *always*, independent of crystal orientations. Moreover, the same result is obtained in experiments which have been performed on cubic crystals which are *twinned*, in which case, were phase memory not the case, the roles of ω_L and ω_T would be exchanged in adjacent domains even without crystal rotation.

In non-cubic systems, in which a preferred axis (or axes) exist in the absence of electromagnetic radiation, this fundamental relationship of the L–T splitting to the propagation direction is more difficult to demonstrate, for the asymmetry of the crystal creates a splitting of its own to which the electromagnetic part discussed in this section only adds or subtracts. This demonstration of the L–T splitting in twinned cubic crystals cannot be unique, however, so that phase memory must be quite general in "single crystals," no matter what their system.

It is of great interest that the interpretation of the L–T splitting we have gained from this microscopic basis is exactly that which resulted from the microscopic basis (Section 4-9). That is to say, it is the anisotropy created by the very existence of a propagation *direction* which, even at $\mathbf{k} = 0$, causes $\omega_{LO} \neq \omega_{TO}$.

5-8 MICROSCOPIC THEORY OF THE DIELECTRIC CONSTANT

In this section it is our aim to construct a molecular theory for Eq. (5-1-2) which expresses the frequency dependence of the principal components of the dielectric constant in the form

$$\varepsilon_\sigma(\nu) = \varepsilon_\sigma(\infty) + \sum_j S_j^{(\sigma)} \nu_j^2 / (\nu_j^2 - \nu^2 - i\gamma_j \nu) \tag{5-8-1}$$

where the sum is over the infrared active fundamental modes polarized in the direction of the σ unit cell axis. The parameter ν_j is the *transverse* frequency for the jth mode, γ_j is a damping (line-width) parameter, and $S_j^{(\sigma)}$ is the strength of the transition. Also, $\varepsilon_\sigma(\infty) = n_\sigma^2(\infty)$, i.e., the $\varepsilon_\sigma(\infty)$ are the squares of the refractive indices at frequencies which are large compared with the vibration frequencies. Such parameters can ordinarily be taken as the refractive indices in the visible region unless the crystal is colored.

Let the total dipole moment of the mth molecule in unit cell $\boldsymbol{\tau}$ be designated as $\mathbf{p}_{\tau m}$; this moment is the sum of an intrinsic moment $\boldsymbol{\mu}_{\tau m}$ which is a function of the molecular normal coordinates, and an induced part which is proportional to the *effective* field $\mathscr{F}_{\tau m}$ at the molecule in question:

$$\mathbf{p}_{\tau m} = \boldsymbol{\mu}_{\tau m} + \boldsymbol{\alpha}_m \mathscr{F}_{\tau m} \tag{5-8-2}$$

in which the proportionality "constant" $\boldsymbol{\alpha}_m$ is the 3×3 matrix of the molecular polarizability due to electronic, as distinct from vibrational, response to the applied field.

The effective field is the sum of the externally applied field $\mathscr{E}_{\tau m}$ which in the present context is due to incident radiation, and to the fields produced at the τ, m molecule by the dipoles of all other molecules in the lattice

$$\mathscr{F}_{\tau m} = \mathscr{E}_{\tau m} + \sum_{\tau', m'} \mathbf{D}_{\tau m, \tau' m'} \mathbf{p}_{\tau' m'} \tag{5-8-3}$$

where $\mathbf{D}$ is the field propagation tensor (5-7-18), that is

$$\mathbf{D}_{\tau m, \tau' m'} = \frac{-1}{r^3}\mathbf{E} + \frac{3}{r^5}\begin{pmatrix} x^2 & xy & xz \\ yx & y^2 & yz \\ zx & zy & z^2 \end{pmatrix} \tag{5-8-4}$$

Our interest in the infrared active modes limits the discussion to total translational symmetry ($k = 0$) which implies that we need consider only the case in which all translationally equivalent dipoles are in phase. Thus, we may write:

$$\begin{aligned} \mathbf{p}_{\tau m} &= \mathbf{p}_m \\ \boldsymbol{\mu}_{\tau m} &= \boldsymbol{\mu}_m \\ \mathscr{E}_{\tau m} &= \mathscr{E} \end{aligned} \tag{5-8-5}$$

Also, the definition of $\mathbf{D}$ in (5-8-4) implies that $\mathbf{D}$ is a function only of the vectorial difference between the positions τ, m and τ', m'. For these reasons, (5-8-3) may be simplified to

$$\mathscr{F}_m = \mathscr{E} + \sum_{\tau', m'} \mathbf{D}_{\tau m, \tau' m'} \mathbf{p}_{m'} = \mathscr{E} + \sum_{m'} \mathbf{D}_{mm'} \mathbf{p}_{m'} \tag{5-8-6}$$

where

$$\mathbf{D}_{mm'} = \sum_{\tau'} \mathbf{D}_{\tau m, \tau' m'} \tag{5-8-7}$$

Such sums are only conditionally convergent; practical methods for their evaluation are discussed in Appendix XI.

It is now possible to eliminate the effective field by combining Eqs. (5-8-2) and (5-8-6) and to solve for the dipole $\mathbf{p}_m$; the result is

$$\mathbf{p}_m = \sum_{m'} \mathbf{B}_{mm'}\boldsymbol{\mu}_{m'} + \sum_{m'} \mathbf{B}_{mm'}\boldsymbol{\alpha}_{m'}\mathscr{E} \tag{5-8-8}$$

in which

$$\mathbf{B} = [\mathbf{E} - \boldsymbol{\alpha}\mathbf{D}]^{-1} \tag{5-8-9}$$

If there are M molecules per unit cell, then the matrix equation (5-8-9) involves square matrices of dimension $3M$; $\boldsymbol{\alpha}$ consists of M 3×3 diagonal blocks, $\boldsymbol{\alpha}_m$.

The polarization per unit volume $\mathscr{P}$, is related to the dielectric tensor according to the expressions from electromagnetic theory,

$$\mathscr{D} = \mathscr{E} + 4\pi\mathscr{P} = \varepsilon\mathscr{E} \tag{5-8-10}$$

From the previous development one sees that

$$\mathscr{P} = \frac{1}{\mathrm{v}}\sum_{m} \mathbf{p}_m = \mathscr{P}_{\mathrm{vib}} + \mathscr{P}_{\mathrm{electronic}} \tag{5-8-11}$$

where v is the unit cell volume and $\mathscr{P}_{\mathrm{electronic}}$ means the terms involving $\boldsymbol{\alpha}_{m'}$ explicitly. Although (5-8-8) shows the dependence of the electronic polarizability on the external field, there is so far no such dependence apparent for the intrinsic dipole term involving $\boldsymbol{\mu}_{m'}$. One can find this dependence by calculating the response of a harmonic oscillator to an external driving term. $\mathscr{P}_{\mathrm{vib}}$ can be expanded in the infrared active normal modes of the crystal, Q_j:

$$\mathscr{P}_{\mathrm{vib}} = \frac{1}{\mathrm{v}}\sum_{j}\sum_{m,m'} \mathbf{B}_{mm'}\frac{\partial\boldsymbol{\mu}_{m'}}{\partial Q_j}Q_j \tag{5-8-12}$$

The coupling energy with the external driving term is

$$-\mathscr{P}_{\mathrm{vib}}\cdot\mathscr{E} = -\mathscr{P}_{\mathrm{vib}}\cdot\mathscr{E}^0 e^{-i\omega t}$$

from which one readily calculates a response function for Q_j of the form

$$Q_j = \sum_{m,m'} \mathbf{B}_{mm'}\frac{\partial\boldsymbol{\mu}_{m'}}{\partial Q_j}\cdot\mathscr{E}(\omega_j^2 - \omega^2 - i\gamma_j\omega)^{-1} \tag{5-8-13}$$

In (5-8-13), ω_j represents the resonance frequency, i.e., the transverse mode frequency of the jth oscillator, and γ_j is a damping constant: note that frequencies here are circular, i.e., in radians sec^{-1}.

A similar result can be obtained from the time dependent quantum mechanics of the harmonic oscillator[44] provided the initial state for a transition has a finite lifetime and its wave mechanical amplitude decays like $e^{-\gamma_j t/2\pi}$.

In the principal axis system of the dielectric constant, (5-8-10) becomes simply

$$\varepsilon_\sigma = 1 + 4\pi\mathscr{P}_\sigma/\mathscr{E}_\sigma \qquad \sigma = x, y, z$$

44. R. Kubo and T. Nagamiya, Eds., *Solid State Physics*, McGraw-Hill, New York, 1969, pages 688–90.

and by combining (5-8-13), (5-8-12), and (5-8-11) and selecting the σ component of $\mathscr{P}$, one finds

$$\varepsilon_\sigma = 1 + \frac{4\pi}{\mathrm{v}}\left\{\left[\sum_{m,m'} \mathbf{B}_{mm'}\boldsymbol{\alpha}_{m'}\right]_{\sigma\sigma} + \sum_j \left[\sum_{m,m'} \mathbf{B}_{mm'}\frac{\partial \boldsymbol{\mu}_{m'}}{\partial Q_j}\right]_\sigma^2 (\omega_k^2 - \omega^2 - i\gamma_j\omega)^{-1}\right\} \tag{5-8-14}$$

This equation, except for the fact that frequency is expressed in radians sec^{-1} rather than cm^{-1}, is identical in form with that given at the beginning of the chapter; by expressing all frequencies appearing in (5-1-2) as ω in the place of ν, one sees that

$$\varepsilon_\sigma(\infty) = 1 + \frac{4\pi}{\mathrm{v}}\left[\sum_{m,m'} \mathbf{B}_{mm'}\boldsymbol{\alpha}_{m'}\right]_{\sigma\sigma} \tag{5-8-15}$$

and

$$S_j^{(\sigma)}\omega_j^2 = \frac{4\pi}{\mathrm{v}}\left[\sum_{m,m'} \mathbf{B}_{mm'}\frac{\partial \boldsymbol{\mu}_{m'}}{\partial Q_j}\right]_\sigma^2 \tag{5-8-16}$$

Equation (5-8-15) may be used to deduce the values of the molecular polarizabilities from the observed refractive indices ($\varepsilon(\infty) = n^2(\infty)$) in the visible region. Often, of course, there are too many parameters and too few observables, but by adopting polarizabilities for a few simple ions such as Li^+ or Na^+ many monatomic and polyatomic molecular ion polarizabilities have been determined assuming, as is implicit throughout this section, that the molecular parameters in the crystal are additive.

Since the strengths S_j can be obtained either from the analysis of reflection band widths or from the limiting integrated absorbance given in (5-5-8), it is evident that the molecular dipole derivatives can be evaluated if the **B** matrix can be obtained. A few remarks are therefore in order about **B**.

Elimination of **p** rather than $\mathscr{F}$ between Eqs. (5-8-2) and (5-8-6) shows that

$$\mathscr{F}_m = \sum_{m'} \mathbf{B}_{mm'}\mathscr{E} + \sum_{m'} (\mathbf{BD})_{mm'}\boldsymbol{\mu}_{m'}$$

Thus, neglecting the molecular moments $\boldsymbol{\mu}_{m'}$ the effective field at molecule m is $\mathscr{E}$ multiplied by the sum

$$\sum_{m'} \mathbf{B}_{mm'} = \sum_{m'} \mathbf{B}_{m'm}$$

In this sense, the **B** sum gives the *effective field ratio,* i.e., $\mathscr{F}_m/\mathscr{E}$. Clearly, if either all $\boldsymbol{\alpha}_{m'}$ or $\boldsymbol{D}_{mm'}$ vanish (the latter case exists at high dilution in a gas), **B** is simply a unit matrix and $\mathscr{F}_m = \mathscr{E}$.

If the molecular polarizabilities and lattice dipole sums are available, **B** can be evaluated. In cubic crystals, like the alkali halides, it is readily shown that **B** is a diagonal matrix with values $(\varepsilon(\infty) + 2)/3$: this is the famous Lorentz–Lorenz field result. In Section 6-5 we shall return to this topic in order to illustrate the evaluation of the dipole derivatives in molecular crystals.

CHAPTER

SIX

MOLECULAR CRYSTALS

6-1 INTERNAL AND EXTERNAL COORDINATES

It is obvious that one could extend the methods employed in Chapter 4 for the analysis of the vibrations of a crystal composed of monatomic units to the case of molecular crystals and ionic crystals containing molecular ions by regarding the contents of the unit cell as individual atoms and neglecting the molecular binding. To do so, however, would fail to take advantage of the wide separation of frequencies of the internal molecular modes as compared with the external modes which can be regarded as essentially translational and rigid rotational motions. Moreover, the comparison of the internal molecular motions with their gas phase counterparts could not lend itself to the perturbation treatment described in Section 1-3. Therefore, throughout this chapter we shall consistently adopt the point of view that the motions of each molecule in the unit cell are classified initially as (i) internal vibrations identical with those in the gas phase molecule, and (ii) rigid translations and rotations of each molecule at its site. The symbol q_k will stand for a normal coordinate of the gas phase molecule and symbols T, R will stand for rigid translations and rotations respectively.

Such a classification of modes implies a certain form for the kinetic and potential energy matrices, which we write as

$$\mathbf{G} = \begin{pmatrix} \mathbf{G}_{HH} & \mathbf{G}_{HL} \\ \mathbf{G}_{LH} & \mathbf{G}_{LL} \end{pmatrix} \quad \text{and} \quad \mathbf{F} = \begin{pmatrix} \mathbf{F}_{HH} & \mathbf{F}_{HL} \\ \mathbf{F}_{LH} & \mathbf{F}_{LL} \end{pmatrix} \tag{6-1-1}$$

in which $\mathbf{G}_{HH}$ is a $3n - f$ dimensional unit matrix and $\mathbf{F}_{HH}$ is a diagonal matrix composed of the eigenvalues ω_k^2.

One must distinguish carefully between two possible definitions of the low frequency (external) coordinates. It is perhaps easier to formulate a reasonable potential function if the low frequency coordinates are defined as rigid translations and rotations of the molecular unit. However, this has the disadvantage that G_{LH} and G_{HL} do not vanish and G_{LL} is not simple. Another choice is to let the low frequency coordinates be the displacements of the center of mass and small rotations of the principal axes of inertia. When this choice is made $\bar{G}_{LH}$ and $\bar{G}_{HL}$ vanish and

$$\bar{\mathbf{G}}_{LL} = \begin{bmatrix} \frac{1}{M} & 0 & 0 & 0 & 0 & 0 \\ 0 & \frac{1}{M} & 0 & 0 & 0 & 0 \\ 0 & 0 & \frac{1}{M} & 0 & 0 & 0 \\ 0 & 0 & 0 & \frac{1}{I_X} & 0 & 0 \\ 0 & 0 & 0 & 0 & \frac{1}{I_Y} & 0 \\ 0 & 0 & 0 & 0 & 0 & \frac{1}{I_Z} \end{bmatrix} \tag{6-1-2}$$

Equation (6-1-2) is, of course, the kinetic energy matrix of a rigid molecule, meaning one in which all internal vibrational coordinates vanish. It therefore follows that the $\bar{G}_{LL}$ must be identical with the result given in Eq. (4-8-2) of Wilson, Decius, and Cross,[1] which in the present notation is

$$\bar{\mathbf{G}}_{LL} = \mathbf{G}_{LL} - \mathbf{G}_{LH}\mathbf{G}_{HH}^{-1}\mathbf{G}_{HL} \tag{6-1-3}$$

This in turn implies that an approximate separation of high and low frequencies can be made in which the high frequencies are determined by

$$|\mathbf{G}_{HH}\mathbf{F}_{HH} - \omega_H^2\mathbf{E}| = 0 \tag{6-1-4}$$

and the low by

$$|\bar{\mathbf{G}}_{LL}\mathbf{F}_{LL} - \omega_L^2\mathbf{E}| = 0 \tag{6-1-5}$$

In other words, in this approximation, the internal mode frequencies are unaffected by the external modes, and the low frequencies may be obtained from the secular equation (6-1-5) which uses the simple form of (6-1-2) together with potential constants (F_{LL}) in the unbarred basis, i.e., that of the rigid displacements.

Higher order approximations may be necessary in some cases. In principle, this is certainly the case where molecules have modes not much higher than the

1. E. B. Wilson, Jr., J. C. Decius, and P. C. Cross, *Molecular Vibrations*, McGraw-Hill, New York, 1955, Chapter 4.

lattice (external) frequencies. One should understand that F_{HH} and G_{HH} refer not to a single molecule, but to all molecules in the lattice; similarly with F_{LL} and G_{LL}. Also, although there are no coupling constants between different molecules in G, this is certainly not the case in F. In the external mode region, the intermolecular coupling constants in F_{LL} are large but most quantitative treatments of the internal modes will make the assumption that the intermolecular coupling terms in F_{HH} are small enough to be handled by perturbation theory.

Even if the approximate separation described in (6-1-1) through (6-1-5) is of insufficient accuracy, the classification of coordinates is fundamental to the symmetry analysis of molecular vibrations in crystals, as we shall see in Section 6-2 below. However, it is important to point out at this stage that molecular crystals are known in which the rotations R_x, R_y, R_z *cannot* be treated as small vibrations. Such crystals include H_2, D_2, CH_4, and possibly others. When the moment of inertia is small, the kinetic energy of rotation as given by $(\hbar^2/2I)J(J+1)$ is large, even for small angular momentum quantum numbers J. Thus it may happen that the energy is sufficiently large that the rotational motion cannot be treated as a small vibration. A further account of this topic may be found in Appendix XII; in the meantime, we proceed with the analysis on the supposition that the T and R motions, like the internal motions, can be treated as small vibrations.

Since, as noted above, there are no intermolecular coupling terms in the **G** matrix, all intermolecular effects are, to the first order, to be found in the **F** matrix. To describe these, a rather cumbersome notation is necessary for generality, but in later examples, redundant indices will be dropped. If the unit cell τ contains more than one molecule, we use m to specify the molecule: all molecules with the same m are translationally equivalent. If the molecular normal mode k is degenerate, then we use α to specify the component of the degeneracy. Thus $q_{k\alpha}(\tau, m)$ identifies a normal coordinate of a certain molecule. The potential energy will then consist of two parts

$$V = V_0 + V' \tag{6-1-6}$$

where

$$2V_0 = \sum_{k\alpha} \omega_{k\alpha}^2 \sum_{\tau,m} q_{k\alpha}^2(\tau, m) \tag{6-1-7}$$

and

$$2V' = \Sigma\Sigma f'_{k\alpha,k'\alpha'}(\tau, m; \tau', m') q_{k\alpha}(\tau, m) q_{k'\alpha'}(\tau', m') \tag{6-1-8}$$

The expression for V_0 shows that we contemplate the lifting of the free molecule degeneracies at the sites, since otherwise $\omega_{k\alpha}^2 = \omega_k^2$. The terms in the intermolecular potential V' are summed over all k, α, k', α' and over all molecule indices for which τ and τ' and/or m and m' are different. The translational symmetry implies that the intermolecular potential constants are functions only of the vector difference between the unit cells:

$$f'_{k\alpha,k'\alpha'}(\tau, m; \tau', m') = f'_{k\alpha,k'\alpha'}(\Delta\tau; m, m') \tag{6-1-9}$$

where

$$\Delta\tau = \tau' - \tau \tag{6-1-10}$$

It is now easy in principle to reduce the problem of the crystal to the dynamics of one unit cell by carrying out the transformation of V_0 and V' which corresponds to the introduction of phonon coordinates:

$$Q_{k\alpha}(\kappa, m) = N^{-3/2} \Sigma\, e^{-i\kappa \cdot \tau}\, q_{k\alpha}(\tau, m) \tag{6-1-11}$$

The term V_0 contributes $\omega_{k\alpha}^2$ to the diagonal elements of the dynamic matrix: the V' terms give rise to

$$F'_{k\alpha,k'\alpha'}(\kappa\,; m, m') = \sum_{\Delta\tau} f'_{k\alpha,k'\alpha'}(\Delta\tau\,; m, m')\, e^{i\kappa \cdot \Delta\tau} \tag{6-1-12}$$

Let us consider a simple example of the form which the dynamical matrix now assumes. Suppose we have two equivalent diatomic molecules per unit cell: in such a case the dynamic matrix assumes the form

$$\begin{pmatrix} \omega_1^2 + F'(\kappa\,; 11) & F'(\kappa\,; 12) \\ F'(\kappa\,; 21) & \omega_1^2 + F'(\kappa\,; 22) \end{pmatrix}$$

Since $F'(\mathbf{0}\,; 11) = F'(\mathbf{0}\,; 22)$ and $F'(\mathbf{0}\,; 12) = F'(\mathbf{0}\,; 21)$ the eigenvalues are

$$\omega^2 = \omega_1^2 + F'(\mathbf{0}\,; 11) \pm F'(\mathbf{0}\,; 12) \tag{6-1-13}$$

In crystals of greater dynamic complexity, the analysis of the splitting of molecular modes depends upon the diagonalization of larger dynamic matrices than the simple one just given as an example. Such analysis relies heavily upon the group theoretical technique of correlation and forms the subject of the next section. In most cases, even for polyatomic molecules, the coupling between *different* molecular normal modes is neglected on the ground that the intermolecular couplings are weak and have a significant first order effect which is limited to the lifting of the degeneracy between identical modes in different molecules. More sophisticated treatments, however, consider the effects of coupling between all the molecular internal and external modes and examples, not limited to $\kappa = 0$, will be given in Section 6-6. The reader should recall from Chapter 4 that even in the "worst" case, i.e., κ corresponding to a general point (C_1) in the Brillouin zone, the size of the dynamical matrix is limited to three times the number of atoms per unit cell.

6-2 SYMMETRY SPECIES IN MOLECULAR CRYSTALS BY CORRELATION

An essential point in the symmetry analysis required in this section is the separate identification of internal and external modes. This is easily accomplished by the technique of correlation. In contrast to the examples of the previous chapter, in which the modes at $\kappa = 0$ were derived by a study of the species of *atomic translations* at the atomic sites followed by correlation of the site groups with the unit cell group, the presence of molecular units demands that the species of the molecular motions at the sites be first identified, the final correlation process being then formally identical with that employed in Chapter 4.

One starts with the molecular unit in the gas phase, and with the aid of Section 2-3, the symmetry species of both internal and external (translational and rotational) modes are identified. Next, one moves to the site occupied by the molecule in the crystal. Unless the crystal is disordered, the site group must either be identical with the free molecule group, or one of its subgroups: the latter situation is encountered in the majority of examples. The species of the modes are correlated from the free molecule to the site. One notes that in this step the number of degrees of freedom is conserved. Finally, the species at the site are correlated with those for the unit cell; this step multiplies the number of degrees of freedom by the number of equivalent sites, which is just the ratio g/h, g being the order of the unit cell group and h that of the site group. If more than one site is occupied, as is always the case for ionic crystals, the process is repeated for each site.

Having identified the symmetry species at $\boldsymbol{\kappa} = 0$ in this way, it is easy by a further application of the correlation technique to describe the behavior of each mode as $\boldsymbol{\kappa}$ becomes finite, although, as we have noted in Section 4-5, on the zone boundary in the non-symmorphic cases, there occur examples which are not isomorphic with the more familiar point groups.

This symmetry analysis at $\boldsymbol{\kappa} = 0$ also lends itself to a very convenient formulation of the intermolecular coupling terms in the dynamic matrix in a form which is independent of any particular model of the intermolecular potential.

This process will now be illustrated with a sufficient variety of examples to cover most of the questions of interpretation which arise in practice.

(a) HCN ($C_{\infty v}$) in Space Group $C_{2v}^{20} = Imm2$

Crystalline HCN at temperatures below 170°K has a structure[2] consisting of linear parallel chains with a body centered, orthorhombic unit cell. The primitive cell is singly occupied and the site, (a) in Table VII-2, has symmetry C_{2v}. In general, single occupancy of the primitive cell implies that the site symmetry is identical with the unit cell symmetry. When this is the case, the final step in correlation naturally is omitted, i.e., it is only necessary to correlate the free molecule modes with the modes under the site. This is readily accomplished, recalling the result of Section 2-6(b) where the XYZ internal modes were classified as ν_1 and ν_3 in Σ^+ and ν_2 in Π of $C_{\infty v}$. The external modes will be described as $T_z = T_\sigma$ in Σ^+ and $(T_x, T_y) = T_\pi$ in Π; also there exist the rotations $(R_x, R_y) = R_\pi$ in Π. The character tables in Appendix I give full details of the translational and rotational species in all groups.

2. R. W. G. Wyckoff, *Crystal Structures*, Interscience, New York, Volume I, 1963, page 160.

By examination of the characters of $C_{\infty v}$ for the selected subgroup $C_{2v} = E, C_2, \sigma, \sigma'$ (or by the use of correlation tables, Appendix IV) one readily discovers that

$$C_{\infty v} \to C_{2v}$$
$$\Sigma^+ \to A_1$$
$$\Pi \to B_1 + B_2$$

The conclusion then is that modes ν_1, ν_3, and T_σ belong to A_1 and that the modes ν_2, T_π, R_π belong to $B_1 + B_2$. In other words, the doubly degenerate modes are split into distinct B_1 and B_2 components; this is the explanation of the observed ν_2 doublet at 828 and 838 cm^{-1} in the crystal.[3] This type of splitting can be understood as a simple consequence of the existence of different bending force constants for HCN bent in the xz compared with the yz plane, as dictated by the non-equivalence of these two directions under orthorhombic symmetry.

More explicitly, the A_1, B_1, and B_2 parts of the dynamic matrix at $\boldsymbol{\kappa} = 0$ could be written

$$\begin{aligned} \omega^2 &= \omega_k^2 + F'_{kk}(0) \qquad (A_1) \qquad k = 1, 3 \\ \omega^2 &= \omega_{21}^2 + F'_{21,21}(0) \qquad (B_1) \\ \omega^2 &= \omega_{22}^2 + F'_{22,22}(0) \qquad (B_2) \end{aligned} \tag{6-2-1}$$

Section 7-2 will discuss means of evaluating the coupling parameters, F'.

(b) CO_2, Space Group $T_h^6 = Pa3$

The free molecule modes are ν_1 in Σ_g^+; ν_2 and T_x, T_y in Π_u, ν_3 and T_z in Σ_u^-; R_x, R_y in Π_g. It is convenient to describe the external modes in the following as $T_z = T_\sigma$, $(T_x, T_y) = T_\pi$, $(R_x, R_y) = R_\pi$. X-ray diffraction studies reveal that there are four molecules per unit cell; the space group is $T_h^6 = Pa3$, whose maximal sites are (a) or (b) of symmetry $C_{3i} = S_6$. Since the order of T_h is $g = 24$ and of S_6 is $h = 6$, there must be four equivalent sites. In T_h (see Table VII-10) every other site has a lower h, so necessarily the number of equivalent sites would be greater than four, and if the CO_2 molecules occupied such sites the number of molecules per cell would be greater than four. In fact the carbon atoms are situated at points of S_6 symmetry, so the first step in the correlation is the identification of the CO_2 species at S_6. Under S_6, one finds ν_1 in A_g, ν_2 and T_π in E_u, ν_3 and T_σ in A_u, and R_π in E_g, by comparison of the species Σ_g^+, Π_u, Σ_u^-, and Π_g of $D_{\infty h}$ with those of S_6. At this stage there are nine degrees of freedom, taking into account the degeneracies.

Finally one may use the correlation table between T_h and S_6, keeping in mind the fact that this step necessarily multiplies the degrees of freedom by four. The

3. R. E. Hoffman and D. F. Hornig, *J. Chem. Phys.*, **17**, 1163 (1949).

whole process, which uses no more information than is contained in the correlation tables (Appendix IV) or the character tables from which the former are derived, is summarized in Table 6-1. Note that the correlation from $D_{\infty h}$ to S_6 is simple, and in particular, there is no splitting of the free molecule degeneracies. Correlation from S_6 to T_h involves one complication, namely the double line connecting the $E(S_6)$ with $F(T_h)$ modes. An important distinction must be made between the splitting of ν_2 predicted here and the simpler site splitting which occurs in HCN. In CO_2 the site symmetry does *not* lift the degeneracy; the three predicted frequencies are a consequence of intermolecular coupling.

Not all the modes given in Table 6-1 under the unit cell group are optical modes. The three acoustic modes must have the symmetry of a translation under T_h; since the species of a translation under T_h is F_u, the acoustic modes are necessarily identified with one of the three degrees of freedom represented by $T_{\sigma F}$, T'_π, and T''_π, leaving 14 distinct frequencies for the optical branches at $\boldsymbol{\kappa} = 0$. Of course not all modes are necessarily active in the infrared or Raman; selection rules for this example will be taken up in section 6-3. Of the 14 optic frequencies, seven are internal. The symmetric stretch yields two frequencies, ν_{1A} and ν_{1F}, the bend yields three frequencies, ν'_{2F}, ν''_{2F}, and ν_{2E}, and the antisymmetric stretch two frequencies ν_{3A} and ν_{3F}. These multiplicities are in the present case entirely due to intermolecular terms in the Hamiltonian.

The form of the dynamic matrix is easy to infer from the analysis just given, noting that the indices m, m' introduced in section 6-1 may now take on the values 1 to 4. For molecular modes 1 and 3, these matrices take the form:

$$\begin{pmatrix} \omega_k^2 + F'_k(1,1) & F'_k(1,2) & F'_k(1,3) & F'_k(1,4) \\ & \omega_k^2 + F'_k(2,2) & F'_k(2,3) & F'_k(2,4) \\ & & \omega_k^2 + F'_k(3,3) & F'_k(3,4) \\ \text{(sym)} & & & \omega_k^2 + F'_k(4,4) \end{pmatrix}$$

Table 6-1 Symmetry of the $\kappa = 0$ modes in CO_2, space group T_h^6

	Free molecule ($D_{\infty h}$)	Site (S_6)	Unit cell (T_h)	
ν_1:	Σ_g^+	A_g	A_g	ν_{1A}
R_π:	Π_g	E_g	F_g	$\nu_{1F}, R'_{\pi F}, R''_{\pi F}$
			E_g	$R_{\pi E}$
ν_3, T_σ	Σ_u^-	A_u	A_u	$\nu_{3A}, T_{\sigma A}$
ν_2, T_π	Π_u	E_u	F_u	$\nu_{3F}, T_{\sigma F}, \nu'_{2F}, \nu''_{2F}, T'_\pi, T''_\pi$
			E_u	$\nu_{2E}, T_{\pi E}$

Here we have compressed the notation $F'_{k\alpha,k'\alpha'}(\boldsymbol{\kappa}; m, m')$ to $F'_k(m, m')$ since the mode is non-degenerate, $k = k'$ and $\boldsymbol{\kappa} = 0$ is understood. By methods already familiar from the analysis of polyatomic molecules, or more formally, from the knowledge that the unit cell group contains operations which send $1 \to 2$ or $1 \to 1$ and $2 \to 3$ etc., it can be demonstrated that (i) all the diagonal elements of F' are equal and (ii) all the off-diagonal elements are equal. Furthermore, the introduction of the appropriate symmetry coordinates for the unit cell reduces this part of the dynamic matrix to

$$\begin{pmatrix} \omega_k^2 + F'_k(1,1) + 3F'_k(1,2) & \text{species: } A_g \text{ for } \nu_1 \\ & A_u \text{ for } \nu_3 \\ \text{and} & \\ \omega_k^2 + F'_k(1,1) - F'_k(1,2) & \text{species: } F_g \text{ for } \nu_1 \\ & F_u \text{ for } \nu_3 \end{pmatrix} \tag{6-2-2}$$

The construction of the dynamical matrix for the degenerate mode ν_2 is more subtle. It was first given in the context of the vibrational exciton method (see Section 6-4) by Hexter.[4] It is advantageous to replace the pair of real, cartesian molecular normal coordinates q_{2x} and q_{2y} by the complex conjugate pair

$$q_{2+} = 2^{-1/2}(q_{2x} + iq_{2y})$$
$$q_{2-} = 2^{-1/2}(q_{2x} - iq_{2y})$$

because these coordinates provide a basis for the separate, though degenerate, representations E_+ and E_- under the local symmetry. Translationally symmetrized coordinates $Q_{2\pm}(\boldsymbol{\kappa}, m)$ may be constructed in the usual way. Abbreviating them in the shorthand notation, for $\boldsymbol{\kappa} = 0$,

$$Q_{2\pm}(\mathbf{0}, m) = m_\pm$$

one finds the convenient transformation properties

$$\begin{aligned} C_{31}^+ 1_+ &= \varepsilon\ 1_+ & C_{31}^+ 1_- &= \varepsilon^* 1_- \\ C_{31}^- 1_+ &= \varepsilon^* 1_+ & C_{31}^- 1_- &= \varepsilon\ 1_- \\ C_{31}^+ 2_+ &= \varepsilon\ 3_+ & C_{31}^+ 2_- &= \varepsilon^* 3_- \\ C_{31}^- 2_+ &= \varepsilon^* 3_+ & C_{31}^- 2_- &= \varepsilon\ 3_- \end{aligned}$$

etc., where $C_{31}^{\pm}$ stand for rotations of $\pm 120°$ around an axis coincident with molecule number 1, and $\varepsilon = \exp(2\pi i/3)$. In consequence of these transformation properties, the dynamical matrix assumes the following form:

4. R. M. Hexter, *J. Chem. Phys.*, **36**, 2285 (1962); **39**, 1608 (1963).

$$
\begin{array}{c} \\ 1_+^* \\ 2_+^* \\ 3_+^* \\ 4_+^* \\ 1_-^* \\ 2_-^* \\ 3_-^* \\ 4_-^* \end{array}
\begin{array}{c} \begin{array}{cccccccc} 1_+ & 2_+ & 3_+ & 4_+ & 1_- & 2_- & 3_- & 4_- \end{array} \\
\begin{pmatrix}
\omega_2^2+A & B & B & B & 0 & C & \varepsilon C & \varepsilon^* C \\
B & \omega_2^2+A & B & B & C & 0 & \varepsilon^* C & \varepsilon C \\
B & B & \omega_2^2+A & B & \varepsilon C & \varepsilon^* C & 0 & C \\
B & B & B & \omega_2^2+A & \varepsilon^* C & \varepsilon C & C & 0 \\
0 & C & \varepsilon^* C & \varepsilon C & \omega_2^2+A & B & B & B \\
C & 0 & \varepsilon C & \varepsilon^* C & B & \omega_2^2+A & B & B \\
\varepsilon^* C & \varepsilon C & 0 & C & B & B & \omega_2^2+A & B \\
\varepsilon C & \varepsilon^* C & C & 0 & B & B & B & \omega_2^2+A
\end{pmatrix} \end{array}
$$

Note that, in this basis of complex coordinates, the dynamical matrix is Hermitian. The elements of the matrix have the meanings

$$A = F'_{2+2+}(m, m)$$

$$B = F'_{2+2+}(m, m') \qquad m \neq m'$$

$$C = F'_{2+2-}(1, 2)$$

Since the + and − basis coordinates behave like distinct one-dimensional representations, they can be symmetrized within the unit cell separately, using the transformation matrix

$$\mathbf{U} = \tfrac{1}{2}\begin{bmatrix} 1 & 1 & 1 & 1 \\ 1 & 1 & -1 & -1 \\ 1 & -1 & 1 & -1 \\ 1 & -1 & -1 & 1 \end{bmatrix} \tag{6-2-3}$$

When this transformation is applied separately to the sets m_+ and m_-, each of the 4×4 blocks of the 8×8 matrix is diagonalized, the + + and − − blocks taking the form

$$\begin{pmatrix} \omega_2^2 + A + 3B & 0 & 0 & 0 \\ 0 & \omega_2^2 + A - B & 0 & 0 \\ 0 & 0 & \omega_2^2 + A - B & 0 \\ 0 & 0 & 0 & \omega_2^2 + A - B \end{pmatrix}$$

while the + − coupling block becomes

$$\begin{pmatrix} 0 & 0 & 0 & 0 \\ 0 & 2Ce^{i\phi_1} & 0 & 0 \\ 0 & 0 & 2Ce^{i\phi_2} & 0 \\ 0 & 0 & 0 & 2Ce^{i\phi_3} \end{pmatrix}$$

Thus the E_u species modes under the unit cell yield

$$\begin{pmatrix} \omega_2^2 + A + 3B & 0 \\ 0 & \omega_2^2 + A + 3B \end{pmatrix}$$

while the triply degenerate F_u species modes are represented by

$$\begin{pmatrix} \omega_2^2 + A - B & 2Ce^{i\phi_\alpha} \\ 2Ce^{-i\phi_\alpha} & \omega_2^2 + A - B \end{pmatrix} \qquad \alpha = 1, 2, 3$$

where the ϕ_α are phase constants which disappear in the final expressions for the eigenvalues. This latter form is readily diagonalized, so that, in summary, one finds the following eigenvalues of the dynamic matrix for ν_2:

$$\begin{aligned} \omega^2 &= \omega_2^2 + A + 3B & &(E_u) \\ \omega^2 &= \omega_2^2 + A - B \pm 2C & &(2F_u) \end{aligned} \tag{6-2-4}$$

One should note that this result is completely general within the framework of the quadratic potential model and the neglect of intermode couplings.

(c) CHI_3, Space Group $C_6^6 = P6_3$

The standard methods of vibrational analysis for free molecules as given in Section 2-6 yield six distinct frequencies for the iodoform molecule of which ν_1, ν_2, and ν_3 belong to A_1, while ν_4, ν_5, and ν_6 are in species E of C_{3v}. The external modes are T_z in A_1, R_z in A_2, and the degenerate pairs (T_x, T_y) and (R_x, R_y) in E. The unit cell contains two molecules which according to Table VII-8 are necessarily on C_3 sites. The appropriate correlation of modes is indicated in Table 6-2 which is quite straightforward. The acoustic modes fall in the A and E_1 species of C_6, so that optic modes in these two species arise exclusively from internal modes and from the rotations.

Since the reduction of the free molecule symmetry from C_{3v} to C_3 at the site in the crystal has no essential effect on the dynamic matrix, for this two molecule per unit cell structure the dynamic matrices take the simple form

$$\begin{pmatrix} \omega_k^2 + F_k'(1,1) & F_k'(1,2) \\ F_k'(1,2) & \omega_k^2 + F_k'(1,1) \end{pmatrix}$$

for all internal modes; for the degenerate modes, the two-by-two blocks like the above are repeated identically. Then for $k = 1, 2, 3$, the symmetry factored dynamic

Table 6-2 Symmetry species at $\kappa = 0$ for iodoform, space group $C_6^6 = P6_3$

Free molecule (C_{3v})		Site (C_3)	Unit cell (C_6)	
$\nu_1, \nu_2, \nu_3, T_\sigma$	A_1	A	A	$\nu_{1A}, \nu_{2A}, \nu_{3A}, T_{\sigma A}, R_{\sigma A}$
R_σ	A_2		B	$\nu_{1B}, \nu_{2B}, \nu_{3B}, T_{\sigma B}, R_{\sigma B}$
$\nu_4, \nu_5, \nu_6, T_\pi, R_\pi$	E	E	E_1	$\nu_{41}, \nu_{51}, \nu_{61}, T_{\pi 1}, R_{\pi 1}$
			E_2	$\nu_{42}, \nu_{52}, \nu_{62}, T_{\pi 2}, R_{\pi 2}$

matrices reduce to

$$\begin{aligned} \omega^2 &= \omega_k^2 + F'_k(1,1) + F'_k(1,2) \qquad \text{species } A \\ \omega^2 &= \omega_k^2 + F'_k(1,1) - F'_k(1,2) \qquad \text{species } B \end{aligned} \tag{6-2-5}$$

while the same formal result applies to modes $k = 4, 5, 6$ in species E_1 and E_2.

(d) Sodium Azide, Space Group $D_{3d}^5 = R\bar{3}m$*

This simple structure will provide our first example of a crystal containing molecular ions, N_3^-, which as "free molecules" would possess $D_{\infty h}$ symmetry and have the same species of modes as CO_2. A number of other interesting substances with the same structure include $NaHF_2$ and $CrHO_2$. The crystal structure can be derived from that of rocksalt by distorting the lattice along a 111 direction in order to accommodate the non-spherical azide ion. The rhombohedral lattice thus obtained is primitive and the unit cell contains one Na^+ ion which may be assigned rhombohedral coordinates 000, and one azide ion whose central nitrogen is at $\frac{1}{2}\frac{1}{2}\frac{1}{2}$. Each of these sites has D_{3d} symmetry, so that the correlation of the sodium motions is trivial—yielding simply the species of translation, $A_{2u} + E_u$, under D_{3d}. The complete correlation analysis is indicated in Table 6-3. Table 6-3 is interpreted to mean that under the unit cell classification, ν_3, T_σ^+, and T_σ^- occur in A_{2u}, while ν_2, T_π^+, and T_π^- occur in E_u. Since there must be acoustic modes in each of these two species, and since such modes could be described approximately as $T_\sigma^+ + T_\sigma^-$ and $T_\pi^+ + T_\pi^-$ respectively, the optic translatory modes will be approximately $T_\sigma^+ - T_\sigma^-$ and $T_\pi^+ - T_\pi^-$.

Since there is only one N_3^- unit per unit cell and the molecular degeneracies are not lifted, the expression for the internal mode frequencies at $\boldsymbol{\kappa} = 0$ assumes the simple form:

$$\omega^2 = \omega_k^2 + F'_k \tag{6-2-6}$$

for $k = 1, 2, 3$.

Table 6-3 Symmetry species of the $\kappa = 0$ vibrations of NaN_3, space group D_{3d}^5

Free azide ion ($D_{\infty h}$)		*Unit cell* (D_{3d})	Sodium ion
ν_1	Σ_g^+	A_{1g}	
R_π	Π_g	E_g	
ν_3, T_σ^-	Σ_u^-	A_{2u}	T_σ^+
ν_2, T_π^-	Π_u	E_u	T_π^+

* This is the phase stable above 18°C.

(e) The Calcite Structure, Space Group $D_{3d}^6 = R\overline{3}c$

This truly classic crystal, whose widespread natural occurrence first revealed the phenomenon of birefringence, is representative of an extensive series of compounds which includes such species as $LiNO_3$ and $NaNO_3$; $MgCO_3$, $CaCO_3$ itself, $MnCO_3$, $FeCO_3$; and $InBO_3$. Such series afford a direct opportunity to test the approximate independence of the internal modes of the anion from the external lattice modes which are substantially changed by variation of the mass of the cation.

The unit cell contains two formula units. The cations are on sites of type (b) which have S_6 symmetry, while the molecular anions occupy sites with a D_3 symmetry (Table VII-7). Of these, (a) may be assigned positions 000, $\frac{1}{2}\frac{1}{2}\frac{1}{2}$ and (b) = $\frac{1}{4}\frac{1}{4}\frac{1}{4}$, $\frac{3}{4}\frac{3}{4}\frac{3}{4}$ in rhombohedral coordinates. The XY_3 anions in an ordered structure must be on sites which are subgroups of D_{3h}, the symmetry of the free molecule; these are the (a) = D_3 sites. The correlation procedure, which by now will be quite familiar to the reader, is summarized in Table 6-4.

The species A_{2u} and E_u of the unit cell contain the acoustic modes. Note that the T_π^+ modes contribute twice ($T_{\pi u}^+$ and $T_{\pi u}^{+\prime}$) to E_u. This may be visualized simply by noting that if S_{bi} is a displacement of a cation, *both* linear combinations of the form $\Delta X_{b1} + \Delta X_{b2}$ and $\Delta X_{b1} - \Delta X_{b2}$ are of species u. It is interesting that only the cation participates in the lattice mode of A_{1u} and only the anions in A_{2g} and E_g. As in the CO_2 example, all internal mode splittings are a consequence of dynamic intermolecular couplings.

The dynamical matrix for the internal modes leads to the eigenvalues

$$\omega^2 = \omega_k^2 + F_k'(1,2) \tag{6-2-7}$$

Table 6-4 Symmetry of the vibrations of the calcite lattice

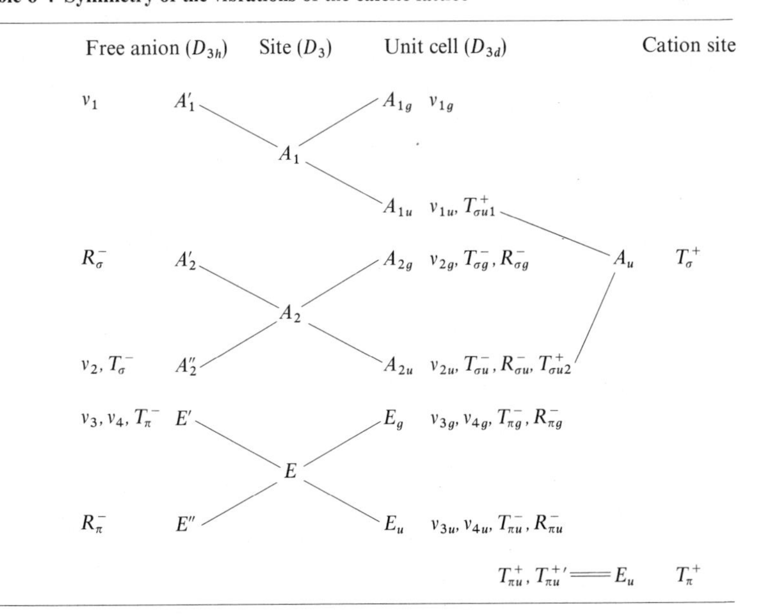

where, for $\omega_k = \omega_1$, the + and − terms correspond with species A_{1g} and A_{1u}, for $\omega_k = \omega_2$, the + and − terms correspond to A_{2u} and A_{2g}, and for k either 3 or 4, to E_g and E_u.

(f) $NaClO_3$, Space Group $T^4 = P2_13$

In this crystal there are four formula units per unit cell, with *both* the Na^+ and the ClO_3^- ions on sites of local C_3 symmetry (a). The unit cell is related to the familiar rocksalt structure; the coordinates of type (a) sites are xxx; $\frac{1}{2}+x, \frac{1}{2}-x, \bar{x}$; $\bar{x}, \frac{1}{2}+x, \frac{1}{2}-x$; and $\frac{1}{2}-x, \bar{x}, \frac{1}{2}+x$. The values of x are 0.0659 for Na^+ and 0.4168 for Cl.[5] The pyramidal C_{3v} molecular ion, ClO_3^-, possesses normal modes ν_1 and ν_2 of type A_1, and ν_3 and ν_4 of type E, where ν_1 and ν_3 arise from bond stretches and ν_2 and ν_4 from bends. The rigid motions of this group are T_z in A_1, R_z in A_2, (T_x, T_y) and (R_x, R_y) in E as shown in Appendix I. By examination of the appropriate correlation tables in Appendix IV, the correlation diagram of Table 6-5 can be constructed. This diagram shows that the internal modes of chlorate give rise to A and F components of ν_1 and ν_2, or to E and $2F$ components of the molecularly degenerate vibrations ν_3 and ν_4.

The external modes consist of two translatory motions, involving both cations and anions and z-axis libration in unit cell species A, two x–y translations and one libration around the x- (or y-) axis in E, and a total of six translatory and three libratory motions in F, of which a single vibration (triply degenerate, of course) is the acoustic mode, leaving a net of five optical translatory modes.

It must be emphasized that natural crystals, strictly speaking, do not possess the idealized symmetry which we have just outlined because the chlorines are randomly distributed between the ^{35}Cl and ^{37}Cl isotopes, the ordinary abundance being about 75 percent of the 35 isotope.

The significant part of the dynamical matrix is governed by symmetry rules identical with those appropriate for CO_2, Eqs. (6-2-2) and (6-2-4), except that there

Table 6-5 Symmetries of the $\kappa = 0$ modes of $NaClO_3$ ($T^4 = P2_13$)

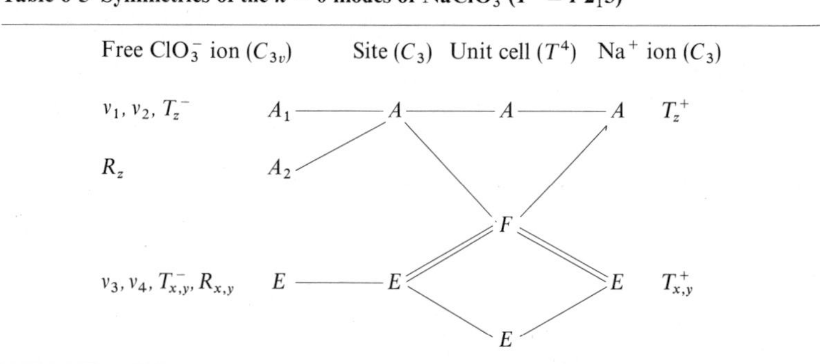

5. R. W. G. Wyckoff, *Crystal Structures*, Interscience, New York, Volume 2, 1964, page 380.

is no distinction between g and u modes. In particular, for the ν_1 and ν_2 modes, the symmetry factoring into A and F type unit cell modes yield

$$\omega^2 = \omega_k^2 + F'_{kk}(1,1) + 3F'_{kk}(1,2) \qquad (A) \tag{6-2-8}$$

and

$$\omega^2 = \omega_k^2 + F'_{kk}(1,1) - F'_{kk}(1,2) \qquad (F) \tag{6-2-9}$$

while for the ν_3 and ν_4 modes, the dynamical matrix reduces to

$$\omega^2 = \omega_k^2 + A + 3B \qquad (E) \tag{6-2-10}$$

$$\omega^2 = \omega_k^2 + A - B \pm 2C \qquad (F) \tag{6-2-11}$$

However, in this case the ν_1 and ν_3 modes are relatively close in energy, and since they each give rise to modes of the F species in the unit cell, it is possible that coupling constants of the type $F'_{13}\,(m, m')$ might have to be considered.

6-3 SELECTION RULES AND INTENSITIES: FUNDAMENTALS

Molecular crystals, like the monatomic systems described in Chapter 4, must obey the selection rule $\boldsymbol{\kappa} \approx 0$ for fundamental transitions, for both external and internal modes. It is therefore expected, and usually observed that such transitions will yield sharp lines in either the Raman or the infrared spectrum, although in the latter case difficulties arise (Section 5-5) if the specimens are too thick.

In contrast, two-phonon transitions, arising from overtones or combinations, will be expected to exhibit greater breadth, since they represent the frequency band $\omega_i(\boldsymbol{\kappa}) + \omega_j(-\boldsymbol{\kappa})$ with $\boldsymbol{\kappa}$ allowed to take on all values in the Brillouin zone. The contrast between these two cases is evident from Fig. 6-1, which illustrates

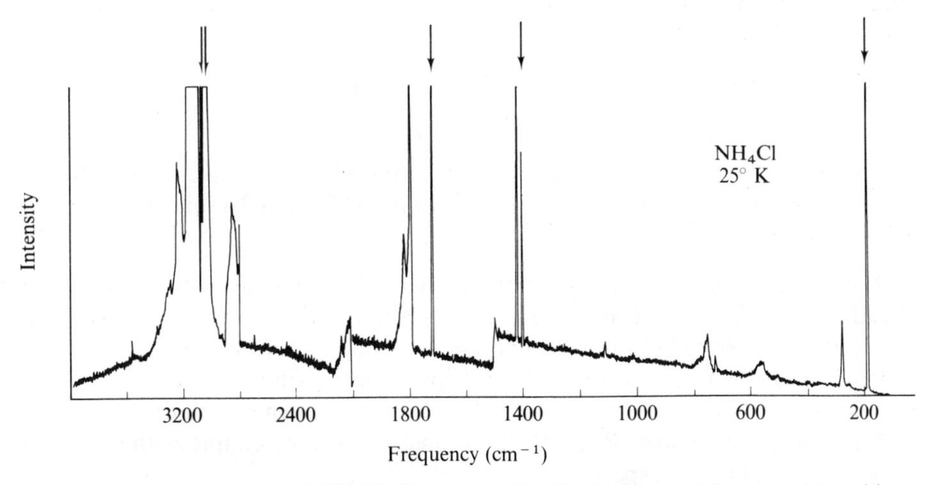

Figure 6-1 Raman spectrum of NH_4Cl. The arrows identify fundamental (transverse) transitions. (After L. R. Fredrickson, Ph.D. Dissertation, Oregon State University, 1977)

the Raman spectrum of NH_4Cl. The sharp lines marked with arrows are fundamentals and the remaining broad features are due to two-phonon processes.

The analysis of combination bands evidently requires a model for the joint density of states and implies a complete solution for the dispersion of the combining branches for all $\boldsymbol{\kappa}$. In all but the very simplest model systems discussed in Chapters 3 and 4, analytical expressions for the dispersion curves are not available, so a rather formidable calculation is implied. Further discussion of multiphonon transitions is postponed until Section 6-6.

The group-theoretical analysis of the allowed transitions in molecular crystals involves no additional principles beyond those set forth in Sections 4-7 through 4-9. Presently we shall give a number of examples. However, the existence of translational and librational modes raises some questions about the physical origins of the intensity. If the molecular unit is an ion, it is clear that the dipole moment of the crystal is in general a function of the translatory displacement just as in the case for a monatomic ion. However the rigid displacement of a neutral molecule does not produce a change in dipole moment, yet both the symmetry theory and experiment afford numerous examples in which external modes of this nature exhibit finite albeit weak infrared intensity.

The intensity of librational modes may be expected to depend upon the existence of an equilibrium dipole moment in the molecule. Since the components of the dipole are required in a laboratory axis system we note the direction cosine transformation:

$$\mu_F = \sum_g \Phi_{Fg}\mu_g \qquad F = X, Y, Z \qquad g = x, y, z \tag{6-3-1}$$

where F refers to laboratory axes and g to molecular axes. In general the elements of Φ must be expressed in terms of the Eulerian angles, but in the librational limit Φ can be written

$$\Phi = \begin{bmatrix} 1 & -R_z & R_y \\ R_z & 1 & -R_x \\ -R_y & R_x & 1 \end{bmatrix} \tag{6-3-2}$$

Suppose the molecule possess only a z component of a dipole moment—either at equilibrium or due to an internal vibration. Then (6-3-1) and (6-3-2) imply the following:

1. A pure librational transition in the R_y mode will couple with X-polarized radiation or a similar transition in the R_x mode will couple with Y-polarized radiation: no pure librational transition involving R_z is possible.
2. If the μ_z arises from an internal vibration, this transition will couple with Z polarized radiation, and combination transitions involving both the μ_z internal vibration together with R_y or R_x librational transitions couple with X- or Y-polarized radiation respectively.

According to the last paragraphs it is thus impossible for a non-polar molecule to exhibit finite infrared librational intensity. Again we must point out that there exist some cases in which the general predictions of the symmetry theory are at variance with the above conclusion. In the point groups D_2, D_3, D_4, D_6, D_{2d}, S_4, T, and O we find cases in which some librational mode (R_x, R_y, R_z) has the same symmetry species as a translation and hence would be predicted to be infrared active—yet in none of these symmetries can the molecule have an equilibrium dipole. Also, in the very simple point group C_3 we find that R_z belongs to the same species as $T_z(A)$, hence the general symmetry theory allows R_z activity as a pure libration although the molecule has no x or y equilibrium moment.

To understand the infrared activity of translational or librational modes of the types discussed in the foregoing paragraphs it is necessary to recall (5-8-2) which expressed the total moment of a molecule in the τth unit cell as

$$\mathbf{p}_{\tau m} = \boldsymbol{\mu}_{\tau m} + \boldsymbol{\alpha}_m \mathscr{F}_{\tau m} \tag{6-3-3}$$

where $\mathscr{F}_{\tau m}$ means the total field at molecule τm due both to other molecules in the crystal and to external radiation. We now see that motions such as translations or certain librations which do not change μ may still change $\mathbf{p}$ and give rise to infrared activity. Crystalline N_2 and CO_2 for example, exhibit infrared active translatory modes and since the polarizability is certainly finite, $\mathbf{p}$ will vary if $\mathscr{F}$ does: the quadrupole moment of such molecules is non-vanishing and produces an $\mathscr{F}$ field which varies for both translational and librational motion.[6] Thus in such crystals we can understand finite infrared intensity for the symmetry allowed translational and librational modes as a consequence of quadrupole-induced dipoles. It is also found[7] that crystalline p-H_2 exhibits infrared absorption due to rotational motion, although in this case the motion is nearly unrestricted by a potential barrier, i.e., it is nowhere near the librational limit.

(a) HCN ($C_{2v}^{20} = Imm2$)

From the analysis in Section 6-2(a), we see that ν_1, ν_3, and T_σ should be infrared active and produce single absorptions having Z polarization in the crystal. Since ν_2, T_π, and R_π give rise both to B_1 and B_2 species, respectively of X and Y polarization, it is expected and observed that ν_2 will be a doublet in a poly-crystalline film. Owing to the strong intermolecular coupling, the B_1 and B_2 librational modes may have rather different infrared frequencies. Since the unit cell is unimolecular, no translational optic modes are expected, i.e., T_σ and T_π are acoustic modes of zero frequency. All modes which are allowed in the infrared spectrum are in this case also allowed in the Raman spectrum, since α'_{XX}, α'_{YY}, α'_{ZZ} are non-vanishing for A_1 modes, α'_{ZX} is non-vanishing for B_1 modes and α'_{ZY} is non-vanishing for B_2 modes (see Appendix I).

6. S. H. Walmsley and J. A. Pople, *Molec. Phys.*, **8**, 345 (1964); see also O. Schnepp, *J. Chem. Phys.*, **46**, 3983 (1967) for the detailed theory of the intensities of the lattice vibrations of N_2 and CO_2.
7. E. J. Allin, W. F. J. Hare, and R. E. MacDonald, *Phys. Rev.*, **98**, 554 (1955); W. F. J. Hare, E. J. Allin, and H. L. Welsh, *Phys. Rev.*, **99**, 1887 (1955).

In principle, the distinction between the B_1 and B_2 components of ν_2 and of R_π could be made with single crystal samples, either in the Raman or infrared. Moreover, each optic mode being infrared active, it should be possible using single crystals to detect separately both the transverse and longitudinal frequencies for all modes, either by oblique reflection or with suitable polarization in the Raman experiment. But, as is the case for most materials which are normally gaseous, there are few single crystal studies. In the absence of single crystal data, it is obviously impossible to assign the B_1 and B_2 components of the ν_2 doublet.

As shown in Table 6-6, the observed Raman frequencies are in good agreement with the infrared frequencies, as one would expect. The Raman lines for ν_1 and ν_3 do however show additional high frequency shoulders which may reasonably be interpreted as the corresponding longitudinal mode frequencies.[8,9] Since dilute solutions of HCN and DCN and vice versa have been reported, it is possible to estimate the coupling constants $F'_{kk}(0)$, $k = 1, 3$ appearing in (6-2-1). Careful studies[10] of the infrared spectra of such mixed isotopic solutions (Section 7-2) have confirmed that, in this hydrogen bonded case, (a) the dipole derivatives for the A_1 modes are significantly changed from their gas phase values, and (b) the dipolar coupling mechanism alone (Section 6-5) does not adequately account for the frequency shifts.

Table 6-6 The infrared and Raman frequencies of crystal HCN

	HCN		HCN/DCN	DCN		DCN/HCN
	Infrared[a]	Raman[b]	Infrared[c]	Infrared[a]	Raman[b]	Infrared[c]
$\nu_2(B_1 + B_2)$	828	827			648	
	838	840			653	
$\nu_1(A_1)$	2098	2098	2097	1885	1885	1887
		~2104			~1900	
$\nu_3(A_1)$	3132	3129	3145	2545	2545	2553
		3150			~2555	

a. R. E. Hoffman and D. F. Hornig, *J. Chem. Phys.*, **17,** 1163 (1949).
b. References 8 and 9.
c. Reference 10.

(b) CO_2 ($T_h^6 = Pa3$)

For this cubic structure according to Appendix I, only the triply degenerate modes of F_u species are allowed to be infrared active, so that single crystals, if they were available, would provide no information from polarization. In contrast, the Raman selection rules allow all three of the species A_g, E_g, and F_g to exhibit activity, with varying polarization properties as shown in Appendix I.

8. M. Pezolet and R. Savoie, *Can. J. Chem.*, **47,** 3041 (1969).
9. M. Pezolet and R. Savoie, *Can. Spectrosc.*, **17,** 39 (1972).
10. H. B. Friedrich and P. F. Krause, *J. Chem. Phys.*, **59,** 4942 (1973).

Since the lowest frequency internal mode at about 650 cm^{-1} is much higher than the highest frequency external mode, around 130 cm^{-1}, we can, as is usually the case, discuss the two types separately.

Internal modes According to Table 6-1, ν_1 may give rise to fundamentals of A_g and F_g symmetry species. Despite the different polarization properties, which persist even for polycrystalline samples, no resolved components have been observed, indicating either that the frequency splitting or that the intensity of the F_g component is negligibly small. However, the Raman spectrum of the solid does exhibit a Fermi resonance of ν_1 with $2\nu_2$, with only very slight shifts relative to the gas phase values. In the frequency range between these two sharp lines there is evidence of a much weaker and broader structure which may well be due to the states of $2\nu_2$ which have the gas phase symmetry species Δ_g and accordingly do not interact with ν_1 in the isolated molecule. The breadth of this feature would be in accord with the solid selection rules which for two-phonon processes allow transitions of the type $\omega(\mathbf{k}) + \omega(-\mathbf{k})$ where $\mathbf{k}$ ranges over the whole Brillouin zone. It is known that there is considerable dispersion in ν_2 (see Section 6-6(e) below).

Table 6-1 shows that the infrared active modes have the following species: ν_2 $(E_u + 2F_u)$ and ν_3 $(A_u + F_u)$. According to (6-2-4), the splitting between the F_u modes is

$$(\omega_2^2 + A - B + 2C) - (\omega_2^2 + A - B - 2C) = 4C$$

where C is the coefficient of $Q_{2+}^*(\mathbf{0}, 1)Q_{2-}(\mathbf{0}, 2)$ in the coupling potential. The allowed modes are observed at 2344 cm^{-1} (ν_3) and 653 and 660 cm^{-1} (ν_2). The overtone and combination spectrum[11] in the infrared exhibits both broad and sharp features, the interpretation of which requires detailed consideration of the intermolecular potential, a topic to which we will return in Section 6-6.

External modes Reference to Section 6-2(b), particularly Table 6-1, reminds us that two types of translatory modes, $T_{\sigma F}$ and T_π, are expected in the infrared spectrum, which have been observed[12] at 68 and 114 cm^{-1}. Note that this is an example of the type discussed above in which the intensity must be attributed to intermolecularly induced dipoles.

Table 6-1 shows that the only external modes expected in the Raman spectrum should be of a libratory type, i.e., of rotational origin. Three are expected to be active, of species $E_g + 2F_g$ and have been observed[13] at 73, 91, and 130 cm^{-1}. Most of these external modes have frequencies which are quite temperature sensitive. A number of interesting models of intermolecular potentials have been applied to the analysis.[14]

11. D. A. Dows and V. Schettino, *J. Chem. Phys.*, **58**, 5009 (1973).
12. A. Anderson and S. H. Walmsley, *Molec. Phys.*, **7**, 583 (1964).
13. J. E. Cahill and G. E. Leroi, *J. Chem. Phys.*, **51**, 1324 (1969).
14. O. Schnepp and N. Jacobi, *Adv. Chem. Phys.*, I. Prigogine and S. A. Rice, Eds., Wiley-Interscience, New York, 1972, Vol. XXII, pages 205–314.

(c) CHI_3 ($C_6^6 = P6_3$)

Although each unit cell of this crystal is doubly occupied, the orientation of the two subsets of molecules is such that the only infrared active fundamentals are those in which the two sets vibrate in phase. Thus, although correlation shows a doubling of the internal modes of each molecule or site occupant, one member of each doublet is infrared inactive. Inspection of Appendix I shows that the same is true in the Raman effect with respect to non-degenerate modes, but not with respect to degenerate modes.

Iodoform crystals are readily grown from acetone solution by slow evaporation. Hexagonal crystals of ~1 cm diameter and several mm thick are readily produced, so that the directional aspects of the selection rules, both in the infrared and Raman, can be tested. In contrast to CO_2, the lowest frequency internal mode (110 cm^{-1}) is not well separated from the highest frequency external mode (51 cm^{-1}), but they are still far enough apart so that there is apparently little interaction at $\mathbf{k} = 0$.

Internal modes There are no surprises in the infrared spectrum at frequencies >100 cm^{-1}. The infrared spectrum conforms to the selection rules (Table 6-2), and the polarization of both fundamentals and binary combinations of internal modes agrees with the simple theory based upon the vector properties of the transition dipole.[15] The spectrum is rich in multiphonon transitions involving both internal and external modes at $k \neq 0$, the most noteworthy of which are the complex sidebands of the CH stretching mode in the 3000 cm^{-1} region. Because of the high frequency of this mode relative to all other fundamentals, little interaction of other modes with it is expected at any **k**. Two-phonon combination bands involving ω_{CH} and each of several lattice modes have been used to determine the dispersion curves of these modes.[16] The results will be discussed in Section 6-6(e).

In the case of the molecular E modes (ν_4, ν_5, ν_6), the difference of the infrared and Raman frequencies is a measure of the intermolecular coupling between the two sublattices, the basis of which is the doubly occupied unit cell, that is $F'_k(1,2)$ of (6-2-5). For example, $\nu_{41}(E_1)$ is active in the infrared (X or Y polarization) while $\nu_{42}(E_2)$ is active only in the Raman (XX, YY, or XY polarizations). Within experimental error, the differences are found to be negligible; hence intermolecular coupling in this crystal is small.

This crystal is not centered; hence, the infrared and Raman spectra are not mutually exclusive. However, the crystal class of iodoform is C_6, one of the piezoelectric classes. We may therefore expect some variation in, say, $\nu_{41}(E_1)$ as the crystal is rotated with respect to the direction of the transition moment of that mode, for unlike $\nu_{42}(E_2)$ it is a mode which develops its own polarization (Section 5-6(c)).

The observed frequencies are listed in Table 6-7.

15. R. M. Hexter and H. Cheung, *J. Chem. Phys.*, **24,** 1186 (1956).
16. N. Neto, O. Oehler, and R. M. Hexter, *J. Chem. Phys.*, **58,** 5661 (1973).

Table 6-7 The infrared and Raman frequencies of crystal CHI_3

	CHI_3	CDI_3	CHI_3
	Infrared		Raman
ν_{1A}	2982	2222	2977
ν_{2A}	425	408	428
ν_{3A}	154	154	152
$R_{\sigma A}$	36	36	33
ν_{41}	1068	786	
ν_{51}	578	556	
ν_{61}	110	110	
$R_{\pi 1}$	24	24	
ν_{42}			1064
ν_{52}			575
ν_{62}			107
$T_{\pi 2}$			(51)*
$R_{\pi 2}$			(11)*

* Calculated.

External modes In the far infrared, only two lattice fundamentals have been observed at 36 and 24 cm^{-1}. These are reasonably assigned as A and E_1, respectively.[16] The Raman spectrum was recorded using the projections $\bar{Z}(XX)Z$, $\bar{Z}(YY)Z$, and $\bar{Z}(XY)Z$, by means of which only A and E_2 modes can be detected. There is a difference. An E_2 mode, if it appears, will do so in all three polarizations, with equal intensity in each; in contrast, an A mode cannot appear in the XY polarization.

Experimentally it is found that only one strong lattice mode appears in the Raman spectrum, at 33 cm^{-1}.[16] Since this is approximately the same frequency as the 36 cm^{-1} absorption in the infrared spectrum, the coincidence suggests its assignment as an A mode. This contradicts the Raman polarization data, as the band appears in the $\bar{Z}(XY)Z$ projection.

(d) NaN_3 ($D_{3d}^5 = R\bar{3}m$)

This is our first example of a molecular ionic crystal: single crystals are readily produced and permit detailed polarization studies. We saw in Section 6-2(d) that the fundamental modes have the following symmetry properties:

$$\nu_1(A_{1g}) \qquad \nu_3, T_\sigma^-, T_\sigma^+(A_{2u})$$

$$R_\pi(E_g) \qquad \nu_2, T_\pi^-, T_\pi^+(E_u)$$

After removing the acoustic modes, which are of species $A_{2u} + E_u$, and which in this case may be simply described as $T_\sigma^+ + T_\sigma^-$ and $T_\pi^+ + T_\pi^-$, there remain optical translatory modes $T_\sigma^+ - T_\sigma^-$ and $T_\pi^+ - T_\pi^-$ of species $A_{2u} + E_u$.

According to Appendix I, the Raman spectrum is expected to exhibit a single line due to ν_1 in any polarization of the type $F(F'F')F''$ and another line due to R_π in either $F(F'F')F''$ (where $F' = X$ or Y) or any crossed polarization such as $F(X\,Y)F''$. Observations[17] show that the region of ν_1 is complicated by a weak Fermi resonance.

Infrared transmission studies have been reported but the samples employed had a thickness >100 μm, thus making it difficult to locate the TO and LO frequencies. Normal transmission through a sufficiently thin XY plate should show a single TO absorption for ν_2 and for the translatory mode $T_{\pi+} - T_{\pi-}$, or through a similar thin plate containing the Z-axis, one would expect ν_3 and $T_\sigma^+ - T_\sigma^-$.

Since the normal habit of the crystal permits the preparation of XY plates, most of the single crystal work has been performed with such an orientation. The ν_2 mode is easily identified: its transverse frequency is 638 cm^{-1}. More confusion surrounds the precise assignment of the ν_3 frequency. This mode is very intense, so that the polycrystalline transmission spectra give an ill-defined result. On the other hand, normal incidence experiments with the XY single crystal plates should not exhibit absorption due to ν_3 whose polarization is Z. Bryant noted two broad bands in transmission which were assigned as $\nu_3 \pm \nu$ where ν is the frequency of the E_g libratory mode; the $\mathbf{k} = 0$ symmetry of such a combination is E_u and thus allowed for normal incidence. In contrast, reflection experiments done at about 45° incidence show only two intense peaks in the region from 600 to 3000 cm^{-1}. These are observed at 641 and 2189 cm^{-1} in TM polarization, but the higher frequency peak disappears in TE polarization.[18] According to Section 5-4(b) these observations (coupled with a detailed analysis of the reflection band shapes) lead to the results: $\nu_{2T} = 638.4$ cm^{-1}, $\nu_{2L} = 642.8$ cm^{-1}, $\nu_{3T} = 2090$ cm^{-1}, and $\nu_{3L} = 2180$ cm^{-1}.

(e) $CaCO_3$ ($D_{3d}^6 = R\bar{3}c$)

Calcite has been the subject of vibrational spectroscopic studies too numerous to review uniformly here. In the internal mode region we were led to expect the following mode symmetries (Section 6-2(e))

ν_1	$A_{1g} + A_{1u}$
ν_2	$A_{2g} + A_{2u}$
ν_3, ν_4	$E_g + E_u$

Since the A_{1u} and A_{2g} modes are inactive in both Raman and infrared spectra, one expects a single Raman line of parallel polarization, i.e., $F(F'F')F''$, for ν_1 and a single Raman line of either parallel polarization or crossed polarization

17. J. I. Bryant, *J. Chem. Phys.*, **40,** 3195 (1964).
18. L. R. Fredrickson and J. C. Decius, *J. Chem. Phys.*, **63,** 2727 (1975).

for ν_3 and ν_4 (limited however to $F' = X$ or Y). In the infrared spectrum, ν_2 is allowed with Z polarization and ν_3 and ν_4 with X, Y polarization.

Many single crystal Raman and infrared transmission and reflection studies have been reported, the transmission studies being perhaps more useful for the study of the weaker combination modes. Figure 6-2 illustrates the Raman effect[19] in the four polarizations, $Z(XX)Y$, $Y(XY)X$, $Y(ZZ)X$, $X(ZX)Y$ and clearly identifies $\nu_{1g} = 1088$ cm^{-1} and $\nu_{4g} = 714$ cm^{-1} in accord with the above analysis.

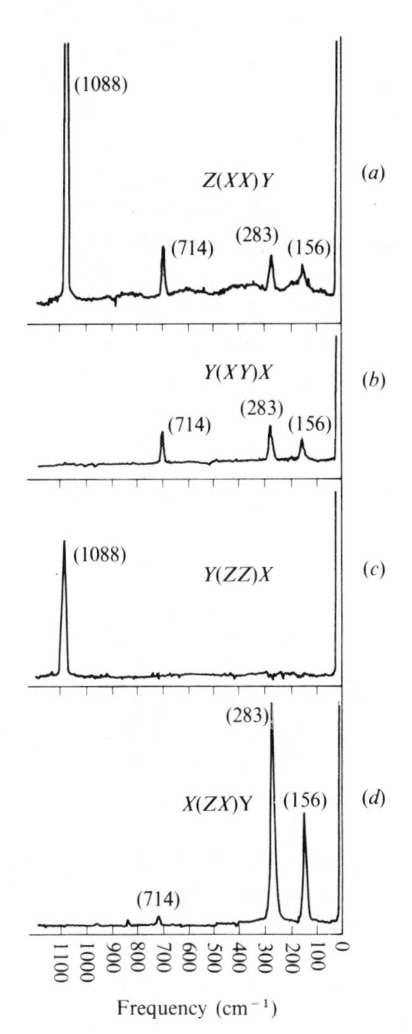

Figure 6-2 Raman spectrum of calcite. The *i(pa)s* notation is explained in section 4-8. (From S. P. S. Porto, J. A. Giordmaine, and T. C. Damen, *Phys. Rev.* **147,** 608 (1966))

19. S. P. S. Porto, J. A. Giordmaine, and T. C. Damen, *Phys. Rev.*, **147,** 608 (1966).

One notes that only ν_1 appears in the $Y(ZZ)X$ polarization in accord with the selection rules as given in Appendix I: the intensity is here proportional to the parameter b^2. In $Z(XX)Y$, the ν_1 intensity is proportional to a^2, and now ν_3 and ν_4 are allowed also with intensities proportional to c^2.

The best estimates of the TO frequencies for $\nu_{2u}(A_{2u})$, $\nu_{3u}(E_u)$, and $\nu_{4u}(E_u)$ come from reflection work and are reported to be $\nu_{2u} = 872\ \text{cm}^{-1}$, $\nu_{3u} = 1407\ \text{cm}^{-1}$, and $\nu_{4u} = 712\ \text{cm}^{-1}$; these are the transverse mode frequencies. Since the crystal is centrosymmetric, no information about the ν_{2u} and ν_{3u} modes can of course be found from Raman experiments, but from various reflection experiments it is found that the longitudinal frequencies are $\nu_{2u}(L) = 890\ \text{cm}^{-1}$, $\nu_{3u}(L) = 1549\ \text{cm}^{-1}$, and $\nu_{4u}(L) = 715\ \text{cm}^{-1}$. The effects of intermolecular coupling on the ν_3 mode are thus evident from the different frequencies for $\nu_{3u}(T)$, $\nu_{3u}(L)$, and ν_{3g}. The much smaller spread of the corresponding values for ν_4 points to much weaker intermolecular coupling.

Referring again to Fig. 6-2, we see the Raman active external modes below $300\ \text{cm}^{-1}$ consist of two lines at $283\ \text{cm}^{-1}$ and $156\ \text{cm}^{-1}$, which appear in $Z(XX)Y$ and $Z(YX)Y$ polarization but not in $X(ZZ)Y$. These must be the modes described as $T_{\pi g}^-$ and $R_{\pi g}^-$ in Table 6-4, but the spectroscopic data above do not determine which modes are translatory and which librational, nor the possible degree of mixing. We recapitulate below the external mode symmetries

$$
\begin{array}{ll}
R_\sigma^- & A_{2g} + A_{2u} \\
T_\sigma^- & A_{2g} + A_{2u} \\
T_\pi^- & E_g + E_u \\
R_\pi^- & E_g + E_u \\
T_\sigma^+ & A_{1u} + A_{2u} \\
T_\pi^+ & 2E_u
\end{array}
$$

Of the g modes, only the E_g species are Raman active and are the frequencies just mentioned. The infrared active species are A_{2u} and E_u: after subtracting the acoustic modes, there remain translatory modes of species $A_{2u} + 2E_u$ and librational modes $A_{2u} + E_u$. Thus a total of two Z polarized and three X, Y polarized infrared modes are expected in the far infrared spectrum. Reflection experiments[20] reveal transverse frequencies of $92\ \text{cm}^{-1}$ and $303\ \text{cm}^{-1}$ which are Z polarized and hence A_{2u}, and of 102, 223, and $297\ \text{cm}^{-1}$ which are X, Y polarized and therefore E_{2u}. It is at present, we believe, an unsettled question as to whether the librational modes $R_\sigma^-(A_{2u})$ and $R_\pi^-(E_u)$ included among the above observations owe their intensities to induction effects or to mixing with the translational modes of the same symmetry. The longitudinal frequencies for these infrared active external modes are 123, 239 (A_{2u}); and 381 (E_u).

In Table 6-8 there is given a summary of the internal mode frequencies for

20. K. H. Hellwege, W. Lesch, M. Plihal, and G. Schaack, *Z. Phys.*, **232**, 61 (1970).

Table 6-8 Internal mode frequencies of crystals belonging to the calcite family (all in cm^{-1})

	ν_{1g}	$\nu_{2u}(T)$	$\nu_{2u}(L)$	ν_{3g}	$\nu_{3u}(T)$	$\nu_{3u}(L)$	ν_{4g}	$\nu_{4u}(T)$	$\nu_{4u}(L)$
$LiNO_3$	1071			1385	1360	1480	735		
$NaNO_3$	1068	838	843	1385	1353	1450			
$MgCO_3$	1096	876	911	1446	1436	1599	736		
$CaCO_3$	1088	871	890	1432	1407	1550	714	712	715
$InBO_3$				1217	1205	1382	640	665	685

five crystals isomorphous with calcite. These frequencies have been observed and assigned via an analysis entirely similar to that just applied to calcite itself. The reader will find it instructive to compare the frequencies of corresponding modes either for identical molecular anions with different cations or for different anions.

(f) $NaClO_3$ ($T^4 = P2_13$)

The structure of this crystal is obviously related to that of CO_2, but differs in the important respect that there is no center of inversion, in consequence of which some modes are simultaneously infrared and Raman active. Because of the detailed information obtainable from single crystal studies, we reproduce in Table 6-9 the essential feature of the selection rules as predicted in Appendix I. Indeed, as is shown in Table 6-9, *all* the infrared allowed modes, of species F, are also allowed in the Raman effect. The polarizability derivatives are written in matrix form, and the relative intensities to be expected for various $i(pa)s$ experiments are evaluated by summing the squares of the parameters in case of degenerate modes. However, for species F modes in the Raman effect, it is important to take into account the breaking of the threefold degeneracy due to the L-T splitting. If one assumes the incident radiation to have a $\mathbf{k}_i$ vector along the Z-axis and the scattered radiation, in a right-angle experiment, to have a wave vector $\mathbf{k}_s$ along the Y-axis, then as explained in Section 5-6 the phonon vector makes an angle of approximately 45° with the incident beam and, in order to evaluate the relative intensities of the T and L modes in the Raman spectrum, one must carry out the transformations given in Eqs. (5-6-9), (5-6-10), and (5-6-11). However, the Raman spectrum of this crystal has been observed[21] using, in addition to right-angle scattering, certain other scattering geometries which have the desirable result of isolating the E modes or the longitudinal F modes. Thus by observing the scattering still at a right angle relative to Z but in a direction lying at 45° relative to either the X- or Y-axes, called Y', and with polarizations in the Y' and X' directions, an experiment designated as $Z(Y'X')Y'$ can be performed, for which each of the fundamental polarizability matrices is subjected

21. C. M. Hartwig, D. L. Rousseau, and S. P. S. Porto, *Phys. Rev.*, **188,** 1328 (1969).

Table 6-9 Selection rules for $NaClO_3$, symmetry T

INFRARED

F modes: $(\partial\mu_X/\partial Q_X) = (\partial\mu_Y/\partial Q_Y) = (\partial\mu_Z/\partial Q_Z) \neq 0$; all others vanish

RAMAN

	$[\partial\boldsymbol{\alpha}/\partial Q]$		INTENSITIES				
	$X\ Y\ Z$	$X'Y'Z$	$Z(XX)Y$	$Z(Y'X')Y'$	$Z(XZ)Y$	$Z(YZ)Y$	$Z(YX)\bar{Z}$
A modes:	$a\begin{vmatrix}1&0&0\\0&1&0\\0&0&1\end{vmatrix}$	$a\begin{vmatrix}1&0&0\\0&1&0\\0&0&1\end{vmatrix}$	a^2	0	0	0	0
E modes:	$b\begin{vmatrix}1&0&0\\0&1&0\\0&0&\bar{2}\end{vmatrix}$	$b\begin{vmatrix}1&0&0\\0&1&0\\0&0&\bar{2}\end{vmatrix}$	$4b^2$	$3b^2$	0	0	0
	$\sqrt{3}b\begin{vmatrix}\bar{1}&0&0\\0&1&0\\0&0&0\end{vmatrix}$	$\sqrt{3}b\begin{vmatrix}0&\bar{1}&0\\1&0&0\\0&0&0\end{vmatrix}$					
F modes (180° scattering):							
$T1 = X$	$c\begin{vmatrix}0&0&0\\0&0&1\\0&1&0\end{vmatrix}$		—	—	—	—	0
$T2 = Y$	$c\begin{vmatrix}0&0&1\\0&0&0\\1&0&0\end{vmatrix}$						
$L = Z$	$c\begin{vmatrix}0&1&0\\1&0&0\\0&0&0\end{vmatrix}$		—	—	—	—	c^2
F modes (90° scattering):							
$T1 = X$	$c\begin{vmatrix}0&0&0\\0&0&1\\0&1&0\end{vmatrix}$	$\frac{c}{\sqrt{2}}\begin{vmatrix}0&0&\bar{1}\\0&0&1\\\bar{1}&1&0\end{vmatrix}$	0	0	$\frac{1}{2}c^2$	c^2	—
$T2 = (Y + Z)/\sqrt{2}$	$\frac{c}{\sqrt{2}}\begin{vmatrix}0&1&1\\1&0&0\\1&0&0\end{vmatrix}$	$\frac{c}{2\sqrt{2}}\begin{vmatrix}\bar{2}&0&1\\0&2&1\\1&1&0\end{vmatrix}$					
$L = (-Y + Z)/\sqrt{2}$	$\frac{c}{\sqrt{2}}\begin{vmatrix}0&1&\bar{1}\\1&0&0\\\bar{1}&0&0\end{vmatrix}$	$\frac{c}{2\sqrt{2}}\begin{vmatrix}\bar{2}&0&\bar{1}\\0&2&\bar{1}\\\bar{1}&\bar{1}&0\end{vmatrix}$	0	0	$\frac{1}{2}c^2$	0	—

to the rotation

$$\frac{\partial \boldsymbol{\alpha}'}{\partial Q} = \mathbf{R} \frac{\partial \boldsymbol{\alpha}}{\partial Q} \tilde{\mathbf{R}}$$

where

$$\mathbf{R} = \begin{pmatrix} 2^{1/2} & -2^{1/2} & 0 \\ 2^{1/2} & 2^{1/2} & 0 \\ 0 & 0 & 1 \end{pmatrix}$$

Such a transformation has no effect upon the diagonal matrix appropriate for A species modes, leaves zeros in the $X'Y'$ position of the $F(T1)$ matrix at the same time that it introduces zeros into the $X'Y'$ position in the $F(T2)$ and $F(L)$ matrices, and has the effect shown in Table 6-9 upon the E matrices. Thus in this setting, only the E modes are expected to exhibit Raman activity for $Y'X'$ polarization.

In order to isolate the longitudinal F modes, a *back-scattering* experiment designated by $Z(YX)\bar{Z}$ can be employed: when this is done the **k** selection rule (quasi-momentum conservation) is simply

$$\mathbf{k}_i = \mathbf{k}_s + \mathbf{k}_p$$

or, since $\mathbf{k}_i$ and $\mathbf{k}_s$ have the same direction but opposite signs and nearly equal magnitudes,

$$\mathbf{k}_p = 2\mathbf{k}_i$$

i.e., the phonon wave vector has the same direction as that of the incident radiation, namely Z. Thus

$$Q_{T1} = Q_X$$

$$Q_{T2} = Q_Y$$

$$Q_L = Q_Z$$

and, as shown in Table 6-9, only the L component of an F mode will be active in the YX polarization, i.e., in $Z(YX)\bar{Z}$.

Although many infrared and Raman studies have been reported, we review here the work of Andermann and Dows[22] which relied entirely upon the reflection spectrum at normal incidence to obtain the frequencies. Although it has been suggested in Sections 5-3 and 5-4 that the transverse and longitudinal frequencies can be deduced by inserting the model equation for the dielectric constant, (5-1-2), into the reflectivity equation (5-4-2), remembering that $n^2 = \varepsilon$, and adjusting the parameters to fit the observed spectrum directly, there is some advantage in first deriving the frequency dependence of the optical constants n' and n'' from the entire reflection spectrum, using the so-called Kramers–Kronig relations.[23] The advantage of this procedure is that it is quite independent of any assumed model for the frequency dependence of the dielectric constant, but such a method has the disadvantage that it requires quite precise measurements of the reflectance R

22. G. Andermann and D. A. Dows, *J. Phys. Chem. Solids*, **28**, 1307 (1967).
23. H. A. Kramers, *Physik. Z.*, **30**, 522 (1929); R. de L. Kronig, *J. Opt. Soc. Amer.*, **12**, 547 (1926).

over a broad spectral region, including parts of the spectrum where R is quite small. Nevertheless, when samples of good optical quality are available, the Kramers–Kronig method in the form developed by Robinson and Price[24] is probably the best available.

In any event, Andermann and Dows successfully employed such an analysis and found the infrared frequencies and strengths which are given in Table 6-10.

These observations may be contrasted with the Raman studies reported by Hartwig, Rousseau, and Porto. As shown in Table 6-9, both the A and E modes are allowed in $Z(XX)Y$, only the E modes in $Z(Y'X')Y'$, both transverse and longitudinal F modes in $Z(XZ)Y$, only the transverse F modes in $Z(YZ)Y$ and only the longitudinal F modes in $Z(YX)\bar{Z}$. On the whole, the agreement between transverse and longitudinal frequencies as found from the two kinds of spectra is satisfactory. The transition strengths have also been evaluated in both experimental reports, and the dipole derivatives, useful in the analysis of the coupling constants, are given in the earlier work.

In order to clarify the relations between the several components of the molecular mode, Fig. 6-3 exhibits the experimental values for the stretching frequencies ν_1 and ν_3. Since except by isotopic dilution it is not feasible to separate terms like $F'_{kk}(1,1)$ in (6-2-2) or A in (6-2-4) from ω_k^2, the uncoupled frequency, Fig. 6-3 refers to values like $\nu_k + a$, b, c, etc., all in units of cm^{-1}, where

$$a = A/8\pi^2 c^2 \nu_k$$

$$b = B/8\pi^2 c^2 \nu_k$$

in virtue of the approximation $(\omega_k^2 + A)^{1/2} \approx \omega_k + A/2\omega_k$.

Table 6-10 Internal mode frequencies of the chlorate ion in $NaClO_3$ in cm^{-1}

Molecular mode	Crystal species A Raman[a]	Crystal species E Raman[a]	Crystal species F Raman[a]	Crystal species F Infrared[b]	Strength
ν_1	937		936(T)	937.5(T)	0.061
			940(L)		
ν_3		957	965(T)	966.4(T)	0.195
			983(L)	983.5(L)	
			988(T)	987.2(T)	0.037
			1030(L)	1027 (L)	
ν_2	618		623(T)	624 (T)	0.051
			629(L)		
ν_4		482	489(T)	481.5(T)	0.037
			490(L)	485.2(L)	

a. Ref. 21.
b. Ref. 22.

24. T. S. Robinson and W. C. Price, *Proc. Phys. Soc.* (*London*), **B65**, 910 (1952); **B66**, 969 (1953).

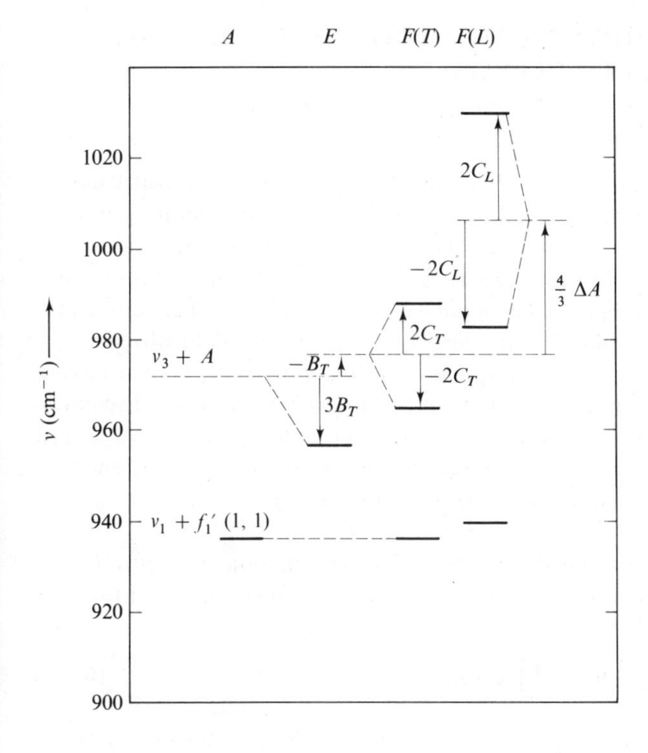

Figure 6-3 Stretching mode (ν_1 and ν_3) frequencies in $NaClO_3$

There is one apparent difficulty in accounting for the manifold of ν_3 levels, since the F modes have both transverse and longitudinal components. The coupling constants A, B, and C of (6-2-4) have both short and long range components, the latter arising from the dipolar coupling. Therefore, in general, one may expect that for molecularly infrared active modes one should write

$$A_L = A_T + \Delta A$$

$$B_L = B_T + \Delta B$$

$$C_L = C_T + \Delta C$$

Since the E_u mode is infrared inactive, it necessarily follows that

$$A_L + 3B_L = A_T + 3B_T$$

i.e., that

$$\Delta A + 3\Delta B = 0$$

One therefore finds that for the F modes, the longitudinal frequencies are given by

$$\omega_3^2 + A_L - B_L \pm 2C_L = \omega_3^2 + A_T - B_T + \tfrac{4}{3}\Delta A \pm 2C_L$$

6-4 EXCITON AND PHONON METHODS OF ANALYZING INTERMOLECULAR COUPLINGS

(a) One Molecule per Unit Cell

Our previous development of crystal dynamics has emphasized the construction of delocalized, phonon normal coordinates (Q) in accordance with symmetry principles, followed by quantization of the Hamiltonian. This is sometimes called the "giant molecule" approach. An alternative scheme, possessing certain advantages to be discussed below, is the *vibrational exciton* technique.[25] This procedure first quantizes the local, molecular modes, and then forms delocalized wave functions for the whole crystal. For simplicity, we first consider a system having only a single molecule per unit cell and, as in Section 6-2, restrict the discussion to a single vibrational degree of freedom; as in Sections 6-1 and 6-2, we regard this approximation as essentially correct when the intermolecular coupling energies are small compared with the differences between the energies of various states of uncoupled molecules.

For any state, one must specify the crystal wave function as a product of wave functions for the individual molecules. For the ground state, such a product is simply

$$\Phi_0 = \prod_{\tau} \varphi_0(q_\tau) = \prod_{\tau} \varphi_\tau \tag{6-4-1}$$

Note the abbreviated notation, φ_τ, which describes the τth molecule in its ground state.

A localized excited state could be designated as

$$\phi_\tau^r = \Phi_0 \varphi_\tau^r / \varphi_\tau \tag{6-4-2}$$

in which r indicates the state of molecular excitation; such wave functions are clearly not symmetrized under the translation group, but can easily be made so by forming the familiar linear combinations

$$\Phi^r(\boldsymbol{\kappa}) = N^{-3/2} \sum_{\tau} e^{-i\boldsymbol{\kappa}\cdot\boldsymbol{\tau}} \phi_\tau^r \tag{6-4-3}$$

If r indicates a state in which a single normal mode is singly excited, the subsequent development of the theory is, as we show below, quite simple and straightforward. In contrast, one must be more careful in the discussion of situations involving multiple excitation: for example, an excitation of $v = 2$ in some molecular mode will produce a state whose energy, for any $\boldsymbol{\kappa}$, will be quite close to the energies of states for the same $\boldsymbol{\kappa}$ characterized by single excitations on two different molecules, of which there are a large number. This sort of problem arises in the analysis of the overtone and combination spectra, a topic to which we return in Section 6-6.

Restricting the discussion to the simple case in which $r = 1$ stands for a single,

25. R. M. Hexter, *J. Chem. Phys.* **33,** 1833 (1960).

fundamental excitation, the evaluation of the energies is straightforward. The transition energy is

$$\langle \Phi^1(\mathbf{0}) | H | \Phi^1(\mathbf{0}) \rangle - \langle \Phi_0 | H | \Phi_0 \rangle$$

where

$$H = H_0 + V'$$

in which H_0 represents the Hamiltonian of the uncoupled molecules, and V' is the same coupling potential as that employed previously, i.e., Eq. (6-1-8). Since V' can make no contribution to the ground state when the molecular wave functions are harmonic oscillator functions, one finds

$$\langle \Phi^1(\mathbf{0}) | H_0 | \Phi^1(\mathbf{0}) \rangle - \langle \Phi_0 | H_0 | \Phi_0 \rangle + \langle \Phi^1(\mathbf{0}) | V' | \Phi^1(\mathbf{0}) \rangle$$

$$= \tfrac{3}{2}\hbar\omega_0 + (N^3 - 1)\frac{\hbar\omega_0}{2} - \frac{N^3\hbar\omega_0}{2} + \tfrac{1}{2}\sum_{\tau}\sum_{\Delta\tau} f'(\Delta\tau)\{\langle \Phi^1(\mathbf{0}) | q_\tau q_{\tau+\Delta\tau} | \Phi^1(\mathbf{0}) \rangle\} \tag{6-4-4}$$

Now for any given pair of unit cells, τ and $\tau + \Delta\tau$, the bracketed integral in (6-4-4) reduces to

$$N^{-3}\{\langle \varphi_\tau^1 \varphi_{\tau+\Delta\tau} | q_\tau q_{\tau+\Delta\tau} | \varphi_\tau \varphi_{\tau+\Delta\tau}^1 \rangle + \langle \varphi_\tau \varphi_{\tau+\Delta\tau}^1 | q_\tau q_{\tau+\Delta\tau} | \varphi_\tau^1 \varphi_{\tau+\Delta\tau} \rangle\} = 2N^{-3}\frac{\hbar}{2\omega_0}$$

so that (6-4-4) becomes

$$\hbar\omega_0 + \frac{\hbar}{2\omega_0}\sum_\tau f'(\Delta\tau) = \hbar\omega_0 + \frac{\hbar}{2\omega_0}F'(\mathbf{0}) \tag{6-4-5}$$

The two contributions to this transition energy are often designated as

$$\hbar\omega_0 = \Delta w + D$$

and

$$I = \frac{\hbar}{2\omega_0}F'(\mathbf{0}) \tag{6-4-6}$$

corresponding to the uncoupled molecule and the shift I due to the coupling.

It is instructive to compare the result given in (6-4-5) with the comparable calculation involving the phonon coordinates, $Q(\boldsymbol{\kappa})$. The dynamical matrix for one oscillator per unit cell is $\omega_0^2 + F'(\boldsymbol{\kappa})$ so that the approximation to the square root of $\omega_0^2 + F'(\mathbf{0})$ yields

$$\hbar\omega \approx \hbar\left[\omega_0 + \frac{1}{2\omega_0}F'(0)\right]$$

showing that the condition for the validity of the exciton result is simply that $F'(\mathbf{0}) \ll \omega_0^2$.

Consider now briefly the exciton description of states whose excitation energy is in the vicinity of $2\hbar\omega_0$. These can arise either by the double excitation of a single molecule and are described by $r = 2$ in (6-4-3), or by the single excitation of two molecules, in which case the symmetrized wave function would be

$$\Phi(\boldsymbol{\kappa}', \boldsymbol{\kappa}'') = N^{-3} C \sum_{\tau', \tau''} e^{-i(\boldsymbol{\kappa}' \cdot \boldsymbol{\tau}' + \boldsymbol{\kappa}'' \cdot \boldsymbol{\tau}'')} \frac{\varphi^1_{\tau'}\, \varphi^1_{\tau''}}{\varphi_{\tau'}\, \varphi_{\tau''}} \Phi_0, \quad C = [\delta(\boldsymbol{\kappa}' - \boldsymbol{\kappa}'') + 1]^{-1/2}$$

Exciton wave functions (like (6-4-3)), in which any amount of excitation is on a single molecule, are called "one-site excitons"; where excitation is distributed over two or more molecules, the wave function is referred to as a "two-site exciton," etc. A two-site exciton transforms with wave vector $\boldsymbol{\kappa}' + \boldsymbol{\kappa}''$ provided the sum is over all $\boldsymbol{\tau}'$ and all $\boldsymbol{\tau}''$ but the terms with $\boldsymbol{\tau}' = \boldsymbol{\tau}''$ do not quite correspond to correct molecular wave functions, since in the case of harmonic oscillator basis functions,

$$\left(\frac{\varphi^1}{\varphi}\right)^2 = 2\gamma q^2$$

but

$$\frac{\varphi^2}{\varphi} = 2^{1/2}[\gamma q^2 - \tfrac{1}{2}]$$

Thus in the purely harmonic case, the appropriate exciton wave function becomes

$$\Phi(\boldsymbol{\kappa}', \boldsymbol{\kappa}'') - N^{-3} C \sum e^{-i(\boldsymbol{\kappa}' + \boldsymbol{\kappa}'') \cdot \boldsymbol{\tau}} \Phi_0 = \Phi(\boldsymbol{\kappa}', \boldsymbol{\kappa}'') - C\, \delta(\boldsymbol{\kappa}' + \boldsymbol{\kappa}'')\Phi_0 \qquad (6\text{-}4\text{-}7)$$

which now contains terms corresponding to two single excitations on different molecules, or double excitations, as given by (6-4-3) with $r = 2$ in the case where $\boldsymbol{\kappa}' + \boldsymbol{\kappa}'' = 0$. In the harmonic case, the $r = 2$ case merges into the two-phonon band for $\boldsymbol{\kappa} = \boldsymbol{\kappa}' + \boldsymbol{\kappa}'' = 0$, but in the presence of anharmonicity, the $r = 2$ level (a one-phonon state) may split away from the two-phonon states, basically described by $\varphi^1_{\tau'} \varphi^1_{\tau''}$ with $\boldsymbol{\tau}' \neq \boldsymbol{\tau}''$.

(b) Several Molecules per Unit Cell

The extension of the exciton method to this case requires only that the locally excited states be distinguished by an index specifying which of the several molecules per cell is excited; using i for this index (analogous to m in Sections 6-1 and 6-2) one has

$$\phi^r_{i\tau} = \Phi_0 \varphi^r_{i\tau} / \varphi_{i\tau}$$

and

$$\Phi^r_i(\boldsymbol{\kappa}) = N^{-3/2} \sum_{\tau} e^{-i\boldsymbol{\kappa} \cdot \boldsymbol{\tau}} \phi^r_{i\tau} \qquad (6\text{-}4\text{-}8)$$

The excitation energies for $\boldsymbol{\kappa} = 0$ are then found by diagonalizing the matrix whose diagonal elements are again given by (6-4-5) and (6-4-6) and whose off-diagonal elements are

$$\sum_{\tau'} I^{11}_{i\tau,j\tau'}$$

where

$$\begin{aligned} I^{11}_{i\tau,j\tau'} &= f'(\tau, i; \tau', j)\langle \varphi^1_{i\tau}\varphi_{j\tau'} | q_{i\tau}q_{j\tau'} | \varphi_{i\tau}\varphi^1_{j\tau'}\rangle \\ &= \frac{\hbar}{2\omega_0} f'(\Delta\tau; ij) \end{aligned} \tag{6-4-9}$$

Then, since

$$F'(\mathbf{0}, ij) = \sum_{\Delta\tau} f'(\Delta\tau; ij)$$

one sees that

$$\sum_{\tau'} I^{11}_{i\tau,j\tau'} = \frac{\hbar}{2\omega_0} F'(0; ij)$$

The diagonalization of this excitation matrix can be accomplished in the usual way, relying mostly upon the unit cell symmetry. For example, with two equivalent molecules per unit cell, one finds eigenvalues

$$\hbar\omega_0 + \frac{\hbar}{2\omega_0} F'(0; 11) \pm \frac{\hbar}{2\omega_0} F'(0; 12)$$

which corresponds with the result found from the dynamical matrix within the range of validity of the square root approximation. The exciton method focuses attention directly upon the most obvious observables, i.e., the multiplet splittings, which are seen to be the eigenvalues of the matrix formed from $\sum_{\tau'} I^{rr}_{i\tau,j\tau'}$. It is not limited, of course, to excitations which are describable as molecular harmonic oscillator states, and therefore is advantageous in the discussion of highly anharmonic states, rotations, etc. (See Appendix XII.)

6-5 ANALYSIS OF MULTIPLETS AND QUANTITATIVE THEORIES OF INTERMOLECULAR POTENTIALS

In this section we present a description of multiplets in fundamentals of molecular crystals, experimental techniques for their observation and measurement, and a discussion of the theories which have been proposed to account for the separation of multiplet components.

It will serve our purposes to have a definite example with which to introduce our discussion. We can illustrate most, if not all, of the features of multiplet splittings using the CH_3Cl crystal. Crystal benzene could also serve well, but the greater complexity, both of the molecule and of the crystal, unnecessarily complicates the description we wish to present.

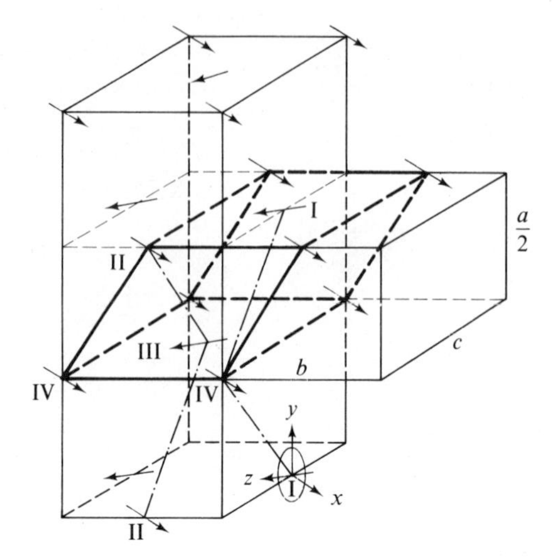

Figure 6-4 Crystal structure of CH_3Cl. (From R. M. Hexter, *J. Chem. Phys.*, **36,** 2285 (1962))

CH_3Cl crystallizes in an orthorhombic space group, $C_{2v}^{12} = Cmc2_1$, with two molecules/cell located at sites $a(C_s)$ of the rhombohedral unit cell.[26] Both the primitive and non-primitive unit cells are illustrated in Fig. 6-4, in which the arrows indicate the C–Cl axes, which lie in the *bc* crystal planes.

Correlation of the molecular, site, and factor groups may be summarized by the diagram shown in Fig. 6-5.

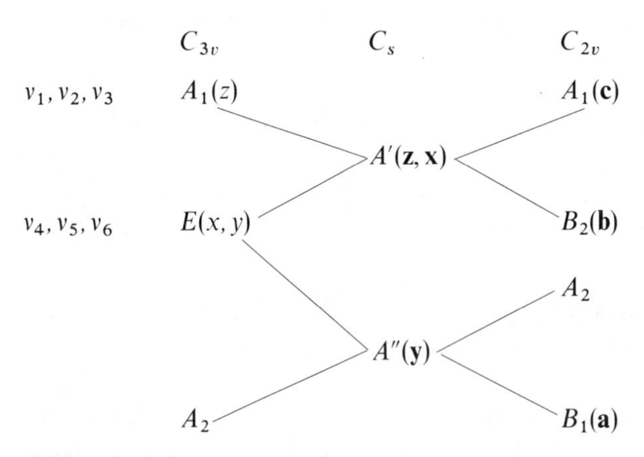

Figure 6-5 Correlation diagram for CH_3Cl crystal. (From R. M. Hexter, *J. Chem. Phys.*, **25,** 1286 (1956))

As we read Fig. 6-5 from left to right, we first note the removal of the two-fold degeneracies of each of the molecular modes ν_4, ν_5, ν_6 due to the fact that

26. R. D. Burbank, *J. Am. Chem. Soc.*, **75,** 1211 (1953).

the site symmetry of each molecule has been reduced from C_{3v} to C_s. No change is predicted for the (molecular) non-degenerate modes ν_1, ν_2, and ν_3; however, we may observe a doublet in place of each of the (formerly) degenerate modes ν_4, ν_5, and ν_6.

The remaining lines on the right-hand side of Fig. 6-5 instruct us that we may build pairs of new symmetry coordinates out of the in-plane components (A' site group states) and the process can be repeated using the A'' components to yield pairs of unit cell group coordinates which belong to A_2 and B_1 of the factor group. This further coupling has been called by various authors "dynamic coupling," "factor group coupling" and "correlation field coupling," in distinction to the crystal field effect of removal of degeneracies under the molecular point group. The latter field also goes variously under the names "local" or "site" or "static" field.

In the absence of data from oriented single crystals there is no way of distinguishing between the factor group components and hence between the two effects, since they are of the same order of magnitude. Moreover, we should view the site effects as only an intermediate step in arriving at the overall correlation of molecular and crystal modes. The reason for this is that static field effects are simply represented by static potentials, and all that is required to remove a degeneracy is a potential of sufficiently low symmetry; there is no unique way of deriving the correct one. On the other hand, something more is known about dynamic effects. By definition these depend upon molecular motion, the constants of which can be derived from molecular spectra. Hence, any theory of multiplets must be developed on the basis of some aspect of molecular dynamics, mechanical or electric.

Multiplet theories must therefore account for all factor group components. In general, each molecular coordinate can give rise to a maximum of $d^\gamma Z$ unit cell modes, where Z is the (molecular) population of the *primitive* unit cell, and d^γ is the dimension of the (molecular) point group representation to which the (molecular) coordinate belongs. Only by some approximation, hopefully one which can be made experimentally feasible, may the theory manifest the site group splittings alone (see Section 7-2).

Since multiplet spectra began to be identified, basically two theories have been developed. One of these, originally due to Dows,[27] identifies the potential which couples the intermolecular vibration as the interaction between pairs of atoms of neighboring molecules whose displacement coordinates figure prominently in the molecular normal coordinates of interest. The other prominent theory, usually called dipole coupling, was originally proposed by Decius[28] and has been elaborated by Hexter.[29] In this theory, the intermolecular coupling arises via the interaction of resonant transition dipoles *throughout the entire crystal.* The

27. D. A. Dows, *J. Chem. Phys.*, **32,** 1342 (1960).
28. J. C. Decius, *J. Chem. Phys.*, **22,** 1941, 1946 (1954); **23,** 1290 (1955).
29. R. M. Hexter, Reference 25, *J. Chem. Phys.*, **36,** 2285 (1962); **37,** 1347 (1962); D. Fox and R. M. Hexter, *J. Chem. Phys.*, **41,** 1125 (1964).

difference between the two theories is therefore twofold: one is based upon the repulsive forces between molecules which are by their nature of short range ($\sim r^{-6}$) while the other, which recognizes that the dipole moment of a molecule changes in the course of a vibration, is much longer in its range ($\sim r^{-3}$).

In order to appreciate the common features and the individual differences of both theories, it is of value to develop a single formalism which can encompass both. The formalism we shall use is more generally applicable to a long-range interaction. We shall use it first, however, to discuss the atom–atom repulsion theory.

We shall find it helpful to begin with a consideration of the non-primitive unit cell illustrated in Fig. 6-4. The common formalism is that of the vibrational exciton, which was presented for unit cells with just one oscillator in Section 6-4. For more densely populated unit cells, an excited state is formed out of linear combinations of the one-site excitons

$$\begin{aligned} \Phi_i^r(\boldsymbol{\kappa}) &= N^{-3/2} \sum_{\tau} \phi_{i\tau}^r e^{-i\boldsymbol{\kappa}\cdot\tau} \\ \phi_{i\tau}^r &= \Phi_0 \varphi_{i\tau}^r / \varphi_{i\tau} \\ \Phi_0 &= \prod_i \prod_{\tau} \varphi_{i\tau} \end{aligned} \tag{6-5-1}$$

The symbol $\varphi_{i\tau}^r$ refers to a molecule in its rth excited state on the ith site $(i = 1, \ldots, h)$ of the τth unit cell. τ is the position vector of the center of that cell. Since our first interest is in internal fundamentals, we may set $\boldsymbol{\kappa} = 0$.

The wave functions (6-5-1) are h-fold degenerate. Linear combinations of the $\boldsymbol{\kappa} = 0$ wave functions can be used as bases for factor group representations by application of the WPO (see Section 2-2). The results are of the form

$$\Psi_\gamma^r = \sum_{i=1}^{h} B_{\gamma i} \Phi_i^r(\mathbf{0}) \tag{6-5-2}$$

where γ is an irreducible representation of the factor group.

The factor group energies are found by diagonalizing the Hamiltonian $H = H^0 + \sum_{k>l} V_{lk}$, where H^0 is the Hamiltonian for the crystal of non-interacting molecules and V_{lk} is the interaction energy of molecules k and l. In the basis of the translationally symmetrized one-site excitons, $\Phi_i^r(\boldsymbol{\kappa})$, and provided r is non-degenerate in the free molecule, the crystal excitation energies can be found by solving the determinant of H, the diagonal elements of which are

$$\Delta w^r + D^r + \sum_{\tau'}{}' I_{i\tau,i\tau'}^{rr} - \Delta E^\gamma$$

and the off-diagonal elements are

$$\sum_{\tau'} I_{i\tau,j\tau'}^{rr}$$

where Δw^r is the excitation energy of the transition in the free molecule,

$$D^r = \sum_{\tau'}{}' \{(\varphi_\tau^r \varphi_{\tau'} | V_{\tau\tau'} | \varphi_\tau^r \varphi_{\tau'}) - (\varphi_\tau \varphi_{\tau'} | V_{\tau\tau'} | \varphi_\tau \varphi_{\tau'})\}$$

and (6-5-3)

$$I_{i\tau,j\tau'}^{rr} = (\varphi_{i\tau}^r \varphi_{j\tau'} | V_{i\tau,j\tau'} | \varphi_{i\tau} \varphi_{j\tau'}^r)$$

Alternatively H can be diagonalized using the unitary transformation $\boldsymbol{B}$. Since we are only interested in *splittings*, or separations of crystal excited states, $\Delta w^r + D^r$ can be subtracted as a shift of the energy scale. The only matrix elements of interest are then the I_{ii} and the I_{ij}. In the factor group basis, these are formed into linear combinations according to

$$I_\gamma^{rr} = \sum_j B_{\gamma i}^* B_{\gamma j} \sum_{\tau'} I_{i\tau,j\tau'}^{rr} \tag{6-5-4}$$

For a non-degenerate molecular wave function φ^r there results just one transition energy belonging to each irreducible representation of the factor group, each separated from the others by the differences among the several I_γ^{rr}. If φ^r belongs to a representation of the molecular point group of dimension d_δ, however, the unitary transformation $\boldsymbol{B}$ block-diagonalizes H, leaving $h\ d_\delta \times d_\delta$ matrices. In the case of CH_3Cl, where the maximum molecular degeneracy is twofold, each 2×2 matrix is of the form

$$\begin{pmatrix} \mathscr{E} + I_\gamma^{xx} & I_\gamma^{xy} \\ I_\gamma^{xy} & \mathscr{E} + I_\gamma^{yy} \end{pmatrix}$$

where

$$\mathscr{E} = \Delta w^r + D - \Delta E \tag{6-5-5}$$

and

$$I_\gamma^{xx} = \sum_j B_{\gamma i}^* B_{\gamma j} \sum_{\tau'} I_{i\tau,j\tau'}^{xx} \tag{6-5-6}$$

x and y, etc., denote the several components of φ^r which are resolved at the site. Relative to the zero shift $\Delta w^r + D$, the roots of the 2×2 matrices are

$$\Delta E_\pm^\gamma = \tfrac{1}{2}(I_\gamma^{xx} + I_\gamma^{yy}) \pm \tfrac{1}{2}\{(I_\gamma^{xx} - I_\gamma^{yy})^2 + 4(I_\gamma^{xy})^2\}^{1/2} \tag{6-5-7}$$

In cases like this there will be *two* distinct excitation energies of crystal states belonging to the same factor group representation, and they are separated by $\{(I_\gamma^{xx} - I_\gamma^{yy})^2 + 4(I_\gamma^{xy})^2\}^{1/2}$. This splitting, which may be viewed as an effect of the crystal field, is a dynamic effect, since it depends upon the transfer of excitation between different molecules.

In order for D^r and $I_{i\tau,j\tau'}^{rr}$ to be finite, V_{ij} must depend explicitly on the product of the molecular normal coordinates $q_i q_j$, provided r represents a single excitation. For example, if V_{ij} is expanded in a power series, for each τ, τ' pair we have

$$V_{ij} = V_{ij}^0 + \frac{\partial V_{ij}}{\partial q_i} q_i + \frac{\partial V_{ij}}{\partial q_j} q_j + \frac{1}{2}\frac{\partial^2 V_{ij}}{\partial q_i^2} q_i^2 + \frac{1}{2}\frac{\partial^2 V_{ij}}{\partial q_j^2} q_j^2 + \frac{\partial^2 V_{ij}}{\partial q_i \, \partial q_j} q_i q_j + \cdots \tag{6-5-8}$$

If we assume, as we did in writing (6-4-9), that only terms like the last contribute to (6-5-3), substitution gives

$$I_{i\tau, j\tau'}^{rr} = \left(\frac{\partial^2 V_{i\tau, j\tau'}}{\partial q_{i\tau} \, \partial q_{j\tau'}}\right) |(\varphi^r | q | \varphi)|^2 \tag{6-5-9}$$

and*

$$D^r = [(\varphi^r | q | \varphi^r) - (\varphi | q | \varphi)](\varphi | q | \varphi) \sum_\tau{}' \left(\frac{\partial^2 V_{\tau\tau'}}{\partial q_\tau \, \partial q_{\tau'}}\right) \tag{6-5-10}$$

The wave functions are assumed to be harmonic oscillator functions (see Section 6-4). The important point is the explicit dependence upon $q_i q_j$, which is characteristic both of the atom–atom repulsion theory as well as the dipole-coupling theory.

Recalling the importance of (6-5-4) (and the several integral sums of (6-5-7)), we note that, as a result of (6-5-9) the factor group states are always split by amounts proportional to the squares of transition moment integrals. *This prediction is common to both theories.*

A further consequence of (6-5-9) is specially relevant to CH_3Cl. The components of the degenerate molecular state were designated φ^x and φ^y because of the special convenience of the cartesian representation in the presence of the site C_s. y is taken $\parallel a$ and x is in the bc plane for all equivalent sites (see Fig. 6-4). The correlation table (Fig. 6-5) indicates that the normal coordinates for a degenerate mode will be q_x for factor group representations A_1 and B_2, and q_y for the A_2 and B_1 representations. Since both $(\varphi^x | q_y | \varphi)$ and $(\varphi^y | q_x | \varphi)$ identically vanish, one of the two excitation energies predicted for each factor group representation by (6-5-7) is zero. This is really a consequence of the use of the non-primitive cell; with the elimination of these states the total number of states agrees with the predictions of the correlation table.

The vanishing of these integrals is peculiar to the crystal structure of CH_3Cl. In another popular class of orthorhombic crystal with four molecules/cell ($D_{2h}^{15} = Pbca$), the site symmetry is C_i. For each equivalent site one of the molecular axes (conventionally called y) may be arbitrarily chosen such that $\mathbf{y} \cdot \mathbf{c} = 0$, but there are no other special trigonometric relationships. Fully eight factor group states result from each degenerate molecular state.

We proceed to evaluate (6-5-7) in the context of the atom–atom interaction theory. The first requirement is a specific form for the intermolecular potential appropriate to this theory. It is generally assumed that the intermolecular potential can be approximated by sums of interactions between non-bonded atom pairs,

* It is to be understood that the molecules τ, τ' are on identical sites i, and are therefore translationally *equivalent*.

each of which has a repulsive and an attractive part. Almost always, the overall atom–atom potential chosen is of the exp-6 type, that is,

$$V_{nm} = -A/r_{nm}^6 + B\exp(-Cr_{nm})$$

in which r_{nm} is the distance between the atom pair, and A, B, and C are parameters. If a particular atom pair is neighboring in a crystal (separated only by their van der Waals radii), only the repulsive part of V_{nm} may be used for that atom pair, but generally all possible non-bonded atom–atom interactions are included. In the expansion of the intermolecular potential in a power series (6-5-8) the terms in $q_n q_m$ are identified as the intermolecular force constants.

$$f_{nm} = \sum_m \partial^2 V_{nm}/\partial q_n\, \partial q_m \tag{6-5-11}$$

where q_n and q_m are the same normal coordinate of different molecules. In the atom–atom repulsion model it is assumed that $V_{nm} = \sum_{i,j} V_{nm}(r_{ij})$ where $r_{ij} = r_{ij}(q_n, q_m)$ is the interatomic distance between the centers of atom i of molecule n and atom j of molecule m. Since

$$\frac{\partial V_{nm}}{\partial q_n} = \sum_{i,j} \frac{dV_{nm}}{dr_{ij}} \frac{\partial r_{ij}}{\partial q_n}$$

and

$$\frac{\partial V_{nm}}{\partial q_m} = \sum_{i,j} \frac{dV_{nm}}{dr_{ij}} \frac{\partial r_{ij}}{\partial q_m} \tag{6-5-12}$$

by differentiation of each of (6-5-12) as a product we obtain

$$\frac{\partial^2 V_{nm}}{\partial q_n\, \partial q_m} = \sum_{i,j} \left[\left(\frac{d^2 V_{nm}}{dr_{ij}^2}\right)\left(\frac{\partial r_{ij}}{\partial q_n}\right)\left(\frac{\partial r_{ij}}{\partial q_m}\right) + \left(\frac{dV_{nm}}{dr_{ij}}\right)\left(\frac{\partial^2 r_{ij}}{\partial q_n\, \partial q_m}\right)\right] \tag{6-5-13}$$

If we knew the *total* intermolecular potential V with respect to all r_{ij}'s, the local minima $dV/dr_{ij} = 0$ would all be identified at the $\{r_{ij}\}$ represented by the equilibrium configuration of the crystal. But $V \neq \Sigma V_{nm}$; that is, the atom–atom repulsive energy is not the total intermolecular potential. Hence, the second term in (6-5-13) cannot be overlooked. Dows has shown that under most circumstances, this term is small compared to the first.[30] We therefore rewrite (6-5-13) as

$$f_{nm} = \frac{\partial^2 V}{\partial q_n\, \partial q_m} = \sum_{i,j} \left(\frac{d^2 V}{dr_{ij}^2}\right)\left(\frac{\partial r_{ij}}{\partial r_i}\right)\left(\frac{\partial r_i}{\partial q_n}\right)\left(\frac{\partial r_{ij}}{\partial r_j}\right)\left(\frac{\partial r_j}{\partial q_m}\right) \tag{6-5-14}$$

From a consideration of the directions of the infinitesimals $d\mathbf{r}_{ij}$ and $d\mathbf{r}_i$, we may identify $(\partial r_{ij}/\partial r_i) = \cos\phi_i$, where ϕ_i is the angle between the displacement coordinate r_i and the interatomic distances r_{ij}. It is important to realize that the f_{nm} are to be evaluated for each normal mode q_n and that the r_i are peculiar to each mode.

The quantities $(\partial r_i/\partial q_n)$ can be evaluated from the relative amplitudes derived in normal coordinate analyses, as shown by Dows.[30] For simplicity, the only r_{ij}

30. D. A. Dows, *J. Chem. Phys.*, **32**, 1342 (1960).

which have been considered are nearest-neighbor hydrogen atoms. Inspection of a map of a CH_3Cl unit cell (and its immediate neighbors) shows that, as defined in (6-5-13), all f_{nm}^{yy} and f_{nm}^{xy} vanish. The same is true of all f_{13} and f_{14}, because the separation of the hydrogen atoms involved are significantly greater than the sum of the van der Waals radii (3Å). In fact, the only finite f's are f_{11}^{xx}, f_{11}^{yy}, and f_{12}^{xx}.* In each case, only first nearest neighbors make any contributions to the lattice sums (6-5-11), a major difference of this model from the dipole-coupling theory. Since all f^{xy} vanish, (6-5-7) becomes

$$\Delta E_{\pm}^{\gamma} = -I_{\gamma}^{xx} \quad \text{or} \quad -I_{\gamma}^{yy} \tag{6-5-15}$$

The four valid factor group excitation energies, corresponding to those predicted by the analysis of the primitive unit cell are:

$$\begin{aligned} \Delta E_{+}^{A_1} &= -(f_{11}^{xx} + f_{12}^{xx}) \\ \Delta E_{+}^{B_2} &= -(f_{11}^{xx} - f_{12}^{xx}) \\ \Delta E_{-}^{A_2} &= -f_{11}^{yy} \\ \Delta E_{-}^{B_1} &= -f_{11}^{yy} \end{aligned} \tag{6-5-16}$$

The factor group splitting is found to be

$$\Delta E_{+}^{A_1} - \Delta E_{+}^{B_2} = -2f_{12}^{xx} \tag{6-5-17}$$

Factor group excitation energies for non-degenerate modes (ν_1, ν_2, and ν_3 of CH_3Cl) were given by (6-5-4), aside from the "shift term," $\Delta w^r + D^r$. In this case, only molecular normal coordinates which transform as q_z are valid and the molecular excited state wave functions are restricted to φ^z. Since q_z (and φ^z) transform exactly like q_x (and φ^x) under C_{2v} (see Fig. 6-5), the first two equations of (6-5-16) apply equally to these modes, provided the force constants (and the associated elements of the matrix of the $\partial r_i/\partial q_n$) are written in terms of the normal coordinates q_z. Thus, the factor group splittings of these modes will be proportional to f_{12}^{zz} instead of f_{12}^{xx}.

Although we have indicated that, in general, "dynamic" crystal field splittings are identified with the separation of states belonging to different representations of the factor group, an *estimate* of the "static site group splitting," $\Delta E_{A'} - \Delta E_{A''}$, can be found by calculating the separation of the *means*

$$\frac{\Delta E_{+}^{A_1} + \Delta E_{+}^{B_2}}{2} - \frac{\Delta E_{-}^{A_2} + \Delta E_{-}^{B_1}}{2} = -(f_{11}^{xx} + f_{11}^{yy}) \tag{6-5-18}$$

* The numbering system is that originally given by Burbank. Following a translation of the origin by $(\frac{1}{2}, \frac{3}{4}, 0)$ the coordinates of the several molecules (as defined by, say, the mid-points of the C—Cl bonds) in the non-primitive unit cell are:

$$\begin{aligned} \text{I} &= (\tfrac{1}{2} \quad 0 \quad \tfrac{1}{2}) \\ \text{II} &= (\tfrac{1}{2} \quad \tfrac{1}{2} \quad 0) \\ \text{III} &= (0 \quad \tfrac{1}{2} \quad \tfrac{1}{2}) \\ \text{IV} &= (0 \quad 0 \quad 0) \end{aligned}$$

We see that even this estimate of a site group splitting points to the importance of *local* anisotropy. Also, by comparison of (6-5-17) and (6-5-18) we see that, as earlier claimed, the two kinds of splitting are of the same order of magnitude.

Evaluation of (6-5-7) in dipole-coupling theory cannot take advantage of the vanishing of one or another element of the F matrix because of a somewhat large distance between two atoms of neighboring molecules. The interaction potential energy is now given by (5-7-6) (more accurately, (5-7-7)). An alternate expression is

$$H_{ij} = -\Sigma' e^{i\mathbf{k}\cdot\mathbf{r}} \left\{ \frac{3}{r^5} (\mathbf{p}_i^* \cdot \mathbf{r})(\mathbf{p}_j \cdot \mathbf{r}) - \frac{(\mathbf{p}_i^* \cdot \mathbf{p}_j)}{r^3} \right\} \tag{6-5-19}$$

where the $\mathbf{p}_i$'s are the resonant dipoles. The summation is over the entire lattice. Along any direction defined by the vector between, say, first nearest neighbor, translationally inequivalent molecules, H_{ij} decays as r^{-3}. Although in one dimension such sums converge reasonably quickly, the discussion of Section 5-7 demonstrated that, due to the long-range nature of the interaction, it is necessary to carry out the summation over the entire crystal.* The actual lattice sum which must be evaluated (5-7-5) is conditionally convergent; that is, it depends upon the order of summation, even at $\mathbf{k} = 0$. Lattice summation techniques are also discussed in Appendix XI. Experience has taught that good convergence is achieved only after summation in direct space over an area of 20×20 unit cells.†

Because of the second scalar product in (6-5-19), in CH_3Cl all $I_\gamma^{xy} = 0$, leading immediately to (6-5-15). In the dipole-coupling model, (6-5-15) is therefore a consequence of the higher symmetry of the coupling potential, as compared to that of the atom–atom repulsion model. The several remaining symmetrized lattice sums have been evaluated using the program developed by Dickmann, and some of the results, both for a non-degenerate and a degenerate mode, are illustrated in Fig. 6-6 for various crystallite shapes (see Section 5-7). The experimental values for these two modes are 5.3 and ≤ 3 cm^{-1}, respectively. It is seen that the splitting of ν_3 can be accounted for in several ways, but not those of ν_6.

Although the transition moment integrals appear explicitly in (6-5-19) as parameters of the dipole-coupling model, it is fundamental to both theories that the factor group splittings are proportional to the squares of the transition moment integrals. It follows that transitions which are both infrared- and Raman-active should manifest factor group splittings in both spectra, to the extent that the integrals (6-5-9) permit.

* It is shown in Section 5-7 that (6-5-19) may be written in such a way (5-7-5) that the transition dipoles may be factored and treated as adjustable parameters. The lattice sum is then one over cartesian components and inverse powers of the intermolecular separations $\mathbf{r}$. In other words, the lattice sum is a property of the *lattice structure*.

† Because of the conditional convergence of the lattice sum, evaluation of (6-5-7) using (5-7-5) is actually carried out for a particular shape (slab-like). Transition energies or factor group splittings for other shapes can be found by adding appropriate elements of the S tensor, as was discussed in Section 5-7 for cubic crystals.

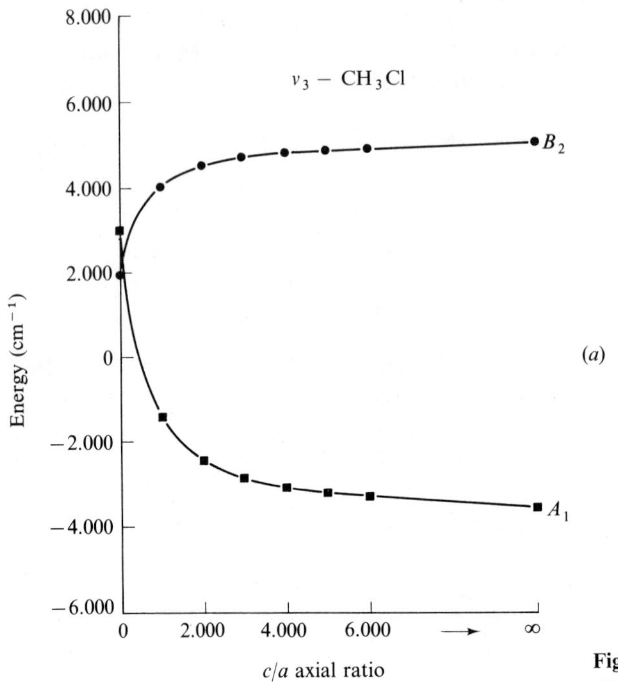

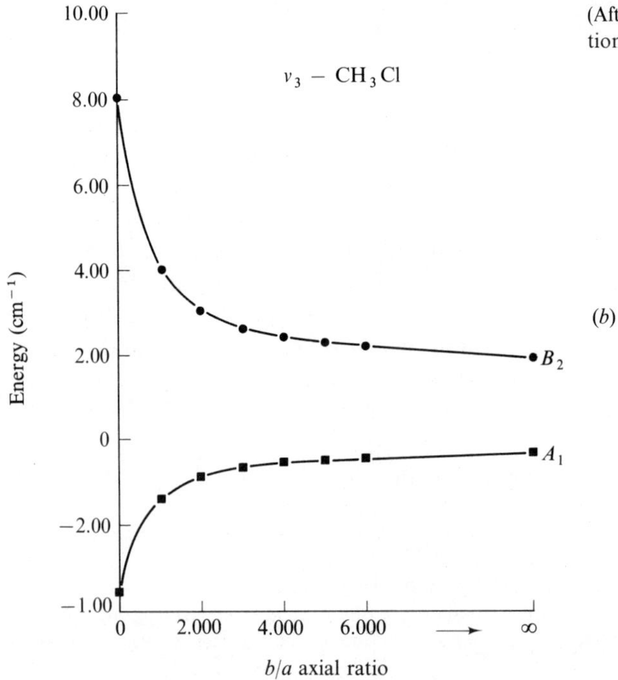

Figure 6-6 (*a*)–(*f*) Multiplet frequencies of ν_3 and ν_6 of CH_3Cl as functions of axial ratios of crystallites along various directions. (After T. Wilczek, Ph.D. Dissertation, University of Minnesota, 1975)

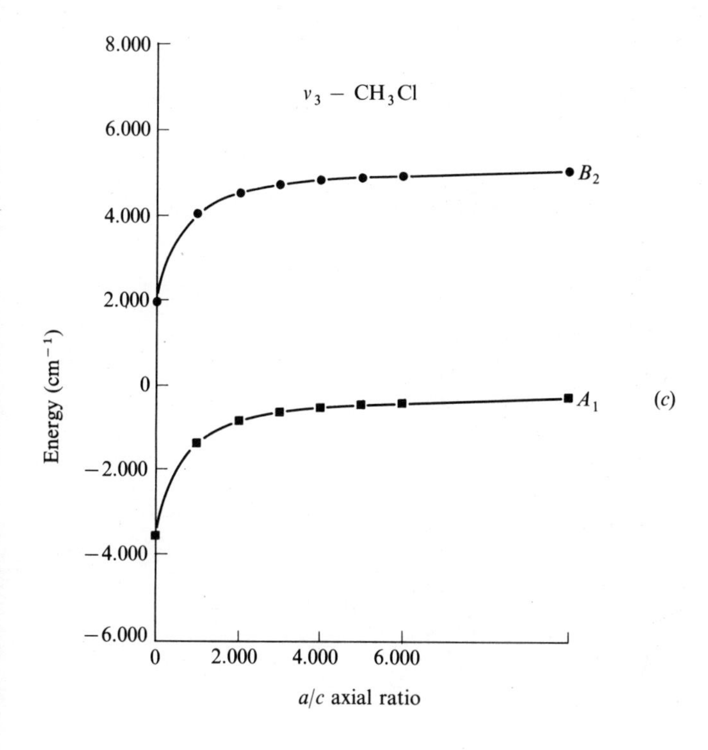

ν_3 – CH_3Cl
8.000
6.000
4.000
2.000
0
−2.000
−4.000
−6.000
Energy (cm^{-1})
0
2.000
4.000
6.000
a/c axial ratio
B_2
A_1

(c)

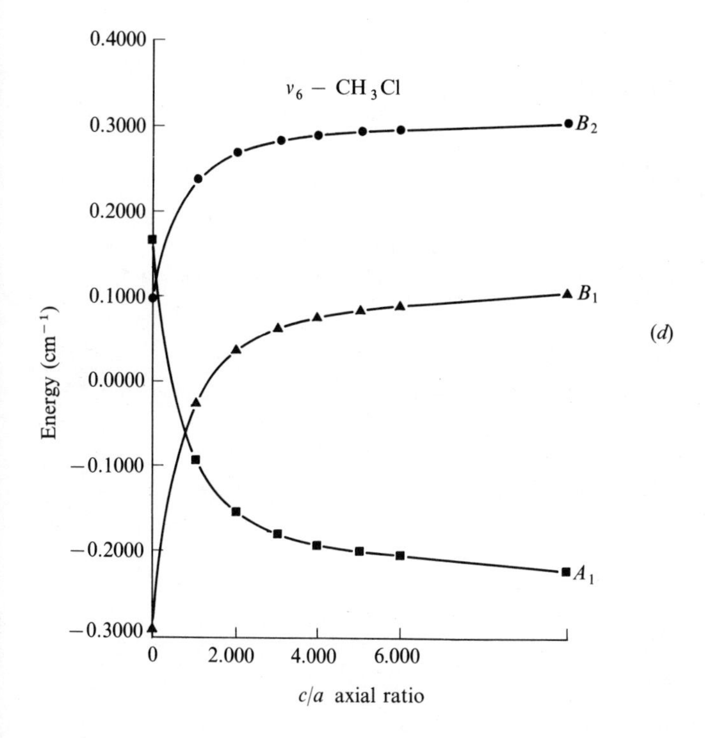

ν_6 – CH_3Cl
0.4000
0.3000
0.2000
0.1000
0.0000
−0.1000
−0.2000
−0.3000
Energy (cm^{-1})
0
2.000
4.000
6.000
c/a axial ratio
B_2
B_1
A_1

(d)

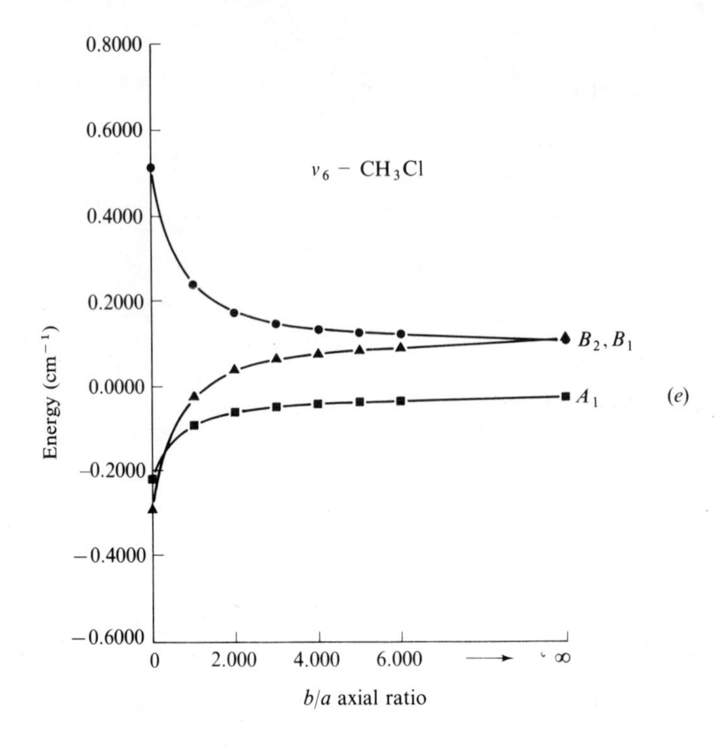
0.8000
0.6000
0.4000
0.2000
0.0000
–0.2000
–0.4000
–0.6000
Energy (cm⁻¹)
ν6 – CH3Cl
B2, B1
A1
0
2.000
4.000
6.000
∞
b/a axial ratio

(e)

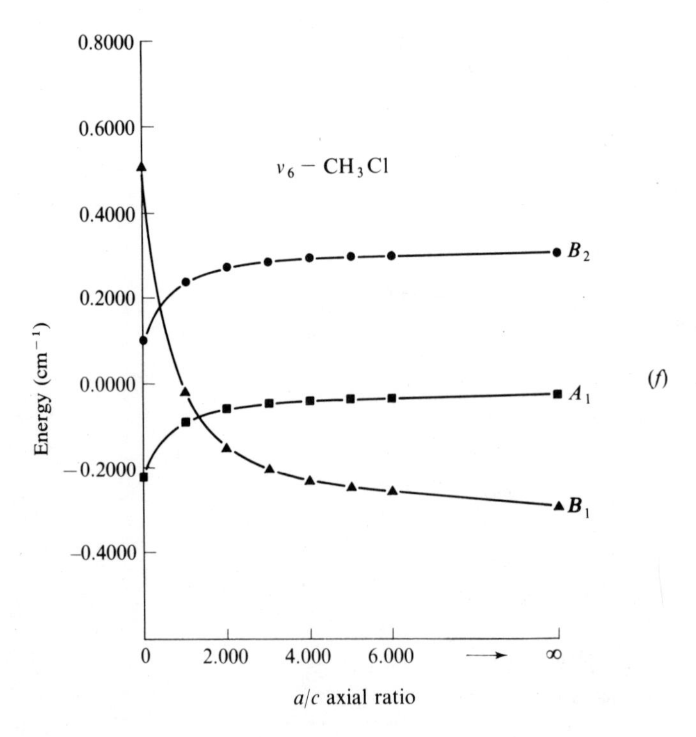
0.8000
0.6000
0.4000
0.2000
0.0000
–0.2000
–0.4000
Energy (cm⁻¹)
ν6 – CH3Cl
B2
A1
B1
0
2.000
4.000
6.000
∞
a/c axial ratio

(f)

The explicit dependence of the splittings upon transition dipoles in the dipole-coupling model leads to certain rules of isotopic invariance, provided the molecular wave functions φ are well-approximated by harmonic oscillator functions. In that case, aside from the geometric factors which appear in (6-5-19), it is easily shown from (6-5-4) and (6-5-9) that

$$\nu\Delta\nu \sim p^2$$

where p is a transition moment integral. Since the geometric factors (i.e., the lattice sum) should be independent of isotopic substitution in the molecule, the ratio $\nu\Delta\nu/\nu^i\Delta\nu^i$ of splittings in, say, fully hydrogenated and fully deuterated crystals should be determined by the intensity ratios p^2/p_i^2 of the analogous bands in the two crystals.

Wilczek has recently compared the two models in the same crystal, benzene, for which the experimental data are quite complete.[31] Calculated and observed splittings for three fundamentals (ν_{11}, ν_{18}, and ν_{19}) are given in Tables 6-11 and 6-12. In the first of these, the calculated splittings (in each case relative to a particular component as origin) were obtained using the dipole coupling model, while in the second, two sets of results due to Williams[32,33] are displayed. It

Table 6-11 Observed and calculated splittings of selected gas-phase infrared-active normal frequencies of C_6H_6 in units of cm^{-1} for the case of a slab-shaped ($c/a = 0$) crystal, using the transition-dipole model

Molecular mode	Crystal mode	$\Delta\nu$ (expt)	$\Delta\nu$ (calc) $(a)^a$	$(b)^b$
	B_{1u}	0.0	0.0	0.0
$\nu_{11}(A_{2u})$	B_{2u}	25.3	1.9	2.7
	B_{3u}	−1.3	15.9	22.3
	B_{1u}	0.0	0.0	0.0
	B_{2u}	−2.0	−0.05	−0.16
	B_{3u}	−0.9	0.08	0.25
$\nu_{18}(E_{1u})$	B_{1u}	4.5	0.71	2.20
	B_{2u}	5.0	0.14	0.43
	B_{3u}	4.0	0.63	1.95
	B_{1u}	0.0	0.0	0.0
	B_{2u}	−0.9	−0.05	−0.23
	B_{3u}	...	0.09	0.41
$\nu_{19}(E_{1u})$	B_{1u}	4.5	0.74	3.33
	B_{2u}	1.7	0.15	0.68
	B_{3u}	2.2	0.65	2.93

a. Calculated using gas-phase dipole derivatives.
b. Calculated using solid state dipole derivatives.

31. T. Wilczek, Jr., Ph.D. Dissertation, University of Minnesota, 1975.
32. D. E. Williams, *J. Chem. Phys.*, **45**, 3770 (1966).
33. D. E. Williams, *J. Chem. Phys.*, **47**, 3680 (1967).

Table 6-12 Observed and calculated splittings of selected gas-phase infrared-active normal frequencies of C_6H_6 in units of cm^{-1} using the atom–atom repulsion model

Molecular mode	Crystal mode	$\Delta\nu$ (expt)	$\Delta\nu$ (calc) (c)[a]	$\Delta\nu$ (calc) (d)[b]
	B_{1u}	0.0	0.0	0.0
$\nu_{11}(A_{2u})$	B_{2u}	25.3	16.0	25.6
	B_{3u}	−1.0	0.6	1.2
	B_{1u}	0.0	0.0	0.0
	B_{2u}	−2.0	0.2	−0.8
	B_{3u}	−0.9	−0.8	−2.2
$\nu_{18}(E_{1u})$	B_{1u}	4.5	4.9	6.3
	B_{2u}	5.0	2.7	4.2
	B_{3u}	4.0	3.9	6.3
	B_{1u}	0.0	0.0	0.0
	B_{2u}	−0.9	−1.6	−2.3
	B_{3u}	...	−0.8	−1.3
$\nu_{19}(E_{1u})$	B_{1u}	4.5	2.9	4.7
	B_{2u}	1.7	0.4	0.4
	B_{3u}	2.2	0.7	1.0

a. Calculated by means of the Williams IVa atom–atom interaction potential (see *J. Chem. Phys.*, **45**, 3770 (1966)).
b. Calculated by means of the Williams IVb atom–atom interaction potential (see *J. Chem. Phys.*, **47**, 3680 (1967)).

should be noted that in this study the linear combinations (6-5-4) belonging to the B_{1u} and B_{3u} representations have opposite polarization to that in the experimental investigation.[34]

While the result for ν_{11} using one of the atom–atom potentials is in almost exact agreement with experiment, it is not clear by inspection which potential is superior for ν_{18} and ν_{19}. Wilczek has defined a correlation coefficient C to aid in this comparison. Using

$$C = 1 - \frac{\sum_i |\Delta\nu_i(\text{expt}) - \Delta\nu_i(\text{calc})|}{\sum_i |\Delta\nu_i(\text{expt})|}$$

the results of Table 6-13 were obtained. According to this definition, a correlation coefficient of unity would correspond to exact agreement between experiment and calculation. From this it would appear that the atom–atom potential is superior for ν_{18}, but for ν_{19}, the dipole coupling model is better.

Of these bands, ν_{11} is the most intense,[35] especially in the solid phase.

34. M. P. Marzocchi, H. Bonadeo and G. Taddei, *J. Chem. Phys.*, **53**, 867 (1970).
35. J. L. Hollenberg and D. A. Dows, *J. Chem. Phys.*, **37**, 1300 (1962); **39**, 495 (1963).

Table 6-13 Correlation coefficients, C, for calculated splittings of the ν_{18} and ν_{19} modes of benzene

ν_{18}	C	ν_{19}	C
Table 6-11, column (*b*)	0.27	Table 6-11, column (*b*)	0.61
Table 6-12, column (*c*)	0.69	Table 6-12, column (*c*)	0.45
Table 6-12, column (*d*)	0.55	Table 6-12, column (*d*)	0.45

Indeed, its intensity is greater than that of the other three infrared-active fundamentals in the gas phase. Under the circumstances, the dipole coupling model is usually superior,[36] but it clearly is not so in this case.

It has sometimes been supposed that dipole–dipole interaction forces may be of dominant importance in ionic crystals, since the molecular ions of a given type are not in contact, being separated from one another by ions of the opposite sign. In these circumstances the longer range dipole forces might be assumed to be the most significant terms in the intermolecular potential, especially if the coupling modes have large infrared intensities and, hence, large transition dipoles.

Investigations of the question have in common the following procedure. The molecular dipole derivatives are evaluated with the aid of (5-8-16), the appropriate strength parameter being available from the analysis of infrared reflection via (5-1-2) and (5-4-2), or in some cases, from direct observation of the transverse and longitudinal mode frequencies in the Raman spectrum. The use of (5-8-16) raises certain questions about the effective field at the molecule in the crystal. In some cases, this has been evaluated with the aid of appropriate lattice dipole sums together with polarizabilities; in others, the Lorentz–Lorenz field, $\mathscr{F} = \mathscr{E}[\varepsilon(\infty) + 2]/3$, is used as an approximation.

Since all the substances to be described below have unit cells containing two or more molecular ions, one expects a number of Raman and infrared frequencies arising from a single molecular mode, as shown in detail for the specific cases of $CaCO_3$ and $NaClO_3$ in Sections 6-2(e) and 6-2(f). The intermolecular coupling constants are then assumed to arise solely from the dipolar coupling, and the frequency differences between the several unit cell modes are evaluated using appropriate lattice sums and the $\partial\mu/\partial q$ parameter, either in the context of the phonon or vibrational exciton procedures.

Such calculations have been carried out for a number of ionic crystals of varying complexity, including LiOH, $Mg(OH)_2$, $CaCO_3$ (calcite) and a number of crystals of the calcite structure (including $LiNO_3$, $MgCO_3$, $InBO_3$), as well as $NaClO_3$, K_2SO_4, and $Ca_{10}(PO_4)_6F_2$ (fluorapatite).

Lithium hydroxide has a bimolecular unit cell of space group D_{4h}^7, with OH^- units on C_{4v} sites. There are accordingly OH stretching modes of species A_{1g} (Raman) and A_{2u} (infrared). The evaluation of $\partial\mu/\partial q$ from the infrared spectrum poses a problem, since the crystals which can conveniently be grown have only a

36. D. Fox and R. M. Hexter, *J. Chem. Phys.*, **41,** 1125 (1964).

large face perpendicular to the C_4 axis, so that at true normal incidence the mode is inactive. However, by using TM off-axis reflectivity[37] from such a face together with Eq. (5-4-4), it was found possible to obtain an estimate of $\partial\mu/\partial q$ which is, in fact, in good agreement with the dipole moment found experimentally for other hydroxides or from electronic quantum mechanics. The frequencies are A_{1g} (Raman) $= 3665\ \mathrm{cm}^{-1}$, A_{2u} $(IR, T) = 3670\ \mathrm{cm}^{-1}$, and A_{2u} $(IR, L) = 3673.5\ \mathrm{cm}^{-1}$. The dipolar theory (details are omitted) would however predict the sequence $\nu(A_{2u}, T) < \nu(A_{1g}) < \nu(A_{2u}, L)$; it is perhaps not surprising that the dipolar coupling theory fails here, since the mode is really very weak.

An opportunity to test the dipole coupling model in calcite and related crystals arises for the molecular modes ν_3 and ν_4, since each gives rise to E_g and E_u components, respectively Raman and infrared active, Section 6-3(3). However, the mode ν_4 is quite weak, so we confine our attention to ν_3. The strengths of this mode have been obtained by several investigators and the necessary effective field factors are available.[38] From the strengths, the dipole derivatives are obtained and subsequently used to calculate the Raman frequency ν_{3g}, the splitting being entirely determined by the coupling constant $F'_k(1, 2)$ of Eq. (6-2-7). The results are summarized in Table 6-14. The agreement with the observed values is only close in the case of $NaNO_3$. The dipole coupling theory consistently predicts that ν_{3g} should occur at about 40 percent of the separation of $\nu_{3u}(L)$ and $\nu_{3u}(T)$, but experiment shows that ν_{3g} is consistently lower and closer to $\nu_{3u}(T)$. It is more probable that this result points to the neglect of short range forces in the calculation, but it should be mentioned that the determination of $\partial\mu/\partial q_3$ may be suspect owing to a possible Fermi resonance of ν_3 with $2\nu_4$.

In the case of $NaClO_3$, Andermann and Dows[39] analyzed the reflectivity spectrum with the aid of Kramers–Kronig integral transform techniques and obtained strengths for the several F species modes which we have previously summarized in Table 6-10. This is a case in which the isolated mode approximation fails because of the proximity of the two F components of ν_3 (and of the F component of ν_1 as well), but even so the analysis in Reference 39 agrees with later work by Hartwig, Rousseau, and Porto[40] in yielding a fivefold difference in the

Table 6-14

	$\nu_{3u}(T)$	$\nu_{3u}(L)$	S_3	$\partial\mu/\partial q_3$	ν_{3g} (calc.)	ν_{3g} (obs.)
$LiNO_3$	1360 cm^{-1}	1480 cm^{-1}	0.55	231 esu $\mathrm{g}^{-1/2}$	1413 cm^{-1}	1385 cm^{-1}
$NaNO_3$	1353	1450	0.37	236	1390	1385
$MgCO_3$	1436	1599	0.71	273	1507	1446
$CaCO_3$	1407	1549	0.58	283	1463	1432

37. R. E. Carlson and J. C. Decius, *J. Chem. Phys.*, **60,** 1251 (1974).
38. R. E. Frech and J. C. Decius, *J. Chem. Phys.*, **54,** 2374 (1971).
39. G. Andermann and D. A. Dows, *J. Phys. Chem. Solids*, **28,** 1307 (1967).
40. C. M. Hartwig, D. L. Rousseau, and S. P. S. Porto, *Phys. Rev.*, **188,** 1328 (1969).

strengths of the two components of ν_3 (see Table 6-10). The expressions for the dielectric strengths (5-8-16), assuming the Lorentz–Lorenz field, become

$$S_1\omega_1^2 = \frac{4\pi}{v}\left(\frac{\varepsilon_\infty + 2}{3}\right)^2 \frac{4}{3}\left(\frac{\partial\mu}{\partial q_1}\right)^2$$

and

$$S_{3a}\omega_{3a}^2 + S_{4a}\omega_{4a}^2 = \frac{4\pi}{v}\left(\frac{\varepsilon_\infty + 2}{3}\right)^2 \frac{8}{3}\left(\frac{\partial\mu}{\partial q_3}\right)^2$$

which yield*

$$\frac{\partial\mu}{\partial q_1} = \pm 120 \text{ esu } g^{-1/2} \quad \text{and} \quad \frac{\partial\mu}{\partial q_3} = \pm 175 \text{ esu } g^{-1/2}$$

As a test of the dipolar coupling potential, Andermann and Dows then used the splitting between ν_{3a} and ν_{3b} as an independent method of evaluating $\partial\mu/\partial q_3 = \pm 163$ (esu/$g^{-1/2}$). The splitting in ν^2 is determined by the coupling constant C of (6-2-11) which depends upon $\partial\mu/\partial q_3$ and appropriate lattice sums which have been discussed by Fox and Hexter[41] employing the exciton formalism.

The crystal K_2SO_4 is orthorhombic and single crystal Raman and infrared studies[42] have identified the numerous allowed components of the sulfate ion fundamentals. The dipole derivatives were evaluated from the reflection band widths, but the dipole coupling model was not found to account successfully for the observed multiplet spacings.

Finally we discuss the case of fluorapatite, $Ca_{10}(PO_4)_6F_2$. This crystal has a C_{6h}^2 (hexagonal) unit cell with six equivalent phosphates on C_s sites. Kravitz, Kingsley, and Elkin[43] studied the polarized Raman and infrared reflection spectra and identified the five Raman allowed and three infrared allowed components of the two molecularly infrared active modes ν_3 and ν_4. By including site splitting parameters as well as the dipole coupling terms deduced from the infrared strengths, a qualitatively successful account of the ν_3 and ν_4 manifolds was achieved, which was improved in a subsequent work[44] by inclusion of the "background polarization," i.e., allowance for the effective fields.

The overall conclusion from this review thus is that the dipole coupling mechanism is in some cases the major contributor to multiplet splittings, but that there are to date no known cases in which it can account quantitatively for these effects, without the inclusion of other, presumably short-range, intermolecular terms.

* These values differ from those reported in Ref. 39 because of the factor $\frac{1}{3}$ in the equations used here.

41. D. Fox and R. M. Hexter, *J. Chem. Phys.*, **41,** 1125 (1964).
42. F. Meserole, J. C. Decius, and R. E. Carlson, *Spectrochim. Acta,* **30A,** 2179 (1974).
43. L. C. Kravitz, J. D. Kingsley, and E. L. Elkin, *J. Chem. Phys.*, **49,** 4600 (1968).
44. J. D. Kingsley, G. D. Mahan, and L. C. Kravitz, *J. Chem. Phys.*, **49,** 4610 (1968).

Relative Intensities of Multiplet Components

In the foregoing, we have discussed the basis of multiplet splittings of pure crystals on the basis of two theories which have been proposed to account for these splittings. We close this section with a short discussion of the relative intensities of the components of a multiplet.

In the case of transitions to a non-degenerate excited state (of a molecule) we have noted that there can be at most Z members of a multiplet. If the unit cell symmetry is high enough, however, some of these components may belong to a representation of the factor group whose dimension is greater than one.[45] We consider first the case where the lack of unit cell symmetry precludes this. If the unit cell has at least orthorhombic symmetry, there will then be one factor group component polarized parallel to each crystallographic axis, and the relative intensities of these will be simply given by the ratios of the squares of the cosines of the transition moments within each molecule and the crystallographic axes. If there are Z molecules/cell, the direction cosines of the transition moment axis are related to one another by simple permutations of their signs, since the factor group operations not in the site group simply permute the molecules among the equivalent sites. The squares of the direction cosines are unchanged.

Alternatively, this result can be appreciated by applying the operations of the factor group not in the site group to a vector, the components of which are the direction cosines of the transition moment axis of one of the Z molecules/cell. The resulting vectors (including the first) are then multiplied by the characters of the same operations and summed, once for each representation of the factor group. The resultant unit cell transition moment, when squared, will give the same result.

Even if the dimension of the factor group representation is greater than one, the use of polarized, plane-parallel radiation will project out only one component of the representation, and its intensity, relative to those of the other components can be found in the same way.

This result is known as the *oriented gas model*.[46] The molecular transition moments are coupled according to one of the perturbation theories which have been proposed (atom–atom interaction or dipole-coupling). However, since the perturbation is only first-order, there can be no quantum mechanical mixing of states and, hence, the relative intensities are given directly by the trigonometric relationships of the molecules with respect to the crystallographic axes, parallel to which it is convenient to polarize the light in most experiments.* Hence, the name of the model.

On the other hand, if the molecular excited state is degenerate, each factor group energy is the root of a $d^\gamma \times d^\gamma$ matrix. First-order perturbation theory for degenerate states may then be used to find the relative intensities of each component belonging to the same representation of the factor group. When $d^\gamma = 2$, the terms

45. H. Winston and R. S. Halford, *J. Chem. Phys.*, **17**, 607 (1949).
46. G. C. Pimentel and A. L. McClellan, *J. Chem. Phys.*, **20**, 270 (1952).

* If this is not done, the relative intensities are further modified by Malus' Law, $I \simeq I_l \cos^2 \theta_l$, where θ_l is the angle between the electric vector and the crystallographic axis l.

I^{xy} in (6-5-7) serve to mix the φ^x and φ^y states. Although the *total* intensity of both components of any one factor group state, relative to that of any other, is still given by the oriented gas model, the *relative* intensities of the two components of each state can be shown to be

$$\frac{I_+^\gamma}{I_-^\gamma} = \frac{1 + 2I_\gamma^{xy}/\Delta_\gamma}{1 - 2I_\gamma^{xy}/\Delta_\gamma} \tag{6-5-20}$$

where

$$\Delta_\gamma = \{(I_\gamma^{xx} - I_\gamma^{yy})^2 + 4(I_\gamma^{xy})^2\}^{1/2} \tag{6-5-21}$$

The relative intensities stated in (6-5-20) have been obtained without consideration of sample shape. In an actual calculation, each lattice sum (I_γ^{xx}, I_γ^{yy}, and I_γ^{xy}) is computed for a range of shapes (needles, spheres, slabs), and the perturbation theory for degenerate states is carried out for each shape, the relative intensities still being computed using (6-5-20).

Similar considerations of the additivity (over the lattice) of *molecular polarizability derivatives* may be used as the basis of predicting the relative intensities of multiplet components in Raman spectra. This is particularly valuable in those cases where a vibrational mode is simultaneously infrared-active, for in that case the multiplet components are split from one another. However, because of the dipole activity, the Raman spectrum will manifest both longitudinal and transverse scattering, often of different relative intensity (see Section 5-6).

In transforming polarizability derivatives from molecular to crystallographic forms, similarity transformations using the matrices of the direction cosines must be carried out, since the polarizability is a tensor. In the analogous transformation of the dipole derivative, only the components of that vector along the crystallographic axes were necessary.

An alternative expression for (6-5-20) is

$$I_\pm^\gamma \propto p^2\left\{x^2 + y^2 \pm \frac{(x^2 - y^2)(I_\gamma^{xx} - I_\gamma^{yy}) \pm 4(xy)I_\gamma^{xy}}{[(I_\gamma^{xx} - I_\gamma^{yy})^2 + 4(I_\gamma^{xy})^2]^{1/2}}\right\} \tag{6-5-22}$$

where x, y are direction cosines between these molecular axes and the crystallographic axis which is the basis of γ. Using (6-5-22) Wilczek has calculated relative intensities for the components of ν_{18} of benzene as a function of crystallite axial ratio.[31]

6-6 COMBINATIONS AND OVERTONES

(a) Selection Rules

This subject has been discussed previously in Sections 2-4 and 4-7. Although no new principles are involved for the case of molecular crystals, it will be convenient to summarize the situation with the aim of distinguishing limiting cases in which spectroscopic activity arises entirely from *electrical* or *mechanical* anharmonicity. For all cases, the general statement in group theoretical terms is that a transition between an initial state $|i\rangle$ and a final state $|f\rangle$ is allowed in the infrared if

$$\Gamma_f^*(\boldsymbol{\kappa}_f) \times \Gamma_\mu(\boldsymbol{\kappa}_\mu) \times \Gamma_i(\boldsymbol{\kappa}_i) \supset \Gamma_1(\mathbf{0})$$

or in the Raman if

$$\Gamma_f^*(\boldsymbol{\kappa}_f) \times \Gamma_\alpha(\boldsymbol{\kappa}_\alpha) \times \Gamma_i(\boldsymbol{\kappa}_i) \supset \Gamma_1(\mathbf{0})$$

Taking the initial state to be the ground state for which $\boldsymbol{\kappa} = 0$, these rules respectively require that the final state have species in common with $\Gamma_\mu(\boldsymbol{\kappa}_\mu)$ or $\Gamma_\alpha(\boldsymbol{\kappa}_\alpha)$. Previously, we have seen that $\boldsymbol{\kappa}_\mu$ and $\boldsymbol{\kappa}_\alpha \simeq 0$, so $\boldsymbol{\kappa}_f \simeq 0$.

For a combination state, the symmetry species for $\Gamma_f(0)$ are determined by formation of the direct product of the species of $Q_j(\boldsymbol{\kappa}_j)$ and $Q_k(\boldsymbol{\kappa}_k)$. It follows that $\boldsymbol{\kappa}_j + \boldsymbol{\kappa}_k = \boldsymbol{\kappa}_f = 0$, and our task is then to discover for each point $\boldsymbol{\kappa}$ in the Brillouin zone, which irreducible representations at $\boldsymbol{\kappa} = 0$, i.e., at the zone center, appear in the reduction of the direct product, $\Gamma_j(\boldsymbol{\kappa}) \times \Gamma_k(\bar{\boldsymbol{\kappa}})$.†

For any $\boldsymbol{\kappa}$ of the types enumerated in Section 4-5 for which the irreducible representations may be formed by symmetrizing the translationally symmetrical coordinates under a point group $H(\boldsymbol{\kappa})$, we have $H(\boldsymbol{\kappa}) = H(\bar{\boldsymbol{\kappa}})$ and the direct product is simply taken between the species of Q_j and Q_k at $H(\boldsymbol{\kappa})$: this is then followed by correlation with the species at $\boldsymbol{\kappa} = 0$.

As an example, let us consider a point $\boldsymbol{\kappa}$ along the threefold axis in the interior of the zone for the T_d^1 space group. Such a $\boldsymbol{\kappa}$ has C_{3v} symmetry; we might want to investigate selection rules for Q_j and Q_k belonging to one or another of the C_{3v} species A_1, A_2, and E. The appropriate direct products in C_{3v} are $A_1 \times A_1 = A_2 \times A_2 = A_1$, $A_1 \times A_2 = A_2$, $A_1 \times E = E$, $A_2 \times E = E$, and $E \times E = A_1 + A_2 + E$. The correlation table between $H(\boldsymbol{\kappa}) = C_{3v}$ and $H(\mathbf{0}) = T_d$ as in Appendix IV shows that

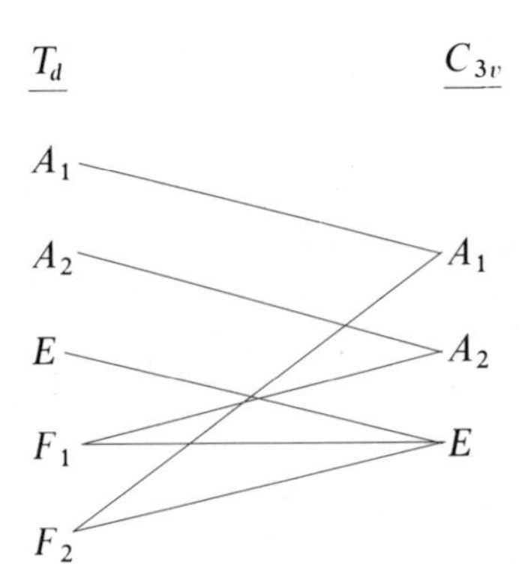

† In this section, $\bar{\boldsymbol{\kappa}} = -\boldsymbol{\kappa}$ must not be confused with the bar used in a different sense in section 4-5.

From this we infer that combinations between all branches except $A_1 \times A_2$ are infrared active, since this is the only combination whose reduction in $C_{3v}(A_2)$ does not correlate with the infrared active species of T_d, namely F_2.

Actually, it is simpler in these circumstances to carry out such an analysis by examining the symmetry species of a combination at $\boldsymbol{\kappa} = 0$ and then to correlate down to C_{3v}. To be specific, consider some combination states of species F_2 at $\boldsymbol{\kappa} = 0$; this is an infrared active combination, which is split into A_1 and E branches for $H(\boldsymbol{\kappa}) = C_{3v}$. Note that all combinations of these branches remain active at C_{3v} as we have just seen; this will quite generally be true for any point in the interior of the zone for combination modes which are active at the center. But, now what about combination modes which are forbidden at $\boldsymbol{\kappa} = 0$? In principle, all of these except A_2 appear to be allowed. This is because $A_2(T_d) \rightarrow A_2(C_{3v}) \rightarrow A_2 + F_1(T_d)$ whereas $A_1(T_d) \rightarrow A_1(C_{3v}) \rightarrow A_1 + F_2(T_d)$, etc. In practice, however, formally allowed combination modes of this latter type have negligible intensities, at least for combinations of two intramolecular modes.

For overtone modes arising from degenerate fundamentals, it is necessary to use the symmetrized Kronecker product when reducing the direct product; see Section 2-4

In the subsequent sections we shall first discuss the structure of combination and overtone bands assuming a completely harmonic potential function, i.e., a *mechanically* harmonic system. In such a case, the intensity arises entirely from *electrical* anharmonicity. In particular, the infrared excitation will require a non-vanishing second derivative of the dipole moment of the form

$$\frac{\partial^2 \mu}{\partial Q_k(\bar{\boldsymbol{\kappa}})\, \partial Q_j(\boldsymbol{\kappa})}$$

while the corresponding Raman intensity requires

$$\frac{\partial^2 \alpha}{\partial Q_k(\bar{\boldsymbol{\kappa}})\, \partial Q_j(\boldsymbol{\kappa})}$$

Considering the infrared case, we assume

$$\mu_{\text{crystal}} = \sum_{\tau} \mu_\tau$$

where μ_τ is the dipole of one molecule. The further assumption that a change in the normal coordinate of a single molecule produces a change in the dipole moment of that molecule but in no other molecule leads to the following expression for the crystal dipole derivatives with respect to phonon coordinates in terms of molecular dipole derivatives with respect to molecular coordinates

$$\frac{\partial \mu}{\partial Q_j(\boldsymbol{\kappa})} = N^{3/2}\, \delta(\boldsymbol{\kappa})\, \frac{\partial \mu}{\partial q_j} \tag{6-6-1}$$

$$\frac{\partial^2 \mu}{\partial Q_k(\bar{\boldsymbol{\kappa}})\, \partial Q_j(\boldsymbol{\kappa})} = \delta(\boldsymbol{\kappa}_j + \boldsymbol{\kappa}_k)\, \frac{\partial^2 \mu}{\partial q_k\, \partial q_j} \tag{6-6-2}$$

in agreement with our oft-asserted rules that for fundamentals $\boldsymbol{\kappa}$ must vanish and for combinations, the sum $\boldsymbol{\kappa}_j + \boldsymbol{\kappa}_k$ must vanish (the above δ's are unity if the vector argument vanishes, zero otherwise). More elaborate expressions may be developed for crystals with several molecules per unit cell, and, naturally, care must be exercised in projecting the components of molecular dipoles onto the crystal axes. Corresponding expressions for the polarizability derivatives are easily obtained.

We noted above that combination transitions of the form $A_1 \times A_2$ are allowed in the infrared spectrum for a T_d unit cell for values of $\boldsymbol{\kappa}_j = \boldsymbol{\kappa} = -\boldsymbol{\kappa}_k \neq 0$, but ordinarily are not observed. We now see from (6-6-2) that such modes, which are molecularly forbidden, cannot appear at any $\boldsymbol{\kappa}$ value according to the above dipole model. If detected, they can only be explained as a consequence of intermolecularly induced dipoles.

In Sections 6-6(b) and (c), we shall develop the consequences of (6-6-2) and discover the expected structure of two-phonon bands subject to the assumption of mechanically harmonic motion.

(b) Dispersion Curves

In Sections 3-2 and 3-3 we derived and discussed the one-photon energies $\omega(\mathbf{k})$ for simple, one-dimensional lattices, and in Section 4-6 we did the same for several three-dimensional, monatomic lattices, as well as for a simple cubic lattice with a basis. One-phonon density-of-states functions for some of these were introduced in Section 4-7(c), as well as their critical points. In at least one case it was pointed out that the *joint* density-of-states function for a summation mode is defined in an analogous manner to that of a one-phonon state, that is,

$$g_{ij}(\omega) = \iint \frac{dS_{ij}}{\nabla_{\mathbf{k}}[\omega_i(\mathbf{k}) + \omega_j(\mathbf{k})]} \tag{6-6-3}$$

It will be readily recognized that the argument of the gradient in the denominator of the integrand of $g_{ij}(\omega)$ is simply the direct sum of the two dispersion surfaces, $\omega_i(\mathbf{k})$ and $\omega_j(\mathbf{k})$. In similar fashion, discussion of difference modes can be facilitated by the construction of the subtractive dispersion surface, $\omega_i(\mathbf{k}) - \omega_j(\mathbf{k})$. Extension of this procedure to ternary, etc., combinations is obvious.

In all such cases, the *overall* dispersion of the combination state is the resultant of that of all of its components. In the special case of a binary combination in which one of the two components is not dispersed, of course, the overall dispersion is just that of the dispersed branch. This situation is frequently encountered in molecular crystals with the combination of internal with external modes. Because of the approximation we call the frequency separation theorem, the internal modes do not have extensive dispersion. This is particularly true for high-frequency molecular fundamental modes, such as hydrogen-stretching frequencies. Consequently, combination of these modes with lattice frequencies gives rise to a new set of states, the dispersion of which is almost exactly that of the external component. In this way, the internal mode is effectively a high frequency amplifier of the external branch, for the external mode density-of-states function is thus shifted to the high-frequency region.

It is worthwhile to reconsider the meaning of a combination state, $\omega_i(\mathbf{0}) + \omega_j(\mathbf{0})$, evaluated at $\mathbf{k} = 0$. We recall that a zone-origin mode corresponds to a motion in which the vibrations of all unit cells are *in phase*. If each unit cell of a crystal is singly occupied, the combination state at $\mathbf{k} = 0$ corresponds to the *molecular combination mode*, $\omega_i + \omega_j$, where i and j are different vibrational degrees of freedom of an individual molecule. On the other hand, we must distinguish between a two-phonon state in which the excitation is that of one molecule and that in which the ith state of one molecule and the jth state of another molecule are simultaneously excited, each with one vibrational quantum.

In the strictly harmonic crystal, these two combination states are degenerate, and in a multiply-occupied unit cell the degeneracy is equal to the cell population. Anharmonicity can remove this degeneracy, but we must be careful to distinguish between *molecular* anharmonic terms $b_{ijk}q_iq_jq_k$, where the q's are coordinates of the same molecule, and crystal anharmonicity in which Q_i, Q_j, and Q_k are molecularly equivalent phonon coordinates which may involve many molecules.

In Section 6-6(d) below, we quantitatively discuss the two effects, and stress their separate contributions to the absorption spectrum of a molecular crystal, especially in the case of $\omega_i = \omega_j$, that is, *overtone bands*.

(c) Density of One- and Two-Phonon States

Equation (6-6-3) expresses the general relationship between the joint density-of-states function $g_{ij}(\omega)$ and the dispersion surface from which it is derived. As we have previously noted in Section 4-7(c), c.p.'s in both component density functions will generally appear in $g_{ij}(\omega)$. However, it may also have additional c.p.'s that are not apparent in the one-phonon densities. For example, since the gradients of $\omega_i(\mathbf{k})$ and $\omega_j(\mathbf{k})$ can be of opposite sign, $\nabla_{\mathbf{k}}[\omega_i(\mathbf{k}) + \omega_j(\mathbf{k})]$ can vanish, thus yielding a c.p. in $g_{ij}(\omega)$ which is in neither $g_i(\omega)$ nor $g_j(\omega)$.

Because of the less rigorous selection rules for overtone and combination bands, as discussed in Section 6-6(a), the existence of c.p.'s at $\mathbf{k} \neq 0$ can be detected. Entire bands become spectroscopically active, their envelopes mapping $g_{ij}(\omega)$. We have earlier described a number of different kinds of c.p.'s in three-dimensional crystals, using Phillips' notation. In each case the singular portion of the dispersion surface gave rise to a contribution to the overall frequency distribution function of the crystal $g(\omega)$ proportional to $(\omega - \omega_0)^{1/2}$, where ω_0 is the frequency at the c.p. Accordingly, if instead of recording the ordinary absorption spectrum of a crystal we were to obtain its frequency-modulated spectrum, $dg(\omega)/d\omega \sim (\omega - \omega_0)^{-1/2}$, the singularities at ω_0 should be especially evident.

This kind of spectroscopy, known as *modulation spectroscopy,* has been especially useful in observing singularities in indirect phonon-aided electronic transitions, such as occur near 1.02 μm in Si.[47] This technique, as applied to

47. I. Balslev, *Solid State Comm.*, **3,** 213 (1965); *Phys. Rev.*, **143,** 636 (1966).

electronic transitions, has been extensively reviewed by Cardona,[48] but it has been little applied to multiphonon spectroscopy. An exception is the recent application of the technique to the related phenomenon of second-order Raman spectra.[49]

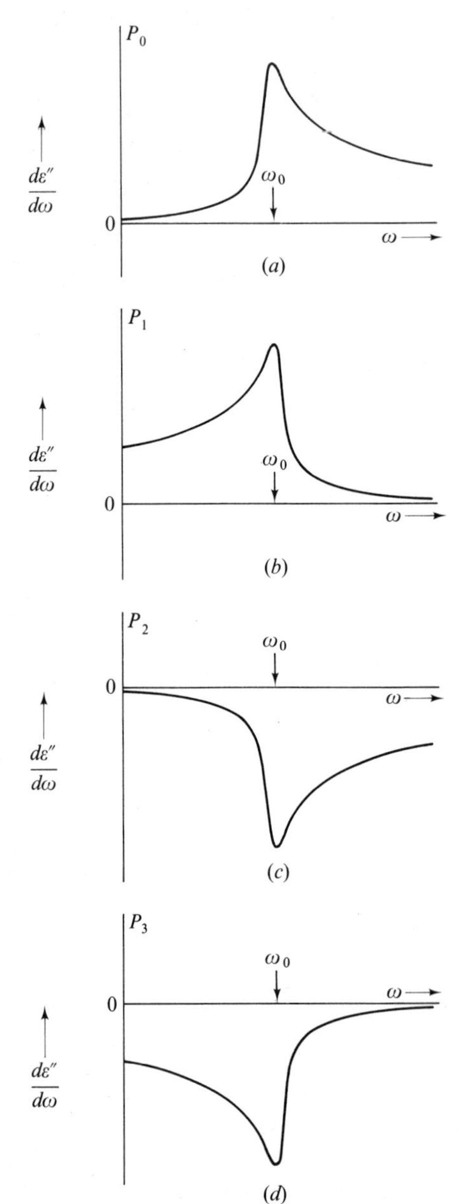

Figure 6-7 Derivatives $d\varepsilon''/d\omega$ of the imaginary part of the dielectric constant with respect to frequency in the neighborhood of analytical critical points P_j. The effect of broadening has been included: (*a*) P_0; (*b*) P_1; (*c*) P_2; (*d*) P_3. (After M. Cardona, *Modulation Spectroscopy*. (Solid State Physics, Supplement **11**, page 95))

48. M. Cardona, *Solid State Physics*, Suppl. 11, Academic Press, New York, 1969.
49. Y. Yacoby and S. Yust, *Surface Science*, **37**, 161 (1973).

The several kinds of c.p.'s in three-dimensional crystals were discussed in Section 4-7(c) in terms of their contributions to the shape of $g(\omega)$, the frequency distribution function, near ω_0. These contributions were illustrated in Fig. 4-15, and consist of the singularity superimposed upon a constant background. As discussed by Cardona,[48] the complex dielectric constant near a three-dimensional c.p. has a similar form, that is

$$\varepsilon \sim b(\omega - \omega_0)^{1/2} + \text{constant}$$

If this function is differentiated with respect to the frequency ω we find

$$\frac{d\varepsilon}{d\omega} = \frac{b}{2}(\omega - \omega_0)^{-1/2} + \frac{db}{d\omega}(\omega - \omega_0)^{1/2}$$

We can neglect the contribution of the second term in the vicinity of ω_0, while the form of the first term, for the imaginary part of ε is illustrated in Fig. 6-7 for the several kinds of singularities P_i.

Clearly this technique has considerable potential in identifying the several kinds of c.p.'s in multiphonon spectra.

(d) Anharmonicity

In order to develop a quantitatively useful discussion of overtone and combination spectra, it is necessary to make several generalizations of our previous treatment of intensities of transitions in molecules (Section 2-6) and crystals (Sections 5-1, 5-3, 5-5, and 5-8). First, we require a description of the dielectric constant which includes contributions from allowed overtones and combinations, instead of being limited to fundamental transitions as in Chapter 5. Also, it will be necessary to investigate the role of mechanical anharmonicity, which leads to a number of complications; this topic will be taken up later in this section.

It is convenient at this point to introduce the *electrical susceptibility* χ which relates the polarization per unit volume, P, to the applied field $\mathscr{E}$:

$$\mathscr{P}_\sigma = \sum_\sigma \chi_{\sigma\sigma'} \mathscr{E}_{\sigma'} \qquad \sigma, \sigma' = x, y, z \tag{6-6-4}$$

We now specialize to principal axes and drop the cartesian indices. Then in view of the relations

$$\mathscr{D} = \varepsilon \mathscr{E} = \mathscr{E} + 4\pi\mathscr{P}$$

it follows that

$$\varepsilon - 1 = 4\pi\chi$$

and further if

$$\varepsilon = \varepsilon' + i\varepsilon'' \quad \text{and} \quad \chi = \chi' + i\chi''$$

then $$\varepsilon' - 1 = 4\pi\chi' \quad \text{and} \quad \varepsilon'' = 4\pi\chi''$$

Thus in order to find the imaginary part of the dielectric constant we study the imaginary part of the susceptibility.

The *fluctuation dissipation theorem* as developed by Kubo[50] establishes a very general relation between the response of a system and the operator, in this case μ, which couples the system to an external force $\mathscr{E}$. Such a relation may in the present case be written as

$$\chi''(\omega) = (2N\mathrm{v}\hbar)^{-1}(1 - e^{-\beta\hbar\omega}) \int_{-\infty}^{+\infty} e^{-i\omega t} \langle \mu(0)\mu(t) \rangle \, dt \qquad \text{(6-6-5)}$$

where $\beta = 1/kT$, v = volume of the unit cell, N is the number of unit cells in the sample, and the expression $\langle \mu(0)\mu(t) \rangle$ represents a statistical average of the dipole correlation function $\mu(0)\mu(t)$. The dipole operators are to be understood in the Heisenberg sense,[51,52] which means that

$$\mu(t) = e^{iHt/h} \cdot \mu(0) \cdot e^{-iHt/h}$$

where H is the complete Hamiltonian of the system, excluding only the part, $-\mu\mathscr{E}$, which couples the system to the external field.

In quantum statistics, the averaging operation for any dynamical variable D is defined by

$$\langle D \rangle = \mathrm{Tr}\, \rho D = \sum_{m,n} \rho_{mn} D_{nm}$$

where ρ is the density matrix. If the states labelled by m or n are eigenstates of H, then ρ is diagonal, and ρ_m simply means the fractional population of the state m which is readily calculated by ordinary Boltzmann statistical mechanics. Also, since

$$\langle p | e^{\pm iHt/h} | q \rangle = \delta_{pq}\, e^{\pm iE_p t/h}$$

one finds that

$$\begin{aligned} \langle \mu(0)\mu(t) \rangle &= \sum_{m,n} \rho_m \mu_{mn}\, e^{iE_n t/\hbar}\, \mu_{nm}\, e^{-iE_m t/\hbar} \\ &= \sum_{m,n} \rho_m |\mu_{nm}|^2\, e^{i\omega_{nm} t} \end{aligned} \qquad \text{(6-6-6)}$$

where

$$\omega_{nm} = (E_n - E_m)/\hbar \qquad \text{(6-6-7)}$$

When the dipole correlation function is written in this way, one is assuming that there are no decay processes, and may expect that the spectrum will consist

50. R. Kubo, *Rep. Prog. Phys.*, **29**, 255 (1966).
51. R. H. Dicke and J. P. Wittke, *Introduction to Quantum Mechanics*, Addison-Wesley Publishing Co., Reading, Mass., 1960, pages 181 and 216.
52. J. M. Ziman, *Elements of Advanced Quantum Theory*, Cambridge University Press, Cambridge, 1969, pages 56–62.

of perfectly sharp lines, i.e., delta functions. In fact this is what the Fourier transform (6-6-5) now yields, namely

$$\chi''(\omega) = (\pi/\hbar N v)(1 - e^{-\beta\hbar\omega}) \sum_{m,n} \rho_m |\mu_{nm}|^2 \delta(\omega_{nm} - \omega) \qquad (6\text{-}6\text{-}8)$$

since

$$\int_{-\infty}^{+\infty} e^{ixt}\, dt = 2\pi\delta(x) \qquad (6\text{-}6\text{-}9)$$

The evaluation of the dipole transition matrix elements μ_{nm} may now be carried out with the aid of an expansion of the dipole in powers of the normal coordinate, which latter operators are conveniently expressed in terms of annihilation and creation operators (Appendix III). For the work which follows we use the abbreviations implicit in the following Taylor's series:

$$\begin{aligned}\mu &= \mu_0 + \sum_j \frac{\partial\mu}{\partial Q_j(0)} Q_j(0) + \frac{1}{2}\sum_{\kappa}\sum_i\sum_j \frac{\partial^2\mu}{\partial Q_i(\kappa)\,\partial Q_j(\bar{\kappa})} Q_i(\kappa)Q_j(\bar{\kappa}) + \cdots \\ &= \mu_0 + \sum_j \mu_j Q_j + \frac{1}{2}\sum_{\kappa}\sum_i\sum_j \mu_{ij}(\kappa)Q_i(\kappa)Q_j(\bar{\kappa}) + \cdots\end{aligned} \qquad (6\text{-}6\text{-}10)$$

Note that we drop the implicit dependence of Q on $\kappa = 0$ in the linear terms, and that the second power terms always involve κ and $\bar{\kappa} = -\kappa$ in accordance with selection rules discussed earlier. It must of course be kept in mind that the summations over i and j are further restricted by the point group selection rules discussed above in Sections 6-3 and 6-6(a).

We next investigate the susceptibility for the harmonic oscillator model. We consider first a fundamental absorption transition for which $n \leftarrow m$ means $v_j + 1 \leftarrow v_j$. Since from Appendix III

$$Q_j = -i(\hbar/2\omega_j)^{1/2}[a_j^+ - a_j]$$

one finds that

$$\mu_{nm} = \mu_j[(v_j + 1)(\hbar/2\omega_j)]^{1/2} \qquad (6\text{-}6\text{-}11)$$

with the aid of Eq. (III-4).

The summation over all initial states, labelled by v_j, and necessary because all transitions of the type $v_j + 1 \leftarrow v_j$ give rise to the same observed frequency, is now easily calculated and yields

$$\sum_{m,n} \rho_m |\mu_{nm}|^2 = (\hbar/2\omega_j)|\mu_j|^2 \Sigma\, \rho_{v_j}(v_j + 1) = (\hbar/2\omega_j)|\mu_j|^2(\bar{v}_j + 1) \qquad (6\text{-}6\text{-}12)$$

in which the bar over v_j indicates the thermal average, which is given by

$$\bar{v}_j + 1 = (1 - e^{-\beta\hbar\omega_j})^{-1} \qquad (6\text{-}6\text{-}13)$$

Now by combining (6-6-8), (6-6-12), and (6-6-13) we find

$$\varepsilon''(\omega) = \left(\frac{4\pi}{N\text{v}}\right)\left(\frac{\pi}{2\omega_j}\right)|\mu_j|^2(1 - e^{-\beta\hbar\omega})(1 - e^{-\beta\hbar\omega_j})^{-1}\,\delta(\omega_j - \omega)$$
$$= \left(\frac{4\pi}{N\text{v}}\right)\left(\frac{\pi}{2\omega_j}\right)|\mu_j|^2\,\delta(\omega_j - \omega) \tag{6-6-14}$$

since the delta function effectively cancels the two temperature dependent factors.

Although this last expression seems to differ from (5-3-10) and (5-8-14), which yield a contribution to ε'' due to a fundamental transition of the form,

$$\varepsilon''(\omega) = \frac{4\pi}{N\text{v}}|\mu_j|^2\gamma_j\omega[(\omega_j^2 - \omega^2)^2 + \gamma_j^2\omega^2]^{-1} \tag{6-6-15}$$

the difference is only apparent, since in order to make the comparison, one must pass to the limit $\gamma_j \to 0$ in (6-6-15). When this limit is evaluated, one finds

$$\lim_{\gamma_j \to 0} \varepsilon''(\omega) = \frac{4\pi}{N\text{v}}|\mu_j|^2\left(\frac{\pi}{2\omega_j}\right)\delta(\omega_j - \omega)$$

which is identical with (6-6-14). Of course, for simplicity in the present discussion we have neglected the effective field factor.

Combination transitions arise from the binary terms in the expansion of the electric dipole. As shown in Appendix III,

$$Q_i(\boldsymbol{\kappa}) = -i[\hbar/2\omega_i(\boldsymbol{\kappa})]^{1/2}[a_i^+(\boldsymbol{\kappa}) - a_i(\bar{\boldsymbol{\kappa}})]$$

and

$$Q_j(\bar{\boldsymbol{\kappa}}) = -i[\hbar/2\omega_j(\bar{\boldsymbol{\kappa}})]^{1/2}[a_j^+(\bar{\boldsymbol{\kappa}}) - a_j(\boldsymbol{\kappa})]$$

so that from (6-6-10) we have the term

$$-\tfrac{1}{2}\mu_{ij}(\boldsymbol{\kappa})\left[\frac{\hbar}{2\omega_i(\boldsymbol{\kappa})}\frac{\hbar}{2\omega_j(\bar{\boldsymbol{\kappa}})}\right]^{1/2}[a_i^+(\boldsymbol{\kappa})a_j^+(\bar{\boldsymbol{\kappa}}) - a_i^+(\boldsymbol{\kappa})a_j(\boldsymbol{\kappa}) - a_i(\bar{\boldsymbol{\kappa}})a_j^+(\bar{\boldsymbol{\kappa}}) + a_i(\bar{\boldsymbol{\kappa}})a_j(\boldsymbol{\kappa})] \tag{6-6-16}$$

Of the four terms involving a and a^+, if $\omega_i(\boldsymbol{\kappa}) > \omega_j(\boldsymbol{\kappa})$, the two of the form $a_i^+(\boldsymbol{\kappa})a_j^+(\bar{\boldsymbol{\kappa}})$ and $-a_i^+(\boldsymbol{\kappa})a_j(\boldsymbol{\kappa})$ will lead to absorption and we confine our attention to just these two terms in what follows. The first term yields a sum band and the second a difference band of frequencies $\omega_i(\boldsymbol{\kappa}) + \omega_j(\bar{\boldsymbol{\kappa}})$ and $\omega_i(\boldsymbol{\kappa}) - \omega_j(\boldsymbol{\kappa})$ respectively. Note that when two phonons are created they must have wave vectors of opposite sign, but when one is created and the other is annihilated, they must have identical $\boldsymbol{\kappa}$ values in order to satisfy the momentum conservation rule.

Proceeding with the calculation of the contribution of the sum band to the dielectric constant, we find that

$$\varepsilon''(\omega) = \frac{\pi^2\hbar}{4N\text{v}}(1 - e^{-\beta\hbar\omega})\sum_{\boldsymbol{\kappa}}\{[|\mu_{ij}(\boldsymbol{\kappa})|^2/\omega_i(\boldsymbol{\kappa})\omega_j(\bar{\boldsymbol{\kappa}})]$$
$$\times(\bar{v}_i(\boldsymbol{\kappa}) + 1)(\bar{v}_j(\bar{\boldsymbol{\kappa}}) + 1)\,\delta[-\omega + \omega_i(\boldsymbol{\kappa}) + \omega_j(\bar{\boldsymbol{\kappa}})]\} \tag{6-6-17}$$
$$= \frac{\pi^2\hbar}{4N\text{v}}\sum_{\boldsymbol{\kappa}}[|\mu_{ij}(\boldsymbol{\kappa})|^2/\omega_i(\boldsymbol{\kappa})\omega_j(\bar{\boldsymbol{\kappa}})](\bar{v}_i(\boldsymbol{\kappa}) + \bar{v}_j(\bar{\boldsymbol{\kappa}}) + 1)\,\delta[-\omega + \omega_i(\boldsymbol{\kappa}) + \omega_j(\bar{\boldsymbol{\kappa}})]$$

where the simplification effected in the last form of (6-6-17) is made possible by the delta function which makes $(1 - e^{-\beta h\omega}) = (1 - e^{-\beta h(\omega_i + \omega_j)})$ along with some elementary algebra. One sees immediately that a combination band, unlike a fundamental, may exhibit temperature dependence.

The result for the difference band is similar, and takes the form

$$\varepsilon''(\omega) = \frac{\pi^2 h}{4N\text{v}} \sum_{\kappa} \{[|\mu_{ij}(\boldsymbol{\kappa})|^2/\omega_i(\boldsymbol{\kappa})\omega_j(\bar{\boldsymbol{\kappa}})](\bar{v}_i(\boldsymbol{\kappa}) - \bar{v}_j(\bar{\boldsymbol{\kappa}}))\,\delta[-\omega + \omega_i(\boldsymbol{\kappa}) - \omega_j(\bar{\boldsymbol{\kappa}})]\} \tag{6-6-18}$$

It may be mentioned here that since many molecular crystals are perforce studied at low temperatures, the simple limits for the sum and difference bands may be appropriate, i.e., as $T \to 0$, $\bar{v}_i + \bar{v}_j + 1 \to 1$ and $\bar{v}_i - \bar{v}_j \to 0$.

For overtone as distinct from combination bands one needs to replace (6-6-16) by

$$-\tfrac{1}{2}\mu_{ii}(\boldsymbol{\kappa}) \frac{h}{2\omega_i(\boldsymbol{\kappa})} [a_i^+(\boldsymbol{\kappa})a_i^+(\bar{\boldsymbol{\kappa}}) - a_i^+(\boldsymbol{\kappa})a_i(\boldsymbol{\kappa}) - a_i(\bar{\boldsymbol{\kappa}})a_i^+(\bar{\boldsymbol{\kappa}}) + a_i(\bar{\boldsymbol{\kappa}})a_i(\boldsymbol{\kappa})]$$

Only the first term contributes to the absorption and leads to the result

$$\varepsilon''(\omega) = \frac{\pi^2 h}{4N\text{v}} \Big\{ 2[|\mu_{ii}(0)|^2/\omega_i^2(0)][2\bar{v}_i(0) + 1]\,\delta[-\omega + 2\omega_i(0)] + \sum_{\kappa \neq 0} [|\mu_{ii}(\boldsymbol{\kappa})|^2/\omega_i^2(\boldsymbol{\kappa})][2\bar{v}_i(\boldsymbol{\kappa}) + 1]\,\delta[-\omega + 2\omega_i(\boldsymbol{\kappa})] \Big\} \tag{6-6-19}$$

in which the distinction between $\boldsymbol{\kappa} = 0$ and $\boldsymbol{\kappa} \neq 0$ arises from the difference in the matrix elements for a one-phonon transition with double excitation $(v_i(0) + 2 \leftarrow v_i(0))$ and the two-phonon transitions.

All of the important results for the ε'' have been obtained previously by many authors, notably in the monograph by Born and Huang.[53] More recently, the elegant and sophisticated methods of many-body theory have been applied; such methods, although beyond the scope of this text, are particularly useful in dealing with the complications which arise when the mechanical anharmonicity is taken into consideration.

Formulas (6-6-17), (6-6-18), and (6-6-19) make very clear the difference between the two-phonon and one-phonon transitions which we have anticipated in Chapter 4 and in the present chapter. The difficulty in applying such formulas lies in the necessity of summing over $\boldsymbol{\kappa}$ which implies a more or less complete knowledge of the phonon dispersion throughout the Brillouin zone.

An approximation which is conceptually very useful for simplifying the description of the two-phonon bands was introduced by Johnson and Cochran.[54] This approximation neglects the dependence of the expression $|\mu_{ij}(\boldsymbol{\kappa})|^2/\omega_i(\boldsymbol{\kappa})\omega_j(\bar{\boldsymbol{\kappa}})$

53. M. Born and K. Huang, *Dynamical Theory of Crystal Lattices*, Oxford University Press, 1954.
54. F. A. Johnson and W. Cochran, Int. Conf. on Semiconductors, Exeter, 1962, page 498.

upon $\boldsymbol{\kappa}$, and attributes the principal intensity variation to the dependence of ε'' upon the delta functions. Then for low temperatures, at which $\bar{v}_i(\boldsymbol{\kappa}) = \bar{v}_j(\bar{\boldsymbol{\kappa}}) = 0$, one may use as the frequency dependent factor,

$$\sum_{\kappa} \delta[-\omega + \omega_i(\boldsymbol{\kappa}) + \omega_j(\bar{\boldsymbol{\kappa}})] = \int_{\mathrm{BZ}} \delta[-\omega + \omega_i(\boldsymbol{\kappa}) + \omega_j(\bar{\boldsymbol{\kappa}})]\, d\boldsymbol{\kappa} = g_{ij}(\omega) \qquad (6\text{-}6\text{-}20)$$

where $g_{ij}(\omega)$ is the joint density of modes, and physically represents the number of modes whose frequencies $\omega = \omega_i(\boldsymbol{\kappa}) + \omega_j(\bar{\boldsymbol{\kappa}})$ lie in the interval ω, $\omega + d\omega$. In the simple non-dispersive case, where $\omega_i(\boldsymbol{\kappa}) = \omega_i$ and $\omega_j(\bar{\boldsymbol{\kappa}}) = \omega_j$, we must clearly have

$$\int_{\mathrm{BZ}} \delta[-\omega + \omega_i(\boldsymbol{\kappa}) + \omega_j(\bar{\boldsymbol{\kappa}})]\, d\boldsymbol{\kappa} = \delta(-\omega + \omega_i + \omega_j) \int_{\mathrm{BZ}} d\boldsymbol{\kappa} = \delta(-\omega + \omega_i + \omega_j) N^3 \qquad (6\text{-}6\text{-}21)$$

since there are exactly N^3 distinct values of $\boldsymbol{\kappa}$ in the Brillouin zone. In this trivial case we therefore find that

$$g_{ij}(\omega) = N^3 \delta(-\omega + \omega_i + \omega_j) \qquad (6\text{-}6\text{-}22)$$

and could proceed to discuss the molecular information available from the dielectric integral $\int \omega\varepsilon''(\omega)\, d\omega$ over the band. Unfortunately, the mechanically harmonic model fails us at this point, and we must now examine the consequences of higher order terms in the potential energy.

The effects of cubic and quartic terms in the potential energy have been studied by a number of workers, of which we shall cite the results of Wallis and Maradudin.[55] First, so far as the fundamental transition is concerned, the anharmonicity introduces temperature and frequency dependent shift, $\Delta(\omega, T)$, and damping, $\Gamma(\omega, T)$, terms which modify the classical formula (5-3-10) so that one has

$$\varepsilon''(\omega) = S_i \omega_i^2 2\Gamma_i \omega_i / [(\omega_i^2 + 2\Delta_i \omega_i - \omega^2)^2 + 4\Gamma_i^2 \omega_i^2] \qquad (6\text{-}6\text{-}23)$$

in place of

$$\varepsilon''(\omega) = S_i \omega_i^2 \gamma \omega / [(\omega_i^2 - \omega^2)^2 + \gamma^2 \omega^2] \qquad (6\text{-}6\text{-}24)$$

Fortunately, both $\Delta_i(\omega, T)$ and $\Gamma_i(\omega, T)$ are small compared with ω_i.

Comparison of (6-6-24) with (6-6-23) shows the limitations of the classical damped harmonic oscillator model introduced in Chapter 5. First, the actual resonant frequency is not precisely ω_i, the harmonic value, but approximately $\omega_i + \Delta_i$; there may therefore be some temperature dependence. Second, putting $2\Gamma_i = \gamma$ does not quite reproduce the classical form, in which γ, itself assumed independent of frequency, always appears multiplied by ω, whereas the present model associates $2\Gamma_i$ with the constant ω_i. Accurate reflection or other measure-

55. R. F. Wallis and A. A. Maradudin, *Phys. Rev.*, **125,** 1277 (1962).

ments designed to obtain ε'' (and ε') are hard to perform if the requirement is large relative accuracy at frequencies ω much different from ω_i. In this way one can understand the successful approximation using the classical equation, but spectra are frequently encountered in which significant deviation from the classical form occurs, and such deviations presumably may be attributed to the different frequency dependence of $2\Gamma_i\omega_i$ as compared with $\gamma\omega$. It is particularly striking that reflection spectra often exhibit dips inconsistent with the classical formula, and that these are associated with maxima in $\Gamma_i(\omega)$ arising, as we shall see presently, from maxima in certain joint density of modes distributions.

We shall only discuss the forms of $\Gamma(\omega, T)$ and $\Delta(\omega, T)$ in the low temperature limit $T \to 0$, referring the reader to the original literature for more detailed equations and derivatives. The damping constant in this limit ($T = 0$ is understood) takes the form

$$\begin{aligned}\Gamma_i(\omega) &= \frac{3\pi}{\hbar} \sum_{\kappa_1 j_1} \sum_{\kappa_2 j_2} \frac{\hbar}{2\omega_i(0)} \frac{\hbar}{2\omega_{j_1}(\boldsymbol{\kappa}_1)} \frac{\hbar}{2\omega_{j_2}(\boldsymbol{\kappa}_2)} \\ &\quad \times |F(0i, \boldsymbol{\kappa}_1 j_1, \boldsymbol{\kappa}_2 j_2)|^2 \, \delta(-\omega + \omega_{j_1}(\boldsymbol{\kappa}_1) + \omega_{j_2}(\boldsymbol{\kappa}_2)) \\ &\simeq \frac{3\pi}{\hbar} \frac{\hbar}{2\omega_i} \sum_{j_1 j_2} \left(\frac{\hbar}{2\omega_{j_1}} \frac{\hbar}{2\omega_{j_2}} \right) |F(0i, j_1, j_2)|^2 g_{ij}(\omega) \end{aligned} \tag{6-6-25}$$

First, we observe that this result could have been obtained approximately from Fermi's celebrated "Golden Rule," which affirms that the rate of a process $m \to n$ is

$$\text{Rate} = \frac{2\pi}{\hbar} |H_{mn}|^2 \rho_f(E_m^\circ) \tag{6-6-26}$$

This has the meaning that the transition rate (actually from a sharp initial state m to a continuum of final states f), depends upon the transition matrix element H_{mn} and the density of final states, ρ_f, whose energy equals that of the initial state. This is the significance of either the sum of the delta functions over $\boldsymbol{\kappa}$ values, or of the approximate form which involves the joint density of modes, $g_{ij}(\omega)$. Note the close similarity to the previous analysis of combination transitions for the harmonic model.

We have assumed that the cubic part of the potential may be written as

$$V_3 = \frac{1}{6} \sum_{\kappa i} \sum_{\kappa_1 j_1} \sum_{\kappa_2 j_2} F(\boldsymbol{\kappa} i\,;\, \boldsymbol{\kappa}_1 j_1\,;\, \boldsymbol{\kappa}_2 j_2) Q_i(\boldsymbol{\kappa}) Q_{j_1}(\boldsymbol{\kappa}_1) Q_{j_2}(\boldsymbol{\kappa}_2) \tag{6-6-27}$$

but since we are interested in the decay of the infrared active phonon whose normal coordinate is $Q_i(0)$, the H_{mn} matrix elements involve only the force constants specified in (6-6-27).

In order to complete this discussion, we need to give an account of the effect of the mechanically anharmonic terms on the intensities of the two-phonon bands. The presence of a term such as $F_{ijk} Q_i(\mathbf{0}) Q_j(\boldsymbol{\kappa}) Q_k(\bar{\boldsymbol{\kappa}})$ in the potential energy mixes the harmonic oscillator states $v_i(\mathbf{0}) v_j(\boldsymbol{\kappa}) v_k(\bar{\boldsymbol{\kappa}})$ and $[v_i(\mathbf{0}) \pm 1][v_j(\boldsymbol{\kappa}) \pm 1][v_k(\bar{\boldsymbol{\kappa}}) \pm 1]$

with the result that the linear, i.e., electrically harmonic term in the dipole moment expansion, $\boldsymbol{\mu}_i Q_i(\mathbf{0})$, will give rise to a non-vanishing matrix element between the states which are described in the harmonic oscillator limit as $v_j(\boldsymbol{\kappa})v_k(\bar{\boldsymbol{\kappa}})$ and $[v_j(\boldsymbol{\kappa}) \pm 1][v_k(\bar{\boldsymbol{\kappa}}) \pm 1]$. Even in small polyatomic molecules, the analogous contributors to combination intensities have rarely been sorted out, so it is not surprising that in the more complex dynamical realm of crystals little is yet known about the relative importance of these two origins of overtone and combination intensity. In crystals, some distinctive and interesting effects arise in consequence of the mechanical anharmonicity. First, consider the limiting case of no dispersion. Then the frequency of the transition $\Delta v(\boldsymbol{\kappa}) = 1$, $\Delta v(\bar{\kappa}) = 1$ for $\boldsymbol{\kappa} \neq 0$ will be precisely twice that of the fundamental transition for which $\Delta v(\mathbf{0}) = 1$. But the frequency of the transition $\Delta v(\mathbf{0}) = 2$ will not be 2ω but instead $2\omega + X$, where X is a (usually) negative anharmonicity parameter. Ron and Hornig[56] described such a case for HCl in the language of molecular transitions, i.e., $\Delta v = 1$ for two different molecules compared with $\Delta v = 2$ for one molecule—an appropriate language in the limit of no dispersion. An infrared study of crystalline CO_2 by Dows and Schettino[57] shows that the combination region $\nu_3 + \nu_1$ and $\nu_3 + 2\nu_2^0$ (recall that ν_1, $2\nu_2^0$ is a Fermi doublet) consists of intense, sharp lines at 3600 and 3708 cm^{-1}, which were assigned as transitions in which a single molecule is excited by $\Delta v_3 = 1$, $\Delta v_1 = 1$ or $\Delta v_3 = 1$, $\Delta v_2 = 2$, thus involving the molecular anharmonicity. But these sharp lines were both accompanied by broader absorption at higher frequencies which is clearly of the two-phonon type, i.e., $\Delta v_3(\boldsymbol{\kappa}) = 1$, $\Delta v_1(\bar{\boldsymbol{\kappa}}) = 1$, since it maps the dispersion of the fundamentals. As we have noted previously (Section 6-4), the case of mechanical anharmonicity may perhaps be discussed more advantageously in the exciton language; in this case, the sharp lines could be described as bound ($\boldsymbol{\kappa} = 0$) biexcitons. Similar effects have been noted in the Raman spectrum of NH_4Cl.[58] The detailed dynamical and intensity analysis of bands of this sort where both dispersion and mechanical anharmonicity play a role can be carried out by applying Green's function methods as in Section 7-1, in this case to the interaction between the sharp state, $v = 2$ in a single molecule, and the band for which $v = 1$ in two different molecules.

(e) Examples

In previous sections the dispersion surfaces of multiphonon states have been discussed. Also, the concept of a joint density function for overtone and combination states was introduced, and some general properties of these functions and their critical points were discussed.

56. A. Ron and D. F. Hornig, *J. Chem. Phys.*, **39**, 1129 (1963).
57. D. A. Dows and V. Schettino, *J. Chem. Phys.*, **58**, 5009 (1973).
58. V. S. Gorelik, G. G. Mitin, and M. M. Sushchinskii, *Sov. Phys. Solid State*, **16**, 1019 (1974). Examples of such sharp lines accompanying two-phonon spectra appear in Fig. 6-1 at 725 and 2800 cm^{-1}.

Dispersion curves have been obtained for only a few molecular crystals. Using simple atom–atom repulsion potentials and/or molecular quadrupole–quadrupole interactions, Schnepp and coworkers have calculated the dispersion of lattice modes of the N_2, CO, and CO_2 crystals, which have similar structures.[59,60] The calculations have been carried out both with and without coupling between translational and librational modes. In either case, the essential features of the derived curves are probably correct, in that the observed frequencies at $\mathbf{k} = 0$ are in reasonable agreement with experiment, which substantiates the reasonableness of the several intermolecular potentials that have been assumed in these calculations.

In Section 6-3(b) the degeneracies at $\mathbf{k} = 0$ were detailed using the CO_2 crystal as an example. At the corner of the simple cubic BZ, the so-called *R*-point, there are two kinds of extra degeneracy, both of which derive from time reversal, as discussed in Appendix X. The calculated dispersion curves also demonstrate the correct degeneracy at the several BZB; thus the symmetrization has been correct.

For this set of crystals multiphonon spectra have been obtained only for CO_2 and, in contrast to the calculations described above, these spectra have been of several combinations of the *internal* modes, $\nu_1 + \nu_2$ and $\nu_1 + \nu_3$.[57,61]

In order to calculate the frequencies of these modes, Dows and Schettino used tensor (5-7-5). A dipole-wave summation is thus called for, which gives rise to problems of conditional convergence and shape dependence at $\mathbf{k} = 0$, as discussed in Section 5-7. These problems were met by explicit summation over spherical samples of 12 Å and 25 Å radius, respectively. As experience with the deWette–Schacher procedure[62] has taught, the results are not very sensitive to this change of radius; they are, however, for changes in shape. This effect can be studied using the continuum (Lorentz) approximation, as discussed in Section 6-5.

Once the dispersion curves for ν_2 and ν_3 have been obtained, the one-phonon density of states in each mode can be obtained by direct sampling throughout the BZ, or more concisely, in one octant of it. If it is assumed that the transition dipoles in the combinations $\nu_1 + \nu_2$ and $\nu_1 + \nu_3$ depend only on the coordinates q_2 or q_3, then the two-phonon frequency distributions should be those of the one-phonon distributions, ν_2 or ν_3, respectively. The width of these distributions agrees with the observed band widths for the several combination bands. Other than this comparison, however, no use has been made of the calculated, two-phonon frequency distribution functions for CO_2. No critical point analysis has been made, for example, although kinks are visible in the two-phonon spectra.

59. *Pa*3 in the case of CO_2, $P2_13$ in the case of N_2 and CO. The difference is the lack of a center of symmetry in the N_2 and CO crystals. (T. H. Jordan, H. W. Smith, W. E. Streib, and W. N. Lipscomb, *J. Chem. Phys.*, **41**, 756 (1969).)
60. These calculations have been thoroughly reviewed by O. Schnepp and N. Jacobi in Reference 14.
61. D. A. Dows and V. Schettino, *Spectrochim. Acta*, **30A**, 1451 (1974).
62. D. B. Dickmann, M.S. Thesis, U.S. Naval Postgraduate School, 1966.

Three other dispersion curve calculations on crystals of small molecules have been carried out. The subject of the first of these is the mineral portlandite, or $Ca(OH)_2$, as studied by Oehler and Günthard,[63] and the second is of CHI_3, studied by Neto, Oehler, and Hexter.[64] The subject of the third calculation is the SiF_4 crystal, recently reported by Schettino.[65] In the first two of these crystals, the spectrum is dominated by a high frequency internal mode, weakly coupled to all other vibrations. Due to the high frequency factoring discussed in Section 6-1, the high frequency internal mode has very little dispersion with **k**, and in combination with other modes it acts as a "frequency amplifier." Thus the lattice frequency region is substantially moved as a unit into a wavelength region that is much more experimentally accessible, especially to high-resolution spectrometers (~ 3 μm). Similar spectra have been observed with crystals of $Mg(OH)_2$ and LiOH.[66-71] Accordingly, there exists an abundance of data for the complex spectra of these crystals in this region.

In the first of these examples, the intermolecular potential consisted of a sum of coulombic terms between point changes e_s at the sites of the Ca, O, and H nuclei, subject to the condition $e_O + e_H = e_{Ca}/2$, and repulsive terms of the form r^{-l} between Ca and O, and between O and H atoms, respectively. The values of l were treated as parameters for each interaction and, after five iterations, were found to be $l_{CaO} = 8.605$ and $l_{OH} = 25.645$, the latter being not very reasonable. Also, the coulombic terms yielded relatively slow convergence of the calculated frequencies, as would be expected. The calculated dispersion curves have zero gradients in more or less the correct frequency regions to account for the complex multiphonon spectra, but it cannot as yet be said that the observed spectra have been accounted for in detail.

In the study of iodoform, the potential function consisted of two parts, an intramolecular part plus an intermolecular portion. The former contained only diagonal force constants plus a stretch-bend interaction term. Trial values were transferred from similar molecules (e.g., CH_2I_2), and these were then refined so as to approximately reproduce the frequencies of the internal modes of this crystal at $\mathbf{k} = 0$. The additional terms in the potential function of the crystal consisted of two sets of $I \cdots I$ and one of $I \cdots H$ nearest-neighbor intermolecular stretching coordinates. Successive refinements were carried out until the best reproduction of both the internal and external frequencies of CHI_3 and CDI_3 at $\mathbf{k} = 0$ was obtained (see Section 6-3(c)).

63. O. Oehler and H. H. Günthard, *J. Chem. Phys.*, **48,** 2036 (1968).
64. N. Neto, O. Oehler, and R. M. Hexter, *J. Chem. Phys.*, **58,** 5661 (1973).
65. V. Schettino, in *Proceedings of the International School of Physics "Enrico Fermi,"* Course LV, *Lattice Dynamics and Intermolecular Forces,* S. Califano, Ed., Academic Press, New York, 1975, page 326.
66. R. T. Mara and G. B. B. M. Sutherland, *J. Opt. Soc. Am.*, **43,** 110 (1953); **46,** 465 (1956).
67. R. M. Hexter, *J. Opt. Soc. Am.*, **48,** 770 (1958).
68. K. A. Wickersheim, *J. Chem. Phys.*, **31,** 870 (1959).
69. R. M. Hexter, *J. Chem. Phys.*, **34,** 941 (1961).
70. R. A. Buchanan, E. L. Kinsey, and H. H. Caspers, *J. Chem. Phys.*, **36,** 2665 (1962).
71. R. M. Hexter, *J. Chem. Phys.*, **38,** 1024 (1963).

Translationally symmetrized coordinates were developed from internal, cartesian coordinates defined within each unit cell in the following manner. Since C_6^6 is non-symmorphic, the transformation is of the form

$$S_\eta(\mathbf{k}) = N^{-3/2} \sum_{R \subset H(\mathbf{k})} \chi^{(\eta)*}_{\{R|\tau_R\}} \{R \mid \tau_R\} s_{i\sigma}(\mathbf{k}) \tag{6-6-28}$$

in which $\{R|\tau_R\}s_{i\sigma}(\mathbf{k})$ stands for the ith internal coordinate into which the translationally symmetrized coordinate $s_{i\sigma}(\mathbf{k})$ is transformed by $\{R \mid \tau_R\}$, an element of $H(\mathbf{k})$. Note that (6-6-28) is a form of (4-5-22). $\chi^{(\eta)}_{\{R|\tau_R\}} = \beta^{(\eta)}(R)\chi^{(l)}_R$ is the character of the ηth irreducible multiplier representation of the group $H(\mathbf{k})$, and $\chi^{(l)}_R$ is the character of the lth irreducible representation of the point group $H_0(\mathbf{k})$, whose operations are $\{R|0\}$. In the case of $\mathbf{k} = (0, 0, \mu)$ $[0 < \mu < \frac{1}{2}]$, $H_0(\mathbf{k}) = C_6$ and the $\beta^{(\eta)}(R)$ can be found from Table 6-15.

The secular determinant, $|\mathbf{GF} - \mathbf{E}\lambda| = 0$, is constructed by writing

$$\mathbf{G} = \mathbf{DD}^\dagger \tag{6-6-29}$$

where $\mathbf{D}$ gives the transformation from mass-weighted cartesian coordinates to the internal coordinates S_j. Coefficients in the S_j are arranged to form the $\mathbf{U}$ matrix, by the use of which the secular determinant is factored to give the roots $\omega_j(\mathbf{k})$. These are then plotted in Figs. 6-8 (*a*), (*b*), (*c*).

Table 6-15 Space group C_6^6—irreducible representations for the groups $H(0, 0, \mu)$
$\Gamma = (0, 0, 0)$; $\Delta = (0, 0, \mu)$; $A = (0, 0, \frac{1}{2})$

		$\{E\|000\}$	$\{C_6\|00\frac{1}{2}\}$	$\{C_3\|000\}$	$\{C_2\|00\frac{1}{2}\}$	$\{\bar{C}_3\|000\}$	$\{\bar{C}_6\|00\frac{1}{2}\}$
Γ:	A	1	1	1	1	1	1
	B	1	-1	1	-1	1	-1
	E_{1+}	1	ε	$-\varepsilon^*$	-1	$-\varepsilon$	ε^* †
	E_{1-}	1	ε^*	$-\varepsilon$	-1	$-\varepsilon^*$	ε †
	E_{2+}	1	$-\varepsilon^*$	$-\varepsilon$	1	$-\varepsilon^*$	$-\varepsilon$ †
	E_{2-}	1	$-\varepsilon$	$-\varepsilon^*$	1	$-\varepsilon$	$-\varepsilon^*$ †
Δ:	A	1	β	1	β	1	β
	B	1	$-\beta$	1	$-\beta$	1	$-\beta$
	E_1'	1	$\beta\varepsilon$	$-\varepsilon^*$	$-\beta$	$-\varepsilon$	$\beta\varepsilon^*$
	E_1''	1	$\beta\varepsilon^*$	$-\varepsilon$	$-\beta$	$-\varepsilon^*$	$\beta\varepsilon$
	E_2'	1	$-\beta\varepsilon^*$	$-\varepsilon$	β	$-\varepsilon^*$	$-\beta\varepsilon$
	E_2''	1	$-\beta\varepsilon$	$-\varepsilon^*$	β	$-\varepsilon$	$-\beta\varepsilon^*$
A:	A	1	i	1	i	1	i †
	B	1	$-i$	1	$-i$	1	$-i$ †
	E_1'	1	$i\varepsilon$	$-\varepsilon^*$	$-i$	$-\varepsilon$	$i\varepsilon^*$ †
	E_2'	1	$-i\varepsilon^*$	$-\varepsilon$	i	$-\varepsilon^*$	$-i\varepsilon$ †
	E_1''	1	$i\varepsilon^*$	$-\varepsilon$	$-i$	$-\varepsilon^*$	$i\varepsilon$ †
	E_2''	1	$-i\varepsilon$	$-\varepsilon^*$	i	$-\varepsilon$	$-i\varepsilon^*$ †

where $\varepsilon = e^{2\pi i/6}$ and $\beta = e^{i\kappa \cdot \tau_R} = e^{\pi i \mu}$

† Degenerate pairs

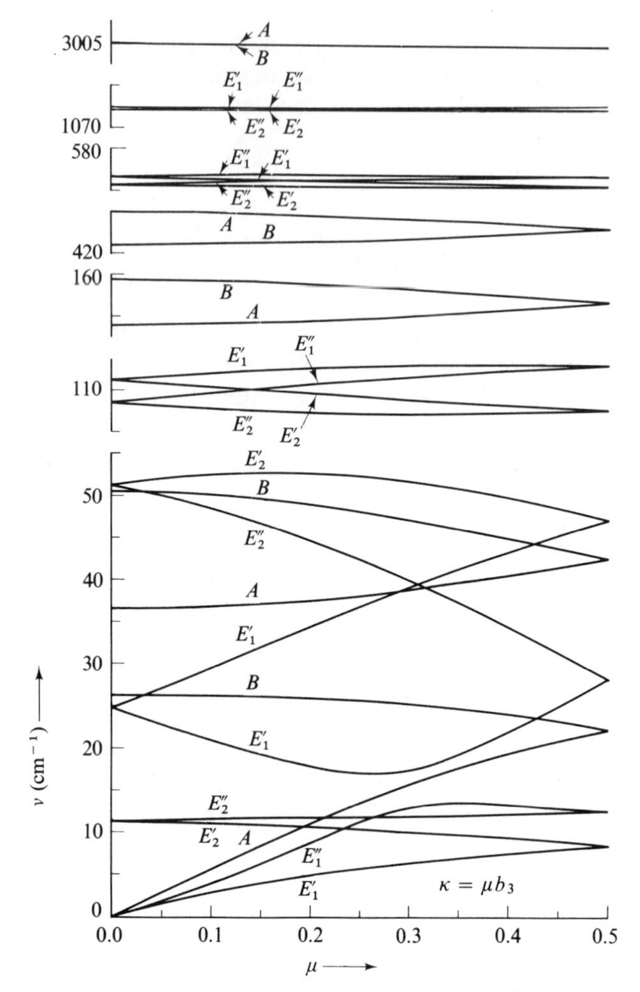

Figure 6-8(*a*) Dispersion curves for $\mathbf{k} = (0, 0, \mu)$ (along the $\Gamma \rightarrow A$ line) in CHI_3 (from Ref. 64)

Numerous avoided crossings may be observed. Without a calculation of the density of states, the extent to which any of these may be singular c.p.'s is unknown. It is also interesting to note the slight mixing of the lattice modes with low-lying internal fundamentals, particularly at $\mathbf{k} \neq 0$. On the other hand, the high-frequency factoring forecast earlier in this section is particularly evident for $\omega_{\text{C—H}}$ which, as a result, shows negligible dispersion.

In addition, numerous *allowed* crossings are to be noted in Fig. 6-8(*a*). This is a result of the lifting of degeneracies at Γ as $\boldsymbol{\kappa}$ becomes finite along the sixfold axis of the crystal. Various states become degenerate again at the top of the BZ ("sticking of branches"). The matter is discussed in detail in Appendix X.

At the point M of the BZ $[\mathbf{k} = (\frac{1}{2}, 0, 0)]$, $H(\mathbf{k}) = C_2$. All phonon states at M are therefore either symmetric or antisymmetric, A or B. Using (4-7-11) we have

$$A \times A = B \times B = A$$

to which only the z-component of $\mathbf{V}_k$ belongs ($\mathbf{z}$ is parallel to the sixfold axis of the crystal). Hence zero gradients for all $\omega(k)$ states are predicted in the plane perpendicular to $\mathbf{z}$, as seen in Fig. 6-8(b) along the line $\Gamma \to M$.

At the point K of the BZ $[\mathbf{k} = (\frac{1}{2}, \frac{1}{2}, 0)]$, $H(\mathbf{k}) = C_3$. All phonon states therefore belong to either the irreducible representations A or E of C_3. Since

$$A \times A = A$$

and

$$E \times E = 2A + E$$

zero gradients are also possible at K, but only for those $\omega(\mathbf{k})$ states which are non-degenerate there (Fig. 6-8(c)).

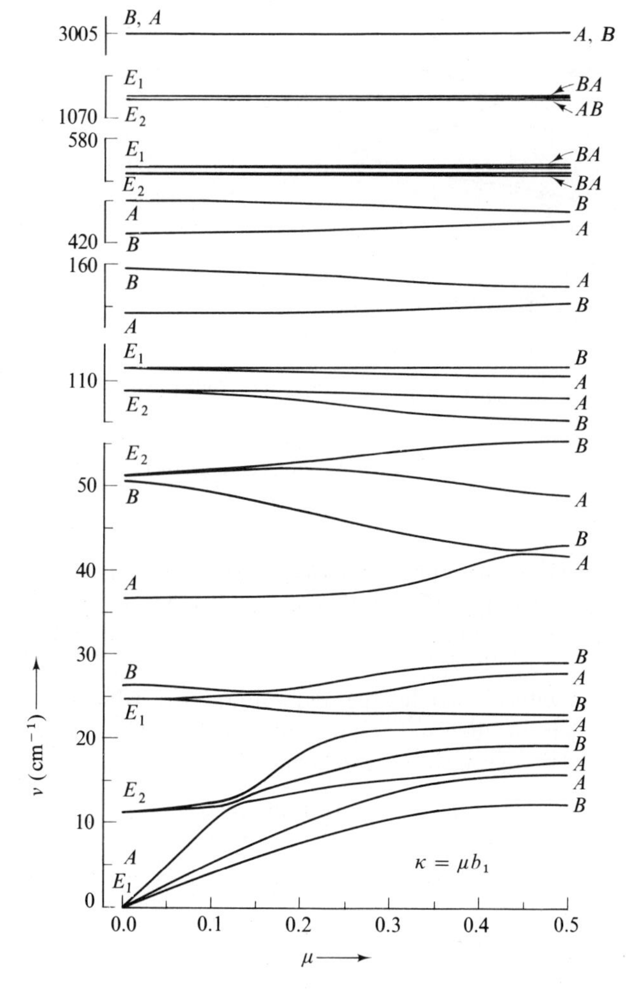

Figure 6-8(b) Dispersion curves for $\mathbf{k} = (\mu, 0, 0)$ (along the $\Gamma \to M$ line) in CHI_3 (from Ref. 64)

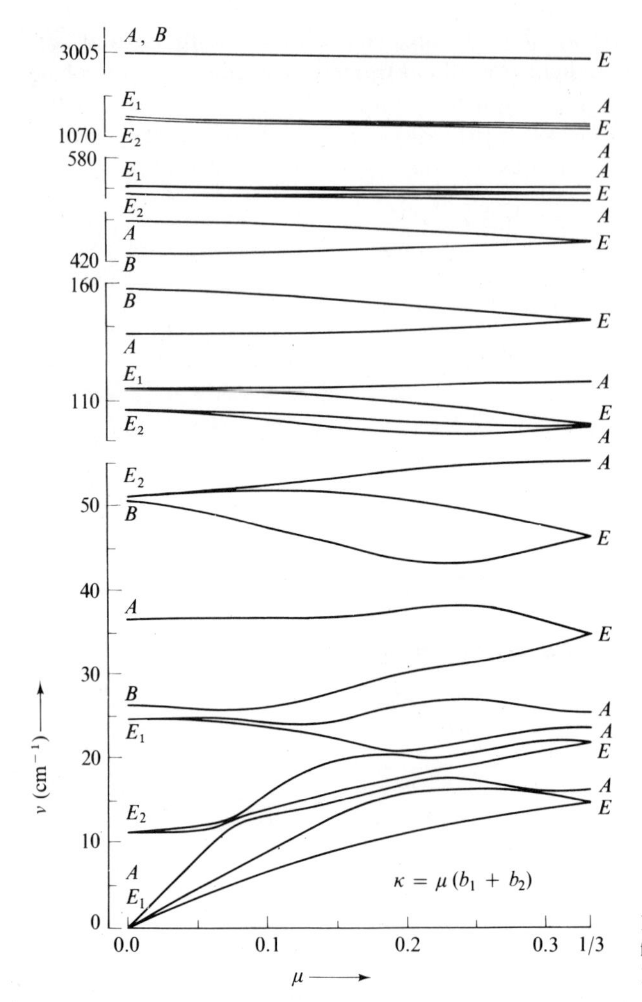

Figure 6-8(*c*) Dispersion curves for $\mathbf{k} = (\mu, \mu, 0)$ (along the $\Gamma \to K$ line) in CHI_3 (from Ref. 64)

Since the density of states can be high for phonon states $\omega(\mathbf{k})$ with zero gradient, some ten frequencies in the CH stretching region (both above and below ω_{CH}) have been assigned to two-phonon transitions, taking place at either the point M or the point K.

By forming direct products with the irreducible representation to which the CH fundamental belongs (at $-\mathbf{k}$ in additive and at $\mathbf{k}$ in subtractive combinations), predictions of the polarization of these transitions can be made. Thus, at point M:

$$A \times A = B \times B = \text{ir}, \parallel \text{polarization; Raman}$$

$$A \times B = \text{ir}, \perp \text{polarization; Raman}$$

and at the point K:

$$A \times A = \text{ir}, \parallel \text{polarization; Raman}$$

Table 6-16 Two-phonon absorption frequencies in the CH stretching region. (From N. Neto, O. Oehler, and R. M. Hexter, *J. Chem. Phys.*, **58**, 5661 (1973))

Calculated	Observed[a]
$2982(A) \pm 22(B) = 3004$ and 2960(ir, $\perp$)	3008(vs, $\perp$) and 2959(vs, $\perp$)
$2982(A) \pm 29(B) = 3011$ and 2953(ir, $\perp$)	3015(vs, $\perp$) and 2952(vs, $\perp$)
$2982(B) \pm 42(A) = 3024$ and 2940(ir, $\perp$)	3021(vs, $\perp$) and 2947(vs, $\perp$)
$2982(B) \pm 43(B) = 3025$ and 2939(ir, $\parallel$)	3020(vw, $\parallel$) and 2946(vw, $\parallel$)
$2982(B) \pm 55(B) = 3037$ and 2927 (ir, $\parallel$)	3046(m, $\parallel$) and 2920(vw, $\parallel$)

a. vs: very strong, vw: very weak.

The results, taken from the calculated dispersion curves, are compared with the observed spectrum in Table 6-16.

The last crystal of small molecules for which dispersion curves have been calculated and which we shall discuss is the SiF_4 crystal, recently published by Schettino.[65] The treatment is similar to that used by Dows and Schettino in their study of CO_2.[57,61] As in that study, the principal result was that the width of the calculated $g(\omega)$ compared favorably with the observed two-phonon band widths. Of greater interest, however, is the fact that Schettino calculated the dispersion curve of ν_3 of SiF_4 for both a spherically-shaped and a slab-shaped crystal, and found that at $\mathbf{k} \neq 0$, the two sets of curves were very much the same. While this is to be expected, there are few other examples of calculated (or experimental) dispersion curves in the literature which demonstrate this fact so well.

The effect is related to the connection between the *LO–TO* splitting obtained from both microscopic and macroscopic theory, as discussed in Section 5-7. In one way or another it has been discussed by a number of authors,[72–81] but most directly and succinctly by Cohen and Keffer.[73] As they show, the dipole field propagation tensor for a sphere of finite dimension for $k \neq 0$ has the form

$$T^{\mu\nu}(\mathbf{k}) = \frac{4\pi}{3}\left(\delta^{\mu\nu} - \frac{3k^{\mu}k^{\nu}}{k^2}\right)\left[1 - \frac{3j_1(kR)}{kR}\right] \qquad (6\text{-}6\text{-}30)$$

72. E. W. Kellerman, *Trans. Roy. Soc.* (*London*), **238,** 513 (1940).
73. M. H. Cohen and F. Keffer, *Phys. Rev.*, **99,** 1128 (1955).
74. H. B. Rosenstock, *Phys. Rev.*, **121,** 416 (1961).
75. A. A. Maradudin and G. H. Weiss, *Phys. Rev.*, **123,** 1968 (1961).
76. T. H. K. Barron, *Phys. Rev.*, **123,** 1995 (1961).
77. D. Fox and R. M. Hexter, *J. Chem. Phys.*, **41,** 1125 (1964).
78. R. E. Frech and J. C. Decius, *J. Chem. Phys.*, **51,** 1536 (1969).
79. E. R. Cowley, *J. Chem. Phys.*, **52,** 5493 (1970).
80. R. E. Frech and J. C. Decius, *J. Chem. Phys.*, **52,** 5494 (1970).
81. S. Aung and H. L. Strauss, *J. Chem. Phys.*, **58,** 2737 (1973).

where $j_1(\rho)$ is the first spherical Bessel function,

$$\frac{j_1(\rho)}{\rho} = \frac{\sin \rho}{\rho^3} - \frac{\cos \rho}{\rho^2}$$

and R is the radius of the sphere.

Equation (6-6-30) can be compared with (5-7-8), its limit as $\mathbf{k} \to 0$. In Section 5-7 we showed how (5-7-26) gives the same result for an infinite crystal as is given by macroscopic theory. Inspection of (6-6-30) shows that the finite sum differs from the infinite sum only for $\mathbf{k}$'s such that $kR \leq 10$. Thus, dipole-wave sums become independent of the origin, and therefore of $\mathbf{k}$, for $kR > 10$.

Since the second term in (6-6-30) approaches unity for $kR < 10$, the field propagation tensor becomes effectively the same as its limiting value at $\mathbf{k} \to 0$, which we have seen gives the same result as a slab-shaped crystal. Once again, $\mathbf{k}$ spoils the symmetry of the cubic crystal, and yields the results of a particular specimen shape, namely, the plane-parallel slab.

CHAPTER

SEVEN

SOLID SOLUTIONS, MATRIX ISOLATION, AND IMPURITY PHENOMENA

In this chapter we recognize that the real systems we deal with are sometimes impure. The introduction of impurities into an otherwise pure sample may be natural, accidental, or intentional. In the first and last cases we include isotopic impurities. As many a molecular scientist has discovered, the presence of isotopes is usually of great advantage because they rarely make for an important difference in chemical properties and, at the same time, the mass difference is usually so slight as to enable analysis of its presence as a small perturbation. Isotope effects are therefore excellent examples of the perturbation approach first discussed in Section 1-3.

Other kinds of impurities can represent more severe modifications of the system. An extreme example would be the isolation of a large molecule, such as benzene, in a rare gas matrix,[1] a substitution which cannot but help to significantly disrupt the host lattice. Intermediate examples may be the substitution of a small tetrahedral molecule for a large inert gas atom—such as CH_4 for Xe—or the substitution of a diatomic for a simple anion in an ionic crystal, particularly where the overall van der Waals radii are similar. In these cases, the primary goal is the study of the motions of the substitute molecule—in isolation from others of its own kind. For example, by this method, the rate of nuclear spin conversion of H_2O at 6.5° has been carefully studied.[2] In contact with other water molecules this would have been impossible. In another example, the restrictions imposed upon free rotation of CN^- as a substitutional impurity in several alkali halides were investigated.[3] In this experiment the CN^- was used as a probe of the effective symmetry of the crystal field.

1. Y. Diamant, R. M. Hexter, and O. Schnepp, *J. Mol. Spectry*, **18**, 158 (1965).
2. H. P. Hopkins, Jr., R. F. Curl, Jr., and K. S. Pitzer, *J. Chem. Phys.*, **48**, 2959 (1968).
3. W. D. Seward and V. Narayanamurti, *Phys. Rev.*, **148**, 463 (1966).

In general, we shall see that experiments carried out with impurities introduced into crystal lattices are designed either to study the motion of the impurity under the conditions imposed by the lattice, or to study the effects of the impurity on the lattice. In the spirit of the perturbation approach, we shall begin with a qualitative study of impurity theory, in which it is supposed that the disturbance of the lattice is slight. Even so, we shall find out that the effects can be significant, such as the induction of $k \neq 0$ modes.

7-1 IMPURITY THEORY

In molecular dynamics the slightest perturbation of which we know is the replacement of one atom by an isotope (Section 2-5). Isotopic substitution changes only G matrix elements, not those of F, although the reduced symmetry attendant upon the substitution may remove degeneracies and thus allow mixings of states which belong to different species in the unsubstituted molecules.

In crystals, too, the first effect of an isotopic substitution is the removal of translational symmetry. Momentum conservation requires that in any transition

$$\mathbf{k}\,(\text{photon}) = \mathbf{k}_f\,(\text{crystal}) - \mathbf{k}_i\,(\text{crystal}) \tag{7-1-1}$$

and translational symmetry yields the *approximate* selection rule,

$$\mathbf{k}\,(\text{photon}) \cong 0 \tag{7-1-2}$$

The destruction of translational symmetry eliminates the effective restriction to small **k**. We thereby immediately recognize a new tool: impurity-activated spectroscopy. Without selection rules, the spectrum will essentially map the crystal's density of phonon states.

But is it true that there are no selection rules? Actually, the lack of any selection rules is itself a selection rule, which like any other, simply states what is *possible*. The question really is: to what extent does an impurity *effectively* spoil the symmetry? The answer may be found by examining the mechanics of an isotopically substituted lattice.

Attempting to emphasize only what is essential to our understanding of real systems, we examine the highly artificial, one-dimensional monatomic lattice with a single isotopic substitution. In the unsubstituted lattice, we have previously seen that the LA frequency of a linear monatomic lattice is given by

$$\begin{aligned}\omega_k^2 &= 2\mu f[1 - \cos(2\pi k/N)] = 4\mu f \sin^2(\pi k/N) \\ &= \omega_{LM}^2 \sin^2(\pi k/N)\end{aligned} \tag{3-2-10}$$

Equation (3-2-10) was obtained by translational symmetrization of the secular equation $|FG - E\lambda| = 0$, the local coordinate having been of the internal variety (i.e., a "bond stretch"). In a related problem in molecular dynamics, a perturbation treatment shows that no frequency can be decreased by a decrease of the mass of any atom, as stated by Rayleigh's Principle.[4,5] Although most discussions of this problem in crystal dynamics are expressed in terms of a local displacement coordinate representation and the equation of motion of the displacement, we shall show here how the matrix representation we have been using can give the same results. We illustrate this using a perturbation calculation, and then indicate how an exact calculation gives essentially the same results.

The perturbation approach to the one-dimensional impurity problem was first carried out by Decius,[6] who treated the several situations in which one or more impurities were substituted sequentially in an otherwise homogeneous, one-dimensional lattice. The general procedure is to call the **G** matrix elements of the unperturbed lattice G while those for the isotopic species are called G'. The intermolecular coupling constants, for first- and second-nearest neighbors, are called F and F', respectively. Thus the **G** matrix, insofar as the affected portion of the lattice is concerned, is diagonal with G' terminal and G otherwise. The **F** matrix elements are $F_{tt} = F$; $F_{t,t\pm 1} = F'$. First-order perturbation theory easily gives the familiar results

$$\begin{aligned} \lambda_1^{(1)} &= GF \\ \lambda_{2\alpha}^{(1)} &= G(F \pm F') \\ \lambda_{3\alpha}^{(1)} &= G(F + \sqrt{2}F'\delta_{\alpha,\alpha'}) \end{aligned} \tag{7-1-3}$$

where $\lambda_{i\alpha}$ is the αth root of the secular equation for a sequence of i impurities. Without F', of course, the α roots are degenerate, for each i.

Second-order perturbation theory may then be used to refine these results. Inspection of the product **GF** for each case shows that (7-1-3) must be modified by adding

$$(u_{\alpha 1}^2 + u_{\alpha n}^2)\frac{(GF')(G'F)}{\lambda_{i\alpha}^{(1)} - G'F}$$

where the $u_{\alpha i}$ are elements of the matrix which diagonalizes the first-order determinant. They are, of course, the coefficients of the local displacement coordinates in the expression for the symmetry coordinates of the *substituted* lattice. The final expressions for the perturbed roots are

$$\begin{aligned} \lambda_1^{(2)} &= GF + \frac{2GG'(F')^2}{(G - G')F} \\ \lambda_{2\alpha}^{(2)} &= G(F + F') + \frac{GG'(F')^2}{G(F + F') - G'F} \\ \lambda_{3\alpha}^{(2)} &= G(F + \sqrt{2}F') + \frac{1}{2}\left[\frac{GG'(F')^2}{G(F + \sqrt{2}F') - G'F}\right] \end{aligned} \tag{7-1-4}$$

4. Sec. 2-5(e).
5. Lord Rayleigh, *Theory of Sound*, Vol. I, Dover, New York, 1945; see pages 88, 92a.
6. J. C. Decius, *J. Chem. Phys.*, **23**, 1290 (1955).

where we have shown only the positive roots because of Rayleigh's theorem.

An exact calculation of these roots has been carried out by Decius, Malan, and Thompson,[7] using a procedure first proposed by Lax for application to lattice vibrations.[8] In this procedure, we first recall the form of the roots of the secular equation with first- and second-nearest neighbor interactions:

$$\lambda_k^0 = G(F + F' e^{2\pi ik/N} + F' e^{-2\pi ik/N}) \tag{3-2-12}$$

We now write the Hamiltonian for the perturbed (impure) lattice as $\mathbf{G}^0\mathbf{F} + \mathbf{F}\mathbf{G}'$. $\mathbf{G}'$ is a diagonal matrix with elements $\Delta G = (G_\alpha - G_\beta)(1 - \delta_{\alpha\beta})$, where α = guest and β = host. The secular equation for the mixed crystal becomes

$$(\mathbf{H}^0 + \mathbf{H}')\mathbf{1} = (\mathbf{G}^0\mathbf{F} + \mathbf{F}\mathbf{G}')\mathbf{1} = \lambda\mathbf{1} \tag{7-1-5}$$

or

$$(\lambda\mathbf{E} - \mathbf{H}^0)^{-1}\mathbf{F}\mathbf{G}'\mathbf{1} = \mathbf{1} \tag{7-1-6}$$

with roots λ.

Defining

$$\mathbf{M} = \frac{\mathbf{F}\mathbf{G}'}{\lambda\mathbf{E} - \mathbf{H}^0} \tag{7-1-7}$$

the solutions of (7-1-6) are given by the system of equations

$$\mathbf{M1} = \mathbf{1} \tag{7-1-8}$$

According to (7-1-7), for the case of a single impurity located at $t = 0$ the elements of $\mathbf{M}$ are

$$M_{tt'} = \delta_{t'0}\left(\frac{\Delta G}{NG_\beta}\right)\sum_k\left(\frac{\lambda_k^0}{\lambda - \lambda_k^0}\right)e^{2\pi ikt/N} \tag{7-1-9}$$

since, in the unperturbed lattice

$$\mathbf{H}^0 = \mathbf{L}^0\mathbf{\Lambda}^0(\mathbf{L}^0)^{-1} \tag{7-1-10}$$

where

$$L_{tk}^0 = \left(\frac{G^0}{N}\right)^{1/2} e^{2\pi ikt/N} \tag{7-1-11}$$

and the elements of Λ^0 are given by (3-2-12). Equation (7-1-8) is then satisfied if

$$M_{00} = \frac{\Delta G}{NG_\beta}\sum_k\left(\frac{\lambda_k^0}{\lambda - \lambda_k^0}\right) = 1 \tag{7-1-12}$$

Since N is very large, the summation can be replaced by an integral, following the change of variables $z = e^{2\pi ik/N}$. Equation (3-2-12) then becomes

7. J. C. Decius, O. G. Malan, and H. W. Thompson, *Proc. Roy. Soc.* (*London*), **275A**, 295 (1963).
8. M. Lax, *Phys. Rev.*, **94**, 1391 (1954).

$$\lambda^0(z) = \lambda_0 + \lambda_1(z + z^{-1}) \tag{7-1-13}$$

where

$$\begin{aligned} \lambda_0 &= GF \\ \lambda_1 &= GF' \end{aligned} \tag{7-1-14}$$

The integral to be evaluated is then

$$\frac{\Delta G}{2\pi i G} \oint \frac{\lambda^0(z)}{z[\lambda - \lambda^0(z)]} dz = 1 \tag{7-1-15}$$

We immediately recognize one pole at $z = 0$; the others are the roots of

$$\lambda - \lambda^0(z_\zeta) = 0 \tag{7-1-16}$$

According to Cauchy's residue theorem, the value of the integral (7-1-15) will be given by its residues. That for the pole at $z = 0$ is found by considering (7-1-13). As $z \to 0$, only the term in z^{-1} blows up. Then

$$\lim_{z \to 0} \frac{\lambda^0(z)}{\lambda - \lambda^0(z)} \to -1$$

since λ is fixed as z is varied. Hence, the residue of the pole at $z = 0$ is $-\Delta G/2\pi i G$.

The roots of (7-1-16) occur in pairs, one inside and one outside the unit circle. According to the contour theorem, we need to consider only those within the contour. Given a function $f(z) = \varphi(z)/\psi(z)$, the residue of a pole at $z = z_0$ is $\varphi(z_0)/\psi(z_0)$. By taking $\varphi(z) = \lambda^0(z)$ in the integral of (7-1-15), we find the other residues to be

$$-\frac{\Delta G}{G} \lambda \sum_\zeta \left[z_\zeta \left(\frac{d\lambda^0(z)}{dz} \right)_{z = z_\zeta} \right]^{-1}$$

As a result, (7-1-15) becomes

$$-\frac{\Delta G}{G} \left\{ 1 + \lambda \sum_\zeta \left[z_\zeta \left(\frac{d\lambda^0(z)}{dz} \right)_{z = z_\zeta} \right]^{-1} \right\} = 1 \tag{7-1-17}$$

Since we restrict our interest to those roots of (7-1-16) which are within the unit circle,

$$z_\zeta = \frac{\lambda - \lambda_0}{\lambda_1} \tag{7-1-18}$$

and the derivative in (7-1-17) is easily evaluated from the restricted form of (7-1-13). The result is

$$(\lambda - \lambda_0)^2 - (\Delta G/G_\alpha)^2 \lambda^2 = 4\lambda_1^2 \tag{7-1-19}$$

whose root is

$$\lambda_{1\alpha} = \frac{FG_\alpha[1 + \{\varepsilon^2 + \eta^2 - \varepsilon^2\eta\}^{1/2}]}{1 + \varepsilon} \tag{7-1-20}$$

where

$$\varepsilon = \frac{(G_\alpha - G_\beta)}{G_\alpha} \quad \text{and} \quad \eta = \frac{2F'}{F} \tag{7-1-21}$$

The results of the exact solution (7-1-20) have been compared with those of the second-order perturbation calculation (7-1-4) in studies of isotopically (^{13}C) substituted carbonates of Ba, Sr, and Ca. The calculated frequencies based upon the two methods are practically the same, although the exact calculation is the more accurate.

The function $(\lambda\mathbf{E} - \mathbf{H}^0)^{-1}$ is in fact the Green's function for the system of oscillators we have been discussing. In the case of lattice vibrations, the Green's function approach has been used by Maradudin and coworkers to obtain impurity mode frequencies in a series of papers beginning in 1955 and summarized in 1971.[9] The procedure outlined here was the first use of the Green's function technique in molecular crystals. Although originally intended for internal modes, it is equally applicable to lattice modes, and is equivalent to the methods discussed by Maradudin *et al.*

In the case of the lattice modes of monatomic crystals, it is a consequence of Rayleigh's theorem that the substitution of a single light isotope *raises* a frequency *out of the band*; viz.,

$$\omega^2 = \frac{\omega_L^2}{1 - \varepsilon^2} \tag{7-1-22}$$

where $\varepsilon = (M - M')/M$ and $\omega_L^2 = 2F_0\mu$, the frequency at the top of the band $(k = N/2)$. Equation (7-1-22) is illustrated in Fig. 7-1. On the other hand, the introduction of a single heavy isotope cannot produce a negative frequency (recall that $\omega(0) = 0$), since the Hamiltonian for the perturbed crystal, like that for the pure, is positive definite.

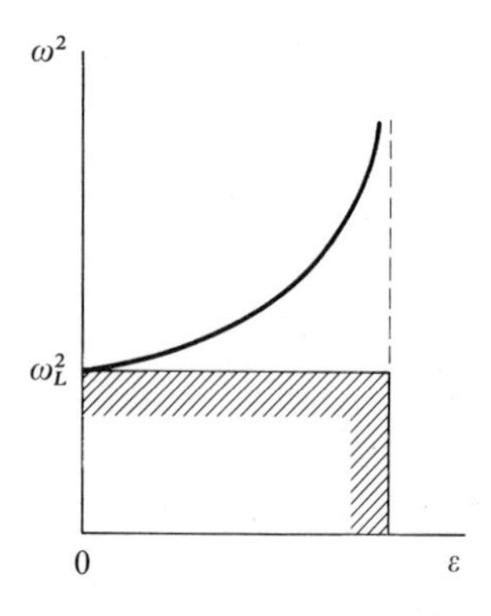

Figure 7-1 The variation of the impurity frequency associated with a light isotope defect in a monatomic linear chain is plotted as a function of the impurity mass. (From R. F. Wallis and A. A. Maradudin, *Progr. Theor. Phys.*, **24**, 1055 (1960))

9. A. A. Maradudin, E. W. Montroll, G. H. Weiss, and I. P. Ipatova, *Theory of Lattice Dynamics in the Harmonic Approximation*, second edition, Suppl. No. 3, *Solid State Physics*, H. Ehrenreich, F. Seitz, and D. Turnbull, Eds., Academic Press, New York, 1971; Section 8.8.

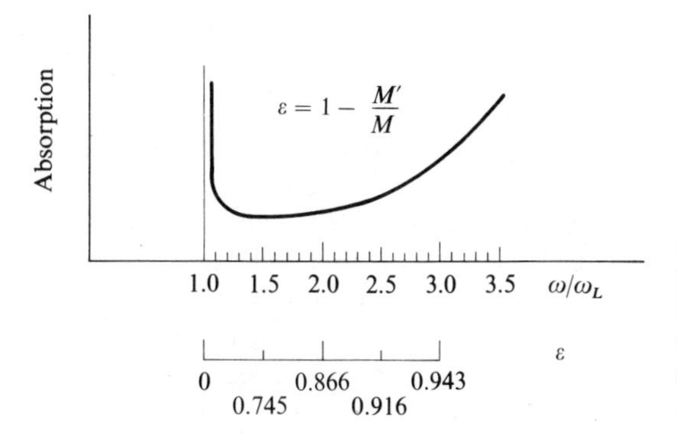

Figure 7-2 Absorption due to the localized vibration mode associated with a light isotope impurity in a diatomic linear chain plotted as a function of the impurity mass. (After R. F. Wallis and A. A. Maradudin, *Progr. Theor. Phys.*, **24,** 1055 (1960))

The impurity mode corresponds to a localized vibration, involving primarily the impurity atom itself. Neighboring atoms do have some amplitude modulated at the impurity mode frequency, but this amplitude dies away quickly with distance from the impurity atom.

The conclusions stated above follow the thorough considerations of Maradudin and coworkers, who have also calculated the relative intensities of the impurity mode as well as those of transitions to the ω_k of the band as a result of the introductions of the impurity atom. The results are shown graphically in Figs. 7-2 and 7-3.

Since no part of the LA is directly accessible by dipole transition in the unsubstituted lattice, and since the introduction of a single impurity atom does not induce substantial spectroscopic intensity at the frequency of the impurity mode, the significant effect of the impurity is the *activation of* ω_L. In real lattices this is also significant, and impurity spectra are useful for the identification of frequencies characteristic of a number of *critical points* (c.p.'s)—i.e., points of high density of states on one or another dispersion surface. We shall discuss this effect more fully below. Before doing so, we complete this discussion of one-dimensional

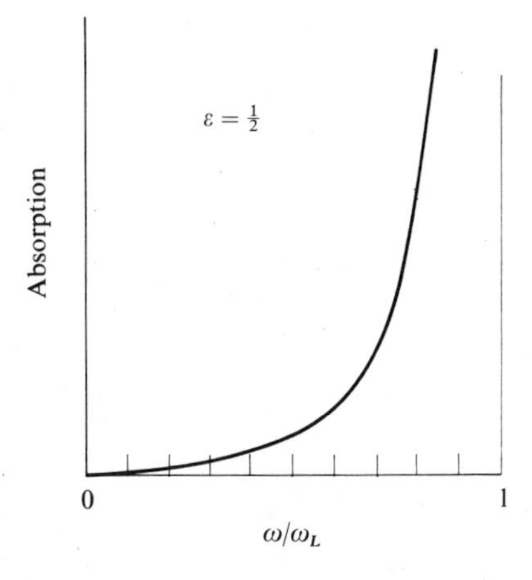

Figure 7-3 The in-band absorption due to a single light isotope impurity in a diatomic linear chain plotted as a function of the frequency. (After R. F. Wallis and A. A. Maradudin, *Progr. Theor. Phys.*, **24,** 1055 (1960))

lattices with an illustration of the effect of isotopic substitution in diatomic lattices. In distinction to the monatomic case, the introduction of a light isotope not only can raise a frequency out of the LA band, but it can do the same out of the LO. Moreover, in the case where the lighter atom of the pair is replaced by a still lighter isotope, the mode in the gap between ω_L and ω_0 (the highest frequency of the LO in the pure crystal) has been *lowered* from the LO band (see Fig. 7-4).

Selection rules for impurity-activated phonon absorption were first discussed by Loudon.[10] The procedure is a straightforward application of correlation. When a substitution for an atom at a lattice site is carried out, the lattice is no longer left invariant by all of the operations of the space group, but instead, only by the operations of the site group. Accordingly, the irreducible representations of the space group become reducible among those of the site group. Actually, it is of considerable economy to restrict this reduction to the irreducible representations of the groups of the wave-vector $H(\mathbf{k})$ at points, lines, or planes of high symmetry in the Brillouin zone, especially because these special points may be critical points in the phonon density of states for reason of symmetry.[11] This reduction is carried out in the usual way, and can be quickly summarized by the use of correlation.

The process will be illustrated with two examples, the fcc and rock salt lattices, both $O_h^5 = Fm3m$. In either case, the site group is O_h. Therefore we need to correlate the irreducible representations of the $H(\mathbf{k})$ at the special points of the space group O_h^5 with those of the point group O_h. The correlation is presented in Table 7-1, which is taken from Ref. 10, except for the addition of point-group representation symbols.*

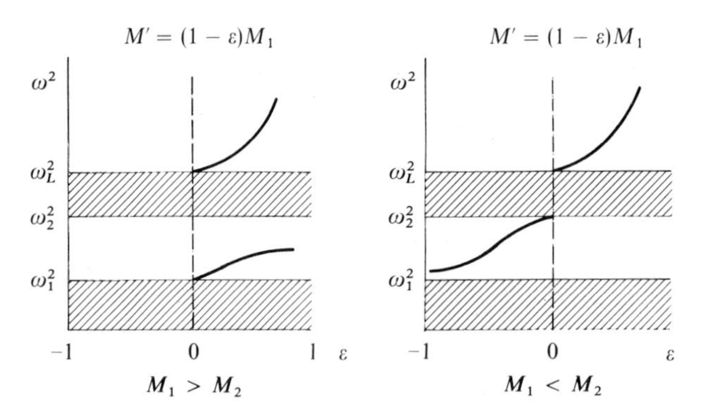

Figure 7-4 The dependence on the impurity mass of the frequencies of the localized vibration modes in a linear diatomic chain when a heavy mass and a light mass atom are replaced by an isotopic impurity, respectively. (From R. F. Wallis and A. A. Maradudin, *Progr. Theor. Phys.*, **24,** 1055 (1960))

10. R. Loudon, *Proc. Phys. Soc.*, **84,** 379 (1964).
11. See Section 4-7(c) and, in particular, Eq. (4-7-11).

* A few minor errors have been corrected.

Table 7-1 Space-group reduction coefficients for the face-centered cubic lattice

O_h			A_{1g}	A_{2g}	E_g	F_{1g}	F_{2g}	A_{1u}	A_{2u}	E_u	F_{1u}	F_{2u}
			Γ_1^+	Γ_2^+	Γ_{12}^+	Γ_{15}^+	Γ_{25}^+	Γ_1^-	Γ_2^-	Γ_{12}^-	Γ_{15}^-	Γ_{25}^-
D_{4h}	X_1^+	A_{1g}	1		1							
	X_2^+	B_{1g}		1	1							
	X_3^+	B_{2g}					1					
	X_4^+	A_{2g}				1						
	X_5^+	E_g				1	1					
	X_1^-	A_{1u}						1		1		
	X_2^-	B_{1u}							1	1		
	X_3^-	B_{2u}										1
	X_4^-	A_{2u}									1	
	X_5^-	E_u									1	1
D_{3d}	L_1^+	A_{1g}	1				1					
	L_2^+	A_{2g}		1		1						
	L_3^+	E_g			1	1	1					
	L_1^-	A_{1u}						1				1
	L_2^-	A_{2u}							1		1	
	L_3^-	E_u								1	1	1
D_{2d}	W_1	A_1	1		1				1	1		
	W_1'	B_2					1				1	
	W_2	A_2				1						1
	W_2'	B_1		1	1			1		1		
	W_3	E				1	1				1	1
C_{4v}	Δ_1	A_1	1		1						1	
	Δ_2	B_1		1	1							1
	Δ_1'	A_2				1		1		1		
	Δ_2'	B_2					1		1	1		
	Δ_5	E				1	1				1	1
C_{3v}	Λ_1	A_1	1				1		1		1	
	Λ_2	A_2		1		1		1				1
	Λ_3	E			1	1	1			1	1	1
C_{2v}	Σ_1	A_1	1		1		1		1	1	1	
	Σ_2	B_2				1	1				1	1
	Σ_3	B_1				1	1				1	1
	Σ_4	A_2		1	1	1		1		1		1
C_{2v}	Z_1	A_1	1	1	2		1				1	1
	Z_2	A_2				1		1	1	2		
	Z_3	B_1				1	1				1	1
	Z_4	B_2				1	1				1	1
C_2	Q_1	A	1	1	2	1	1	1	1	2	1	1
	Q_2	B				2	2				2	2

The representations of the $H(\mathbf{k})$ to which the several phonon branches belong have already been tabulated (Table 4-5). Since the localized vibrations of an impurity atom at a lattice site must belong to F_{1u} of O_h, and since the acoustic branches at any $H(\mathbf{k})$ must belong to the representations of a vector, we see from

Table 7-1 that an impurity at a lattice site can activate direct absorption by the acoustic branch of an fcc at any special point in the BZ, just as was found analytically in one dimension.

It was noted in Section 4-6 that, although the $H(\mathbf{k})$ at certain Brillouin zone boundary points as well as at its center include the inversion operation, the interior of any BZ cannot, since $i\mathbf{k} \neq \mathbf{k}$. It follows that the g, u distinction is lost inside the BZ, although it may be restored at the zone boundaries. In NaCl, the acoustic and optic branches are both u at Γ (see Table 4-5) and, as we learned in Section 4-9, the optic mode is split into an LO and two (degenerate) TO modes by the local electric field created by the motion of the ions themselves. Along the direction $\Lambda = k(111)$, the u-character is lost, since $G(\Lambda) = C_{3v}$. In the interior of the zone, then, along this direction, the acoustic and optic branches can mix. (See the discussion of this mixing in Section 3-3.) At the zone boundary (L), $H(\mathbf{k}) = D_{3d}$. In a particular, assumed interaction model,[12] the mixing is such that at L, the LA and TA of one sublattice are g, while the LO and TO remain u. In this model, then, an impurity at either a Na^+ or a Cl^- site cannot induce spectroscopic activity of the acoustic branches at L.

Substitution of halides by hydride ions in the NaCl lattice creates an impurity known as the U-center. Absorption due to these fall in the region 700–1000 cm^{-1}, depending on the lattice.[13] Side bands of U-centers have been observed and assigned as combination bands of the lattice modes and the localized modes. Since both transform as F_{1u} at Γ, the combination belongs to the direct product $F_{1u} \times F_{1u} = A_{1g} + E_g + F_{1g} + F_{2g}$. The correlation of each of these with the representations of the $H(\mathbf{k})$ at the special points can be found in Table 7-1. In order for the combination to be spectroscopically active, the correlated species must be that of the acoustic or optic branches at $H(\mathbf{k})$. Because of the particular mixing at L described earlier, some of the observed sidebands can be due to the acoustic branches at that special point.

Other examples have been discussed by Loudon.[10]

7-2 ISOTOPIC SOLID SOLUTIONS

A special case of impurity spectra of great importance for the elucidation of site symmetry and the intermolecular potential function is that of an isotopic impurity in a molecular crystal. It was first emphasized by Hrostowski and Pimentel[14] and by Hiebert and Hornig[15] that for the case of molecules containing hydrogen isotopes, the isotopic shift for at least some of the modes is large compared with frequency shifts due to intermolecular couplings. Thus, in the notation of (7-1-20)

12. See Fig. 3-11, p. 93.
13. See A. E. Hughes, *Phys. Rev.*, **173**, 860 (1968) and references there cited.
14 H. J. Hrostowski and G. C. Pimentel, *J. Chem. Phys.*, **19**, 661 (1951).
L. Hiebert and D. F. Hornig, *J. Chem. Phys.*, **20**, 918 (1952).

and (7-1-21) one could take the limit in which $\eta/\varepsilon \to 0$, since η represents the intermolecular coupling and ε the isotopic shift, leading to

$$\lambda_{1\alpha} \to FG_{\alpha}$$

For the more general case than that of coupling within linear chains, this limiting case corresponds to the frequencies designated in Chapter 6, e.g. (6-2-1), (6-2-2), (6-2-4), (6-2-5), etc., as ω_k^2, which neglect the intermolecular force constants F_k'.

A simple example of this sort of observation is afforded by the crystal $NaHF_2$ which is isomorphic with NaN_3 (see Sections 6-2(d) and 6-3(d)). Rush, Schroeder, and Melveger[16] studied $NaHF_2$ (and KHF_2) with deuterium content ranging from 2 to 80 percent by infrared transmission spectroscopy employing polycrystalline samples. This work was extended by R. D. Cooke[17] to deuterium concentrations as high as 96 percent, and to the observation of reflection spectra whose analysis yields the longitudinal as well as transverse frequencies. The frequencies for the isotopically dilute species are quite close to the longitudinal mode frequencies, which is at variance with the prediction of a dipole coupling model which would require the isolated frequencies to fall about two-thirds of the way between the transverse and longitudinal frequencies of the respective pure crystals.

Another simple crystal structure with a unimolecular unit cell which has been studied using dilute isotopic impurities is hydrogen cyanide.[18] Here it is found that for HCN at low concentrations in DCN, the weaker of the two stretching modes (ν_1) appears at a slightly lower frequency (2097) than in the pure HCN crystal (2098), whereas ν_3 exhibits a shift of the opposite sign, occurring at 3145 in isolation, but at 3132 in pure HCN (see Table 6-6). These results are apparently inconsistent with a dipolar coupling model, but the shifts between the frequencies of the pure and isotopically isolated crystal provide a measure of the coupling constants as defined in (6-2-1).

Potassium acid fluoride has two molecules per unit cell on D_{2h} sites in the D_{4h} crystal. The bending mode is split at the site so that the doublet persists in the isotopically dilute sample, the location of the isolated frequencies being quite similar to those found in $NaHF_2$, i.e., rather closer to the longitudinal than the transverse frequency.

Crystals which exhibit vibrational multiplets, but which consist of molecules occupying just one set of equivalent sites, may be expected to exhibit a collapse of the multiplet into a singlet on going to the limit of low isotopic concentrations, provided the mode is non-degenerate. Hornig and Hiebert[19] have made a quantitative analysis of HCl–DCl and HBr–DBr mixed crystals over a wide range of H/D ratios. In the pure crystals doublets are observed in HCl at 2706 and 2747 cm^{-1} and at 1966 and 1993 in DCl (92.5 percent D). These doublets are

16. J. J. Rush, L. W. Schroeder, and A. J. Melveger, *J. Chem. Phys.*, **56**, 2793 (1972).
17. R. D. Cooke, Ph.D. Thesis, Oregon State University, 1973.
18. H. B. Friedrich and P. F. Krause, *J. Chem. Phys.*, **59**, 4942 (1973).
19. D. F. Hornig and G. L. Hiebert, *J. Chem. Phys.*, **27**, 752 (1957).

replaced by single peaks at approximately 2725 (7.5 percent HCl in DCl) and 1983 (3.5 percent DCl in HCl).

In more complex crystals, two or more non-equivalent sites may be occupied. In such cases, the isotopically dilute sample should reveal, even for non-degenerate molecular modes, separate frequencies corresponding to the different sites. The very complex spectrum of PH_3 below 30°K may be such an example.[20]

For the case of a simple site, the difference between the frequency of the isotopically isolated molecule and the frequency of the same mode in the isotopically pure crystal provides a direct measure of the intermolecular coupling constant.

If the molecule contains more than one hydrogen, further complications may occur, depending upon the relation between the molecular and the site symmetry. In an orientationally ordered crystal, the site symmetry is necessarily a subgroup of the molecular symmetry (for a molecule of homogeneous isotopic composition). As an elementary example, consider an AB_2 molecule with C_{2v} symmetry. The possible sites are C_{2v}, C_2, C_s, and C_1. If the molecule ABB* (where B and B* are different isotopes) is located at a C_{2v} or C_2 site, the potential function experienced by the molecule is obviously unchanged if the positions of B and B* are interchanged, whereas on a C_s^* or C_1 site, the two possible orientations of the molecule are inequivalent, in consequence of which the spectrum of the isolated molecule should yield doublets. As another example, any molecule with threefold symmetry, say AB_3 (C_{3v}), may occupy a C_{3v}, C_3, C_s, or C_1 site. The mixed isotopic form AB_2B^* on C_{3v} and C_3 sites is independent of orientation, but has, respectively, two and three distinct orientations on C_s and C_1 sites. In the C_s case, two of the orientations, namely those with B in the plane of symmetry, are equivalent, so the expected doublet should have a two-to-one intensity ratio.

An early example in which the orientation effect may have been observed is that of NH_4NCS in which the ammonium ion is situated at a C_1 position. NH_3D^+ accordingly has four distinct orientations and $NH_2D_2^+$ has six. There is some evidence in the infrared spectrum[21] that the modes of the species NH_3D^+, $NH_2D_2^+$, and NHD_3^+ occur with these rather high multiplicities, but further work is needed in which one isotopic species is isolated at low concentrations in a second isotopic species as the host. Such work has recently been initiated by Oxton, Knop, and Falk.[22]

Another example is provided by studies of crystal benzene,[23] where the samples consisted of a low concentration of the isotopic guest in a pure isotopic host. Here, the site is C_i in a D_{2h}^{15} unit cell. When C_6H_6 is isolated in C_6D_6, the molecular E_u modes appear as doublets with splittings of the order of 4–8 cm^{-1} and the

* This conclusion assumes that C_s is the molecular plane; the former conclusion applies if the molecule is perpendicular to the plane of the site.

20. M. D. Francia and E. R. Nixon, *J. Chem. Phys.*, **58**, 1061 (1973).
21. L. L. Oden and J. C. Decius, *Spectrochim. Acta*, **20**, 667 (1964).
22. I. A. Oxton, O. Knop, and M. Falk, *Can. J. Chem.*, **53**, 2675 (1975).
23. E. R. Bernstein, *J. Chem. Phys.*, **50**, 4842 (1969).

non-degenerate modes appear as singlets. For the isotopically heterogeneous species such as C_6H_5D, ortho, meta, and para $C_6H_4D_2$, etc., a number of possibilities arise. For C_6H_5D, there would be six distinguishable orientations at a C_1 site, but this number is reduced to three at a C_i site. In fact, several of the modes of C_6H_5D are observed to be triplets of roughly equal intensity, although others are only seen as doublets with roughly two-to-one intensity ratios, the former being out-of-plane vibrations and the latter in-plane. This observation has been interpreted as evidence that the site symmetry, strictly C_i, approximates C_{2h}.

A general group theoretical analysis of the (maximum) number of orientational components to be expected has been given by Kopelman.[24]

Isotopic substitution in general may be expected to lower the molecular symmetry. If the unsubstituted molecule has the symmetry group M of order m, and the substituted molecule the symmetry group M^* of order m^*, then the number of distinguishable configurations of the substituted molecule is m/m^*. In abstract group theory, m/m^* is the number of cosets of M determined by M^*, i.e., every operation of M belongs to a coset X_iM^*, where $X_i = 1, 2, \ldots, m/m^*$:

$$M = \sum_{i=1}^{m/m^*} X_i M^* \qquad (X_1 = E)$$

If the site in the crystal has the trivial symmetry C_1, each of the above set of m/m^* configurations will have a distinct set of frequencies. However, for higher site symmetries S, if any operation of S exists which transforms one configuration into another, then those two configurations will have identical sets of frequencies, or as Kopelman puts it: "We get it (the number of physically different species) by striking out from the cosets... any coset differing from another only by an element of S."

The language for this analysis in abstract group theory[25] involves the concept of a *double coset*: The group M can be expressed as

$$M = \sum_j SY_jM^*$$

A particular term in the sum can be shown to consist of a number of single cosets of the form X_iM^*. The groups S and $Y_jM^*Y_j^{-1}$ will contain a number of common operations, forming a group which we designate as D_j. Then, the number of cosets X_iM^* contained in SY_jM^* is equal to the ratio s/d_j where s is the order of S and d_j is the order of D_j. These numbers are important in determining the statistical weights of the distinguishable configurations.

To illustrate, we return to our earlier example of an AB_2B^* molecule at a C_s site. We have $M = C_{3v}$, $M^* = C_s$, $S = C_s$. The single coset expansion of M is

$$C_{3v} = EC_s + C_3C_s + C_3^2C_s$$

24. R. Kopelman, *J. Chem. Phys.*, **47**, 2631 (1967).
25. A. Speiser, *Die Theorie der Gruppen von Endlicher Ordnung*, Dover, New York, 1945.

while the double coset expansion becomes

$$C_{3v} = C_s E C_s + C_s C_3 C_s$$

The groups D_j are C_s and E so that $s/d_j = 1$ and 2.

For 1, 3, 5 deuterated benzene ($M^* = D_{3h}$) at a C_i site, the double coset expansion consists of a single term

$$D_{6h} = C_i D_{3h}$$

so no orientational splitting is expected.

It should be emphasized that Kopelman's original discussion considered only the distinct *orientations* as physically inequivalent. But, if the site symmetry S includes no improper operations (i.e., i, σ, or S_n), then enantiomorphic molecules in principle should yield distinct spectra at such a site, which is the reason why we find the maximum number of distinct configurations to be m/m^* rather than the ratio of the orders of the rotational subgroups of M and M^* which is another way of stating Kopelman's result. In the simplest possible terms "*d*" and "*l*" molecules at, say, a "*d*" site, are diastereoisomers and may have different spectra.

It must also be pointed out that the number of distinguishable configurations at the site is not necessarily equal to the number of fine structure components, since it may happen that a degeneracy which persists through the isotopic substitution $M \to M^*$ is lifted for one or more of the site configurations. Thus, Oxton, Knop, and Falk[26] in a study of NH_3D^+ in NH_4ReO_4 (S_4 site) found that ν_1 is a singlet, while ν_{4E} (i.e., the doubly degenerate bending mode under C_{3v}) is a doublet: in this case, all four configurations are indistinguishable, but the *effective symmetry*, defined as the group of operations common to $M^* = C_{3v}$ and $S = S_4$ is only C_1, i.e., $C_{3v} \cap S_4 = C_1$.

It has been reported that orientational splittings, when observed, are comparable in magnitude with site splittings of degenerate modes. It is easy to understand this with the aid of a simple but quite realistic model. Consider, for example, the frequency problem for the three stretching modes of an AH_3 molecule (C_{3v} symmetry). The FG matrix methods described in Chapter 2 yield the following result, if the interaction with the lower frequency (bending) modes is neglected:

$$\omega_{A_1}^2 = (f + 2f')G_{A_1}$$
$$\omega_E^2 = (f - f')G_E$$

This is the free molecule result. If now a site splitting is introduced, it may be regarded as a small perturbation of the force constants; for a C_s site this could be expressed as $f_1 = f + \Delta f$, $f_2 = f_3 = f + \Delta f'$ (f' above is the coupling force constant between two different bonds). The effect of the Δf perturbation is easy to estimate for the AH_3 molecule. The frequencies become

$$\omega^2 \simeq \omega_{A_1}^2 + \tfrac{1}{3}(\Delta f + 2\Delta f')G_{A_1}$$

26. I. A. Oxton, O. Knop, and M. Falk, *Can. J. Chem.*, **54**, 892 (1976).

$$\omega^2 \simeq \omega_E^2 + \tfrac{1}{3}(2\Delta f + \Delta f')G_E$$

$$\omega^2 \simeq \omega_E^2 + \Delta f G_E$$

i.e., the split in the E mode is $\frac{1}{3}(\Delta f - \Delta f')G_E$.

Now consider the AH_2D molecule: the AD stretch modes are widely separated from the AH stretches, so we can easily approximate the AD stretch frequencies as

$$\omega_s^2 = (f + \Delta f)G_D$$

and

$$\omega_a^2 = (f + \Delta f')G_D$$

so that

$$\omega_s^2 - \omega_a^2 = (\Delta f - \Delta f')G_D$$

Here s signifies the orientation in which the isotopic molecular symmetry plane coincides with the site symmetry plane and a signifies the other case, in which the effective symmetry is only C_1. Although several approximations have been made, the essential point has been established, namely that both the site and orientation splits are proportional to $\Delta f - \Delta f'$.

7-3 INERT GAS MATRICES

Matrix isolation was proposed in the same year by Norman and Porter[27] and by Whittle, Dows, and Pimentel.[28] Most of its applications have had to do with the trapping, identification, and the study of the chemistry of small polyatomic (mostly triatomic) free radicals, whose lifetime and concentration, when they are produced in the gas phase are usually too small to permit accurate study by infrared spectroscopy. In recent years, however, the procedure has also been well utilized to study the spectra of stable but non-volatile inorganic compounds. It is not the purpose of this section to provide a comprehensive review of the subject.[29]

27. I. Norman and G. Porter, *Nature,* **174,** 508 (1954).
28. E. Whittle, D. A. Dows, and G. C. Pimentel, *J. Chem. Phys.,* **22,** 1943 (1954).
29. The subject has been well reviewed. See:
    - G. C. Pimentel in *Formation and Trapping of Free Radicals,* A. M. Bass and H. P. Broida, Eds., Academic Press, New York, 1960, pages 69–116.
    - G. C. Pimentel, *Pure Appl. Chem.,* **4,** 61 (1962).
    - G. C. Pimentel and S. W. Charles, *Pure Appl. Chem.,* **7,** 111 (1963).
    - M. E. Jacox and D. E. Milligan, *Appl. Optics,* **3,** 873 (1964).
    - T. S. Hermann and S. R. Harvey, *Appl. Specty,* **23,** 435 (1969). (Literature review with 612 references.)
    - D. E. Milligan and M. E. Jacox in *Physical Chemistry: An Advanced Treatise,* Vol. 4: *Molecular Properties,* H. Eyring, Ed., Academic Press, New York, 1969, chapter 4.
    - *Vibrational Spectroscopy of Trapped Species,* H. E. Hallam, Ed., Wiley, New York (1973).
    - A. J. Downs and S. C. Peake, *Mol. Spectros.* (Specialist Periodical Reports), **1,** 523 (1973). (Literature review through July 1972 and bibliography with 436 references).
    - B. M. Chadwick, *Mol. Spectros.* (Specialist Periodical Reports), **3,** 281 (1975). (Literature review July 1972 to September 1974 and bibliography with 374 references.)

Instead, we shall only mention the aspects of the technique as they relate to the general subject of this chapter—solid solutions and impurity phenomena.

Rare gas matrices are not solid solutions. Indeed, careful studies of the phase diagrams of Ar and N_2 and Ar and CO show that, although at low temperatures true solid solutions do form, near the melting point of Ar a few percent of either molecule causes segregation. It is not well known at what sites of the rare gas lattices the guests reside—that is, whether the sites are substitutional or interstitial and, if the latter, of what kind (tetrahedral or octahedral). In point of fact, some rare gas matrices are not even crystalline.[30] This is somewhat expected since most molecules, even small diatomics, have larger diameters than the diameters of the substitutional and interstitial sites in the solid rare gases.

Fundamentally, the reason matrix isolation is used is to achieve what the name says: *isolation*—from either other molecules of the same kind or those which may chemically react with the isolate. At the same time, it is desired that there be little or no change in the molecule's vibrational parameters from those it would manifest in the gas phase. It is chiefly for this reason that the host is chosen to be a rare gas. Since even the crystalline modifications of condensed rare gases are so highly symmetrical, the principal forces which can shift the vibrational levels of the molecule to be isolated are expected to be of the dispersion type and, since their angular parts should spatially average to zero, the so-called matrix shift should be relatively small.

In many cases the shifts are only of a few cm^{-1}, presumably because intramolecular bond energies are large compared to the binding energies of the matrix as well as the energy which binds the molecule to the matrix, and because vibrational amplitudes are small compared to lattice spacings.[31] Where they are larger ($\sim -60\ cm^{-1}$), they have been attributed to a familiar mélange of multipole interactions, and these attributions have formed the bases of calculations of the shifts in particular cases.[32] The important fact is that the shifts are sufficiently small and uniform so that the spectrum of the isolated molecule is identifiable. Thus, the spacings of vibrational progressions are usually changed but little from their values in the gas phase.[33,34]

Of greater concern to the matrix isolator are four other general effects, any of which may spoil the identification of the trapped species:

1. Aggregation
2. Multiple trapping sites
3. Molecular rotation
4. Coupling with lattice vibrations

30. H. S. Peiser in *Formation and Trapping of Free Radicals*, A. M. Bass and H. P. Broida, Eds., Academic Press, New York, 1960, pages 301–26.
31. O. Schnepp, Fourth International Symposium on Free Radicals, Washington, DC, 1962, Paper F'I.
32. See, for example, M. L. Linevsky, *J. Chem. Phys.*, **34,** 587 (1961).
33. K. Dressler and O. Schnepp, *J. Chem. Phys.*, **42,** 2482 (1965).
34. J. Roncin, N. Damany, and J. Romand, *J. Mol. Spectry*, **22,** 154 (1967).

The first of these is generally addressed by recourse to a greater "*M*/*R*"—the ratio of the concentration of *m*atrix atoms to *r*eactive species molecules. Alternatively, aggregation may manifest itself as a function of time—that is, say, while the spectrum is being recorded. Diffusion is responsible for this phenomenon, and the solution is to repeat the experiment at a lower temperature. In some of his earliest work on this subject, Pimentel showed that the ratio of the diffusion temperature to the melting point is generally 0.5, but never less than 0.25. Since the melting point of Ar is 34°K, this means that most matrix isolation is carried out with liquid hydrogen or helium as the cryogenate.

An important technique in matrix isolation is a so-called warm-up, cool-down experiment, in which controlled diffusion is encouraged in order to cause aggregation which may enable further identification of the isolate. In the case of stable molecules, aggregates are *n*-mers of the original isolate, whose spectra are related to that of the pure crystals, as originally discussed by Hornig and Hiebert.[35] In some cases, even with stable molecules, the heat of aggregation may be sufficient to raise the temperature of the matrix and thus permit further diffusion. Further aggregation results, and, thus, a chain reaction leading ultimately to loss of the sample from the cold window upon which it has been sprayed or evaporated. (Because the heat of vaporization of He is so small, runaway aggregation can also lead to abrupt loss of refrigerant from the experimental cryostat.)

Multiple trapping sites have been implicated in vibrational investigations, such as that of Leroi *et al.* on CO in Ar,[36] although one of the two bands observed was assigned to aggregates. In other studies different trapping sites are established by parallel behavior upon warm-up, with accompanying development of a new band due to aggregate formation.[37] In a study of the electronic spectrum of C_6H_6 in a variety of matrices, there was no question of two sites, since two *parallel* progressions ($^1B_{2u} \leftarrow {}^1A_{1g}$) of sharp lines were observed.[38,39]

Another source of a multiple "line spectrum" is due to rotation of the trapped molecule in the matrix. Again because the diameters of even substitutional sites are so small and molecular diameters are (relatively) so large, the phenomenon is restricted to only a few cases, such as HF[40] in the several rare gases, H_2O[41] and NH_3[42] in Ar and CH_4 in Xe, Kr and Ar.[43] In all cases except that of HF, the rotation is hindered to some extent. The case of HCl in Ar is complex, there being simultaneous evidence for restricted rotation, aggregation, rotation–translation coupling and complexation with N_2 impurity.[44]

35. D. F. Hornig and G. L. Hiebert, *J. Chem. Phys.*, **27,** 752 (1957).
36. G. E. Leroi, G. E. Ewing, and G. C. Pimentel, *J. Chem. Phys.*, **40,** 2298 (1964).
37. W. Charles and K. O. Lee, *Trans. Faraday Soc.*, **61,** 614 (1965).
38. Y. Diamant, R. M. Hexter, and O. Schnepp, *J. Mol. Spectry,* **18,** 158 (1965).
39. R. B. Merrithew, G. V. Marusak, and C. E. Blount, *J. Mol. Science,* **25,** 269 (1968).
40. M. T. Bowers, G. I. Herley, and W. H. Flygare, *J. Chem. Phys.*, **45,** 3399 (1966).
41. M. E. Jacox and D. E. Milligan, *Appl. Optics,* **3,** 873 (1964).
42. A. Cabana, G. B. Savitsky, and D. F. Hornig, *J. Chem. Phys.*, **39,** 2942 (1963).
43. R. L. Redington and D. E. Milligan, *J. Chem. Phys.*, **37,** 2162 (1962).
44. D. E. Milligan, R. M. Hexter, and K. Dressler, *J. Chem. Phys.*, **34,** 1009 (1961).

From the point of view of this chapter coupling with lattice vibrations would be most interesting. Although the expectations of Section 7-1 have been demonstrated,[45] the identification of a localized impurity mode "in the gap," which is both the counterpart to the activation of the acoustic branch as well as equivalent to a coupling with a lattice mode, has yet to be a definitive assignment in the vibrational spectrum of a molecule in matrix isolation. In electronic spectra, such as that of benzene,[46] so-called phonon-assisted transitions are a commonplace. Their absence in vibrational spectra may be related to the absence of combination tones between counter-ions in such crystals as NH_4NO_3.[47]

In the absence of these effects, spectra of molecules in matrix isolation are relatively simple, and more important, reliable. Gone are the necessities of locating band origins, pressurizing with an inert gas in order to obtain accurate intensities for concentration measurement, deconvoluting overlapping bands, etc. Instead, the spectrum is a collection of *sharp,* well-separated transitions, the intensity of each being accurately gauged by its peak height. In fact, the spectra are even sharper than those of pure crystals, due to the fact that they are truly absorption spectra, uncomplicated by reflection phenomena (see Section 5-4).

In Section 2-8 we saw that, in the gas phase, the natural lifetimes of vibrationally excited states are $\sim 10^{-2}$ sec, leading to natural widths of $\sim 10^{-8}$ cm^{-1}. Relativistic and collisional processes act to broaden the natural widths of vibrational lines to $\sim 10^{-2}$ cm^{-1}. The analog of gas-phase collisional processes in any condensed state is some kind of coupling of the oscillator to the "lattice." That such couplings exist has been amply demonstrated, although no quantitative studies have been made of relaxation processes of species in matrix isolation.[48] Theoretical consideration has been given to relaxation processes in pure crystals, some aspects of which were discussed in Section 6-6(d). Suffice it to say here that, so long as the several effects previously discussed are effectively disposed, vibrational transitions in matrices have bandwidths dictated by the instrumental slits and hence are well-suited for analytical purposes.[49]

These several matters are well illustrated in a study of the infrared spectrum of trimethylamine.[50] The spectrum of that compound in the 2800–3000 cm^{-1} region has been studied in the gas phase, in the neat solid and in matrix isolation. The contrasting resolution in the last two cases is instructive, particularly with respect to the effectiveness of matrix isolation in resolving nearly-degenerate vibrational transitions.

45. G. O. Jones and J. M. Woodfine, *Proc. Phys. Soc. (London),* **86,** 101 (1965).
46. B. Meyer, *Low Temperature Spectroscopy,* American Elsevier Publishing Co., New York, 1971, page 59 and Fig. 3.2.
47. R. Newman and R. S. Halford, *J. Chem. Phys.,* **18,** 1276, 1291 (1950).
48. M. Gouterman, *J. Chem. Phys.,* **36,** 2846 (1962). See, however, H. Dubost and R. Charneau, *Chem. Phys.,* **12,** 407 (1976).
49. M. M. Rochkind, *Science,* **160,** 196 (1968).
50. T. D. Goldfarb and B. N. Khare, *J. Chem. Phys.,* **46,** 3379 (1967).

7-4 IONIC MATRICES

In much the same way that condensed inert gases have proved to be useful hosts for the isolation of neutral molecules (and radicals), the alkali halides have been employed as host matrices for the spectroscopic study of various small molecular ions. Many simple molecular ionic species have been studied in this way by means of infrared absorption, and a smaller number have been observed in the Raman effect. In many cases, the studies have comprised a variety of hosts, including as alkali cations, Na^+, K^+, Rb^+, Cs^+ and as halide anions, Cl^-, Br^-, I^-. Some molecular ions capable of substituting for halide or alkali ions are indicated in Fig. 7-5, which shows approximate dimensions to scale.

In contrast to the case of matrix isolation with inert gas matrices, the alkali halide hosts are stable over a wide range of temperatures, from 0°K up to the melting point, approximately 1000°K. Even without heating above room temperature, it is usually possible to discern a large number of "hot bands," i.e., upper stage transitions, which provide more extensive information about the molecular potential function than can be learned from the fundamentals alone. In some cases, this information has been used to obtain the spectroscopic parameters ν_i and X_{ij} in the expression for the vibrational levels of the anharmonic polyatomic molecule,

$$E = \sum_j h\nu_j(v_j + \tfrac{1}{2}) + \sum_{i<j}\sum hX_{ij}\left(v_i + \frac{d_i}{2}\right)\left(v_j + \frac{d_j}{2}\right) \tag{7-4-1}$$

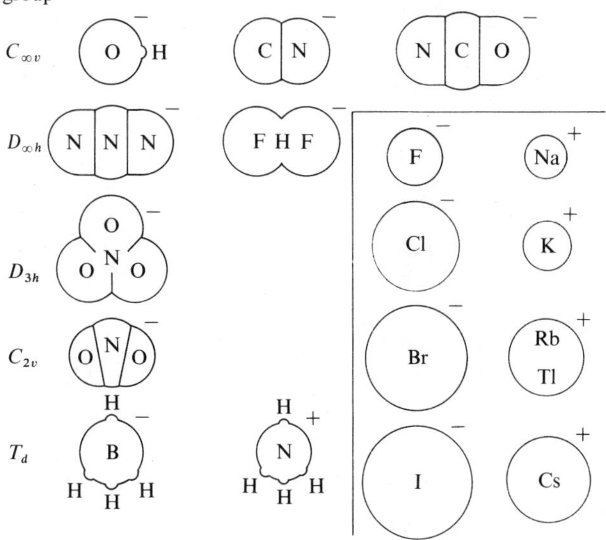

Figure 7-5 Diagrams illustrating the size and shape of the ions that have been studied. (From W. C. Price, W. F. Sherman, and G. R. Wilkinson, *Spectrochim. Acta*, **16**, 663 (1960))

while in other cases programmed computers have been employed to fit the energy levels directly from an anharmonic potential function including some or all of the cubic and quartic constants allowed by symmetry in a Taylor series expansion of the vibrational potential energy. The latter type of analysis automatically takes into account the possibilities of various Fermi resonances.

The possibility of such anharmonic analyses depends upon knowledge of a relatively large number of overtone and combination levels; as we have seen previously, this is, in part, possible due to the detection of hot bands, but it also requires relatively large sample densities in the radiation beam. This has been achieved by using thick host crystals—up to one or two centimeters—while still maintaining a low enough volume concentration in the host to keep the guest molecules isolated from one another. Also, the detection of transitions is facilitated by the fact that for many small, typically triatomic molecular ions, the line width is very small (often limited by the spectrometer resolution), so that the peak intensity is very large.

Samples for these studies have been prepared by all of the following methods:

1. Diffusion of the guest into a single crystal of the host.
2. Coprecipitation of alkali halide and the desired impurity from water, followed by pelletting.
3. Pelletting the powder obtained by grinding the cooled mass obtained from a fusion of the alkali halide host and the guest.
4. Formation of a single crystal from a melt containing the desired impurity.
5. Chemical reaction in a pellet containing reactants which yield the desired product upon heating.

(a) The Diatomic Species, OH^- and CN^-

The single vibrational mode of these species was first observed in a number of alkali halides by Price, Sherman, and Wilkinson.[51] For CN^-, they found a variation from 2105 cm^{-1} in NaCl to 2053 in CsI. A striking temperature dependence of the half width was noted: at 10°K, the width was less than 0.6 cm^{-1}, but over 20 cm^{-1} at 300°K. Many studies have been aimed at the discovery of the orientation and potential function restricting rotation; these are reviewed by Narayanamurti and Pohl.[52] The most revealing experiments involve the application of stress or d.c. electric fields at temperatures below 5°K. The hydroxide ion apparently has six equivalent minima in (100) directions. On the other hand, Lüty[53] has observed an infrared doublet for CN^- in KCl at temperatures below 5°K whose response to applied stress is only consistent with 111 orientation of CN^-. For such an orientation, the ground rotational state is eightfold degenerate

51. W. C. Price, W. F. Sherman, and G. R. Wilkinson, *Spectrochim. Acta*, **16**, 663 (1960); *Proc. Roy. Soc. London*, A**255**, 5 (1960). These workers also studied many of the other molecular ions discussed in this Section, including NCO^-, N_3^-, HF_2^-, NO_3^-, NH_4^+, and BH_4^-.
52. V. Narayanamurti and R. O. Pohl, *Rev. Mod. Phys.*, **42**, 201 (1970).
53. F. Lüty, *Phys. Rev.*, **B10**, 3677 (1974).

for very high barriers, but is split into $A_{1g} + F_{1u} + F_{2g} + A_{1u}$ states under the O_h group when tunnelling is allowed. The observed splitting is 2Δ, where Δ is the constant spacing between the levels $A_{1g} < F_{1u} < F_{2g} < A_{1u}$.

(b) Linear Triatomic Anions

The related species BO_2^-, N_3^-, HF_2^-, and NCO^- have been very extensively studied in alkali halide matrices. All of these are 20-electron molecules—isoelectronic with CO_2, and apparently of such dimensions as to fit substitutionally in place of a halide ion.

The first member of this family to be reported was FHF^-, whose two infrared active fundamentals ν_2 and ν_3 were observed in KCl, KBr, KI, and NaBr by Ketelaar, Haas, and van der Elsken.[54] The ν_3 mode is extremely broad in pure KHF_2, but is significantly narrowed and shifted in the alkali halides. The cyanate ion has been the subject of many studies beginning in 1958;[55] it is an example in which a fairly complete anharmonic potential function has been evaluated.[56] The number of observed transitions for the cyanate ion is rather large for a condensed phase species. For example, 125 lines have been observed for NCO^- in KCl. These arise from 80 vibrational levels in seven molecular isotopic species, since C, N, O each have at least two isotopic species which are detectable at natural abundance or with some enrichment.

The two infrared active fundamentals of the azide ion were studied in four alkali halides by Bryant and Turrell,[57] whose work also included careful measurements of the shifts of these frequencies with temperature. Bryant and Turrell, as well as other workers have analyzed the influence of the host matrix on the frequencies and thereby obtained "free" molecule, i.e., hypothetical gas phase frequencies. The "site" effect is found to be mainly a "blue" shift, which arises from the net effect of induction, coulombic, and short range (atom–atom repulsion) forces.

In some ways the most interesting case is that of BO_2^- which is not known to exist as a simple linear triatomic structure outside the alkali halide environment. In compounds such as $NaBO_2$, $Ca(BO_2)_2$, etc., metaborate invariably forms cyclic or linear polymers. The infrared spectrum of BO_2^- monomer was first noted as an impurity in alkali halides used as optical components in infrared spectrometers and identified by R. S. McDonald.[58] Further transitions were reported by Morgan and Staats,[59,60] by Hisatsune and Suarez[61] and by other workers (see

54. J. A. A. Ketelaar, C. Haas, and J. van der Elsken, *J. Chem. Phys.*, **24**, 624 (1956).
55. A. Maki and J. C. Decius, *J. Chem. Phys.*, **28**, 1003 (1958); A. Maki and J. C. Decius, *J. Chem. Phys.*, **31**, 772 (1959); J. C. Decius, J. L. Jacobson, W. F. Sherman, and G. R. Wilkinson, *J. Chem. Phys.*, **43**, 2180 (1965); V. Schettino and I. C. Hisatsune, *J. Chem. Phys.*, **52**, 9 (1970).
56. D. F. Smith, Jr., J. Overend, J. C. Decius, and D. J. Gordon, *J. Chem. Phys.*, **58**, 1636 (1973).
57. J. I. Bryant and G. C. Turrell, *J. Chem. Phys.*, **37**, 1069 (1962).
58. R. S. McDonald, *Spectrochim. Acta*, **15**, 773 (1959).
59. H. W. Morgan and P. A. Staats, *J. Appl. Phys.*, **33**, 364 (1962).
60. H. W. Morgan and P. A. Staats, *Spectrochim. Acta*, **28A**, 600 (1972).
61. I. C. Hisatsune and N. H. Suarez, *Inorg. Chem.*, **3**, 168 (1964).

Ref. 60 for further references and for interesting comments on the intensity and ubiquity of this species). Studies of the hot bands, $\nu_1 + \nu_3 - \nu_3$, $2\nu_2 + \nu_3 - \nu_3$, $\nu_1 - \nu_2$, and $2\nu_2 - \nu_2$, have been used to locate the infrared inactive ν_1 and $2\nu_2$ states which were subsequently detected in the Raman effect. An anharmonic potential function has been evaluated for BO_2^-.[62]

(c) NO_2^- and NO_3^-

These anions were amongst the earliest to be studied isolated in the alkali halides.[63] Since melt grown crystals invariably contain both species, we discuss them together. The matrix shifts were reported by Price, Sherman, and Wilkinson,[64] who noted comparatively broad bands. Narayanamurti, Seward, and Pohl[65] observed the infrared spectra in the potassium halides at temperatures down to 2°K and resolved numerous satellite components which they interpreted as: (i) slightly hindered rotation around the least axis of inertia in NO_2^-; (ii) highly hindered rotation (i.e. libration) around the symmetry axis; (iii) libration in the case of nitrate. However, a somewhat different interpretation has been advanced by Bonn, Metselaar, and van der Elsken.[66] According to Narayanamurti *et al.*, the ν_1 mode of NO_3^- is observed at 15°K. This could be due to (a) natural abundance of an oxygen isotope, or (b) to the fact that the rigorous definition of the site symmetry for a non-rotating impurity (high barrier limit) requires that the site be a subgroup of the crystal field symmetry (presumably O_h) and of the free molecule symmetry (D_{3h}). The largest such subgroups are D_3 or C_{3v}, corresponding to different orientations of the nitrate around its threefold axis. Under D_3, ν_1 is infrared forbidden, but allowed under C_{3v}.

(d) NH_4^+ and BH_4^-

These tetrahedral hydrides, isoelectronic with methane, have been studied by many workers. Both species appear to have a significant amount of Fermi interaction between ν_3, $\nu_2 + \nu_4$, and $2\nu_4$, which is easier to analyze in matrix isolation than in crystal ammonium halides, where $\nu_2 + \nu_4$ and $2\nu_4$, become two-phonon bands. In the NaCl type lattices, Vedder and Hornig[67] found a large number of components—particularly at low temperatures in the ν_4 region—which they attributed to hindered rotation. In contrast Sherman and Smulovitch[68] reported that NH_4^+ in CsBr has a ν_3 spectrum which can be fully described by invoking combinations with two external modes of about 142 and 315 cm^{-1}, respectively

62. D. F. Smith, Jr., *Spectrochim. Acta,* **30A,** 875 (1974).
63. I. Maslakowez, *Z. Physik,* **51,** 696 (1928).
64. W. C. Price, W. F. Sherman, and G. R. Wilkinson, *Spectrochim. Acta,* **16,** 663 (1960).
65. V. Narayanamurti, W. D. Seward, and R. O. Pohl, *Phys. Rev.,* **148,** 481 (1966).
66. R. Bonn, R. Metselaar, and J. van der Elsken, *J. Chem. Phys.,* **46,** 1988 (1967).
67. W. Vedder and D. F. Hornig, *J. Chem. Phys.,* **35,** 1560 (1961).
68. W. F. Sherman and P. P. Smulovitch, *J. Chem. Phys.,* **52,** 5187 (1970).

of translational and librational characters. This discrepancy may be real, and due to the difference between the host lattices, but it is also possible that the rotational fine structure reported by Vedder and Hornig from data at liquid helium temperatures was not resolved in the higher temperature (100°K) experiments of Sherman and Smulovitch.

(e) The Tetrahedral Oxyanions, XO_4^{-n}

Perchlorate ion, ClO_4^-, substitutes readily for iodide, less easily for bromide, still less for chloride, as would be expected from the stereochemistry. The spectrum is consistent with T_d selection rules. However, the aliovalent ions ($n > 1$), rarely exhibit a T_d spectrum. Sulfate, for example, appears to obey C_{2v} site selection rules, since $\nu_3(F_2)$ is split into three components, and $\nu_1(A_1)$ is active in the infrared spectrum. This has been attributed to the fact that SO_4^{2-} complexes with divalent cationic impurities, such as Ca^{2+}, Ba^{2+}, etc., on an adjacent alkali ion site, producing a distortion along one of the sulfate C_2 axes, and leading thereby to C_{2v} selection rules.

Although in the foregoing we have mentioned the rotatory modes of the impurity in some cases, it should be emphasized that all ionic impurities in the alkali halides are expected to produce impurity modes of translational origin. These may occur in the vicinity of the gap between the acoustic and optic modes, or above the latter, depending upon the relative masses of the host and guest which it displaces. A great many such modes have been noted in combination with the internal modes, particularly for the BO_2^-, NCO^-, etc., family.[69]

69. M. A. Cundell and W. F. Sherman, *Phys. Rev.*, **168,** 1007 (1968).

APPENDIX

I

CHARACTER TABLES, SYMMETRY SPECIES OF TRANSLATION, ROTATION, AND FORMS OF THE DIPOLE ($\boldsymbol{\mu}' = \partial\boldsymbol{\mu}/\partial Q$) AND POLARIZABILITY ($\boldsymbol{\alpha}' = \partial\boldsymbol{\alpha}/\partial Q$) DERIVATIVES FOR THE CRYSTALLOGRAPHIC POINT GROUPS

The tables are arranged according to the crystal system, i.e., *triclinic* (C_1, C_i); *monoclinic* (C_2, C_s, C_{2h}); *orthorhombic* (C_{2v}, D_2, D_{2h}); *trigonal* (*hexagonal*) ($C_3, C_{3i}, C_{3v}, D_3, D_{3d}$); *tetragonal* ($C_4, S_4, C_{4h}, C_{4v}, D_4, D_{2d}, D_{4h}$); *hexagonal* ($C_6, C_{3h}, C_{6h}, D_6, C_{6v}, D_{3h}, D_{6h}$; and *cubic* ($T, T_h, T_d, O, O_h$).

In the groups C_3, C_4, C_6, C_{3i}, C_{3h}, C_{4h}, C_{6h}, S_4, T, and T_h, *separably degenerate* irreducible representations occur. The degeneracy is twofold (E species) and the characters of the one-dimensional representations into which these species may be separated are complex conjugates, and are shown in braces. When components of **T** (translation) or **R** (rotation) transform as such representations, we indicate real cartesian components which are related to the E_+ and E_- species according to the transformation $E_+ = X + iY$ and $E_- = X - iY$. When there is a unique axis, this is always taken as Z.

The right-hand columns indicate the independent parameters in the dipole derivative ($\boldsymbol{\mu}' = \partial\boldsymbol{\mu}/\partial Q$) and polarizability derivative ($\boldsymbol{\alpha}' = \partial\boldsymbol{\alpha}/\partial Q$) tensors. Thus in the point group C_3 we see that $\partial\mu_X/\partial Q_X = \partial\mu_Y/\partial Q_Y = l$ and $\partial\mu_Z/\partial Q_A = n$, where Q_X and Q_Y are any two normal coordinates which transform like the real representation formed from E_+ and E_-, and Q_A is any totally symmetric normal coordinate. Also in this point group one finds that

$$\frac{\partial\boldsymbol{\alpha}}{\partial Q_A} = \begin{bmatrix} a & 0 & 0 \\ 0 & a & 0 \\ 0 & 0 & b \end{bmatrix} \quad \frac{\partial\boldsymbol{\alpha}}{\partial Q_X} = \begin{bmatrix} c & d & e \\ d & -c & f \\ e & f & 0 \end{bmatrix} \quad \text{and} \quad \frac{\partial\boldsymbol{\alpha}}{\partial Q_Y} = \begin{bmatrix} d & -c & -f \\ -c & -d & e \\ -f & e & 0 \end{bmatrix}$$

For the most part these conventions agree with those adopted by Loudon[1] who first assembled tables showing in full detail the parametric relations for the polarizability derivatives.

The symmetry species notation employed here was apparently introduced by Placzek.[2] It has been used in several earlier works in molecular spectroscopy, notably by G. Herzberg,[3] recommended by R. S. Mulliken in a report of the Joint Commission for Spectroscopy,[4] and has for several decades become part of the *lingua franca* of molecular physicists.[5]

In the following character tables, many of the groups which are expressible as direct products of a smaller group with i are indicated in abridged form. Since i commutes with everything, one simply appends g or u to the symmetry species of the smaller group, e.g., $D_{3d} = D_3 \times i$ has species A_{1g}, A_{2g}, E_g; A_{1u}, A_{2u}, E_u. In general if $G = H \times i$, then for any operation R in H, the characters for R and $R \times i$ are identical in representation Γ_g of G with the character of R in representation Γ of H; for the Γ_u representation of G, again the character of R in G is the same as for R in H, but the character of $R \times i$ is the negative of that of R.

The vectors **T** and **R** and the parameters of $\boldsymbol{\mu}'$ and $\boldsymbol{\alpha}'$ have the same form in $G = H \times i$ as they do in H, with the additional specification that **T** and $\boldsymbol{\mu}'$ are in u species (representations) while **R** and $\boldsymbol{\alpha}'$ are in g species.

Triclinic: C_1, C_i

C_i	E	i		μ'_X	μ'_Y	μ'_Z	α'_{XX}	α'_{YY}	α'_{ZZ}	α'_{XY}	α'_{XZ}	α'_{YZ}
A_g	1	1	**R**	0	0	0	a	b	c	d	e	f
A_u	1	−1	**T**	l	m	n	0	0	0	0	0	0

Monoclinic: C_2, C_s, C_{2h}

C_2	E	C_2										
A	1	1	T_Z; R_Z	0	0	n	a	b	c	d	0	0
B	1	−1	T_X, T_Y; R_X, R_Y	l	m	0	0	0	0	0	e	f

C_s	E	σ										
A'	1	1	T_X, T_Y; R_Z	l	m	0	a	b	c	d	0	0
A''	1	−1	T_Z; R_X, R_Y	0	0	n	0	0	0	0	e	f

1. R. Loudon, *Adv. in Phys.*, **13**, 423 (1964); **14**, 621 (1965).
2. G. Placzek, *Handb. d. Radiologie*, **VI**, [2] 209 (1934).
3. G. Herzberg, *Infrared and Raman Spectra of Polyatomic Molecules*, D. Van Nostrand Co., Inc., New York, 1945.
4. R. S. Mulliken, *J. Chem. Phys.*, **23**, 1997 (1955).
5. J. L. Warren, *Rev. Mod. Phys.*, **40**, 38 (1968) compares various systems of notation.

Monoclinic—*continued*

C_{2h}	E	C_2	i	σ_h		μ'_X	μ'_Y	μ'_Z	α'_{XX}	α'_{YY}	α'_{ZZ}	α'_{XY}	α'_{XZ}	α'_{YZ}
A_g	1	1	1	1	R_Z	0	0	0	a	b	c	d	0	0
B_g	1	−1	1	−1	R_X, R_Y	0	0	0	0	0	0	0	e	f
A_u	1	1	−1	−1	T_Z	0	0	n	0	0	0	0	0	0
B_u	1	−1	−1	1	T_X, T_Y	l	m	0	0	0	0	0	0	0

Orthorhombic: C_{2v}, D_2, D_{2h}

C_{2v}	E	C_2	σ_{XZ}	σ_{YZ}										
A_1	1	1	1	1	T_Z	0	0	n	a	b	c	0	0	0
A_2	1	1	−1	−1	R_Z	0	0	0	0	0	0	d	0	0
B_1	1	−1	1	−1	$T_X; R_Y$	l	0	0	0	0	0	0	e	0
B_2	1	−1	−1	1	$T_Y; R_X$	0	m	0	0	0	0	0	0	f

D_2	E	C_{2Z}	C_{2Y}	C_{2X}										
A	1	1	1	1		0	0	0	a	b	c	0	0	0
B_1	1	1	−1	−1	$T_Z; R_Z$	0	0	n	0	0	0	d	0	0
B_2	1	−1	1	−1	$T_Y; R_Y$	0	m	0	0	0	0	0	e	0
B_3	1	−1	−1	1	$T_X; R_X$	l	0	0	0	0	0	0	0	f

D_{2h}	E	C_{2Z}	C_{2Y}	C_{2X}	i	σ_{XY}	σ_{ZX}	σ_{YZ}											
A_g	1	1	1	1	1	1	1	1		0	0	0	a	b	c	0	0	0
B_{1g}	1	1	−1	−1	1	1	−1	−1	R_Z	0	0	0	0	0	0	d	0	0
B_{2g}	1	−1	1	−1	1	−1	1	−1	R_Y	0	0	0	0	0	0	0	e	0
B_{3g}	1	−1	−1	1	1	−1	−1	1	R_X	0	0	0	0	0	0	0	0	f
A_u	1	1	1	1	−1	−1	−1	−1		0	0	0	0	0	0	0	0	0
B_{1u}	1	1	−1	−1	−1	−1	1	1	T_Z	0	0	n	0	0	0	0	0	0
B_{2u}	1	−1	1	−1	−1	1	−1	1	T_Y	0	m	0	0	0	0	0	0	0
B_{3u}	1	−1	−1	1	−1	1	1	−1	T_X	l	0	0	0	0	0	0	0	0

Trigonal (hexagonal): $C_3, C_{3i} = S_6, C_{3v}, D_3, D_{3d}$ $\qquad \varepsilon = e^{2\pi i/3}$

C_3	E	C_3^+	C_3^-										
A	1	1	1	$T_Z; R_Z$	0	0	n	a	a	b	0	0	0
E_+	$\{1$	ε	$\varepsilon^*\}$	$T_X; R_X$	l	0	0	c	$-c$	0	d	e	f
E_-	$\{1$	ε^*	$\varepsilon\}$	$T_Y; R_Y$	0	l	0	d	$-d$	0	$-c$	$-f$	e

$C_{3i} = C_3 \times i$

T, $\boldsymbol{\mu}'$ in A_u, E_u
R, $\boldsymbol{\alpha}'$ in A_g, E_g

Trigonal—*continued*

C_{3v}	E	$2C_3$	$3\sigma_v$		μ'_X	μ'_Y	μ'_Z	α'_{XX}	α'_{YY}	α'_{ZZ}	α'_{XY}	α'_{XZ}	α'_{YZ}
A_1	1	1	1	T_Z	0	0	n	a	a	b	0	0	0
A_2	1	1	-1	R_Z	0	0	0	0	0	0	0	0	0
E	2	-1	0	$T_X; R_Y$	l	0	0	0	0	0	c	d	0
				$T_Y; R_X$	0	l	0	c	$-c$	0	0	0	d

D_3	E	$2C_3$	$3C_2$		μ'_X	μ'_Y	μ'_Z	α'_{XX}	α'_{YY}	α'_{ZZ}	α'_{XY}	α'_{XZ}	α'_{YZ}
A_1	1	1	1		0	0	0	a	a	b	0	0	0
A_2	1	1	-1	$T_Z; R_Z$	0	0	n	0	0	0	0	0	0
E	2	-1	0	$T_X; R_X$	l	0	0	c	$-c$	0	0	0	d
				$T_Y; R_Y$	0	l	0	0	0	0	$-c$	$-d$	0

$D_{3d} = D_3 \times i = \{E, 2C_3, 3C_2; i, 2S_6, 3\sigma_d\}$

$\mathbf{T}, \boldsymbol{\mu}'$ in A_{2u}, E_u

$\mathbf{R}, \boldsymbol{\alpha}'$ in A_{1g}, E_g

Tetragonal: C_4, S_4, C_{4h}, C_{4v}, D_4, D_{2d}, D_{4h}

C_4	E	C_4^+	C_2	C_4^-										
S_4	E	S_4^+	C_2	S_4^-		μ'_X	μ'_Y	μ'_Z	α'_{XX}	α'_{YY}	α'_{ZZ}	α'_{XY}	α'_{XZ}	α'_{YZ}
A	1	1	1	1	$T_Z(C_4); R_Z$	0	0	$n(C_4)$	a	a	b	0	0	0
B	1	-1	1	-1	$T_Z(S_4)$	0	0	$n(S_4)$	c	$-c$	0	d	0	0
E_+	$\{1$	i	-1	$-i\}$	$T_X; R_{X,Y}$	l	0	0	0	0	0	0	e	f
E_-	$\{1$	$-i$	-1	$i\}$	$T_Y; R_{Y,X}$	0	l	0	0	0	0	0	$\pm f^*$	$\mp e^*$

* upper sign S_4, lower sign C_4

$C_{4h} = C_4 \times i = \{E, C_4^+, C_2, C_4^-; i, S_4^+, \sigma_h, S_4^-\}$

$\mathbf{T}, \boldsymbol{\mu}'$ in A_u, E_u; $\mathbf{R}$ in A_g, E_g; $\boldsymbol{\alpha}'$ in A_g, B_g, E_g

C_{4v}	E	$2C_4$	C_2	$2\sigma_v$	$2\sigma_d$										
D_4	E	$2C_4$	C_2	$2C'_2$	$2C''_2$										
D_{2d}	E	$2S_4$	C_2	$2C'_2$	$2\sigma_d$		μ'_X	μ'_Y	μ'_Z	α'_{XX}	α'_{YY}	α'_{ZZ}	α'_{XY}	α'_{XZ}	α'_{YZ}
A_1	1	1	1	1	1	$T_Z(C_{4v})$	0	0	$n(C_{4v})$	a	a	b	0	0	0
A_2	1	1	1	-1	-1	$T_Z(D_4); R_Z$	0	0	$n(D_4)$	0	0	0	0	0	0
B_1	1	-1	1	1	-1		0	0	0	c	$-c$	0	0	0	0
B_2	1	-1	1	-1	1	$T_Z(D_{2d})$	0	0	$n(D_{2d})$	0	0	0	d	0	0
E	2	0	-2	0	0	$T_X; R_{Y,X,Y}$	l	0	0	0	0	0	0	e*	0*
						$T_Y; R_{X,Y,X}$	0	l	0	0	0	0	0	0*	e*

* for C_{4v}; for D_4 replace by $\begin{pmatrix} 0 & e \\ -e & 0 \end{pmatrix}$; for D_{2d} by $\begin{pmatrix} 0 & e \\ e & 0 \end{pmatrix}$

$D_{4h} = D_4 \times i = \{E, 2C_4, C_2, 2C'_2, 2C''_2; i, 2S_4, \sigma_h, 2\sigma_v, 2\sigma_d\}$

$\mathbf{T}, \boldsymbol{\mu}'$ in A_{2u}, E_u; $\mathbf{R}$ in A_{2g}, E_g; $\boldsymbol{\alpha}'$ in $A_{1g}, B_{1g}, B_{2g}, E_g$.

Hexagonal: C_6, C_{3h}, C_{6h}, D_6, C_{6v}, D_{3h}, D_{6h} $\qquad \varepsilon = e^{2\pi i/6}$

C_6	E	C_6^+	C_3^+	C_2	C_3^-	C_6^-		μ'_X	μ'_Y	μ'_Z	α'_{XX}	α'_{YY}	α'_{ZZ}	α'_{XY}	α'_{XZ}	α'_{YZ}
A	1	1	1	1	1	1	$T_Z; R_Z$	0	0	n	a	a	b	0	0	0
B	1	−1	1	−1	1	−1		0	0	0	0	0	0	0	0	0
E_{1+}	1	ε	$-\varepsilon^*$	−1	$-\varepsilon$	ε^*	$T_X; R_X$	l	0	0	0	0	0	0	c	d
E_{1-}	1	ε^*	$-\varepsilon$	−1	$-\varepsilon^*$	ε	$T_Y; R_Y$	0	l	0	0	0	0	0	$-d$	c
E_{2+}	1	$-\varepsilon^*$	$-\varepsilon$	1	$-\varepsilon^*$	$-\varepsilon$		0	0	0	e	$-e$	0	f	0	0
E_{2-}	1	$-\varepsilon$	$-\varepsilon^*$	1	$-\varepsilon$	$-\varepsilon^*$		0	0	0	f	$-f$	0	$-e$	0	0

C_{3h}	E	C_3^+	C_3^-	σ_h	S_3^+	S_3^-	$\varepsilon = e^{2\pi i/3}$									
A'	1	1	1	1	1	1	R_Z	0	0	0	a	a	b	0	0	0
E'_+	1	ε	ε^*	1	ε	ε^*	T_X	l	0	0	c	$-c$	0	d	0	0
E'_-	1	ε^*	ε	1	ε^*	ε	T_Y	0	l	0	d	$-d$	0	$-c$	0	0
A''	1	1	1	−1	−1	−1	T_Z	0	0	n	0	0	0	0	0	0
E''_+	1	ε	ε^*	−1	$-\varepsilon$	$-\varepsilon^*$	R_X	0	0	0	0	0	0	0	e	f
E''_-	1	ε^*	ε	−1	$-\varepsilon^*$	$-\varepsilon$	R_Y	0	0	0	0	0	0	0	$-f$	e

$C_{6h} = C_6 \times i = \{E, C_6^+, C_3^+, C_2, C_3^-, C_6^-; i, S_3^-, S_6^-, \sigma_h, S_6^+, S_3^+\}$
$\mathbf{T}, \boldsymbol{\mu}'$ in A_u, E_{1u}; $\mathbf{R}$ in A_g, E_{1g}; $\boldsymbol{\alpha}'$ in A_g, E_{1g}, E_{2g}

D_6 C_{6v}	E E	$2C_6$ $2C_6$	$2C_3$ $2C_3$	C_2 C_2	$3C'_2$ $3\sigma_v$	$3C''_2$ $3\sigma_d$										
A_1	1	1	1	1	1	1	$T_Z(C_{6v})$	0	0	n	a	a	b	0	0	0
A_2	1	1	1	1	−1	−1	$T_Z(D_6)$; R_Z	0	0	n	0	0	0	0	0	0
B_1	1	−1	1	−1	1	−1		0	0	0	0	0	0	0	0	0
B_2	1	−1	1	−1	−1	1		0	0	0	0	0	0	0	0	0
E_1	2	1	−1	−2	0	0	$T_X; R_{X,Y}$	l	0	0	0	0	0	0	0*	c*
							$T_Y; R_{Y,X}$	0	l	0	0	0	0	0	$-c$*	0*
E_2	2	−1	−1	2	0	0		0	0	0	0	0	0	d	0	0
								0	0	0	d	$-d$	0	0	0	0

* for D_6; for C_{6v} replace by $\begin{pmatrix} c & 0 \\ 0 & c \end{pmatrix}$

D_{3h}	E	$2C_3$	$3C_2$	σ_h	$2S_3$	$3\sigma_v$										
A'_1	1	1	1	1	1	1		0	0	0	a	a	b	0	0	0
A'_2	1	1	−1	1	1	−1	R_Z	0	0	0	0	0	0	0	0	0
E'	2	−1	0	2	−1	0	T_X	l	0	0	0	0	0	c	0	0
							T_Y	0	l	0	c	$-c$	0	0	0	0
A''_1	1	1	1	−1	−1	−1		0	0	0	0	0	0	0	0	0
A''_2	1	1	−1	−1	−1	1	T_Z	0	0	n	0	0	0	0	0	0
E''	2	−1	0	−2	1	0	R_X	0	0	0	0	0	0	0	0	d
							R_Y	0	0	0	0	0	0	0	$-d$	0

$D_{6h} = D_6 \times i = \{E, 2C_6, 2C_3, C_2, 3C'_2, 3C''_2; i, 2S_3, 2S_6, \sigma_h, 3\sigma_d, 3\sigma_v\}$
$\mathbf{T}, \boldsymbol{\mu}'$ in A_{2u}, E_{1u}; $\mathbf{R}$ in A_{2g}, E_{1g}; $\boldsymbol{\alpha}'$ in A_{1g}, E_{1g}, E_{2g}

Cubic: T, T_h, O, T_d, O_h $\qquad \varepsilon = e^{2\pi i/3}$

T	E	$4C_3^+$	$4C_3^-$	$3C_2$		μ'_X	μ'_Y	μ'_Z	α'_{XX}	α'_{YY}	α'_{ZZ}	α'_{XY}	α'_{XZ}	α'_{YZ}
A	1	1	1	1		0	0	0	a	a	a	0	0	0
E_+	1	ε	ε^*	1		0	0	0	b	b	$-2b$	0	0	0
E_-	1	ε^*	ε	1		0	0	0	$-3^{1/2}b$	$3^{1/2}b$	0	0	0	0
F	3	0	0	-1	T_X, R_X	l	0	0	0	0	0	0	0	c
					T_Y, R_Y	0	l	0	0	0	0	0	c	0
					T_Z, R_Z	0	0	l	0	0	0	c	0	0

$T_h = T \times i = \{E, 4C_3^+, 4C_3^-, 3C_2; i, 4S_6^+, 4S_6^-, 3\sigma_h\}$
$\qquad \mathbf{T}, \boldsymbol{\mu}'$ in F_u; $\mathbf{R}$ in F_g; $\boldsymbol{\alpha}'$ in A_g, E_g, F_g

O	E	$8C_3$	$3C_2$	$6C_4$	$6C'_2$										
T_d	E	$8C_3$	$3C_2$	$6S_4$	$6\sigma_d$		μ'_X	μ'_Y	μ'_Z	α'_{XX}	α'_{YY}	α'_{ZZ}	α'_{XY}	α'_{XZ}	α'_{YZ}
A_1	1	1	1	1	1		0	0	0	a	a	a	0	0	0
A_2	1	1	1	-1	-1		0	0	0	0	0	0	0	0	0
E	2	-1	2	0	0		0	0	0	b	b	$-2b$	0	0	0
							0	0	0	$-3^{1/2}b$	$3^{1/2}b$	0	0	0	0
F_1	3	0	-1	1	-1	$T_X(O)$; R_X	$l(O)$	0	0	0	0	0	0	0	0
						$T_Y(O)$; R_Y	0	$l(O)$	0	0	0	0	0	0	0
						$T_Z(O)$; R_Z	0	0	$l(O)$	0	0	0	0	0	0
F_2	3	0	-1	-1	1	$T_X(T_d)$	$l(T_d)$	0	0	0	0	0	0	0	c
						$T_Y(T_d)$	0	$l(T_d)$	0	0	0	0	0	c	0
						$T_Z(T_d)$	0	0	$l(T_d)$	0	0	0	c	0	0

$O_h = O \times i = \{E, 8C_3, 3C_2, 6C_4, 6C'_2; i, 8S_6, 3\sigma_h, 6S_4, 6\sigma_d\}$
$\qquad \mathbf{T}, \boldsymbol{\mu}'$ in F_{1u}; $\mathbf{R}$ in F_{1g}; $\boldsymbol{\alpha}'$ in A_{1g}, E_g, F_{2g}

Linear Molecules: $C_{\infty v}$, $D_{\infty h}$

$C_{\infty v}$	E	$2C_\infty^\phi$	…	$\infty\sigma_v$		μ'_X	μ'_Y	μ'_Z	α'_{XX}	α'_{YY}	α'_{ZZ}	α'_{XY}	α'_{XZ}	α'_{YZ}
Σ^+	1	1	…	1	T_Z	0	0	n	a	a	b	0	0	0
Σ^-	1	1	…	-1	R_Z	0	0	0	0	0	0	0	0	0
Π	2	$2\cos\phi$	…	0	T_X; R_Y	l	0	0	0	0	0	0	c	0
					T_Y; R_X	0	l	0	0	0	0	0	0	c
Δ	2	$2\cos 2\phi$	…	0		0	0	0	0	0	d	0	0	0
						0	0	0	d	$-d$	0	0	0	0
…	…	…	…	…		…	…	…	…		…	…	…	…

$D_{\infty h} = C_{\infty v} \times i = \{E, 2C_\infty^\phi, \ldots \infty\sigma_v; i, 2S_\infty^\phi, \ldots \infty C_2\}$
$\qquad \mathbf{T}, \boldsymbol{\mu}'$ in Σ_u^+, Π_u; $\mathbf{R}$ in Σ_g^-, Π_g; $\boldsymbol{\alpha}'$ in Σ_g^+, Π_g, Δ_g

APPENDIX

II

SYMMETRY ANALYSIS IN CARTESIAN COORDINATES

Since we wanted the reader to acquire in Section 2-3 a better sense of the form of normal vibration modes than can be grasped from a cartesian description, we used internal coordinates, but pointed out that redundancies, i.e., linear relations amongst internal coordinates, may complicate the choice of a kinetically complete set, by which we mean a set capable of describing every possible molecular deformation.

A cartesian basis set does not suffer from this difficulty: all that is required is to choose three cartesian displacements for each atom, and finally to subtract three translational and two or three rotational degrees of freedom accordingly as the molecule is linear or non-linear. We will now subject such a set of coordinates to symmetry analysis in order to find the symmetry species of all the normal coordinates.

The matrices which represent the transformation of the $3N$ cartesian coordinates can only have diagonal elements associated with those atoms in the molecule which are *unshifted* by the symmetry operation in question. Only such atoms need to be considered in evaluating the character. Further, since the character is invariant to any change in the orientation of the cartesian axes, one may choose this orientation in whatever is the most convenient manner. We shall adopt the z-axis as that about which any rotation is performed. Then the 3×3 matrix representing the transformation of the cartesian coordinates of an unshifted atom assumes the form:

$$\begin{pmatrix} \cos\theta_R & -\sin\theta_R & 0 \\ \sin\theta_R & \cos\theta_R & 0 \\ 0 & 0 & \pm 1 \end{pmatrix} \qquad \text{(II-1)}$$

in which θ_R is the angle of rotation and the choice of signs in the zz element depends upon whether the operation is purely rotational, i.e., the operations E, C_n

or whether the operation includes reflection, namely, σ, i, and S_n operations. One speaks of the purely rotational operations as *proper*, and those involving reflection as *improper*; the + sign applies to the proper and the − to the improper operations.

The trace of the matrix in (II-1) is $\pm 1 + 2\cos\theta_R$, and only the *unshifted* atoms can contribute to the character of the $3N$-dimensional representation, so that

$$\chi_R(3N) = u_R(\pm 1 + 2\cos\theta_R) \tag{II-2}$$

where u_R is the number of atoms *unshifted* by operation R. The decomposition formula (2-2-27) can now be used to find the numbers of normal modes of each symmetry species, i.e.,

$$\eta^{(\gamma)}(3N) = \frac{1}{g}\sum_R \chi_R^{(\gamma)*}\chi_R(3N) \tag{II-3}$$

For the example of XY_3 with D_{3h} symmetry we summarize the steps in this sort of calculation in Table II-1.

In the upper part of the table we have simply transcribed the character table for the irreducible representations of this group. The last three lines exhibit the application of the discussion given above. Then the use of (II-3) enables one to calculate the quantities $\eta^{(\gamma)}$ which are tabulated in the next to last column of the table, i.e.,

$$\Gamma(3N) = A_1' + A_2' + 3E' + 2A_2'' + E''$$

In order to remove the translational and rotational modes it is only necessary to identify their symmetry species which we indicated for each group in Appendix I by the symbols T and R. Note that in the degenerate species T_x, T_y or R_x, R_y count as one pair. Thus, one arrives at the last column, describing the $3N - 6$ vibrational degrees of freedom as

$$\Gamma(\text{Vib.}) = A_1' + 2E' + A_2''$$

Table II-1 Symmetry analysis of the XY_3 molecule

D_{3h}	E	$2C_3$	$3C_2$	σ_h	$2S_3$	$3\sigma_v$		$\eta^{(\gamma)}(3N)$	$\eta^{(\gamma)}(\text{Vib.})$
A_1'	1	1	1	1	1	1		1	1
A_2'	1	1	−1	1	1	−1	R_z	1	0
E'	2	−1	0	2	−1	0	T_x, T_y	3	2
A_1''	1	1	1	−1	−1	−1		0	0
A_2''	1	1	−1	−1	−1	1	T_z	2	1
E''	2	−1	0	−2	1	0	R_x, R_y	1	0
u_R	4	1	2	4	1	2			
$\pm 1 + 2\cos\theta_R$	3	0	−1	1	−2	1			
$\chi_R(3N)$	12	0	−2	4	−2	2			

It may be noted that the character of the three translational degrees of freedom, T_x, T_y, T_z is the trace of (II-1). A rotation R_x, R_y, R_z is another kind of vector, namely a vector product of two three-dimensional vectors. Sometimes the two types are called polar (T) and axial (R). It is easily demonstrated that the character of any axial vector is

$$\chi_R = (1 \pm 2 \cos \theta_R)$$

where the sign alternatives are the same as used previously, + for the proper and − for improper symmetry operations.

A much more rapid way of effecting the same calculation uses the correlation tables. Each atom in the molecule belongs to a symmetrically equivalent set, and any one member of such a set occupies a site in the molecule, meaning that its position is invariant under a subgroup of operations from the molecular group. The three-dimensional cartesian representation of an atom at its site can be ascertained by searching for the species of a translation in the character table for the appropriate subgroup. The contribution of all atoms in a symmetrically equivalent set may be found with the aid of a correlation table. The overall $3N$ cartesian representation is simply the sum of such contributions from all equivalent sets.

For $XY_3(D_{3h})$, the site of X is D_{3h} itself, but that of Y is C_{2v}. Thus the X atom contributes $A_2'' + E'$ and the $3Y$ atoms contribute the species with which the translational species of C_{2v}, namely $A_1 + B_1 + B_2$, correlate on passing to D_{3h}. These are found in Appendix IV to be

C_{2v}	D_{3h}
A_1	$\rightarrow A_1' + E'$
B_1	$\rightarrow A_2'' + E''$
B_2	$\rightarrow A_2' + E'$

Thus

$$\Gamma(3N) = A_1' + A_2' + 3E' + 2A_2'' + E''$$

in agreement with the result found above using a lengthier calculation.

The principle which underlies this convenient method is contained in (2-5-5), where it was shown how to apply the decomposition formula in finding the correlation between a group G and one of its subgroups H. There, the intent was to relate the symmetry species of a parent molecule belonging to group G with those of an isotopic derivative of lower symmetry H. In the present problem we note that $\chi_R(3N)$ can be expressed as a sum over atoms α; each atom contributes the character of a translation if R is in its site group H_α and zero otherwise. Thus (II-3) becomes

$$\eta^{(\gamma)}(3N) = \sum_\alpha \frac{1}{g} \sum_{R \subset H_\alpha} \chi_R^{(\gamma)*} \chi_R(T) \tag{II-4}$$

But although the site groups H_α which leave equivalent atoms invariant are not identical, they are related by the formula

$$H_{\alpha'} = XH_\alpha X^{-1} \tag{II-5}$$

where $H_{\alpha'}$ means the subgroup which leaves α' invariant and X is any operation which sends $\alpha \rightarrow \alpha'$. Thus they are conjugate and yield exactly the same character sums over $R \subset H_\alpha$ in (II-4). Moreover there are necessarily g/h_α atoms in an equivalent set, so that (II-4) becomes

$$\begin{aligned} \eta^{(\gamma)}(3N) &= \sum_{\substack{\text{(non-equivalent} \\ \text{sets)}}} \left(\frac{1}{g}\frac{g}{h_\alpha}\right) \sum_{R \subset H_\alpha} \chi_R^{(\gamma)*} \chi_R(T) \\ &= \frac{1}{h_\alpha} \sum_{\substack{\text{(non-equivalent} \\ \text{sets)}}} \sum_{R \subset H_\alpha} \chi_R^{(\gamma)*} \chi_R(T) \end{aligned} \tag{II-6}$$

Aside from the sum over non-equivalent atomic sets, (II-6) now is precisely the correlation formula (2-5-5) between certain symmetry species of H_α, namely those describing translation and symmetry species of G.

APPENDIX

III

HARMONIC OSCILLATOR STATE FUNCTIONS: CREATION AND ANNIHILATION OPERATORS

In the harmonic approximation, the state function for a molecule with f degrees of freedom ($f = 3n - 5$ for linear and $3n - 6$ for non-linear molecules) is a product of independent oscillator functions:

$$|v\rangle = |v_1\rangle |v_2\rangle \cdots |v_f\rangle$$

We proceed to consider the most important properties of these individual state functions (kets in the Dirac language). We define the dimensionless operator (remembering that molecular normal coordinates Q have dimensions $(\text{mass})^{1/2} \times \text{length}$)

$$a = (2\hbar\omega)^{-1/2}(P - i\omega Q) \tag{III-1}$$

and call it the *annihilation operator* because it has the important property that

$$a|v\rangle = -iv^{1/2}|v - 1\rangle \tag{III-2}$$

Its complex conjugate

$$a^+ = (2\hbar\omega)^{-1/2}(P + i\omega Q) \tag{III-3}$$

(where P and Q are real) is a *creation operator*, since

$$a^+|v\rangle = i(v + 1)^{1/2}|v + 1\rangle \tag{III-4}$$

The reader unfamiliar with this formalism may consult any of the recent texts in quantum mechanics, or may verify (III-3) and (III-4) with the aid of recursion formulas for the harmonic oscillator functions given analytically in the introductory textbooks.

The creation operator a^+ may be used to express any ket in terms of the ground state: by repeated application of Eq. (III-4) one finds

$$|v\rangle = i^{-v}(v!)^{-1/2}(a^+)^v|0\rangle \tag{III-5}$$

Expressions such as (III-5) will be useful in discussing selection rules and intensities for transitions.

For a bra as distinct from a ket, one must observe the general rule that conjugation is required, so that if, from (III-4)

$$|v+1\rangle = i^{-1}(v+1)^{-1/2}a^{+}|v\rangle$$

then

$$\langle v+1| = i(v+1)^{-1/2}\langle v|a$$

etc.

Since the bras taken together with the kets satisfy the orthonormality rule

$$\langle v|v'\rangle = \delta_{vv'}$$

it is easy to compute the various transition elements needed, for example, in the intensity theory. Thus if one requires

$$\langle v'|Q|v\rangle$$

one first expresses

$$Q = (\hbar/2\omega)^{1/2} i(a - a^{+}) \tag{III-6}$$

and then uses

$$(a - a^{+})|v\rangle = -iv^{1/2}|v-1\rangle - i(v+1)^{1/2}|v+1\rangle$$

to obtain

$$\begin{aligned}\langle v'|Q|v\rangle &= (\hbar/2\omega)^{1/2}[v^{1/2}\langle v'|v-1\rangle + (v+1)^{1/2}\langle v'|v+1\rangle] \\ &= (\hbar/2\omega)^{1/2}[v^{1/2}\delta_{v',v-1} + (v+1)^{1/2}\delta_{v',v+1}]\end{aligned}$$

Higher powers of Q and the various powers of P can be developed from (III-6) and the corresponding relation

$$P = (\hbar\omega/2)^{1/2}(a + a^{+}) \tag{III-7}$$

Such relations are useful, for example, in the discussion of anharmonicity, where it is necessary to evaluate matrix elements involving cubic and quartic terms in Q.

In the above discussion, it has been assumed that P and Q are real, which is usually the case in molecular dynamics. When P and Q are complex, which invariably occurs when they are phonon operators, the appropriate generalization consists in defining

$$a = (2\hbar\omega)^{-1/2}(P - i\omega Q^{*}) \tag{III-8}$$

and

$$a^{+} = (2\hbar\omega)^{-1/2}(P^{*} + i\omega Q) \tag{III-9}$$

Operators such as (III-8) and (III-9) obey the simple commutation rule

$$[a, a^{+}] = 1$$

and since the Hamiltonian is

$$H = \frac{\hbar\omega}{2}(aa^+ + a^+a)$$

the eigenvalues are just those found in the case of real P and Q, namely

$$E = \hbar\omega(v + 1/2)$$

This treatment makes it unnecessary to appeal to the detailed analytic forms of the harmonic oscillator wave functions, but the reader who desires some reassurance that appropriate wave functions can be constructed may derive consolation from a little further development.

Although it has been shown that translational symmetrization leads to a diagonalized Hamiltonian which then admits quantization, one must be careful to remember that the normal coordinates are in general complex but that the standard treatment of the harmonic oscillator wave equation is usually written in terms of a real coordinate variable. This situation causes no difficulty inasmuch as the universal $+\boldsymbol{\kappa}$, $-\boldsymbol{\kappa}$ degeneracy implies a potential energy of the form

$$V = 1/2 \sum_{\boldsymbol{\kappa}} \omega^2(\boldsymbol{\kappa}) Q^*(\boldsymbol{\kappa}) Q(\boldsymbol{\kappa})$$

in which $\omega(+\boldsymbol{\kappa}) = \omega(-\boldsymbol{\kappa})$. If we introduce the real and imaginary parts of Q as the dynamic variables,

$$Q(\boldsymbol{\kappa}) = x(\boldsymbol{\kappa}) + iy(\boldsymbol{\kappa})$$

and make use of the relation

$$Q^*(\boldsymbol{\kappa}) = x(\boldsymbol{\kappa}) - iy(\boldsymbol{\kappa}) = Q(-\boldsymbol{\kappa}) = x(-\boldsymbol{\kappa}) + iy(-\boldsymbol{\kappa})$$

we can express the potential energy as

$$V = 1/2 \sum_{\boldsymbol{\kappa}} \omega^2(\boldsymbol{\kappa})[x^2(\boldsymbol{\kappa}) + y^2(\boldsymbol{\kappa})]$$

The ordinary harmonic oscillator functions are then appropriate solutions using the real variables $x(\boldsymbol{\kappa})$ and $y(\boldsymbol{\kappa})$ and we can represent states for a given $\boldsymbol{\kappa}$ with the aid of the two quantum numbers v_x and v_y, i.e., $|v_x v_y\rangle$. Since these states do not belong to a single symmetry species of the translational group, i.e., they are not eigenfunctions of the "crystal momentum," it is useful to form linear combinations with this desirable property. A few of these are shown in Table III-1, which indicates the pair of quantum numbers $v(\boldsymbol{\kappa})$ and $v(-\boldsymbol{\kappa})$, the resultant translational symmetry, and the appropriate linear combination of the $|v_x v_y\rangle$ states. As an example, consider $v(\boldsymbol{\kappa}) = 2$, $v(-\boldsymbol{\kappa}) = 0$: such a function must contain $Q^2(\boldsymbol{\kappa})$ in order that it transform like $2\boldsymbol{\kappa}$, and therefore has the polynomial form

$$Q^2(\kappa) = (x + iy)^2 = x^2 - y^2 + 2ixy$$

Table III-1 Relation between real and complex harmonic oscillator wave functions for phonons

Quantum numbers for complex phonon coordinates			
$v(\boldsymbol{\kappa})$	$v(-\boldsymbol{\kappa})$	Resultant $\boldsymbol{\kappa}$	$\Sigma a_{v_x v_y} \lvert v_x v_y\rangle$
0	0	0	$\lvert 00\rangle$
1	0	$\boldsymbol{\kappa}$	$2^{-1/2}[\lvert 10\rangle + i\lvert 01\rangle]$
0	1	$-\boldsymbol{\kappa}$	$2^{-1/2}[\lvert 10\rangle - i\lvert 01\rangle]$
2	0	$2\boldsymbol{\kappa}$	$4^{-1/2}[\lvert 20\rangle - \lvert 02\rangle + 2^{1/2}i\lvert 11\rangle]$
1	1	0	$2^{-1/2}[\lvert 20\rangle + \lvert 02\rangle]$
0	2	$-2\boldsymbol{\kappa}$	$4^{-1/2}[\lvert 20\rangle - \lvert 02\rangle - 2^{1/2}i\lvert 11\rangle]$

But since

$$\lvert 20\rangle = \psi_{00}(2^2 \cdot 2!)^{-1/2}(4\gamma x^2 - 2)$$
$$\lvert 11\rangle = \psi_{00}(2^1 \cdot 1!)^{-1}4\gamma xy$$
$$\lvert 02\rangle = \psi_{00}(2^2 2!)^{-1/2}(4\gamma y^2 - 2)$$

it follows that the (normalized) linear combination must be as stated in the table. Similarly the state $v(\boldsymbol{\kappa}) = v(-\boldsymbol{\kappa}) = 1$, which has a resultant $\boldsymbol{\kappa} = 0$, must be of the form $Q(\boldsymbol{\kappa})Q(-\boldsymbol{\kappa}) + \text{const.}$ which is $x^2 + y^2 + \text{const.}$; this is easily seen to involve equal components of the states $\lvert 20\rangle$ and $\lvert 02\rangle$, proportional to $\gamma Q(\boldsymbol{\kappa})Q(-\boldsymbol{\kappa}) - 1/2$.

We conclude by remarking that this analysis may seem to have doubled the degrees of freedom: this is, of course, not the case, simply because the real phonon variables $x(\boldsymbol{\kappa})$ and $y(\boldsymbol{\kappa})$ obey the relations

$$x(\boldsymbol{\kappa}) = x(-\boldsymbol{\kappa})$$

and

$$y(\boldsymbol{\kappa}) = -y(-\boldsymbol{\kappa})$$

APPENDIX

IV

CORRELATION BETWEEN POINT GROUPS, ETC.

The method of correlation of symmetry species between a group and its subgroups is so useful a tool as noted frequently in Chapters 4 and 6, that it is worthwhile to tabulate these relations in some detail. So far as the crystallographic point groups (32 in all) are concerned, every group is a subgroup either of O_h or of D_{6h}. This makes it possible to list all such correlations in two tables, as has been done by Poulet and Mathieu.[1] Of course in using a correlation table in such a form, one must be careful to note that if G_1 and G_2 are both subgroups, say of O_h, it does not necessarily follow that G_2 is a subgroup of G_1 or vice versa.

In the present appendix we have followed the format of Wilson, Decius, and Cross[2] which uses a somewhat larger space in order to indicate all subgroups of a given group with the exception of O_h, for which all the maximal subgroup correlations are exhibited, requiring the reader to perform the overall correlation in two steps.

Other group correlations are required, of which the most common ones involve the linear molecule, of symmetry $D_{\infty h}$ or $C_{\infty v}$. Here we may encounter the following possibilities:

1. $D_{\infty h}$ molecular symmetry. The largest possible crystallographic site symmetry is D_{6h}; any subgroup of D_{6h} is, of course, also a possible site symmetry and therefore of possible interest. By giving the correlation between $D_{\infty h}$ and

1. H. Poulet and J.-P. Mathieu, *Spectres de Vibration et Symétrie des Cristaux*, Gordon and Breach, Paris, 1970.
2. E. B. Wilson, Jr., J. C. Decius, and P. C. Cross, *Molecular Vibrations*, McGraw-Hill, New York, 1955, Appendix X-14.

D_{6h} or D_{4h}, we enable the reader to carry out the overall correlation from a linear symmetric $D_{\infty h}$ molecule to any possible site in two stages.

2. $C_{\infty v}$ molecular symmetry. Here the maximal site symmetry is C_{6v} or C_{4v}, and the remaining procedure is completely analogous to case (a).

Many direct correlations to lower order subgroups for the linear molecules are given by Herzberg.[3] For example, the CO_2 molecule occupies an S_6 site in the crystal: the sequential correlation $D_{\infty h} \rightarrow D_{6h} \rightarrow S_6$ by using Tables IV-2 and IV-1 yields: $\Sigma_g^+ \rightarrow A_{1g} \rightarrow A_g$, or $\Pi_u \rightarrow E_{1u} \rightarrow E_u$, etc. Similarly, HCN is on a C_{2v} site in the low temperature crystal, and one finds for $C_{\infty v} \rightarrow C_{6v} \rightarrow C_{2v}$, $\Sigma^+ \rightarrow A_1 \rightarrow A_1$, $\Pi \rightarrow E_1 \rightarrow B_1 + B_2$, $\Delta \rightarrow E_2 \rightarrow A_1 + A_2$.

As noted in Chapter 4, the application of symmetry to the factoring of the dynamic matrix throughout the Brillouin zone relies heavily upon correlations between $G = H(\mathbf{0})$ and $H(\boldsymbol{\kappa})$; for the *symmorphic* space groups this can be effected essentially with the use of information from this appendix. For the *non-symmorphic* space groups, again, with some precautions about phase factors, one needs only the point groups for $\boldsymbol{\kappa}$ in the interior of the zone, since $H(\boldsymbol{\kappa})$ is still isomorphic with the point groups. However, for $\boldsymbol{\kappa}$ at points of symmetry on the surface of the Brillouin zone, it frequently happens that $H(\boldsymbol{\kappa})$ is not isomorphic with a point group. This will be the case in those situations where, if $\{R \mid \tau_R\}$ and $\{S \mid \tau_S\}$ are any two factor group operations in $H(\boldsymbol{\kappa})$ such that,

$$S^{-1}\boldsymbol{\kappa} = \boldsymbol{\kappa} + \bar{\boldsymbol{\kappa}}_S$$

the quantity

$$\exp(i\boldsymbol{\kappa}_S \cdot \boldsymbol{\tau}_R) \neq 1$$

for some S and R. Examples mentioned earlier included the point $\frac{1}{2}0\frac{1}{2}$ (X) in the diamond structure $(O_h^7 = Fd3m)$. For these cases, the reader must be referred to other sources.[4–10] A few examples are given in Appendix X for cases discussed in this text, namely O_h^7 (diamond), C_6^6 (CHI_3), T_h^6 (CO_2), and D_{3d}^6 (calcite).

3. G. Herzberg, *Electronic Spectra of Polyatomic Molecules*, D. Van Nostrand Co., Inc., Princeton, N.J., 1966, Appendix IV, Table 59, page 576.
4. C. J. Bradley and A. P. Cracknell, *The Mathematical Theory of Symmetry in Solids*, Clarendon Press, Oxford, 1972.
5. G. Ya. Lyubarskii, *The Application of Group Theory in Physics*, Pergamon Press, New York, 1960.
6. O. V. Kovalev, *Irreducible Representations of the Space Groups*, Gordon and Breach, New York, 1965.
7. J. Zak, A. Casher, M. Glück, and Y. Gur, *The Irreducible Representations of Space Groups*, W. A. Benjamin, Inc., New York, 1969.
8. S. C. Miller and W. F. Love, *Tables of Irreducible Representations of Space Groups and Co-Representations of Magnetic Space Groups*, Pruett Press, Boulder, Colorado, 1967.
9. J. C. Slater, *Quantum Theory of Molecules and Solids*, McGraw-Hill Book Co., New York, 1965, Volume II, Appendix 3.
10. J. L. Warren, *Rev. Mod. Phys.*, **40**, 38 (1968).

Table IV-1 Correlation of symmetry for the crystallographic point groups

C_4	C_2
A	A
B	A
E	$2B$

C_6	C_3	C_2
A	A	A
B	A	B
E_1	E	$2B$
E_2	E	$2A$

D_2	C_2	C_2	C_2
A	A	A	A
B_1	A	B	B
B_2	B	A	B
B_3	B	B	A

D_3	C_3	C_2
A_1	A	A
A_2	A	B
E	E	$A + B$

D_4	C_4	C_2	C_2' C_2	C_2'' C_2
A_1	A	A	A	A
A_2	A	A	B	B
B_1	B	A	A	B
B_2	B	A	B	A
E	E	$2B$	$A + B$	$A + B$

D_6	C_6	C_2' D_3	C_2'' D_3	D_2	C_3	C_2	C_2' C_2	C_2'' C_2
A_1	A	A_1	A_1	A	A	A	A	A
A_2	A	A_2	A_2	B_1	A	A	B	B
B_1	B	A_1	A_2	B_2	A	B	A	B
B_2	B	A_2	A_1	B_3	A	B	B	A
E_1	E_1	E	E	$B_2 + B_3$	E	$2B$	$A + B$	$A + B$
E_2	E_2	E	E	$A + B_1$	E	$2A$	$A + B$	$A + B$

C_{2v}	C_2	$\sigma(zx)$ C_s	$\sigma(yz)$ C_s
A_1	A	A'	A'
A_2	A	A''	A''
B_1	B	A'	A''
B_2	B	A''	A'

C_{3v}	C_3	C_s
A_1	A	A'
A_2	A	A''
E	E	$A' + A''$

C_{4v}	C_4	σ_v C_{2v}	σ_d C_{2v}	C_2	σ_v C_s	σ_d C_s
A_1	A	A_1	A_1	A	A'	A'
A_2	A	A_2	A_2	A	A''	A''
B_1	B	A_1	A_2	A	A'	A''
B_2	B	A_2	A_1	A	A''	A'
E	E	$B_1 + B_2$	$B_1 + B_2$	$2B$	$A' + A''$	$A' + A''$

C_{6v}	C_6	σ_v C_{3v}	σ_d C_{3v}	$\sigma_v \to \sigma(zx)$ C_{2v}	C_3	C_2	σ_v C_3	σ_d C_3
A_1	A	A_1	A_1	A_1	A	A	A'	A'
A_2	A	A_2	A_2	A_2	A	A	A''	A''
B_1	B	A_1	A_2	B_1	A	B	A'	A''
B_2	B	A_2	A_1	B_2	A	B	A''	A'
E_1	E_1	E	E	$B_1 + B_2$	E	$2B$	$A' + A''$	$A' + A''$
E_2	E_2	E	E	$A_1 + A_2$	E	$2A$	$A' + A''$	$A' + A''$

C_{2h}	C_2	C_s	C_i	C_{3h}	C_3	C_s
A_g	A	A'	A_g	A'	A	A'
B_g	B	A''	A_g	E'	E	$2A'$
A_u	A	A''	A_u	A''	A	A''
B_u	B	A'	A_u	E''	E	$2A''$

C_{4h}	C_4	S_4	C_{2h}	C_2	C_s	C_i
A_g	A	A	A_g	A	A'	A_g
B_g	B	B	A_g	A	A'	A_g
E_g	E	E	$2B_g$	$2B$	$2A''$	$2A_g$
A_u	A	B	A_u	A	A''	A_u
B_u	B	A	A_u	A	A''	A_u
E_u	E	E	$2B_u$	$2B$	$2A'$	$2A_u$

C_{6h}	C_6	C_{3h}	S_6	C_{2h}	C_3	C_2	C_s	C_i
A_g	A	A'	A_g	A_g	A	A	A'	A_g
B_g	B	A''	A_g	B_g	A	B	A''	A_g
E_{1g}	E_1	E''	E_g	$2B_g$	E	$2B$	$2A''$	$2A_g$
E_{2g}	E_2	E'	E_g	$2A_g$	E	$2A$	$2A'$	$2A_g$
A_u	A	A''	A_u	A_u	A	A	A''	A_u
B_u	B	A'	A_u	B_u	A	B	A'	A_u
E_{1u}	E_1	E'	E_u	$2B_u$	E	$2B$	$2A'$	$2A_u$
E_{2u}	E_2	E''	E_u	$2A_u$	E	$2A$	$2A''$	$2A_u$

D_{2h}	D_2	$C_2(z)$ C_{2v}	$C_2(y)$ C_{2v}	$C_2(x)$ C_{2v}	$C_2(z)$ C_{2h}	$C_2(y)$ C_{2h}	$C_2(x)$ C_{2h}
A_g	A	A_1	A_1	A_1	A_g	A_g	A_g
B_{1g}	B_1	A_2	B_2	B_1	A_g	B_g	B_g
B_{2g}	B_2	B_1	A_2	B_2	B_g	A_g	B_g
B_{3g}	B_3	B_2	B_1	A_2	B_g	B_g	A_g
A_u	A	A_2	A_2	A_2	A_u	A_u	A_u
B_{1u}	B_1	A_1	B_1	B_2	A_u	B_u	B_u
B_{2u}	B_2	B_2	A_1	B_1	B_u	A_u	B_u
B_{3u}	B_3	B_1	B_2	A_1	B_u	B_u	A_u

D_{2h} (cont.)	$C_2(z)$ C_2	$C_2(y)$ C_2	$C_2(x)$ C_2	$\sigma(xy)$ C_s	$\sigma(zx)$ C_s	$\sigma(yz)$ C_s
A_g	A	A	A	A'	A'	A'
B_{1g}	A	B	B	A'	A''	A''
B_{2g}	B	A	B	A''	A'	A''
B_{3g}	B	B	A	A''	A''	A'
A_u	A	A	A	A''	A''	A''
B_{1u}	A	B	B	A''	A'	A'
B_{2u}	B	A	B	A'	A''	A'
B_{3u}	B	B	A	A'	A'	A''

D_{3h}	C_{3h}	D_3	C_{3v}	$\sigma_h \to \sigma_v(zy)$ C_{2v}	C_3	C_2	σ_h C_s	σ_v C_s
A'_1	A'	A_1	A_1	A_1	A	A	A'	A'
A'_2	A'	A_2	A_2	B_2	A	B	A'	A''
E'	E'	E	E	$A_1 + B_2$	E	$A + B$	$2A'$	$A' + A''$
A''_1	A''	A_1	A_2	A_2	A	A	A''	A''
A''_2	A''	A_2	A_1	B_1	A	B	A''	A'
E''	E''	E	E	$A_2 + B_1$	E	$A + B$	$2A''$	$A' + A''$

D_{4h}	D_4	$C'_2 \to C'_2$ D_{2d}	$C''_2 \to C'_2$ D_{2d}	C_{4v}	C_{4h}	C'_2 D_{2h}	C''_2 D_{2h}	C_4	S_4
A_{1g}	A_1	A_1	A_1	A_1	A_g	A_g	A_g	A	A
A_{2g}	A_2	A_2	A_2	A_2	A_g	B_{1g}	B_{1g}	A	A
B_{1g}	B_1	B_1	B_2	B_1	B_g	A_g	B_{1g}	B	B
B_{2g}	B_2	B_2	B_1	B_2	B_g	B_{1g}	A_g	B	B
E_g	E	E	E	E	E_g	$B_{2g} + B_{3g}$	$B_{2g} + B_{3g}$	E	E
A_{1u}	A_1	B_1	B_1	A_2	A_u	A_u	A_u	A	B
A_{2u}	A_2	B_2	B_2	A_1	A_u	B_{1u}	B_{1u}	A	B
B_{1u}	B_1	A_1	A_2	B_2	B_u	A_u	B_{1u}	B	A
B_{2u}	B_2	A_2	A_1	B_1	B_u	B_{1u}	A_u	B	A
E_u	E	E	E	E	E_u	$B_{2u} + B_{3u}$	$B_{2u} + B_{3u}$	E	E

D_{4h} (cont.)	C'_2 D_2	C''_2 D_2	C_2, σ_v C_{2v}	C_2, σ_d C_{2v}	C'_2 C_{2v}	C''_2 C_{2v}
A_{1g}	A	A	A_1	A_1	A_1	A_1
A_{2g}	B_1	B_1	A_2	A_2	B_1	B_1
B_{1g}	A	B_1	A_1	A_2	A_1	B_1
B_{2g}	B_1	A	A_2	A_1	B_1	A_1
E_g	$B_2 + B_3$	$B_2 + B_3$	$B_1 + B_2$	$B_1 + B_2$	$A_2 + B_2$	$A_2 + B_2$
A_{1u}	A	A	A_2	A_2	A_2	A_2
A_{2u}	B_1	B_1	A_1	A_1	B_2	B_2
B_{1u}	A	B_1	A_2	A_1	A_2	B_2
B_{2u}	B_1	A	A_1	A_2	B_2	A_2
E_u	$B_2 + B_3$	$B_2 + B_3$	$B_1 + B_2$	$B_1 + B_2$	$A_1 + B_1$	$A_1 + B_1$

D_{4h} (cont.)	C_2 C_{2h}	C'_2 C_{2h}	C''_2 C_{2h}	C_2 C_2	C'_2 C_2	C''_2 C_2
A_{1g}	A_g	A_g	A_g	A	A	A
A_{2g}	A_g	B_g	B_g	A	B	B
B_{1g}	A_g	A_g	B_g	A	A	B
B_{2g}	A_g	B_g	A_g	A	B	A
E_g	$2B_g$	$A_g + B_g$	$A_g + B_g$	$2B$	$A + B$	$A + B$
A_{1u}	A_u	A_u	A_u	A	A	A
A_{2u}	A_u	B_u	B_u	A	B	B
B_{1u}	A_u	A_u	B_u	A	A	B
B_{2u}	A_u	B_u	A_u	A	B	A
E_u	$2B_u$	$A_u + B_u$	$A_u + B_u$	$2B$	$A + B$	$A + B$

D_{4h} (cont.)	σ_h C_s	σ_v C_s	σ_d C_s	C_i
A_{1g}	A'	A'	A'	A_g
A_{2g}	A'	A''	A''	A_g
B_{1g}	A'	A'	A''	A_g
B_{2g}	A'	A''	A'	A_g
E_g	$2A''$	$A' + A''$	$A' + A''$	$2A_g$
A_{1u}	A''	A''	A''	A_u
A_{2u}	A''	A'	A'	A_u
B_{1u}	A''	A''	A'	A_u
B_{2u}	A''	A'	A''	A_u
E_u	$2A'$	$A' + A''$	$A' + A''$	$2A_u$

D_{6h}	D_6	C'_2 D_{3h}	C''_2 D_{3h}	C_{6v}	C_{6h}	C''_2 D_{3d}	C'_2 D_{3d}	$\sigma_h \to \sigma(xy)$ $\sigma_v \to \sigma(yz)$ D_{2h}	C_6	C_{3h}	C'_2 D_3	C''_2 D_3	σ_v C_{3v}	σ_d C_{3v}	S_6	D_2
A_{1g}	A_1	A'_1	A'_1	A_1	A_g	A_{1g}	A_{1g}	A_g	A	A'	A_1	A_1	A_1	A_1	A_g	A
A_{2g}	A_2	A'_2	A'_2	A_2	A_g	A_{2g}	A_{2g}	B_{1g}	A	A'	A_2	A_2	A_2	A_2	A_g	B_1
B_{1g}	B_1	A''_1	A''_2	B_2	B_g	A_{2g}	A_{1g}	B_{2g}	B	A''	A_1	A_2	A_2	A_1	A_g	B_2
B_{2g}	B_2	A''_2	A''_1	B_1	B_g	A_{1g}	A_{2g}	B_{3g}	B	A''	A_2	A_1	A_1	A_2	A_g	B_3
E_{1g}	E_1	E''	E''	E_1	E_{1g}	E_g	E_g	$B_{2g} + B_{3g}$	E_1	E''	E	E	E	E	E_g	$B_2 + B_3$
E_{2g}	E_2	E'	E'	E_2	E_{2g}	E_g	E_g	$A_g + B_{1g}$	E_2	E'	E	E	E	E	E_g	$A + B_1$
A_{1u}	A_1	A''_1	A''_1	A_2	A_u	A_{1u}	A_{1u}	A_u	A	A''	A_1	A_1	A_2	A_2	A_u	A
A_{2u}	A_2	A''_2	A''_2	A_1	A_u	A_{2u}	A_{2u}	B_{1u}	A	A''	A_2	A_2	A_1	A_1	A_u	B_1
B_{1u}	B_1	A'_1	A'_2	B_1	B_u	A_{2u}	A_{1u}	B_{2u}	B	A'	A_1	A_2	A_1	A_2	A_u	B_2
B_{2u}	B_2	A'_2	A'_1	B_2	B_u	A_{1u}	A_{2u}	B_{3u}	B	A'	A_2	A_1	A_2	A_1	A_u	B_3
E_{1u}	E_1	E'	E'	E_1	E_{1u}	E_u	E_u	$B_{2u} + B_{3u}$	E_1	E'	E	E	E	E	E_u	$B_2 + B_3$
E_{2u}	E_2	E''	E''	E_2	E_{2u}	E_u	E_u	$A_u + B_{1u}$	E_2	E''	E	E	E	E	E_u	$A + B_1$

D_{6h} (cont.)	C_2 C_{2v}	C'_2 C_{2v}	C''_2 C_{2v}	C_2 C_{2h}	C'_2 C_{2h}	C''_2 C_{2h}	C_3
A_{1g}	A_1	A_1	A_1	A_g	A_g	A_g	A
A_{2g}	A_2	B_1	B_1	A_g	B_g	B_g	A
B_{1g}	B_1	A_2	B_2	B_g	A_g	B_g	A
B_{2g}	B_2	B_2	A_2	B_g	B_g	A_g	A
E_{1g}	$B_1 + B_2$	$A_2 + B_2$	$A_2 + B_2$	$2B_g$	$A_g + B_g$	$A_g + B_g$	E
E_{2g}	$A_1 + A_2$	$A_1 + B_1$	$A_1 + B_1$	$2A_g$	$A_g + B_g$	$A_g + B_g$	E
A_{1u}	A_2	A_2	A_2	A_u	A_u	A_u	A
A_{2u}	A_1	B_2	B_2	A_u	B_u	B_u	A
B_{1u}	B_2	A_1	B_1	B_u	A_u	B_u	A
B_{2u}	B_1	B_1	A_1	B_u	B_u	A_u	A
E_{1u}	$B_1 + B_2$	$A_1 + B_1$	$A_1 + B_1$	$2B_u$	$A_u + B_u$	$A_u + B_u$	E
E_{2u}	$A_1 + A_2$	$A_2 + B_2$	$A_2 + B_2$	$2A_u$	$A_u + B_u$	$A_u + B_u$	E

D_{6h} (cont.)	C_2 C_2	C'_2 C_2	C''_2 C_2	σ_h C_s	σ_d C_s	σ_v C_s	C_i
A_{1g}	A	A	A	A'	A'	A'	A_g
A_{2g}	A	B	B	A'	A''	A''	A_g
B_{1g}	B	A	B	A''	A'	A''	A_g
B_{2g}	B	B	A	A''	A''	A'	A_g
E_{1g}	$2B$	$A + B$	$A + B$	$2A''$	$A' + A''$	$A' + A''$	$2A_g$
E_{2g}	$2A$	$A + B$	$A + B$	$2A'$	$A' + A''$	$A' + A''$	$2A_g$
A_{1u}	A	A	A	A''	A''	A''	A_u
A_{2u}	A	B	B	A''	A'	A'	A_u
B_{1u}	B	A	B	A'	A''	A'	A_u
B_{2u}	B	B	A	A'	A'	A''	A_u
E_{1u}	$2B$	$A + B$	$A + B$	$2A'$	$A' + A''$	$A' + A''$	$2A_u$
E_{2u}	$2A$	$A + B$	$A + B$	$2A''$	$A' + A''$	$A' + A''$	$2A_u$

D_{2d}	S_4	$C_2 \rightarrow C_2(z)$ D_2	C_{2v}	C_2 C_2	C'_2 C_2	C_s
A_1	A	A	A_1	A	A	A'
A_2	A	B_1	A_2	A	B	A''
B_1	B	A	A_2	A	A	A''
B_2	B	B_1	A_1	A	B	A'
E	E	$B_2 + B_3$	$B_1 + B_2$	$2B$	$A + B$	$A' + A''$

D_{3d}	D_3	C_{3v}	S_6	C_3	C_{2h}	C_2	C_s	C_i
A_{1g}	A_1	A_1	A_g	A	A_g	A	A'	A_g
A_{2g}	A_2	A_2	A_g	A	B_g	B	A''	A_g
E_g	E	E	E_g	E	$A_g + B_g$	$A + B$	$A' + A''$	$2A_g$
A_{1u}	A_1	A_2	A_u	A	A_u	A	A''	A_u
A_{2u}	A_2	A_1	A_u	A	B_u	B	A'	A_u
E_u	E	E	E_u	E	$A_u + B_u$	$A + B$	$A' + A''$	$2A_u$

S_4	C_2
A	A
B	A
E	$2B$

S_6	C_3	C_i
A_g	A	A_g
E_g	E	$2A_g$
A_u	A	A_u
E_u	E	$2A_u$

T	D_2	C_3	C_2
A	A	A	A
E	$2A$	E	$2A$
F	$B_1 + B_2 + B_3$	$A + E$	$A + 2B$

T_h	T	D_{2h}	S_6	D_2
A_g	A	A_g	A_g	A
E_g	E	$2A_g$	E_g	$2A$
F_g	F	$B_{1g} + B_{2g} + B_{3g}$	$A_g + E_g$	$B_1 + B_2 + B_3$
A_u	A	A_u	A_u	A
E_u	E	$2A_u$	E_u	$2A$
F_u	F	$B_{1u} + B_{2u} + B_{3u}$	$A_u + E_u$	$B_1 + B_2 + B_3$

T_h (cont.)	C_{2v}	C_{2h}	C_3	C_2	C_s	C_i
A_g	A_1	A_g	A	A	A'	A_g
E_g	$2A_1$	$2A_g$	E	$2A$	$2A'$	$2A_g$
F_g	$A_2 + B_1 + B_2$	$A_g + 2B_g$	$A + E$	$A + 2B$	$A' + 2A''$	$3A_g$
A_u	A_2	A_u	A	A	A''	A_u
E_u	$2A_2$	$2A_u$	E	$2A$	$2A''$	$2A_u$
F_u	$A_1 + B_1 + B_2$	$A_u + 2B_u$	$A + E$	$A + 2B$	$2A' + A''$	$3A_u$

T_d	T	D_{2d}	C_{3v}	S_4	D_2
A_1	A	A_1	A_1	A	A
A_2	A	B_1	A_2	B	A
E	E	$A_1 + B_1$	E	$A + B$	$2A$
F_1	F	$A_2 + E$	$A_2 + E$	$A + E$	$B_1 + B_2 + B_3$
F_2	F	$B_2 + E$	$A_1 + E$	$B + E$	$B_1 + B_2 + B_3$

T_d (cont.)	C_{2v}	C_3	C_2	C_s
A_1	A_1	A	A	A'
A_2	A_2	A	A	A''
E	$A_1 + A_2$	E	$2A$	$A' + A''$
F_1	$A_2 + B_1 + B_2$	$A + E$	$A + 2B$	$A' + 2A''$
F_2	$A_1 + B_1 + B_2$	$A + E$	$A + 2B$	$2A' + A''$

O	T	D_4	D_3	C_4	$3C_2$ D_2	$C_2, 2C'_2$ D_2
A_1	A	A_1	A_1	A	A	A
A_2	A	B_1	A_2	B	A	B_1
E	E	$A_1 + B_1$	E	$A + B$	$2A$	$A + B_1$
F_1	F	$A_2 + E$	$A_2 + E$	$A + E$	$B_1 + B_2 + B_3$	$B_1 + B_2 + B_3$
F_2	F	$B_2 + E$	$A_1 + E$	$B + E$	$B_1 + B_2 + B_3$	$A + B_2 + B_3$

O (cont.)	C_3	C_2	C_2
A_1	A	A	A
A_2	A	A	B
E	E	$2A$	$A + B$
F_1	$A + E$	$A + 2B$	$A + 2B$
F_2	$A + E$	$A + 2B$	$2A + B$

O_h	O	T_d	T_h	D_{4h}	D_{3d}
A_{1g}	A_1	A_1	A_g	A_{1g}	A_{1g}
A_{2g}	A_2	A_2	A_g	B_{1g}	A_{2g}
E_g	E	E	E_g	$A_{1g} + B_{1g}$	E_g
F_{1g}	F_1	F_1	F_g	$A_{2g} + E_g$	$A_{2g} + E_g$
F_{2g}	F_2	F_2	F_g	$B_{2g} + E_g$	$A_{1g} + E_g$
A_{1u}	A_1	A_2	A_u	A_{1u}	A_{1u}
A_{2u}	A_2	A_1	A_u	B_{1u}	A_{2u}
E_u	E	E	E_u	$A_{1u} + B_{1u}$	E_u
F_{1u}	F_1	F_2	F_u	$A_{2u} + E_u$	$A_{2u} + E_u$
F_{2u}	F_2	F_1	F_u	$B_{2u} + E_u$	$A_{1u} + E_u$

Table IV-2 Correlation of symmetry for linear molecules

$D_{\infty h}$	D_{6h}	D_{4h}	$D_{\infty h}$	D_{6h}	D_{4h}	$C_{\infty v}$	C_{6v}	C_{4v}
Σ_g^+	A_{1g}	A_{1g}	Σ_u^+	A_{2u}	A_{2u}	Σ^+	A_1	A_1
Σ_g^-	A_{2g}	A_{2g}	Σ_u^-	A_{1u}	A_{1u}	Σ^+	A_2	A_2
Π_g	E_{1g}	E_g	Π_u	E_{1u}	E_u	Π	E_1	E
Δ_g	E_{2g}	$B_{1g} + B_{2g}$	Δ_u	E_{2u}	$B_{1u} + B_{2u}$	Δ	E_2	$B_1 + B_2$
Φ_g	$B_{1g} + B_{2g}$	E_g	Φ_u	$B_{1u} + B_{2u}$	E_u	Φ	$B_1 + B_2$	E

APPENDIX

V

THE 230 SPACE GROUPS

In this appendix the generating elements of all 230 space groups are listed, in order of their number in the International system. The Bravais lattice and Schöflies symbol of each is also listed. It is to be noted that in defining the generating elements, the choices of origin and unit cell orientation are each not unique. The generating elements listed conform to the choices which are for the most part consistent with those given in Volume 1 of the ITXRC. Point group operation symbols in the cubic and hexagonal classes are defined in BC, Table 1-4 (see Reference 1, p. 364). (Note that I is used to denote the inversion operation instead of i, as is used elsewhere in the text.)

International number	International symbol	Bravais lattice	Schönflies symbol	Generating elements
1	$P1$	Γ_t	C_1^1	$\{E\|000\}$
2	$P\bar{1}$	Γ_t	C_i^1	$\{I\|000\}$
3	$P2$	Γ_m	C_2^1	$\{C_{2z}\|000\}$
4	$P2_1$	Γ_m	C_2^2	$\{C_{2z}\|00\frac{1}{2}\}$
5	$B2$	Γ_m^b	C_2^3	$\{C_{2z}\|000\}$
6	Pm	Γ_m	C_s^1	$\{\sigma_z\|000\}$
7	Pb	Γ_m	C_s^2	$\{\sigma_z\|\frac{1}{2}00\}$
8	Bm	Γ_m^b	C_s^3	$\{\sigma_z\|000\}$
9	Bb	Γ_m^b	C_s^4	$\{\sigma_z\|\frac{1}{2}00\}$
10	$P2/m$	Γ_m	C_{2h}^1	$\{C_{2z}\|000\}, \{I\|000\}$
11	$P2_1/m$	Γ_m	C_{2h}^2	$\{C_{2z}\|00\frac{1}{2}\}, \{I\|000\}$
12	$B2/m$	Γ_m^b	C_{2h}^3	$\{C_{2z}\|000\}, \{I\|000\}$
13	$P2/b$	Γ_m	C_{2h}^4	$\{C_{2z}\|\frac{1}{2}00\}, \{I\|000\}$
14	$P2_1/b$	Γ_m	C_{2h}^5	$\{C_{2z}\|\frac{1}{2}0\frac{1}{2}\}, \{I\|000\}$
15	$B2/b$	Γ_m^b	C_{2h}^6	$\{C_{2z}\|\frac{1}{2}00\}, \{I\|000\}$
16	$P222$	Γ_0	D_2^1	$\{C_{2x}\|000\}, \{C_{2y}\|000\}$
17	$P222_1$	Γ_0	D_2^2	$\{C_{2x}\|000\}, \{C_{2y}\|00\frac{1}{2}\}$
18	$P2_12_12$	Γ_0	D_2^3	$\{C_{2x}\|\frac{1}{2}\frac{1}{2}0\}, \{C_{2y}\|\frac{1}{2}\frac{1}{2}0\}$
19	$P2_12_12_1$	Γ_0	D_2^4	$\{C_{2x}\|\frac{1}{2}\frac{1}{2}0\}, \{C_{2y}\|\frac{1}{2}0\frac{1}{2}\}$
20	$C222_1$	Γ_0^b	D_2^5	$\{C_{2x}\|000\}, \{C_{2y}\|00\frac{1}{2}\}$
21	$C222$	Γ_0^b	D_2^6	$\{C_{2x}\|000\}, \{C_{2y}\|000\}$
22	$F222$	Γ_0^f	D_2^7	$\{C_{2x}\|000\}, \{C_{2y}\|000\}$
23	$I222$	Γ_0^v	D_2^8	$\{C_{2x}\|000\}, \{C_{2y}\|000\}$
24	$I2_12_12_1$	Γ_0^v	D_2^9	$\{C_{2x}\|\frac{1}{2}\frac{1}{2}0\}, \{C_{2y}\|\frac{1}{2}0\frac{1}{2}\}$

International number	International symbol	Bravais lattice	Schönflies symbol	Generating elements
25	*Pmm*2	Γ_0	C_{2v}^1	$\{\sigma_x\|000\}, \{\sigma_y\|000\}$
26	*Pmc*2_1	Γ_0	C_{2v}^2	$\{\sigma_x\|000\}, \{\sigma_y\|00\tfrac{1}{2}\}$
27	*Pcc*2	Γ_0	C_{2v}^3	$\{\sigma_x\|00\tfrac{1}{2}\}, \{\sigma_y\|00\tfrac{1}{2}\}$
28	*Pma*2	Γ_0	C_{2v}^4	$\{\sigma_x\|0\tfrac{1}{2}0\}, \{\sigma_y\|0\tfrac{1}{2}0\}$
29	*Pca*2_1	Γ_0	C_{2v}^5	$\{\sigma_x\|0\tfrac{1}{2}\tfrac{1}{2}\}, \{\sigma_y\|0\tfrac{1}{2}0\}$
30	*Pnc*2	Γ_0	C_{2v}^6	$\{\sigma_x\|\tfrac{1}{2}0\tfrac{1}{2}\}, \{\sigma_y\|\tfrac{1}{2}0\tfrac{1}{2}\}$
31	*Pmn*2_1	Γ_0	C_{2v}^7	$\{\sigma_x\|000\}, \{\sigma_y\|0\tfrac{1}{2}\tfrac{1}{2}\}$
32	*Pba*2	Γ_0	C_{2v}^8	$\{\sigma_x\|\tfrac{1}{2}\tfrac{1}{2}0\}, \{\sigma_y\|\tfrac{1}{2}\tfrac{1}{2}0\}$
33	*Pna*2_1	Γ_0	C_{2v}^9	$\{\sigma_x\|\tfrac{1}{2}\tfrac{1}{2}\tfrac{1}{2}\}, \{\sigma_y\|\tfrac{1}{2}\tfrac{1}{2}0\}$
34	*Pnn*2	Γ_0	C_{2v}^{10}	$\{\sigma_x\|\tfrac{1}{2}\tfrac{1}{2}\tfrac{1}{2}\}, \{\sigma_y\|\tfrac{1}{2}\tfrac{1}{2}\tfrac{1}{2}\}$
35	*Cmm*2	Γ_0^b	C_{2v}^{11}	$\{\sigma_x\|000\}, \{\sigma_y\|000\}$
36	*Cmc*2_1	Γ_0^b	C_{2v}^{12}	$\{\sigma_x\|000\}, \{\sigma_y\|00\tfrac{1}{2}\}$
37	*Ccc*2	Γ_0^b	C_{2v}^{13}	$\{\sigma_x\|00\tfrac{1}{2}\}, \{\sigma_y\|00\tfrac{1}{2}\}$
38	*Amm*2	Γ_0^b	C_{2v}^{14}	$\{\sigma_x\|000\}, \{\sigma_y\|000\}$
39	*Abm*2	Γ_0^b	C_{2v}^{15}	$\{\sigma_x\|\tfrac{1}{2}0\tfrac{1}{2}\}, \{\sigma_y\|\tfrac{1}{2}0\tfrac{1}{2}\}$
40	*Ama*2	Γ_0^b	C_{2v}^{16}	$\{\sigma_x\|0\tfrac{1}{2}0\}, \{\sigma_y\|0\tfrac{1}{2}0\}$
41	*Aba*2	Γ_0^b	C_{2v}^{17}	$\{\sigma_x\|\tfrac{1}{2}\tfrac{1}{2}\tfrac{1}{2}\}, \{\sigma_y\|\tfrac{1}{2}\tfrac{1}{2}\tfrac{1}{2}\}$
42	*Fmm*2	Γ_0^f	C_{2v}^{18}	$\{\sigma_x\|000\}, \{\sigma_y\|000\}$
43	*Fdd*2	Γ_0^f	C_{2v}^{19}	$\{\sigma_x\|\tfrac{3}{4}\tfrac{3}{4}\tfrac{3}{4}\}, \{\sigma_y\|\tfrac{3}{4}\tfrac{3}{4}\tfrac{3}{4}\}$
44	*Imm*2	Γ_0^v	C_{2v}^{20}	$\{\sigma_x\|000\}, \{\sigma_y\|000\}$
45	*Iba*2	Γ_0^v	C_{2v}^{21}	$\{\sigma_x\|\tfrac{1}{2}\tfrac{1}{2}0\}, \{\sigma_y\|\tfrac{1}{2}\tfrac{1}{2}0\}$
46	*Ima*2	Γ_0^v	C_{2v}^{22}	$\{\sigma_x\|\tfrac{1}{2}0\tfrac{1}{2}\}, \{\sigma_y\|\tfrac{1}{2}0\tfrac{1}{2}\}$
47	*Pmmm*	Γ_0	D_{2h}^1	$\{C_{2x}\|000\}, \{C_{2y}\|000\}, \{I\|000\}$
48	*Pnnn*	Γ_0	D_{2h}^2	$\{C_{2x}\|000\}, \{C_{2y}\|000\}, \{I\|\tfrac{1}{2}\tfrac{1}{2}\tfrac{1}{2}\}$
49	*Pccm*	Γ_0	D_{2h}^3	$\{C_{2x}\|00\tfrac{1}{2}\}, \{C_{2y}\|00\tfrac{1}{2}\}, \{I\|000\}$
50	*Pban*	Γ_0	D_{2h}^4	$\{C_{2x}\|000\}, \{C_{2y}\|000\}, \{I\|\tfrac{1}{2}\tfrac{1}{2}0\}$
51	*Pmma*	Γ_0	D_{2h}^5	$\{C_{2x}\|0\tfrac{1}{2}0\}, \{C_{2y}\|000\}, \{I\|000\}$
52	*Pnna*	Γ_0	D_{2h}^6	$\{C_{2x}\|\tfrac{1}{2}0\tfrac{1}{2}\}, \{C_{2y}\|\tfrac{1}{2}\tfrac{1}{2}\tfrac{1}{2}\}, \{I\|000\}$
53	*Pmna*	Γ_0	D_{2h}^7	$\{C_{2x}\|000\}, \{C_{2y}\|0\tfrac{1}{2}\tfrac{1}{2}\}, \{I\|000\}$
54	*Pcca*	Γ_0	D_{2h}^8	$\{C_{2x}\|0\tfrac{1}{2}\tfrac{1}{2}\}, \{C_{2y}\|00\tfrac{1}{2}\}, \{I\|000\}$
55	*Pbam*	Γ_0	D_{2h}^9	$\{C_{2x}\|\tfrac{1}{2}\tfrac{1}{2}0\}, \{C_{2y}\|\tfrac{1}{2}\tfrac{1}{2}0\}, \{I\|000\}$
56	*Pccn*	Γ_0	D_{2h}^{10}	$\{C_{2x}\|0\tfrac{1}{2}\tfrac{1}{2}\}, \{C_{2y}\|\tfrac{1}{2}0\tfrac{1}{2}\}, \{I\|000\}$
57	*Pbcm*	Γ_0	D_{2h}^{11}	$\{C_{2x}\|\tfrac{1}{2}00\}, \{C_{2y}\|\tfrac{1}{2}0\tfrac{1}{2}\}, \{I\|000\}$
58	*Pnnm*	Γ_0	D_{2}^{12}	$\{C_{2x}\|\tfrac{1}{2}\tfrac{1}{2}\tfrac{1}{2}\}, \{C_{2y}\|\tfrac{1}{2}\tfrac{1}{2}\tfrac{1}{2}\}, \{I\|000\}$
59	*Pmmn*	Γ_0	D_{2h}^{13}	$\{C_{2x}\|\tfrac{1}{2}\tfrac{1}{2}0\}, \{C_{2y}\|\tfrac{1}{2}\tfrac{1}{2}0\}, \{I\|\tfrac{1}{2}\tfrac{1}{2}0\}$
60	*Pbcn*	Γ_0	D_{4h}^{14}	$\{C_{2x}\|\tfrac{1}{2}\tfrac{1}{2}0\}, \{C_{2y}\|00\tfrac{1}{2}\}, \{I\|000\}$
61	*Pbca*	Γ_0	D_{2h}^{15}	$\{C_{2x}\|\tfrac{1}{2}\tfrac{1}{2}0\}, \{C_{2y}\|\tfrac{1}{2}0\tfrac{1}{2}\}, \{I\|000\}$
62	*Pnma*	Γ_0	D_{2h}^{16}	$\{C_{2x}\|\tfrac{1}{2}\tfrac{1}{2}\tfrac{1}{2}\}, \{C_{2y}\|\tfrac{1}{2}00\}, \{I\|000\}$
63	*Cmcm*	Γ_0^b	D_{2h}^{17}	$\{C_{2x}\|000\}, \{C_{2y}\|00\tfrac{1}{2}\}, \{I\|000\}$
64	*Cmca*	Γ_0^b	D_{2h}^{18}	$\{C_{2x}\|000\}, \{C_{2y}\|\tfrac{1}{2}\tfrac{1}{2}\tfrac{1}{2}\}, \{I\|000\}$
65	*Cmmm*	Γ_0^b	D_{2h}^{19}	$\{C_{2x}\|000\}, \{C_{2y}\|000\}, \{I\|000\}$
66	*Cccm*	Γ_0^b	D_{2h}^{20}	$\{C_{2x}\|00\tfrac{1}{2}\}, \{C_{2y}\|00\tfrac{1}{2}\}, \{I\|000\}$
67	*Cmma*	Γ_0^b	D_{2h}^{21}	$\{C_{2x}\|000\}, \{C_{2y}\|\tfrac{1}{2}\tfrac{1}{2}0\}, \{I\|000\}$
68	*Ccca*	Γ_0^b	D_{2h}^{22}	$\{C_{2x}\|\tfrac{1}{2}\tfrac{1}{2}\tfrac{1}{2}\}, \{C_{2y}\|00\tfrac{1}{2}\}, \{I\|000\}$
69	*Fmmm*	Γ_0^f	D_{2h}^{23}	$\{C_{2x}\|000\}, \{C_{2y}\|000\}, \{I\|000\}$
70	*Fddd*	Γ_0^f	D_{2h}^{24}	$\{C_{2x}\|000\}, \{C_{2y}\|000\}, \{I\|\tfrac{3}{4}\tfrac{3}{4}\tfrac{3}{4}\}$
71	*Immm*	Γ_0^v	D_{2h}^{25}	$\{C_{2x}\|000\}, \{C_{2y}\|000\}, \{I\|000\}$
72	*Ibam*	Γ_0^v	D_{2h}^{26}	$\{C_{2x}\|\tfrac{1}{2}\tfrac{1}{2}0\}, \{C_{2y}\|\tfrac{1}{2}\tfrac{1}{2}0\}, \{I\|000\}$
73	*Ibca*	Γ_0^v	D_{2h}^{27}	$\{C_{2x}\|\tfrac{1}{2}\tfrac{1}{2}0\}, \{C_{2y}\|\tfrac{1}{2}0\tfrac{1}{2}\}, \{I\|000\}$
74	*Imma*	Γ_0^v	D_{2h}^{28}	$\{C_{2x}\|000\}, \{C_{2y}\|0\tfrac{1}{2}\tfrac{1}{2}\}, \{I\|000\}$

International number	International symbol	Bravais lattice	Schönflies symbol	Generating elements
75	$P4$	Γ_q	C_4^1	$\{C_{4z}^+\|000\}$
76	$P4_1$	Γ_q	C_4^2	$\{C_{4z}^+\|00\frac{1}{4}\}$
77	$P4_2$	Γ_q	C_4^3	$\{C_{4z}^+\|00\frac{1}{2}\}$
78	$P4_3$	Γ_q	C_4^4	$\{C_{4z}^+\|00\frac{3}{4}\}$
79	$I4$	Γ_q^v	C_4^5	$\{C_{4z}^+\|000\}$
80	$I4_1$	Γ_q^v	C_4^6	$\{C_{4z}^+\|\frac{3}{4}\frac{1}{4}\frac{1}{2}\}$
81	$P\bar{4}$	Γ_q	S_4^1	$\{S_{4z}^+\|000\}$
82	$I\bar{4}$	Γ_q^v	S_4^2	$\{S_{4z}^+\|000\}$
83	$P4/m$	Γ_q	C_{4h}^1	$\{C_{4z}^+\|000\}, \{I\|000\}$
84	$P4_2/m$	Γ_q	C_{4h}^2	$\{C_{4z}^+\|00\frac{1}{2}\}, \{I\|000\}$
85	$P4/n$	Γ_q	C_{4h}^3	$\{C_{4z}^+\|\frac{1}{2}\frac{1}{2}0\}, \{I\|\frac{1}{2}\frac{1}{2}0\}$
86	$P4_2/n$	Γ_q	C_{4h}^4	$\{C_{4z}^+\|\frac{1}{2}\frac{1}{2}\frac{1}{2}\}, \{I\|\frac{1}{2}\frac{1}{2}\frac{1}{2}\}$
87	$I4/m$	Γ_q^v	C_{4h}^5	$\{C_{4z}^+\|000\}, \{I\|000\}$
88	$I4_1/a$	Γ_q^v	C_{4h}^6	$\{C_{4z}^+\|\frac{3}{4}\frac{1}{4}\frac{1}{2}\}, \{I\|\frac{3}{4}\frac{1}{4}\frac{1}{2}\}$
89	$P422$	Γ_q	D_4^1	$\{C_{4z}^+\|000\}, \{C_{2x}\|000\}$
90	$P42_12$	Γ_q	D_4^2	$\{C_{4z}^+\|\frac{1}{2}\frac{1}{2}0\}, \{C_{2x}\|\frac{1}{2}\frac{1}{2}0\}$
91	$P4_122$	Γ_q	D_4^3	$\{C_{4z}^+\|00\frac{1}{4}\}, \{C_{2x}\|00\frac{1}{2}\}$
92	$P4_12_12$	Γ_q	D_4^4	$\{C_{4z}^+\|\frac{1}{2}\frac{1}{2}\frac{1}{4}\}, \{C_{2x}\|\frac{1}{2}\frac{1}{2}\frac{3}{4}\}$
93	$P4_222$	Γ_q	D_4^5	$\{C_{4z}^+\|00\frac{1}{2}\}, \{C_{2x}\|000\}$
94	$P4_22_12$	Γ_q	D_4^6	$\{C_{4z}^+\|\frac{1}{2}\frac{1}{2}\frac{1}{2}\}, \{C_{2x}\|\frac{1}{2}\frac{1}{2}\frac{1}{2}\}$
95	$P4_322$	Γ_q	D_4^7	$\{C_{4z}^+\|00\frac{3}{4}\}, \{C_{2x}\|00\frac{1}{2}\}$
96	$P4_32_12$	Γ_q	D_4^8	$\{C_{4z}^+\|\frac{1}{2}\frac{1}{2}\frac{3}{4}\}, \{C_{2x}\|\frac{1}{2}\frac{1}{2}\frac{1}{4}\}$
97	$I422$	Γ_q^v	D_4^9	$\{C_{4z}^+\|000\}, \{C_{2x}\|000\}$
98	$I4_122$	Γ_q^v	D_4^{10}	$\{C_{4z}^+\|\frac{3}{4}\frac{1}{4}\frac{1}{2}\}, \{C_{2x}\|\frac{3}{4}\frac{1}{4}\frac{1}{2}\}$
99	$P4mm$	Γ_q	C_{4v}^1	$\{C_{4z}^+\|000\}, \{\sigma_x\|000\}$
100	$P4bm$	Γ_q	C_{4v}^2	$\{C_{4z}^+\|000\}, \{\sigma_x\|\frac{1}{2}\frac{1}{2}0\}$
101	$P4_2cm$	Γ_q	C_{4v}^3	$\{C_{4z}^+\|00\frac{1}{2}\}, \{\sigma_x\|00\frac{1}{2}\}$
102	$P4_2nm$	Γ_q	C_{4v}^4	$\{C_{4z}^+\|\frac{1}{2}\frac{1}{2}\frac{1}{2}\}, \{\sigma_x\|\frac{1}{2}\frac{1}{2}\frac{1}{2}\}$
103	$P4cc$	Γ_q	C_{4v}^5	$\{C_{4z}^+\|000\}, \{\sigma_x\|00\frac{1}{2}\}$
104	$P4nc$	Γ_q	C_{4v}^6	$\{C_{4z}^+\|000\}, \{\sigma_x\|\frac{1}{2}\frac{1}{2}\frac{1}{2}\}$
105	$P4_2mc$	Γ_q	C_{4v}^7	$\{C_{4z}^+\|00\frac{1}{2}\}, \{\sigma_x\|000\}$
106	$P4_2bc$	Γ_q	C_{4v}^8	$\{C_{4z}^+\|00\frac{1}{2}\}, \{\sigma_x\|\frac{1}{2}\frac{1}{2}0\}$
107	$I4mm$	Γ_q^v	C_{4v}^9	$\{C_{4z}^+\|000\}, \{\sigma_x\|000\}$
108	$I4cm$	Γ_q^v	C_{4v}^{10}	$\{C_{4z}^+\|000\}, \{\sigma_x\|\frac{1}{2}\frac{1}{2}0\}$
109	$I4_1md$	Γ_q^v	C_{4v}^{11}	$\{C_{4z}^+\|\frac{3}{4}\frac{1}{4}\frac{1}{2}\}, \{\sigma_x\|000\}$
110	$I4_1cd$	Γ_q^v	C_{4v}^{12}	$\{C_{4z}^+\|\frac{3}{4}\frac{1}{4}\frac{1}{2}\}, \{\sigma_x\|\frac{1}{2}\frac{1}{2}0\}$
111	$P\bar{4}2m$	Γ_q	D_{2d}^1	$\{S_{4z}^+\|000\}, \{C_{4x}\|000\}$
112	$P\bar{4}2c$	Γ_q	D_{2d}^2	$\{S_{4z}^+\|000\}, \{C_{2x}\|00\frac{1}{2}\}$
113	$P\bar{4}2_1m$	Γ_q	D_{2d}^3	$\{S_{4z}^+\|000\}, \{C_{2x}\|\frac{1}{2}\frac{1}{2}0\}$
114	$P\bar{4}2_1c$	Γ_q	D_{2d}^4	$\{S_{4z}^+\|000\}, \{C_{2x}\|\frac{1}{2}\frac{1}{2}\frac{1}{2}\}$
115	$P\bar{4}m2$	Γ_q	D_{2d}^5	$\{S_{4z}^+\|000\}, \{C_{2a}\|000\}$
116	$P\bar{4}c2$	Γ_q	D_{2d}^6	$\{S_{4z}^+\|000\}, \{C_{2a}\|00\frac{1}{2}\}$
117	$P\bar{4}b2$	Γ_q	D_{2d}^7	$\{S_{4z}^+\|000\}, \{C_{2a}\|\frac{1}{2}\frac{1}{2}0\}$
118	$P\bar{4}n2$	Γ_q	D_{2d}^8	$\{S_{4z}^+\|000\}, \{C_{2a}\|\frac{1}{2}\frac{1}{2}\frac{1}{2}\}$
119	$I\bar{4}m2$	Γ_q^v	D_{2d}^9	$\{S_{4z}^+\|000\}, \{C_{2a}\|000\}$
120	$I\bar{4}c2$	Γ_q^v	D_{2d}^{10}	$\{S_{4z}^+\|000\}, \{C_{2a}\|\frac{1}{2}\frac{1}{2}0\}$
121	$I\bar{4}2m$	Γ_q^v	D_{2d}^{11}	$\{S_{4z}^+\|000\}, \{C_{2x}\|000\}$
122	$I\bar{4}2d$	Γ_q^v	D_{2d}^{12}	$\{S_{4z}^+\|000\}, \{C_{2x}\|\frac{3}{4}\frac{1}{4}\frac{1}{2}\}$

International number	International symbol	Bravais lattice	Schönflies symbol	Generating elements
123	$P4/mmm$	Γ_q	D_{4h}^{1}	$\{C_{4z}^{+}\|000\}, \{C_{2x}\|000\}, \{I\|000\}$
124	$P4/mcc$	Γ_q	D_{4h}^{2}	$\{C_{4z}^{+}\|000\}, \{C_{2x}\|00\tfrac{1}{2}\}, \{I\|000\}$
125	$P4/nbm$	Γ_q	D_{4h}^{3}	$\{C_{4z}^{+}\|\tfrac{1}{2}00\}, \{C_{2x}\|0\tfrac{1}{2}0\}, \{I\|000\}$
126	$P4/nnc$	Γ_q	D_{4h}^{4}	$\{C_{4z}^{+}\|000\}, \{C_{2x}\|000\}, \{I\|\tfrac{1}{2}\tfrac{1}{2}\tfrac{1}{2}\}$
127	$P4/mbm$	Γ_q	D_{4h}^{5}	$\{C_{4z}^{+}\|000\}, \{C_{2x}\|\tfrac{1}{2}\tfrac{1}{2}0\}, \{I\|000\}$
128	$P4/mnc$	Γ_q	D_{4h}^{6}	$\{C_{4z}^{+}\|000\}, \{C_{2x}\|\tfrac{1}{2}\tfrac{1}{2}\tfrac{1}{2}\}, \{I\|000\}$
129	$P4/nmm$	Γ_q	D_{4h}^{7}	$\{C_{4z}^{+}\|\tfrac{1}{2}\tfrac{1}{2}0\}, \{C_{2x}\|\tfrac{1}{2}\tfrac{1}{2}0\}, \{I\|\tfrac{1}{2}\tfrac{1}{2}0\}$
130	$P4/ncc$	Γ_q	D_{4h}^{8}	$\{C_{4z}^{+}\|\tfrac{1}{2}\tfrac{1}{2}0\}, \{C_{2x}\|\tfrac{1}{2}\tfrac{1}{2}\tfrac{1}{2}\}, \{I\|\tfrac{1}{2}\tfrac{1}{2}0\}$
131	$P4_2/mmc$	Γ_q	D_{4h}^{9}	$\{C_{4z}^{+}\|00\tfrac{1}{2}\}, \{C_{2x}\|000\}, \{I\|000\}$
132	$P4_2/mcm$	Γ_q	D_{4h}^{10}	$\{C_{4z}^{+}\|00\tfrac{1}{2}\}, \{C_{2x}\|00\tfrac{1}{2}\}, \{I\|000\}$
133	$P4_2/nbc$	Γ_q	D_{4h}^{11}	$\{C_{4z}^{+}\|\tfrac{1}{2}\tfrac{1}{2}\tfrac{1}{2}\}, \{C_{2x}\|00\tfrac{1}{2}\}, \{I\|\tfrac{1}{2}\tfrac{1}{2}\tfrac{1}{2}\}$
134	$P4_2/nnm$	Γ_q	D_{4h}^{12}	$\{C_{4z}^{+}\|\tfrac{1}{2}\tfrac{1}{2}\tfrac{1}{2}\}, \{C_{2x}\|000\}, \{I\|\tfrac{1}{2}\tfrac{1}{2}\tfrac{1}{2}\}$
135	$P4_2/mbc$	Γ_q	D_{4h}^{13}	$\{C_{4z}^{+}\|00\tfrac{1}{2}\}, \{C_{2x}\|\tfrac{1}{2}\tfrac{1}{2}0\}, \{I\|000\}$
136	$P4_2/mnm$	Γ_q	D_{4h}^{14}	$\{C_{4z}^{+}\|\tfrac{1}{2}\tfrac{1}{2}\tfrac{1}{2}\}, \{C_{2x}\|\tfrac{1}{2}\tfrac{1}{2}\tfrac{1}{2}\}, \{I\|000\}$
137	$P4_2/nmc$	Γ_q	D_{4h}^{15}	$\{C_{4z}^{+}\|\tfrac{1}{2}\tfrac{1}{2}\tfrac{1}{2}\}, \{C_{2x}\|\tfrac{1}{2}\tfrac{1}{2}\tfrac{1}{2}\}, \{I\|\tfrac{1}{2}\tfrac{1}{2}\tfrac{1}{2}\}$
138	$P4_2/ncm$	Γ_q	D_{4h}^{16}	$\{C_{4z}^{+}\|\tfrac{1}{2}\tfrac{1}{2}\tfrac{1}{2}\}, \{C_{2x}\|\tfrac{1}{2}\tfrac{1}{2}0\}, \{I\|\tfrac{1}{2}\tfrac{1}{2}\tfrac{1}{2}\}$
139	$I4/mmm$	Γ_q^v	D_{4h}^{17}	$\{C_{4z}^{+}\|000\}, \{C_{2x}\|000\}, \{I\|000\}$
140	$I4/mcm$	Γ_q^v	D_{4h}^{18}	$\{C_{4z}^{+}\|000\}, \{C_{2x}\|\tfrac{1}{2}\tfrac{1}{2}0\}, \{I\|000\}$
141	$I4_1/amd$	Γ_q^v	D_{4h}^{19}	$\{C_{4z}^{+}\|0\tfrac{1}{2}0\}, \{C_{2x}\|000\}, \{I\|000\}$
142	$I4_1/acd$	Γ_q^v	D_{4h}^{20}	$\{C_{4z}^{+}\|0\tfrac{1}{2}0\}, \{C_{2x}\|\tfrac{1}{2}\tfrac{1}{2}0\}, \{I\|000\}$
143	$P3$	Γ_h	C_{3}^{1}	$\{C_{3}^{+}\|000\}$
144	$P3_1$	Γ_h	C_{3}^{2}	$\{C_{3}^{+}\|00\tfrac{1}{3}\}$
145	$P3_2$	Γ_h	C_{3}^{3}	$\{C_{3}^{+}\|00\tfrac{2}{3}\}$
146	$R3$	Γ_{rh}	C_{3}^{4}	$\{C_{3}^{+}\|000\}$
147	$P\bar{3}$	Γ_h	C_{3i}^{1}	$\{S_{6}^{+}\|000\}$
148	$R\bar{3}$	Γ_{rh}	C_{3i}^{2}	$\{S_{6}^{+}\|000\}$
149	$P312$	Γ_h	D_{3}^{1}	$\{C_{3}^{+}\|000\}, \{C'_{21}\|000\}$
150	$P321$	Γ_h	D_{3}^{2}	$\{C_{3}^{+}\|000\}, \{C''_{21}\|000\}$
151	$P3_112$	Γ_h	D_{3}^{3}	$\{C_{3}^{+}\|00\tfrac{1}{3}\}, \{C'_{21}\|00\tfrac{1}{3}\}$
152	$P3_121$	Γ_h	D_{3}^{4}	$\{C_{3}^{+}\|00\tfrac{1}{3}\}, \{C''_{21}\|00\tfrac{2}{3}\}$
153	$P3_212$	Γ_h	D_{3}^{5}	$\{C_{3}^{+}\|00\tfrac{2}{3}\}, \{C'_{21}\|00\tfrac{2}{3}\}$
154	$P3_221$	Γ_h	D_{3}^{6}	$\{C_{3}^{+}\|00\tfrac{2}{3}\}, \{C''_{21}\|00\tfrac{1}{3}\}$
155	$R32$	Γ_{rh}	D_{3}^{7}	$\{C_{3}^{+}\|000\}, \{C''_{21}\|000\}$
156	$P3m1$	Γ_h	C_{3v}^{1}	$\{C_{3}^{+}\|000\}, \{\sigma_{v1}\|000\}$
157	$P31m$	Γ_h	C_{3v}^{2}	$\{C_{3}^{+}\|000\}, \{\sigma_{d1}\|000\}$
158	$P3c1$	Γ_h	C_{3v}^{3}	$\{C_{3}^{+}\|000\}, \{\sigma_{v1}\|00\tfrac{1}{2}\}$
159	$P31c$	Γ_h	C_{3v}^{4}	$\{C_{3}^{+}\|000\}, \{\sigma_{d1}\|00\tfrac{1}{2}\}$
160	$R3m$	Γ_{rh}	C_{3v}^{5}	$\{C_{3}^{+}\|000\}, \{\sigma_{v1}\|000\}$
161	$R3c$	Γ_{rh}	C_{3v}^{6}	$\{C_{3}^{+}\|000\}, \{\sigma_{v1}\|\tfrac{1}{2}\tfrac{1}{2}\tfrac{1}{2}\}$
162	$P\bar{3}1m$	Γ_h	D_{3d}^{1}	$\{S_{6}^{+}\|000\}, \{\sigma_{d1}\|000\}$
163	$P\bar{3}1c$	Γ_h	D_{3d}^{2}	$\{S_{6}^{+}\|000\}, \{\sigma_{d1}\|00\tfrac{1}{2}\}$
164	$P\bar{3}m1$	Γ_h	D_{3d}^{3}	$\{S_{6}^{+}\|000\}, \{\sigma_{v1}\|000\}$
165	$P\bar{3}c1$	Γ_h	D_{3d}^{4}	$\{S_{6}^{+}\|000\}, \{\sigma_{v1}\|00\tfrac{1}{2}\}$
166	$R\bar{3}m$	Γ_{rh}	D_{3d}^{5}	$\{S_{6}^{+}\|000\}, \{\sigma_{v1}\|000\}$
167	$R\bar{3}c$	Γ_{rh}	D_{3d}^{6}	$\{S_{6}^{+}\|000\}, \{\sigma_{v1}\|\tfrac{1}{2}\tfrac{1}{2}\tfrac{1}{2}\}$

International number	International symbol	Bravais lattice	Schönflies symbol	Generating elements
168	$P6$	Γ_h	C_6^1	$\{C_6^+ \mid 000\}$
169	$P6_1$	Γ_h	C_6^2	$\{C_6^+ \mid 00\frac{1}{6}\}$
170	$P6_5$	Γ_h	C_6^3	$\{C_6^+ \mid 00\frac{5}{6}\}$
171	$P6_2$	Γ_h	C_6^4	$\{C_6^+ \mid 00\frac{1}{3}\}$
172	$P6_4$	Γ_h	C_6^5	$\{C_6^+ \mid 00\frac{2}{3}\}$
173	$P6_3$	Γ_h	C_6^6	$\{C_6^+ \mid 00\frac{1}{2}\}$
174	$P\bar{6}$	Γ_h	C_{3h}^1	$\{S_3^+ \mid 000\}$
175	$P6/m$	Γ_h	C_{6h}^1	$\{C_6^+ \mid 000\}, \{\sigma_h \mid 000\}$
176	$P6_3/m$	Γ_h	C_{6h}^2	$\{C_6^+ \mid 00\frac{1}{2}\}, \{\sigma_h \mid 00\frac{1}{2}\}$
177	$P622$	Γ_h	D_6^1	$\{C_6^+ \mid 000\}, \{C'_{21} \mid 000\}$
178	$P6_122$	Γ_h	D_6^2	$\{C_6^+ \mid 00\frac{1}{6}\}, \{C''_{21} \mid 000\}$
179	$P6_522$	Γ_h	D_6^3	$\{C_6^+ \mid 00\frac{5}{6}\}, \{C''_{21} \mid 000\}$
180	$P6_222$	Γ_h	D_6^4	$\{C_6^+ \mid 00\frac{1}{3}\}, \{C'_{21} \mid 000\}$
181	$P6_422$	Γ_h	D_6^5	$\{C_6^+ \mid 00\frac{2}{3}\}, \{C'_{21} \mid 000\}$
182	$P6_322$	Γ_h	D_6^6	$\{C_6^+ \mid 00\frac{1}{2}\}, \{C''_{21} \mid 000\}$
183	$P6mm$	Γ_h	C_{6v}^1	$\{C_6^+ \mid 000\}, \{\sigma_{v1} \mid 000\}$
184	$P6cc$	Γ_h	C_{6v}^2	$\{C_6^+ \mid 000\}, \{\sigma_{v1} \mid 00\frac{1}{2}\}$
185	$P6_3cm$	Γ_h	C_{6v}^3	$\{C_6^+ \mid 00\frac{1}{2}\}, \{\sigma_{v1} \mid 00\frac{1}{2}\}$
186	$P6_3mc$	Γ_h	C_{6v}^4	$\{C_6^+ \mid 00\frac{1}{2}\}, \{\sigma_{v1} \mid 000\}$
187	$P\bar{6}m2$	Γ_h	D_{3h}^1	$\{S_3^+ \mid 000\}, \{\sigma_{v1} \mid 000\}$
188	$P\bar{6}c2$	Γ_h	D_{3h}^2	$\{S_3^+ \mid 00\frac{1}{2}\}, \{\sigma_{v1} \mid 00\frac{1}{2}\}$
189	$P\bar{6}2m$	Γ_h	D_{3h}^3	$\{S_3^+ \mid 000\}, \{\sigma_{d1} \mid 000\}$
190	$P\bar{6}2c$	Γ_h	D_{3h}^4	$\{S_3^+ \mid 00\frac{1}{2}\}, \{\sigma_{d1} \mid 00\frac{1}{2}\}$
191	$P6/mmm$	Γ_h	D_{6h}^1	$\{C_6^+ \mid 000\}, \{C'_{21} \mid 000\}, \{I \mid 000\}$
192	$P6/mcc$	Γ_h	D_{6h}^2	$\{C_6^+ \mid 000\}, \{C'_{21} \mid 00\frac{1}{2}\}, \{I \mid 000\}$
193	$P6_3/mcm$	Γ_h	D_{6h}^3	$\{C_6^+ \mid 00\frac{1}{2}\}, \{C'_{21} \mid 000\}, \{I \mid 000\}$
194	$P6_3/mmc$	Γ_h	D_{6h}^4	$\{C_6^+ \mid 00\frac{1}{2}\}, \{C'_{21} \mid 00\frac{1}{2}\}, \{I \mid 000\}$
195	$P23$	Γ_c	T^1	$\{C_{2z} \mid 000\}, \{C_{2x} \mid 000\}, \{C_{31}^+ \mid 000\}$
196	$F23$	Γ_c^f	T^2	$\{C_{2z} \mid 000\}, \{C_{2x} \mid 000\}, \{C_{31}^+ \mid 000\}$
197	$I23$	Γ_c^v	T^3	$\{C_{2z} \mid 000\}, \{C_{2x} \mid 000\}, \{C_{31}^+ \mid 000\}$
198	$P2_13$	Γ_c	T^4	$\{C_{2z} \mid \frac{1}{2}0\frac{1}{2}\}, \{C_{2x} \mid \frac{1}{2}\frac{1}{2}0\}, \{C_{31}^+ \mid 000\}$
199	$I2_13$	Γ_c^v	T^5	$\{C_{2z} \mid \frac{1}{2}0\frac{1}{2}\}, \{C_{2x} \mid \frac{1}{2}\frac{1}{2}0\}, \{C_{31}^+ \mid 000\}$
200	$Pm3$	Γ_c	T_h^1	$\{C_{2z} \mid 000\}, \{C_{2x} \mid 000\}, \{C_{31}^+ \mid 000\}, \{I \mid 000\}$
201	$Pn3$	Γ_c	T_h^2	$\{C_{2z} \mid 000\}, \{C_{2x} \mid 000\}, \{C_{31}^+ \mid 000\}, \{I \mid \frac{1}{2}\frac{1}{2}\frac{1}{2}\}$
202	$Fm3$	Γ_c^f	T_h^3	$\{C_{2z} \mid 000\}, \{C_{2x} \mid 000\}, \{C_{31}^+ \mid 000\}, \{I \mid 000\}$
203	$Fd3$	Γ_c^f	T_h^4	$\{C_{2z} \mid 000\}, \{C_{2x} \mid 000\}, \{C_{31}^+ \mid 000\}, \{I \mid \frac{1}{4}\frac{1}{4}\frac{1}{4}\}$
204	$Im3$	Γ_c^v	T_h^5	$\{C_{2z} \mid 000\}, \{C_{2x} \mid 000\}, \{C_{31}^+ \mid 000\}, \{I \mid 000\}$
205	$Pa3$	Γ_c	T_h^6	$\{C_{2z} \mid \frac{1}{2}0\frac{1}{2}\}, \{C_{2x} \mid \frac{1}{2}\frac{1}{2}0\}, \{C_{31}^+ \mid 000\}, \{I \mid 000\}$
206	$Ia3$	Γ_c^v	T_h^7	$\{C_{2z} \mid \frac{1}{2}0\frac{1}{2}\}, \{C_{2x} \mid \frac{1}{2}\frac{1}{2}0\}, \{C_{31}^+ \mid 000\}, \{I \mid 000\}$
207	$P432$	Γ_c	O^1	$\{C_{2z} \mid 000\}, \{C_{2x} \mid 000\}, \{C_{2a} \mid 000\}, \{C_{31}^+ \mid 000\}$
208	$P4_232$	Γ_c	O^2	$\{C_{2z} \mid 000\}, \{C_{2x} \mid 000\}, \{C_{2a} \mid \frac{1}{2}\frac{1}{2}\frac{1}{2}\}, \{C_{31}^+ \mid 000\}$
209	$F432$	Γ_c^f	O^3	$\{C_{2z} \mid 000\}, \{C_{2x} \mid 000\}, \{C_{2a} \mid 000\}, \{C_{31}^+ \mid 000\}$
210	$F4_132$	Γ_c^f	O^4	$\{C_{2z} \mid 000\}, \{C_{2x} \mid 000\}, \{C_{2a} \mid \frac{1}{4}\frac{1}{4}\frac{1}{4}\}, \{C_{31}^+ \mid 000\}$
211	$I432$	Γ_c^v	O^5	$\{C_{2z} \mid 000\}, \{C_{2x} \mid 000\}, \{C_{2a} \mid 000\}, \{C_{31}^+ \mid 000\}$
212	$P4_332$	Γ_c	O^6	$\{C_{2z} \mid \frac{1}{2}0\frac{1}{2}\}, \{C_{2x} \mid \frac{1}{2}\frac{1}{2}0\}, \{C_{2a} \mid \frac{1}{4}\frac{3}{4}\frac{3}{4}\}, \{C_{31}^+ \mid 000\}$
213	$P4_132$	Γ_c	O^7	$\{C_{2z} \mid \frac{1}{2}0\frac{1}{2}\}, \{C_{2x} \mid \frac{1}{2}\frac{1}{2}0\}, \{C_{2a} \mid \frac{3}{4}\frac{1}{4}\frac{1}{4}\}, \{C_{31}^+ \mid 000\}$
214	$I4_132$	Γ_c^v	O^8	$\{C_{2z} \mid \frac{1}{2}0\frac{1}{2}\}, \{C_{2x} \mid \frac{1}{2}\frac{1}{2}0\}, \{C_{2a} \mid \frac{1}{2}00\}, \{C_{31}^+ \mid 000\}$

International number	International symbol	Bravais lattice	Schönflies symbol	Generating elements
215	$P\bar{4}3m$	Γ_c	T_d^1	$\{C_{2z}\|000\}, \{C_{2x}\|000\}, \{\sigma_{da}\|000\}, \{C_{31}^+\|000\}$
216	$F\bar{4}3m$	Γ_c^f	T_d^2	$\{C_{2z}\|000\}, \{C_{2x}\|000\}, \{\alpha_{da}\|000\}, \{C_{31}^+\|000\}$
217	$I\bar{4}3m$	Γ_c^v	T_d^3	$\{C_{2z}\|000\}, \{C_{2x}\|000\}, \{\sigma_{da}\|000\}, \{C_{31}^+\|000\}$
218	$P\bar{4}3n$	Γ_c	T_d^4	$\{C_{2z}\|000\}, \{C_{2x}\|000\}, \{\sigma_{da}\|\frac{1}{2}\frac{1}{2}\frac{1}{2}\}, \{C_{31}^+\|000\}$
219	$F\bar{4}3c$	Γ_c^f	T_d^5	$\{C_{2z}\|000\}, \{C_{2x}\|000\}, \{\sigma_{da}\|\frac{1}{2}\frac{1}{2}\frac{1}{2}\}, \{C_{31}^+\|000\}$
220	$I\bar{4}3d$	Γ_c^v	T_d^6	$\{C_{2z}\|\frac{1}{2}0\frac{1}{2}\}, \{C_{2x}\|\frac{1}{2}\frac{1}{2}0\}, \{\sigma_{da}\|\frac{1}{2}00\}, \{C_{31}^+\|000\}$
221	$Pm3m$	Γ_c	O_h^1	$\{C_{2z}\|000\}, \{C_{2x}\|000\}, \{C_{2a}\|000\}, \{C_{31}^+\|000\}, \{I\|000\}$
222	$Pn3n$	Γ_c	O_h^2	$\{C_{2z}\|000\}, \{C_{2x}\|000\}, \{C_{2a}\|000\}, \{C_{31}^+\|000\}, \{I\|\frac{1}{2}\frac{1}{2}\frac{1}{2}\}$
223	$Pm3n$	Γ_c	O_h^3	$\{C_{2z}\|000\}, \{C_{2x}\|000\}, \{C_{2a}\|\frac{1}{2}\frac{1}{2}\frac{1}{2}\}, \{C_{31}^+\|000\}, \{I\|000\}$
224	$Pn3m$	Γ_c	O_h^4	$\{C_{2z}\|000\}, \{C_{2x}\|000\}, \{C_{2a}\|\frac{1}{2}\frac{1}{2}\frac{1}{2}\}, \{C_{31}^+\|000\}, \{I\|\frac{1}{2}\frac{1}{2}\frac{1}{2}\}$
225	$Fm3m$	Γ_c^f	O_h^5	$\{C_{2z}\|000\}, \{C_{2x}\|000\}, \{C_{2a}\|000\}, \{C_{31}^+\|000\}, \{I\|000\}$
226	$Fm3c$	Γ_c^f	O_h^6	$\{C_{2z}\|000\}, \{C_{2x}\|000\}, \{C_{2a}\|\frac{1}{2}\frac{1}{2}\frac{1}{2}\}, \{C_{31}^+\|000\}, \{I\|000\}$
227	$Fd3m$	Γ_c^f	O_h^7	$\{C_{2z}\|000\}, \{C_{2x}\|000\}, \{C_{2a}\|\frac{1}{4}\frac{1}{4}\frac{1}{4}\}, \{C_{31}^+\|000\}, \{I\|\frac{1}{4}\frac{1}{4}\frac{1}{4}\}$
228	$Fd3c$	Γ_c^f	O_h^8	$\{C_{2z}\|000\}, \{C_{2x}\|000\}, \{C_{2a}\|\frac{1}{4}\frac{1}{4}\frac{1}{4}\}, \{C_{31}^+\|000\}, \{I\|\frac{3}{4}\frac{3}{4}\frac{3}{4}\}$
229	$Im3m$	Γ_c^v	O_h^9	$\{C_{2z}\|000\}, \{C_{2x}\|000\}, \{C_{2a}\|000\}, \{C_{31}^+\|000\}, \{I\|000\}$
230	$Ia3d$	Γ_c^v	O_h^{10}	$\{C_{2z}\|\frac{1}{2}0\frac{1}{2}\}, \{C_{2x}\|\frac{1}{2}\frac{1}{2}0\}, \{C_{2a}\|\frac{1}{2}00\}, \{C_{31}^+\|000\}, \{I\|000\}$

APPENDIX

VI

FACTOR GROUP THEOREMS

We begin this appendix by reviewing standard definitions and theorems of abstract group theory, and proceed to use them to firmly establish references to them in Sections 4-2 and 4-5.

A subset $\{H\}$ of a group **G** that is itself a group using the same binary operation as that in **G** is called a subgroup **H** of **G**. A group always has at least two subgroups—the group itself and the trivial group consisting of just the identity. These two are called *improper* subgroups; all others are called *proper*.

If **H** is a subgroup of **G** and R is any element of **G**, the subset **H**R is a *right coset* of **H**; similarly, R**H** is a *left coset*. R is called the *coset representative*. Every element of **G** is in some, say, left coset, but no element is in more than one. Each left coset contains the same number of elements—the order of **H**, h. The number of left cosets of **H** is called the index t of **H** in **G**, where

$$t = g/h \qquad \text{(VI-1)}$$

where g and h are the orders of **G** and **H**, respectively.

We now come to the singularly most important kind of subgroup in the theory of space groups—the invariant subgroup. If **H** is a subgroup of **G** such that GHG^{-1} is in **H** (all G in **G**, all H in **H**), then **H** is called an *invariant subgroup* of **G**. **H** is an invariant subgroup if and only if (iff) all right and left cosets of **H** are the same. Also, **H** can be shown to be an invariant subgroup iff it is invariant under all mappings carried out by conjugation, that is, if **H** is composed of entire classes of **G**.

If **H** is an invariant subgroup of **G**, its left cosets form a group whose multiplication law is

$$(G_i\mathbf{H})(G_j\mathbf{H}) = (G_iG_j)\mathbf{H} \qquad \text{(VI-2)}$$

This new group is called the *factor group* **G/H** of order t. *The identity element of* **G/H** *is the subgroup* **H** *itself.*

$$(G_iH)(G_iH)^{-1} = (G_iG_i^{-1})\mathbf{H} = E\mathbf{H} = \mathbf{H} \qquad \text{(VI-3)}$$

As an example, consider the group $\mathbf{C}_{3v}$ with subgroup $\mathbf{C}_3$, illustrated in the figure

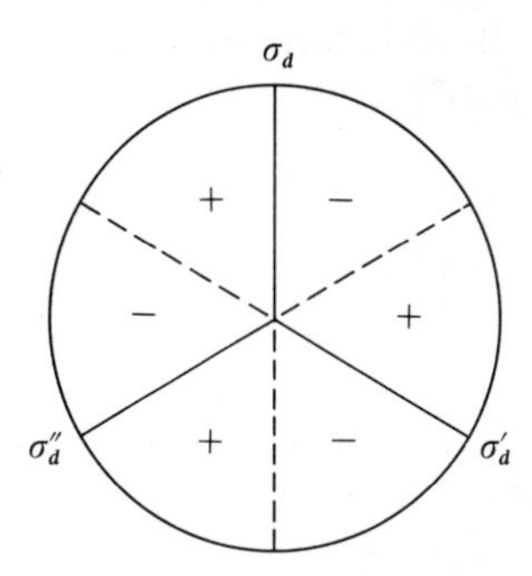

Figure A6-1 The group $\mathbf{C}_{3v}$

The factor group $\mathbf{C}_{3v}/\mathbf{C}_3$ has index 2, and it is isomorphic with $\mathbf{C}_s$. The identity element of $\mathbf{C}_{3v}/\mathbf{C}_3$ (the invariant subgroup $\mathbf{C}_3$ itself) simply serves to move one trisection into another. The other operation, $\sigma\mathbf{C}_3 = \sigma$, interchanges positive and negative areas.

Similarly, in space groups, the subgroup $\mathbf{T} = \{E\,|\,\tau\}$ of all translations of the Bravais lattice is easily seen to be an invariant subgroup. For one thing, it is composed of an entire *class* of the space group. The factor group **G/T** is therefore isomorphic with the point group which leaves the unit cell invariant, i.e., the unit cell group, and the identity element of **G/T** is the set of all translations, $\{E\,|\,\tau\}$. Just as in the example above, the factor group operations are effective within one unit cell (a trisection), and the identity operation preserves the lattice.

Since the elements of the unit cell group are point group operations, the representations of the invariant subgroup $\mathbf{T} = \{E\,|\,\tau\}$ must be such that for all τ,

$$e^{-i\kappa\cdot\tau} = 1 \qquad \text{(VI-4)}$$

We immediately see why the unit cell group is the appropriate group for $\boldsymbol{\kappa} = 0$.

Elsewhere in the Brillouin zone, $\boldsymbol{\kappa} \neq 0$, there will still be some translations τ_κ for which

$$e^{-i\kappa\cdot\tau_\kappa} = 1 \qquad \text{(VI-5)}$$

The translations $\{E\,|\,\tau_\kappa\}$ also constitute an invariant subgroup $\mathbf{G}_\kappa/\mathbf{T}_\kappa$ of the space group for which the set of translations $\{T_\kappa\}$ constitutes the identity element. We denote $\mathbf{G}_\kappa$ the *group of the wave vector,* called elsewhere $H(\boldsymbol{\kappa})$.

In general, then, a space group **G** can be decomposed into an invariant subgroup consisting of a group of translations, **T** and its, say, left cosets:

$$\mathbf{G} = \{E\,|\,\mathbf{0}\}\mathbf{T} + \{R_1\,|\,\tau_{R_1}\}\mathbf{T} + \{R_2\,|\,\tau_{R_2}\}\mathbf{T} + \cdots + \{R_n\,|\,\tau_{R_n}\}\mathbf{T} \qquad \text{(VI-6)}$$

The coset representatives $\{R_i | \tau_{R_i}\}$ are space group operations. For a symmorphic group, all $\tau_{R_i} = 0$ but for non-symmorphic groups at least one $\tau_{R_i} \neq 0$.

Because the invariant subgroup **T** is composed of an entire class of **G**,

$$\{R | \tau_R\}\{E | \tau\}\{R | \tau_R\}^{-1} = \{\mathbf{E} | R\tau\} \tag{VI-7}$$

Since **T** is self-conjugate, $R\tau$ is a translation of the Bravais lattice, as is τ. The operation R is a point operation, a member of a group isomorphic with $\mathbf{G}_\kappa/\mathbf{T}_\kappa$. It is also a subgroup of the point group of the Bravais lattice, that is, a subgroup of one of the seven crystal systems (see Section 4-2).

APPENDIX

VII

LIST OF SITES IN THE 230 SPACE GROUPS

The analysis of the symmetry species of normal vibrations in crystals as developed in Chapter 6, has relied heavily on the method of correlation of the site symmetry of the molecular units with the unit cell (space group) symmetry. Appendix IV shows these correlations in detail. The present appendix gives a tabulation of all distinct sites in each of the 230 space groups. Both Schönflies and Mauguin-Hermann notations are indicated respectively in the first and second columns. Since most crystallographic investigations employing X-ray or neutron diffraction techniques report atomic positions in terms of the Wyckoff notation for a set of equivalent positions, the sites which exist in a given space group are indicated by small letters in the Wyckoff notation.[1] For example, in the space group C_{2h}^1, ($P2/m$), there are eight distinct sites each of symmetry C_{2h} and containing only a single point denoted by the letters *a* through to *h*, four distinct sites, each consisting of two equivalent points of symmetry C_2 designated by the letters *i* through to *l*, two further distinct sites indicated by the letters *m* and *n*, each comprising two equivalent points of symmetry C_s, and, finally, the general position *o* with four equivalent points having no symmetry (C_1). Detailed geometric coordinates may be found in the International Tables for X-Ray Crystallography.[2]

1. R. W. G. Wyckoff, *The Analytical Expression of the Results of the Theory of Space Groups*, Carnegie Institution of Washington, Second Edition, 1930.
2. *International Tables for Crystallography*, Vol. I, N. F. M. Henry and K. Lonsdale, Eds., Kynoch Press, Birmingham, 1952.

The number of equivalent points constituting a site within the primitive unit cell is the ratio of the order of the unit cell group to the order of the site group. The International Tables, however, define the number of equivalent sites per crystallographic unit cell: when this cell is primitive as indicated by the symbol P in the Hermann-Mäuguin notation, the two unit cells, primitive and crystallographic, are identical by definition. However, the crystallographic unit cells in other cases indicated by A, B, C (centered in opposite faces), I (body centered) are larger by a factor of 2 and F (face centered) unit cells by a factor of 4 than the primitive unit cell. Thus, for example, the space group D_2^7 ($F222$) is reported in the International Tables to contain 16 equivalent points of site k, symmetry C_1 (the general position), whereas the primitive unit cell which is needed in the analysis of the symmetry of the normal vibrations contains only 4 equivalent points at site k, this being the ratio of the order of the unit cell group D_2^7 which is 4 to the order of the site symmetry C_1 which is unity. It is, of course, important to bear this distinction in mind when noting the number of molecules per unit cell which is usually stated per crystallographic cell in the X-ray diffraction literature.

Table VII-1 Site symmetries in the space groups C_1, C_i, C_2^n, C_s^n, C_{2h}^n, D_2^n

		Site symmetry					
		D_2	C_{2h}	C_2	C_s	C_i	C_1
Space group							
C_1^1	$P1$						*a*
C_i^1	$P\bar{1}$					*a–h*	*i*
C_2^1	$P2$			*a–d*			*e*
C_2^2	$P2_1$						*a*
C_2^3	$B2$			*a, b*			*c*
C_s^1	Pm				*a, b*		*c*
C_s^2	Pb						*a*
C_s^3	Bm				*a*		*b*
C_s^4	Bb						*a*
C_{2h}^1	$P2/m$		*a–h*	*i–l*	*m, n*		*o*
C_{2h}^2	$P2_1/m$				*e*	*a–d*	*f*
C_{2h}^3	$B2/m$		*a–d*	*g, h*	*i*	*e, f*	*j*
C_{2h}^4	$P2/b$			*e, f*		*a–d*	*g*
C_{2h}^5	$P2_1/b$					*a–d*	*e*
C_{2h}^6	$B2/b$			*e*		*a–d*	*f*
D_2^1	$P222$	*a–h*		*i–t*			*u*
D_2^2	$P222_1$			*a–d*			*e*
D_2^3	$P2_12_12$			*a, b*			*c*
D_2^4	$P2_12_12_1$						*a*
D_2^5	$C222_1$			*a, b*			*c*
D_2^6	$C222$	*a–d*		*e–k*			*l*
D_2^7	$F222$	*a–d*		*e–j*			*k*
D_2^8	$I222$	*a–d*		*e–j*			*k*
D_2^9	$I2_12_12_1$			*a–c*			*d*

Table VII-2 Site symmetries in the space groups C_{2v}^n

		Site symmetry			
		C_{2v}	C_2	C_s	C_1
Space group					
C_{2v}^1	*Pmm*2	*a–d*		*e–h*	*i*
C_{2v}^2	*Pmc*2$_1$			*a, b*	*c*
C_{2v}^3	*Pcc*2		*a–d*		*e*
C_{2v}^4	*Pma*2		*a, b*	*c*	*d*
C_{2v}^5	*Pca*2$_1$				*a*
C_{2v}^6	*Pnc*2		*a, b*		*c*
C_{2v}^7	*Pmn*2$_1$			*a*	*b*
C_{2v}^8	*Pba*2		*a, b*		*c*
C_{2v}^9	*Pna*2$_1$				*a*
C_{2v}^{10}	*Pnn*2		*a, b*		*c*
C_{2v}^{11}	*Cmm*2	*a, b*	*c*	*d, e*	*f*
C_{2v}^{12}	*Cmc*2$_1$			*a*	*b*
C_{2v}^{13}	*Ccc*2		*a–c*		*d*
C_{2v}^{14}	*Amm*2	*a, b*		*c–e*	*f*
C_{2v}^{15}	*Abm*2		*a, b*	*c*	*d*
C_{2v}^{16}	*Ama*2		*a*	*b*	*c*
C_{2v}^{17}	*Aba*2		*a*		*b*
C_{2v}^{18}	*Fmm*2	*a*	*b*	*c, d*	*e*
C_{2v}^{19}	*Fdd*2		*a*		*b*
C_{2v}^{20}	*Imm*2	*a, b*		*c, d*	*e*
C_{2v}^{21}	*Iba*2		*a, b*		*c*
C_{2v}^{22}	*Ima*2		*a*	*b*	*c*

Table VII-3 Site symmetries in the space groups D_{2h}^n

		Site symmetry							
		D_{2h}	C_{2h}	D_2	C_{2v}	C_i	C_2	C_s	C_1
Space group									
D_{2h}^1	*Pmmm*	*a–h*			*i–t*			*u–z*	*α*
D_{2h}^2	*Pnnn*			*a–d*		*e, f*	*g–l*		*m*
D_{2h}^3	*Pccm*		*a–d*	*e–h*			*i–p*	*q*	*r*
D_{2h}^4	*Pban*			*a–d*		*e, f*	*g–l*		*m*
D_{2h}^5	*Pmma*		*a–d*		*e, f*		*g, h*	*i–k*	*l*
D_{2h}^6	*Pnna*					*a, b*	*c, d*		*e*
D_{2h}^7	*Pmna*		*a–d*				*e–g*	*h*	*i*
D_{2h}^8	*Pcca*					*a, b*	*c–e*		*f*
D_{2h}^9	*Pbam*		*a–d*				*e, f*	*g, h*	*i*
D_{2h}^{10}	*Pccn*					*a, b*	*c, d*		*e*
D_{2h}^{11}	*Pbcm*					*a, b*	*c*	*d*	*e*
D_{2h}^{12}	*Pnnm*		*a–d*				*e, f*	*g*	*h*
D_{2h}^{13}	*Pmmn*				*a, b*	*c, d*		*e, f*	*g*
D_{2h}^{14}	*Pbcn*					*a, b*	*c*		*d*
D_{2h}^{15}	*Pbca*					*a, b*			*c*
D_{2h}^{16}	*Pnma*					*a, b*		*c*	*d*
D_{2h}^{17}	*Cmcm*		*a, b*		*c*	*d*	*e*	*f, g*	*h*
D_{2h}^{18}	*Cmca*		*a, b*			*c*	*d, e*	*f*	*g*
D_{2h}^{19}	*Cmmm*	*a–d*	*e, f*		*g–l*		*m*	*n–q*	*r*
D_{2h}^{20}	*Cccm*		*c–f*	*a, b*			*g–k*	*l*	*m*
D_{2h}^{21}	*Cmma*		*c–f*	*a, b*	*g*		*h–l*	*m, n*	*o*
D_{2h}^{22}	*Ccca*			*a, b*		*c, d*	*e–h*		*i*
D_{2h}^{23}	*Fmmm*	*a, b*	*c–e*	*f*	*g–i*		*j–l*	*m–o*	*p*
D_{2h}^{24}	*Fddd*			*a, b*		*c, d*	*e–g*		*h*
D_{2h}^{25}	*Immm*	*a–d*			*e–j*	*k*		*l–n*	*o*
D_{2h}^{26}	*Ibam*		*c, d*	*a, b*		*e*	*f–i*	*j*	*k*
D_{2h}^{27}	*Ibca*					*a, b*	*c–e*		*f*
D_{2h}^{28}	*Imma*		*a–d*		*e*		*f, g*	*h, i*	*j*

Table VII-4 Site symmetries in the space groups C_4^n, S_4^n, C_{4h}^n, D_4^n

		Site symmetry									
		D_4	C_{4h}	D_2	C_{2h}	S_4	C_4	C_i	C_2	C_s	C_1
Space group											
C_4^1	$P4$						*a, b*		*c*		*d*
C_4^2	$P4_1$										*a*
C_4^3	$P4_2$								*a–c*		*d*
C_4^4	$P4_3$										*a*
C_4^5	$I4$						*a*		*b*		*c*
C_4^6	$I4_1$								*a*		*b*
S_4^1	$P\bar{4}$					*a–d*			*e–g*		*h*
S_4^2	$I\bar{4}$					*a–d*			*e, f*		*g*
C_{4h}^1	$P4/m$		*a–d*		*e, f*		*g, h*		*i*	*j, k*	*l*
C_{4h}^2	$P4_2/m$				*a–d*	*e, f*			*g–i*	*j*	*k*
C_{4h}^3	$P4/n$					*a, b*	*c*	*d, e*	*f*		*g*
C_{4h}^4	$P4_2/n$					*a, b*		*c, d*	*e, f*		*g*
C_{4h}^5	$I4/m$		*a, b*		*c*	*d*	*e*	*f*	*g*	*h*	*i*
C_{4h}^6	$I4_1/a$					*a, b*		*c, d*	*e*		*f*
D_4^1	$P422$	*a–d*		*e, f*			*g, h*		*i–o*		*p*
D_4^2	$P42_12$			*a, b*			*c*		*d–f*		*g*
D_4^3	$P4_122$								*a–c*		*d*
D_4^4	$P4_12_12$								*a*		*b*
D_4^5	$P4_222$			*a–f*					*g–o*		*p*
D_4^6	$P4_22_12$			*a, b*					*c–f*		*g*
D_4^7	$P4_322$								*a–c*		*d*
D_4^8	$P4_32_12$								*a*		*b*
D_4^9	$I422$	*a, b*		*c, d*			*e*		*f–j*		*k*
D_4^{10}	$I4_122$			*a, b*					*c–f*		*g*

Table VII-5 Site symmetries in the space groups C_{4v}^n and D_{2d}^n

		Site symmetry								
		C_{4v}	C_4	D_{2d}	D_2	S_4	C_{2v}	C_2	C_s	C_1
Space group										
C_{4v}^1	$P4mm$	*a, b*					*c*		*d–f*	*g*
C_{4v}^2	$P4bm$		*a*				*b*		*c*	*d*
C_{4v}^3	$P4_2cm$						*a, b*	*c*	*d*	*e*
C_{4v}^4	$P4_2nm$						*a*	*b*	*c*	*d*
C_{4v}^5	$P4cc$		*a, b*					*c*		*d*
C_{4v}^6	$P4nc$		*a*					*b*		*c*
C_{4v}^7	$P4_2mc$						*a–c*		*d, e*	*f*
C_{4v}^8	$P4_2bc$							*a, b*		*c*
C_{4v}^9	$I4mm$	*a*					*b*		*c, d*	*e*
C_{4v}^{10}	$I4cm$		*a*				*b*		*c*	*d*
C_{4v}^{11}	$I4_1md$						*a*		*b*	*c*
C_{4v}^{12}	$I4_1cd$							*a*		*b*
D_{2d}^1	$P\bar{4}2m$			*a–d*	*e, f*		*g, h*	*i–m*	*n*	*o*
D_{2d}^2	$P\bar{4}2c$				*a–d*	*e, f*		*g–m*		*n*
D_{2d}^3	$P\bar{4}2_1m$					*a, b*	*c*	*d*	*e*	*f*
D_{2d}^4	$P\bar{4}2_1c$					*a, b*		*c, d*		*e*
D_{2d}^5	$P\bar{4}m2$			*a–d*			*e–g*	*h, i*	*j, k*	*l*
D_{2d}^6	$P\bar{4}c2$				*a, b*	*c, d*		*e–i*		*j*
D_{2d}^7	$P\bar{4}b2$				*c, d*	*a, b*		*e–h*		*i*
D_{2d}^8	$P\bar{4}n2$				*c, d*	*a, b*		*e–h*		*i*
D_{2d}^9	$I\bar{4}m2$			*a–d*			*e, f*	*g, h*	*i*	*j*
D_{2d}^{10}	$I\bar{4}c2$				*a, d*	*b, c*		*e–h*		*i*
D_{2d}^{11}	$I\bar{4}2m$			*a, b*	*c*	*d*	*e*	*f–h*	*i*	*j*
D_{2d}^{12}	$I\bar{4}2d$					*a, b*		*c, d*		*e*

Table VII-6 Site symmetries in the space groups D_{4h}^n

		Site symmetry														
		D_{4h}	D_{2h}	C_{4v}	D_4	C_{4h}	D_{2d}	C_{2v}	C_{2h}	D_2	C_4	S_4	C_2	C_s	C_i	C_1
Space group																
D_{4h}^1	$P4/mmm$	*a–d*	*e, f*	*g, h*				*i–o*						*p–t*		*u*
D_{4h}^2	$P4/mcc$				*a, c*	*b, d*			*e*	*f*	*g, h*		*i–l*	*m*		*n*
D_{4h}^3	$P4/nbm$				*a, b*		*c, d*	*h*	*e, f*		*g*		*i–l*	*m*		*n*
D_{4h}^4	$P4/nnc$				*a, b*					*c*	*e*	*d*	*g–j*		*f*	*k*
D_{4h}^5	$P4/mbm$		*c, d*			*a, b*		*f–h*			*e*			*i–k*		*l*
D_{4h}^6	$P4/mnc$					*a, b*			*c*	*d*	*e*		*f, g*	*h*		*i*
D_{4h}^7	$P4/nmm$			*c*			*a, b*	*f*	*d, e*				*g, h*	*i, j*		*k*
D_{4h}^8	$P4/ncc$									*a*	*c*	*b*	*e, f*		*d*	*g*
D_{4h}^9	$P4_2/mmc$		*a–d*				*e, f*	*g–m*					*n*	*o–q*		*r*
D_{4h}^{10}	$P4_2/mcm$		*a, c*				*b, d*	*g–j*	*f*	*e*			*k–m*	*n, o*		*p*
D_{4h}^{11}	$P4_2/nbc$									*a–c*		*d*	*f–j*		*e*	*k*
D_{4h}^{12}	$P4_2/nnm$						*a, b*	*g*	*e, f*	*c, d*			*h–l*	*m*		*n*
D_{4h}^{13}	$P4_2/mbc$								*a, c*	*d*		*b*	*e–g*	*h*		*i*
D_{4h}^{14}	$P4_2/mnm$		*a, b*					*e–g*	*c*			*d*	*h*	*i, j*		*k*
D_{4h}^{15}	$P4_2/nmc$						*a, b*	*c, d*					*f*	*g*	*e*	*h*
D_{4h}^{16}	$P4_2/ncm$							*e*	*c, d*	*a*		*b*	*f–h*	*i*		*j*
D_{4h}^{17}	$I4/mmm$	*a, b*	*c*	*e*			*d*	*g–j*	*f*				*k*	*l–n*		*o*
D_{4h}^{18}	$I4/mcm$		*d*		*a*	*c*	*b*	*g, h*	*e*		*f*		*i, j*	*k, l*		*m*
D_{4h}^{19}	$I4_1/amd$						*a, b*	*e*	*c, d*				*f, g*	*h*		*i*
D_{4h}^{20}	$I4_1/acd$									*b*		*a*	*d–f*		*c*	*g*

Table VII-7 Site symmetries of the space groups C_3^n, C_{3i}^n, D_3^n, C_{3v}^n, D_{3d}^n

		Site symmetry									
		D_{3d}	D_3	C_{3i}	C_{3v}	C_{2h}	C_3	C_i	C_2	C_s	C_1
Space group											
C_3^1	$P3$						*a–c*				*d*
C_3^2	$P3_1$										*a*
C_3^3	$P3_2$										*a*
C_3^4	$R3$						*a*				*b*
C_{3i}^1	$P\bar{3}$			*a, b*			*c, d*	*e, f*			*g*
C_{3i}^2	$R\bar{3}$			*a, b*			*c*	*d, e*			*f*
D_3^1	$P312$		*a–f*				*g–i*		*j, k*		*l*
D_3^2	$P321$		*a, b*				*c, d*		*e, f*		*g*
D_3^3	$P3_112$								*a, b*		*c*
D_3^4	$P3_121$								*a, b*		*c*
D_3^5	$P3_212$								*a, b*		*c*
D_3^6	$P3_221$								*a, b*		*c*
D_3^7	$R32$		*a, b*				*c*		*d, e*		*f*
C_{3v}^1	$P3m1$				*a–c*					*d*	*e*
C_{3v}^2	$P31m$				*a*		*b*			*c*	*d*
C_{3v}^3	$P3c1$						*a–c*				*d*
C_{3v}^4	$P31c$						*a, b*				*c*
C_{3v}^5	$R3m$				*a*					*b*	*c*
C_{3v}^6	$R3c$						*a*				*b*
D_{3d}^1	$P\bar{3}1m$	*a, b*	*c, d*		*e*	*f, g*	*h*		*i, j*	*k*	*l*
D_{3d}^2	$P\bar{3}1c$		*a, c, d*	*b*			*e, f*	*g*	*h*		*i*
D_{3d}^3	$P\bar{3}m1$	*a, b*			*c, d*	*e, f*			*g, h*	*i*	*j*
D_{3d}^4	$P\bar{3}c1$		*a*	*b*			*c, d*	*e*	*f*		*g*
D_{3d}^5	$R\bar{3}m$	*a, b*			*c*	*d, e*			*f, g*	*h*	*i*
D_{3d}^6	$R\bar{3}c$		*a*	*b*			*c*	*d*	*e*		*f*

Table VII-8 Site symmetries in the space groups C_6^n, C_{3h}^n, C_{6h}^n, D_6^n, C_{6v}^n, D_{3h}^n

		Site symmetry																
		C_{6h}	D_6	D_{3h}	C_{6v}	D_3	C_6	C_{3h}	C_{3i}	C_{3v}	C_{2h}	D_2	C_{2v}	C_3	C_2	C_s	C_i	C_1
Space group																		
C_6^1	$P6$						*a*							*b*	*c*			*d*
C_6^2	$P6_1$																	*a*
C_6^3	$P6_5$																	*a*
C_6^4	$P6_2$														*a, b*			*c*
C_6^5	$P6_4$														*a, b*			*c*
C_6^6	$P6_3$													*a, b*				*c*
C_{3h}^1	$P\bar{6}$							*a–f*						*g–i*		*j, k*		*l*
C_{6h}^1	$P6/m$	*a, b*					*e*	*c, d*			*f, g*			*h*	*i*	*j, k*		*l*
C_{6h}^2	$P6_3/m$							*a, c, d*	*b*					*e, f*		*h*	*g*	*i*
D_6^1	$P622$		*a, b*			*c, d*	*e*					*f, g*		*h*	*i–m*			*n*
D_6^2	$P6_122$														*a, b*			*c*
D_6^3	$P6_522$														*a, b*			*c*
D_6^4	$P6_222$											*a–d*			*e–j*			*k*
D_6^5	$P6_422$											*a–d*			*e–j*			*k*
D_6^6	$P6_322$					*a–d*								*e, f*	*g, h*			*i*
C_{6v}^1	$P6mm$				*a*					*b*			*c*			*d, e*		*f*
C_{6v}^2	$P66c$						*a*							*b*	*c*			*d*
C_{6v}^3	$P6_3cm$									*a*				*b*		*c*		*d*
C_{6v}^4	$P6_3mc$									*a, b*						*c*		*d*
D_{3h}^1	$P\bar{6}m2$			*a–f*						*g–i*			*j, k*			*l–n*		*o*
D_{3h}^2	$P\bar{6}c2$					*a, c, e*		*b, d, f*						*g–i*	*j*	*k*		*l*
D_{3h}^3	$P\bar{6}2m$			*a, b*				*c, d*		*e*			*f, g*	*h*		*i–k*		*l*
D_{3h}^4	$P\bar{6}2c$					*a*		*b–d*						*e, f*	*g*	*h*		*i*

Table VII-9 Site symmetries in the space groups D_{6h}^n

Space group		Site symmetry																	
		D_{6h}	D_{3d}	D_{3h}	D_6	C_{6v}	C_{6h}	D_{2h}	D_3	C_{3h}	C_{3v}	C_6	D_2	C_{2h}	C_{2v}	C_3	C_2	C_s	C_1
D_{6h}^1	$P6/mmm$	a, b		c, d		e		f, g			h				i–m			n–q	r
D_{6h}^2	$P6/mcc$				a		b		c	d		e	f	g		h	i–k	l	m
D_{6h}^3	$P6_3/mcm$		b	a					d	c		e		f	g	h	i	j, k	l
D_{6h}^4	$P6_3/mmc$		a	b–d							e, f			g	h		i	j, k	l

Table VII-10 Site symmetries in the space groups T^n, T_h^n, O^n

		Site symmetry															
		O	T_h	T	D_{2h}	D_4	D_3	C_{3i}	D_2	C_4	C_{2h}	C_{2v}	C_3	C_2	C_s	C_i	C_1
Space group																	
T^1	$P23$			*a, b*					*c, d*				*e*	*f–i*			*j*
T^2	$F23$			*a–d*									*e*	*f, g*			*h*
T^3	$I23$			*a*					*b*				*c*	*d, e*			*f*
T^4	$P2_13$												*a*				*b*
T^5	$I2_13$												*a*	*b*			*c*
T_h^1	$Pm3$		*a, b*		*c, d*							*e–h*	*i*		*j, k*		*l*
T_h^2	$Pn3$			*a*				*b, c*	*d*				*e*	*f, g*			*h*
T_h^3	$Fm3$		*a, b*	*c*							*d*	*e*	*f*	*g*	*h*		*i*
T_h^4	$Fd3$			*a, b*				*c, d*					*e*	*f*			*g*
T_h^5	$Im3$		*a*		*b*			*c*				*d, e*	*f*		*g*		*h*
T_h^6	$Pa3$							*a, b*					*c*				*d*
T_h^7	$Ia3$							*a, b*					*c*	*d*			*e*
O^1	$P432$	*a, b*				*c, d*				*e, f*			*g*	*h–j*			*k*
O^2	$P4_232$			*a*			*b, c*		*d–f*				*g*	*h–l*			*m*
O^3	$F432$	*a, b*		*c*					*d*	*e*			*f*	*g–i*			*j*
O^4	$F4_132$			*a, b*			*c, d*						*e*	*f, g*			*h*
O^5	$I432$	*a*				*b*	*c*		*d*	*e*			*f*	*g–i*			*j*
O^6	$P4_332$						*a, b*						*c*	*d*			*e*
O^7	$P4_132$						*a, b*						*c*	*d*			*e*
O^8	$I4_132$						*a, b*		*c, d*				*e*	*f–h*			*i*

Table VII-11 Site symmetries in the space groups T_d^n and O_h^n

		Site symmetry																						
		O_h	O	T_h	T_d	D_{4h}	T	D_{3d}	D_4	D_{2h}	D_{2d}	C_{4h}	C_{4v}	D_3	C_{3i}	C_{3v}	D_2	S_4	C_4	C_{2v}	C_3	C_2	C_s	C_1
Space group																								
T_d^1	$P\bar{4}3m$				*a, b*						*c, d*					*e*				*f, g*		*h*	*i*	*j*
T_d^2	$F\bar{4}3m$				*a–d*											*e*				*f, g*			*h*	*i*
T_d^3	$I\bar{4}3m$				*a*						*b*					*c*		*d*		*e*		*f*	*g*	*h*
T_d^4	$P\bar{4}3n$						*a*										*b*	*c, d*			*e*	*f–h*		*i*
T_d^5	$F\bar{4}3c$						*a, b*											*c, d*			*e*	*f, g*		*h*
T_d^6	$I\bar{4}3d$																	*a, b*			*c*	*d*		*e*
O_h^1	$Pm3m$	*a, b*				*c, d*							*e, f*			*g*				*h–j*			*k–m*	*n*
O_h^2	$Pn3n$		*a*						*b*						*c*			*d*	*e*		*f*	*g, h*		*i*
O_h^3	$Pm3n$			*a*						*b*	*c, d*			*e*						*f–h*	*i*	*j*	*k*	*l*
O_h^4	$Pn3m$				*a*			*b, c*			*d*					*e*	*f*			*g*		*h–j*	*k*	*l*
O_h^5	$Fm3m$	*a, b*			*c*					*d*			*e*			*f*				*g–i*			*j, k*	*l*
O_h^6	$Fm3c$		*a*	*b*							*c*	*d*							*f*	*e*	*g*	*h*	*i*	*j*
O_h^7	$Fd3m$				*a, b*			*c, d*								*e*				*f*		*h*	*g*	*i*
O_h^8	$Fd3c$						*a*							*b*	*c*			*d*			*e*	*f, g*		*h*
O_h^9	$Im3m$	*a*				*b*		*c*			*d*		*e*			*f*				*g, h*		*i*	*j, k*	*l*
O_h^{10}	$Ia3d$													*b*	*a*		*c*	*d*			*e*	*f, g*		*h*

APPENDIX

VIII

RECIPROCAL LATTICES AND SPACE GROUPS; RECIPROCITY OF SITES AND SPECIAL POINTS

Reciprocal space is the space spanned by the reciprocal lattice vectors, defined by

$$\bar{\mathbf{k}} = m_1\mathbf{b}_1 + m_2\mathbf{b}_2 + m_3\mathbf{b}_3$$

$$\mathbf{a}_i \cdot \mathbf{b}_j = \delta_{ij}$$

$$\mathbf{b}_i = \frac{\mathbf{a}_j \times \mathbf{a}_k}{\mathbf{a}_i \cdot (\mathbf{a}_j \times \mathbf{a}_k)}$$

where $\mathbf{a}_j$ is a primitive vector of the real lattice, and the m_i are integers.

The choice of the $\mathbf{a}_i$ is to a certain extent arbitrary and conventional. Each unit cell must contain at least one lattice point, and the entire solid must be mapped by the $\mathbf{a}_i$ without any vacancies. In the standard works on crystallography, rectangular, rhombohedral, or hexagonal parallelepipeds are generally chosen so as to most simply conform with the translational symmetry of one or another of the 14 Bravais lattices. This is not necessary, however, and it does obscure the fact that a cell can display the *point* symmetry of a Bravais lattice. For example, the usual shape and form of the body-centered cubic lattice, $O_h^9(Im3m)$, does not readily manifest the fact that the point symmetry along one of the cube diagonals is C_{3v}. This difficulty is overcome by the use of the so-called Wigner–Seitz cell[1] whose boundary planes are the perpendicular bisectors of the lines connecting nearest-neighbor and next-to-nearest neighbor lattice points, etc., until the planes so generated are no longer contained within the fundamental Bravais cell. The Wigner–Seitz cells for the bc and fc cubes are illustrated in Figs. VIII-1 and VIII-2.

1. E. Wigner and F. Seitz, *Phys. Rev.*, **43,** 804 (1933).

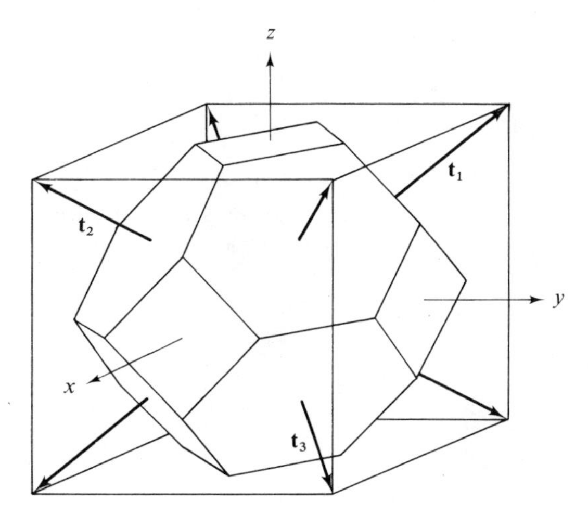

Figure VIII-1 Wigner–Seitz cell for the body-centered cubic structure. (After J. C. Slater, *Quantum Theory of Molecules and Solids,* Vol. 2, McGraw-Hill, New York, 1965, page 17)

For all monatomic crystals, the primitive cell is a rhombohedron. We distinguish in notation between the primitive translations of a real lattice in a cartesian basis, $\mathbf{t}_i$, and those in a rhombohedral basis, τ_i. The same distinction can be made in reciprocal space: **k**-vectors are reserved for use in cartesian k-space but they are denoted by $\boldsymbol{\kappa}$ in a rhombohedral basis.

In bcc, the $\mathbf{t}_i$ and the τ_i are related by

$$\mathbf{t} = \frac{a}{2}\begin{pmatrix} -1 & 1 & 1 \\ 1 & -1 & 1 \\ 1 & 1 & -1 \end{pmatrix}\tau$$

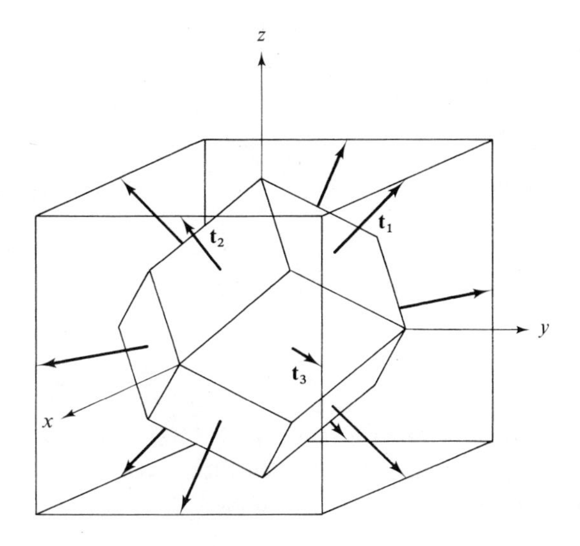

Figure VIII-2 Wigner–Seitz cell for the face-centered cubic structure. (After J. C. Slater, *Quantum Theory of Molecules and Solids,* Vol. 2, McGraw-Hill, New York, 1965, page 19)

while in fcc the transformation is

$$\mathbf{t} = \frac{a}{2}\begin{pmatrix} 0 & 1 & 1 \\ 1 & 0 & 1 \\ 1 & 1 & 0 \end{pmatrix}\boldsymbol{\tau}$$

Because of Eq. (4-1-3), the transformations preserve the scalar product, $2\pi\mathbf{k}\cdot\mathbf{t} = \boldsymbol{\kappa}\cdot\boldsymbol{\tau}$. From this fact it follows that the bcc and the fcc lattices are reciprocal to each other. In fcc,

$$\boldsymbol{\kappa} = \frac{2\pi}{a}\begin{pmatrix} 0 & 1 & 1 \\ 1 & 0 & 1 \\ 1 & 1 & 0 \end{pmatrix}\mathbf{k}$$

while in bcc,

$$\boldsymbol{\kappa} = \frac{2\pi}{a}\begin{pmatrix} -1 & 1 & 1 \\ 1 & -1 & 1 \\ 1 & 1 & -1 \end{pmatrix}\mathbf{k}$$

This reciprocity is not restricted to these cubic cases. Similar relationships in the other crystal systems yield the following rules for reciprocation for the Bravais lattices.

1. Cubic system:
   (*a*) $P \leftrightarrow P$
   (*b*) $F \leftrightarrow I$
2. Hexagonal:
   (*a*) $P \leftrightarrow P$
3. Rhombohedral (trigonal):
   (*a*) $R \leftrightarrow R$
4. Tetragonal:
   (*a*) $P \leftrightarrow P$
   (*b*) $I \leftrightarrow I$
5. Orthorhombic:
   (*a*) $P \leftrightarrow P$
   (*b*) $C \leftrightarrow C$
   (*c*) $A \leftrightarrow A$
   (*d*) $I \leftrightarrow F$
6. Monoclinic:
   (*a*) $P \leftrightarrow P$
   (*b*) $C \leftrightarrow C$
7. Triclinic:
   (*a*) $P \leftrightarrow P$

A consequence of the fundamental reciprocity between the BZ's and the Wigner–Seitz cells of the Bravais lattices is that within each there exist certain special positions, including the origin, which have non-trivial point symmetry. In direct lattices, these positions are called *sites* and they have the following significance.

As discussed in Sections 4-1 and 4-2, translational symmetry makes it possible to analyze important portions of a crystal's motion in terms of the motion at $\mathbf{k} = 0$. Since $\mathbf{k} = 0$ corresponds to the motion with all unit cells in phase, we in effect need analyze the motion of just one unit cell, and the point group G is then referred to as the *unit cell group*.

In a few cases, crystals have only one molecule (or atom) per unit cell. The space groups of such crystals are necessarily symmorphic and the lattice point occupied by the molecule must be at the center of the Wigner–Seitz cell. The molecule will then be left invariant in orientation and position by all operations of the factor group G. The subgroup of G which in general accomplishes this is defined as the *site group* S for the occupied lattice point and that point is denoted a *site*. In this case, $S = G$.

Usually a crystal will have more than one molecule (or atom) per cell. No more than one of these, of course, can be at the center of the Wigner–Seitz cell. Indeed, if the space group is non-symmorphic, the unit cell cannot be singly occupied. In these cases, the unit cell occupants occur in sets, each member of which is left invariant in position and orientation by the operations of at least one subgroup of G, and the remaining operations of G not in S interchange members of the set. In all cases, S is the intersection of G and the molecular point group and $\mathcal{O}(G)/\mathcal{O}(S) = Z$, the unit cell population. It follows that in non-symmorphic space groups, the *maximal site* (the lattice point whose site group is of highest order among all possible sites of the space group) must be of lower order than that of G and the center of the Wigner–Seitz cell will be unoccupied.

It is then possible to carry out a preliminary (and sometimes sufficient) point group analysis of the molecular motions, using S. Furthermore, *correlation* (see Section 4-4) of the representations of S with those of G will yield much information about the possible multiplicities of $\mathbf{k} = 0$ modes which arise *via* intermolecular coupling (Section 6-5). Correlation of S with the molecular point group will indicate which degenerate modes may be split by crystal field effects, since S describes the symmetry of the intermolecular potential energy surface in the vicinity of the site.

In reciprocal space the sites are called the *special points* of the BZ and the site groups are denoted by $H(\mathbf{k})$, the group of the $\mathbf{k}$-vector. The subgroup $H(\mathbf{k})$ is defined such that $R\mathbf{k} = \mathbf{k} + \bar{\mathbf{k}}$, where $\bar{\mathbf{k}}$ is a lattice translation in reciprocal space and R is a point operation of $H(\mathbf{k})$. In the interior of the BZ, $\bar{\mathbf{k}} = 0$ and $H(\mathbf{k})$ consists only of those operations which leave $\mathbf{k}$ invariant. At certain boundary points, however, the more general rule holds. If R is in G but not in $H(\mathbf{k})$, $R\mathbf{k} = \mathbf{k}'$. The set of all distinct $\mathbf{k}'$ constitutes the *star* of $\mathbf{k}$.

In real space, a space group operation $\{R \mid \tau_R\}$ has the effect of adding the translation τ_R to the effect of the point operation:

$$\{R \mid \tau_R\}\mathbf{r} = \{R \mid \mathbf{0}\}\mathbf{r} + \{E \mid \tau_R\}$$

In reciprocal space the analogous transformation $R\mathbf{k} = \mathbf{k} + \bar{\mathbf{k}}$ takes place.

However $\bar{\mathbf{k}}$ has the property that $e^{2\pi i\bar{\mathbf{k}}\cdot\mathbf{t}} = 1$ so that a pair of vectors $\mathbf{k}'$, $\mathbf{k}$ related by

$$\mathbf{k}' = \mathbf{k} + \bar{\mathbf{k}}$$

are said to be equivalent, since their effect in symmetrization is equivalent. Hence all considerations of $\mathbf{k}$ may be restricted to the BZ, whose boundaries are the $\bar{\mathbf{k}}$.

Because of these properties, *all reciprocal lattices are symmorphic,* and we may find their special points from a knowledge of the sites of the symmorphic space groups plus a few further details of space group reciprocation. For example, if a space group is primitive and symmorphic, its sites are the special points of its reciprocal lattice, since $P \leftrightarrow P$ always. An example is $Pm3m$. On the other hand, if both F and I exist (as in the cubic and orthorhombic systems) and the space group is symmorphic, then the sites of an F group are the special points of the I group with the same factor group, e.g., $Fm3m \leftrightarrow Im3m$, as is well known.* A less well-known example is $C_{2v}^{20} \leftrightarrow C_{2v}^{18}$ ($Imm2 \leftrightarrow Fmm2$).

If a space group is non-symmorphic, its reciprocal lattice has the special points of the reciprocal of the "parent" *symmorphic* space group. Thus $O_h^7 \to O_h^9$ ($Fd3m \to Im3m$), $D_{6h}^4 \to D_{6h}^1$ ($P6_3/mmc \to P6/mmm$), and $C_{2h}^5 \to C_{2h}^1$ ($P2_1/b \to P2/m$).

There are a few special precautions to lattice reciprocation which arise among certain hexagonal and tetragonal groups. In the hexagonal system, for factor groups D_3, C_{3v}, D_{3d}, and D_{3h}, there arise *pairs* of *symmorphic* space groups ($P312$, $P321$; $P3m1$, $P31m$; $P\bar{3}1m$, $P\bar{3}m1$; and $P\bar{6}m2$, $P\bar{6}2m$) in which the members of each pair differ only in the two ways in which the primitive translation vectors can be arranged relative to the dihedral axes or vertical planes. Reciprocation interchanges these axes or planes. Hence D_{3d}^2 ($P\bar{3}1c$), whose parent symmorphic group is $D_{3d}^1(P\bar{3}1m)$ has $D_{3d}^3(P\bar{3}m1)$ as its reciprocal, etc. Note (from Appendix VII) that these reciprocal pairs may be easily identified by the occurrence, in the International symbol for each pair of space groups, of a permutation in the order in which the dyad, monad, and/or mirror plane are indicated.

Since the majority of space groups are non-symmorphic, we reiterate our recipe using the examples D_{3h}^4 ($P\bar{6}2c$) and D_{3h}^2 ($P\bar{6}c2$). Their symmorphic parents are respectively $P\bar{6}2m$ and $P\bar{6}m2$. *These* are reciprocal. The reciprocal space groups of $P\bar{6}2c$ and $P\bar{6}c2$ are therefore, respectively, $P\bar{6}m2$ and $P\bar{6}2m$.

In the tetragonal system there are two such sets of symmorphic pairs: $P\bar{4}2m$, $P\bar{4}m2$ and $I\bar{4}m2$, $I\bar{4}2m$. Only the members of the body-centered pair are reciprocal, however; in the primitive pairs, each space group is *self-reciprocal.*

* First noted by L. Brillouin, in *Wave Propagation in Periodic Structures,* McGraw-Hill, New York, 1946, page 153.

APPENDIX

IX

BRILLOUIN ZONES AND SPECIAL POINTS

In this appendix we illustrate the Brillouin zones and special points of the space groups used as examples in Sections 6-2 and 6-3. These are as follows:

Crystal	Space group: Schönflies symbol	Space group: International symbol	Bravais lattice	Special conditions
HCN	C_{2v}^{20}	$Imm2$	Γ_0^v	$b > c > a$
CO_2	T_h^6	$Pa3$	Γ_c	
CHI_3	C_6^6	$P6_3$	Γ_h	
NaN_3	D_{3d}^5	$R\bar{3}m$	Γ_{rh}	$a < 2^{1/2}c$
$CaCO_3$	D_{3d}^6	$R\bar{3}c$	Γ_{rh}	$a < 2^{1/2}c$
$NaClO_3$	T^4	$P2_13$	Γ_c	

In each case, we illustrate the BZ with its special points, tabulate the lattice parameters of the crystal, and identity $H(\mathbf{k})$ for each special point.

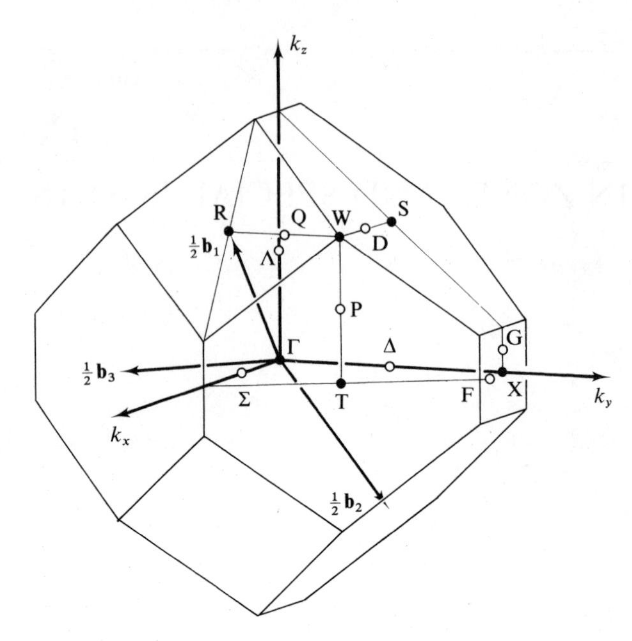

Figure IX-1 The Brillouin zone for the HCN crystal (space group $C_{2v}^{20} = Imm2$). The Bravais lattice is Γ_0^v, with $b > c > a$.

$H(\mathbf{k})$:

$\Gamma, X = C_{2v}$	$a = 4.13$ Å
$R, S = C_s$	$b = 4.85$
$T, W = C_2$	$c = 4.34$

(After C. J. Bradley and A. P. Cracknell, *The Mathematical Theory of Symmetry in Solids,* Clarendon Press, Oxford, 1972, page 99)

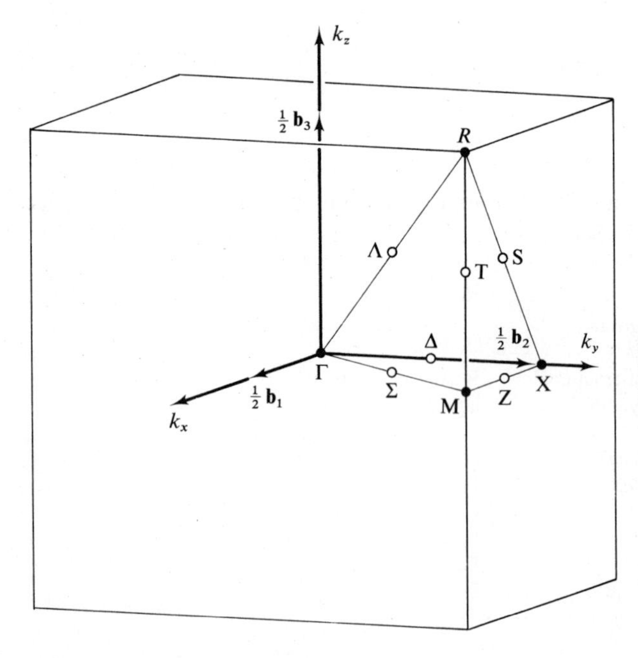

Figure IX-2 The Brillouin zone for CO_2 and $NaClO_3$ (space groups $T_h^6 = Pa3$ and $T^4 = P2_13$). The Bravais lattice is Γ_c.

$H(\mathbf{k})$:

$\Gamma, R = T_h(T)$	$a = 5.575\ (6.570)$ Å
$\Lambda = C_3$	
$\Sigma, \Delta = C_{2v}$	
$M, X = D_{2h}$	
$Z, T = C_{2v}$	
$S = C_s$	

(After C. J. Bradley and A. P. Cracknell, *The Mathematical Theory of Symmetry in Solids,* Clarendon Press, Oxford, 1972, page 108)

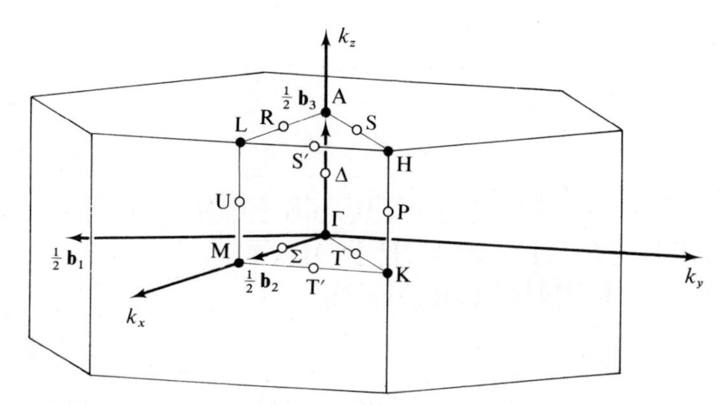

Figure IX-3 The Brillouin zone for CHI_3. (Space group $C_6^6 = P6_3$.) The Bravais lattice is Γ_h.

$H(\mathbf{k})$:		
	$\Gamma, A, \Delta = C_6$	$a = 6.83$ Å
	$H, P, K = C_3$	$c = 7.53$
	$L, U, M = C_2$	

(After C. J. Bradley and A. P. Cracknell, *The Mathematical Theory of Symmetry in Solids,* Clarendon Press, Oxford, 1972, page 107)

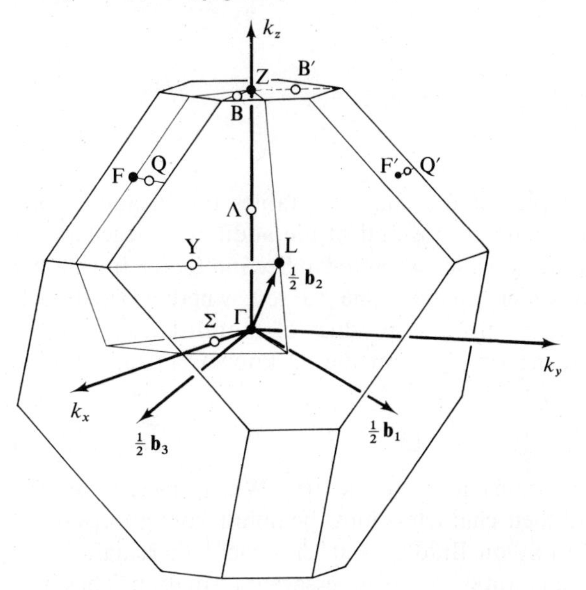

Figure IX-4 The Brillouin zone for NaN_3 and calcite. (Space groups $D_{3d}^5 = R\bar{3}m$ and $D_{3d}^6 = R\bar{3}c$.) The Bravais lattice is $\Gamma_{rh}\left(a_0 < \sqrt{2}c_0\right)$.

$H(\mathbf{k})$:	NaN_3	$CaCO_3$
$\Gamma, Z = D_{3d}$	$a =$ 3.637 Å	$a =$ 4.99008 Å
$\Lambda = C_{3v}$	$c =$ 15.209	$c =$ 17.05951
$L, F = C_{2h}$		
$B, \Sigma, Q, Y = C_2$		

(After C. J. Bradley and A. P. Cracknell, *The Mathematical Theory of Symmetry in Solids,* Clarendon Press, Oxford, 1972, page 107)

APPENDIX

X

IRREDUCIBLE REPRESENTATIONS FOR SOME SELECTED BRILLOUIN ZONE POINTS IN NON-SYMMORPHIC GROUPS

Here we shall give a few examples of the character tables for Brillouin zone boundary points in non-symmorphic groups and of the additional degeneracies produced by time reversal symmetry, which we noted in Section 6-6(e). In Chapter 4 a criterion was developed which enables one to test whether irreducible representations can be constructed using a product of a translational and a rotational factor, the latter arising simply from the 32 known crystallographic point groups. This test was expressed as

$$e^{i\boldsymbol{\kappa}_R \cdot \boldsymbol{\tau}_S} = 1 \qquad \text{(X-1)}$$

where R and S are any two operations in the group $H(\boldsymbol{\kappa})$. When this test fails, the irreducible representations and their characters must be found from groups other than the 32 point groups. We rely on Bradley and Cracknell[1] (hereinafter BC) for a complete analysis and tabulation of all the necessary information. They give the character tables of some 95 abstract groups of orders ≤ 192. However, their work includes both single-valued and double-valued representations, the latter being required for the treatment of electron spin, but not in the context of the present work, so that not all these abstract groups are required for the analysis of vibrations in crystals. In fact, the largest group encountered is of order 96. Moreover, only part of the irreducible representations of these groups are required,

1. C. J. Bradley and A. P. Cracknell, *The Mathematical Theory of Symmetry in Solids,* Clarendon Press, Oxford, 1972.

since only those whose basis elements are eigenfunctions of the translational group are needed. We have anticipated this fact in Chapter 4 by initiating the construction of symmetry coordinates by translational symmetrization. These are called "small" representations and BC further show for these representations whether time reversal will cause additional degeneracy.

The operations defining the space groups transform the space coordinates of the atoms. The Hamiltonian is however invariant not only to these operations, but also to time reversal. Examination of the time-dependent Schrödinger equation,

$$H\psi = i\hbar \frac{\partial \psi}{\partial t}$$

shows that, for real H, changing the sign of t is equivalent to taking the complex conjugate of ψ. This in turn implies that in the time-independent Schrödinger equation, the spatial wave functions ψ and ψ^* must have the same energy eigenvalues. Since the wave functions ψ are basis functions for the irreducible representations of the space group, it follows that complex conjugate representations Γ and Γ^* are degenerate. For point groups this degeneracy is recognized immediately for representations indicated by E_+ and E_- in C_3, T, etc.

In the space groups it is necessary to consider the effects of both the rotational and translational operations, and the analysis as shown originally by Wigner[2] requires a determination of the relation between Γ and Γ^*:

1. Γ is equivalent to a real representation
2. Γ is equivalent to Γ^* but not to a real representation
3. Γ is not equivalent to Γ^*

For case (1) time reversal produces no extra degeneracy; for case (2), two different levels, both belonging to Γ become degenerate; for case (3), two different representations become degenerate. The test to distinguish these cases was devised by Frobenius and Schur for any group of order g:

$$\begin{aligned} \sum_R \chi(R^2) &= g \quad \text{case (1)} \\ &= -g \quad \text{case (2)} \\ &= 0 \quad \text{case (3)} \end{aligned}$$

Herring[3] adapted this test to the space groups, so that the required summation could be abbreviated to a summation, not over the full space group, but just to a selection of operations R such that $R\boldsymbol{\kappa}$ is equivalent to $-\boldsymbol{\kappa}$. It may be mentioned that extra degeneracy due to time reversal may occur in the interior as well as the boundary of the Brillouin zone, for example in T_d^2 along a twofold axis.

We shall now discuss a number of interesting examples with the aid of Bradley and Cracknell's work.

2. E. P. Wigner, *Nachr. Kgl. Ges. Wiss. Göttingen Math-Physik Kl.*, 546 (1932).
3. C. Herring, *Phys. Rev.*, **52**, 361 (1937).

$O_h^7(Fd3m)$

The symmorphic space group reciprocal to O_h^7 is O_h^9, whose sites (Appendix VII) and BSW conventional zone symbols include (a) = $O_h(\Gamma)$, (b) = $D_{4h}(X)$, (c) = $D_{3d}(L)$, (d) = $D_{2d}(W)$, together with interior points and regions of lower symmetry on the zone boundary. See Fig. 4-3(b) for identification of these points. The point L offers no problems: it satisfies the test (X-1). This is shown in the BC tables by the fact that the irreducible representations at L are those of a direct product of D_{3d} with a translational group.

But the points X and W cannot be so represented; BC (page 384) shows that the irreducible representations at X are those numbered 10, 11, 13, and 14 of the abstract group G_{32}^2, while those at W are representations 11 and 12 of G_{32}^4. Moreover, as shown in BC, time reversal symmetry does not in these cases produce any additional degeneracy. We shall not here reproduce the full character tables for the points, but only a portion thereof pertaining to the point group C_{4v}. The reason for making this simplified selection is that we merely need to ascertain the degeneracy of the phonon branches on the zone boundary, having developed in Chapter 4 a method of factoring the dynamic matrix at all interior points. Thus we can apply simple correlation methods to anticipate the behavior of the branches at the boundary points, in the present case X, which can be reached from $O_h(\Gamma)$ via $C_{4v}(\Delta)$. The necessary portions of the character table and resultant correlation are given in Table X-1.

Table X-1 Part of the character table at X and correlation of symmetry species in O_h^7

$X(G_{32}^2)$	E	$2C_4$	C_4^2	$2\sigma_v$	$2\sigma_d$	i
$R_{10} = E_1$	2	0	2	0	2	0
$R_{11} = E_2$	2	0	−2	0	0	0
$R_{13} = E_3$	2	0	2	0	−2	0
$R_{14} = E_4$	2	0	−2	0	0	0

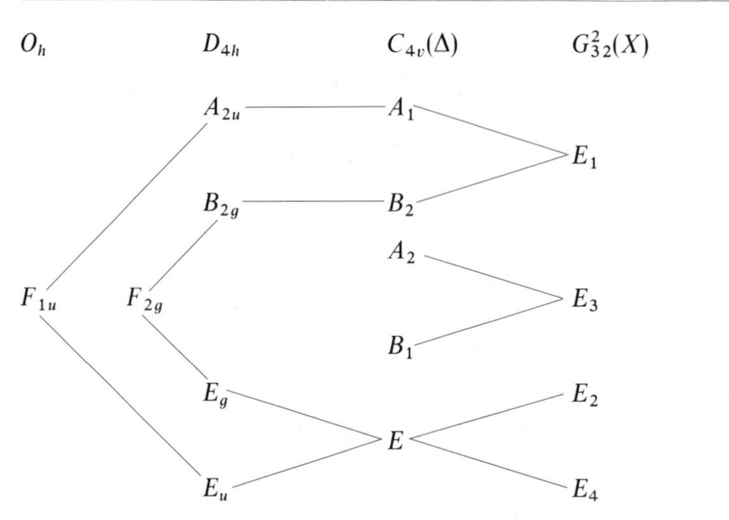

For the particular case of diamond, we noted in Chapter 4 that the acoustic and optic branches, belonging respectively to F_{1u} and F_{2g} at $\kappa = 0$ correlate with $A_1 + E$ and with $B_2 + E$ for $\kappa = \Delta(C_{4v})$. Table X-1 now shows us that the longitudinal branches A_1 and B_2 in the interior become degenerate, of species E_1, at the boundary point X. Also, the transverse branches ($2E$ under C_{4v}) give rise to symmetry species E_2 and E_4 at X. These cannot, however, be classified as g and u, because, as shown by BC in the full character table for G_{32}^2, the characters of all four relevant species E_1, E_2, E_3, and E_4 are zero under the inversion. The difference between this and a symmorphic case (such as O_h^5) is also illustrated in Table X-1 which shows that part of the correlation of C_{4v} with D_{4h} which would apply to modes of species F_{1u} and F_{2g} at the origin of the zone.

$T_h^6(Pa3)$

This space group was encountered in Section 6-2(b) in the discussion of carbon dioxide. The symmorphic group reciprocal to T_h^6 is T_h^1 whose important sites (Appendix VII) are (a) = $T_h(\Gamma)$, (b) = $T_h(R)$, (c) = $D_{2h}(M)$, and (d) = $D_{2h}(X)$; see Fig. IX-2. In T_h^6, none of the three zone boundary points R, M, and X satisfies the test (X-1).

First we consider the point R (the cube corner) which may be approached via Λ, which is a Wyckoff site (i) of symmetry C_3. As shown in BC (page 377), the character table for point R is based upon a group of order 48, with six relevant representations, all of which are doubly degenerate. Bradley and Cracknell have proposed a reasonable and appropriate notation for all symmetry species of all the space groups, but there are a few points of conflict with other conventions. We have adhered to the convention prevalent amongst vibrational spectroscopists whereby F (rather than T as used extensively by electronic spectroscopists) is used to indicate a real triply degenerate representation. For complex conjugate pairs of individual dimensions 1, 2, 3, BC propose ${}^1E$, ${}^2E$; ${}^1F$, ${}^2F$; and ${}^1H$, ${}^2H$. Since there is no danger of confusing the super prefixes with spin multiplicity, we shall, with only a minor inconsistency, defer to the BC convention in discussing the zone boundary symmetry, but adhere to E_+ and E_- (or E' and E'') for complex conjugate species, or F for triply degenerate species at interior points.

Before excerpting a small portion of the character table for the point R from BC, we note also that of the six relevant representations, all have increased degeneracy because of time reversal symmetry. In the case of ${}^1F_g$ and ${}^2F_g$ and of ${}^1F_u$ and ${}^2F_u$, the increased degeneracy is of the familiar kind in which the characters of the pairs are complex conjugate. But in the cases of E_g and E_u, the situation is rather different: time reversal causes *two different* states of species E_g to be degenerate; similarly with E_u. Thus at R *all* energies are fourfold degenerate. The excerpted character table and correlation of R with C_3 is given in Table X-2.

To see the implication for the dispersion of the internal modes in CO_2, recall that ν_1 yields $A_g + F_g$, ν_2 yields $E_u + 2F_u$, and ν_3 yields $A_u + F_u$ at $\kappa = O(T_h)$. Along the line Λ in the interior, under C_3 symmetry, ν_1 becomes $2A + E_+ + E_-$,

Table X-2 Part of the character table at R and correlation in T_h^6 ($\omega = \exp(2\pi i/3)$)

R	E	C_3^+	C_3^-	i
E_g	2	-1	-1	2
${}^2F_g$	2	$-\omega$	$-\omega^*$	2
${}^1F_g$	2	$-\omega^*$	$-\omega$	2
E_u	2	-1	-1	-2
${}^2F_u$	2	$-\omega$	$-\omega^*$	-2
${}^1F_u$	2	$-\omega^*$	$-\omega$	-2

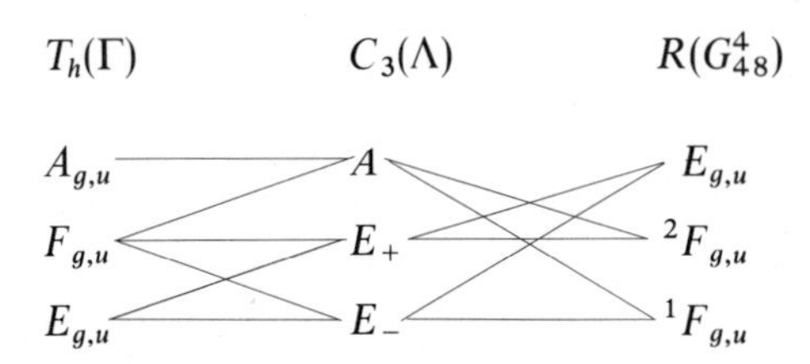

as does ν_3, while ν_2 becomes $2A + 3E_+ + 3E_-$. From the diagram in Table X-2, it is evident that the two stretching modes ν_1 and ν_3 will yield ${}^2F + {}^1F$, while the bending mode must yield $2E + {}^2F + {}^1F$. The rationale behind this conclusion is that one must obtain complete representations, i.e., two components of each of the species E, ${}^2F_1$, and ${}^1F$, and that the internal modes are too widely separated to mix strongly with one another. The question of the g, u labels of each mode may be resolved by the observation that the inversion i is a member of both T_h and $R = G_{48}^4$ and commutes with all the elements in each group. Therefore, although the branches lose their g, u labels in the interior, provided there is no strong intermode mixing, we may expect finally that ν_1 becomes ${}^2F_g + {}^1F_g$, ν_3 becomes ${}^2F_u + {}^1F_u$, and ν_2 becomes $2E_u + {}^2F_u + {}^1F_u$ at R. Note that this implies a single, fourfold degenerate level for ν_1 and for ν_3, but two fourfold levels ($2E_u$ and ${}^2F_u + {}^1F_u$) for ν_2.

A similar analysis using BC shows that at the zone boundary point X only two doubly degenerate representations exist, which combine the branches at $\Delta(C_{2v})$ in the interior according to the rule $A_1 + B_1(C_{2v}) \rightarrow E'(X)$ and $A_2 + B_2(C_{2v}) \rightarrow E''(X)$. At the point M it is also true that only two doubly degenerate representations exist, but in this case these representations are complex conjugate so that time reversal produces effectively a fourfold degeneracy, which in fact exists everywhere on the edges of the cube which describes the Brillouin zone.

$C_6^6(P6_3)$

The case of this non-symmorphic space group is somewhat different from the examples previously discussed. The symmorphic reciprocal group is C_6^1, which has sites (a) $= C_6(\Gamma, \Delta, A)$, (b) $= C_3(K, P, H)$, and (c) $= C_2(M, U, L)$: see Appendix VII and Fig. IX-3. Now although in the zone boundary the criterion $e^{i\kappa_R \cdot \tau_S} = 1$ is everywhere satisfied, since all $\kappa_R = 0$, which means that the representatives may be constructed using a simple product of the translational factor $e^{i\kappa \cdot \tau_R}$ and a point group representation, some care must be exercised in describing the degeneracies. We consider the path $\Gamma \to \Delta \to A$: everywhere along this path we have $\chi\{R\,|\,\tau_R\} = e^{i\kappa \cdot \tau_R}\chi_R^{(l)}$ where the $\chi_R^{(l)}$ are the characters of the cyclic, Abelian group, C_6. For the generating element $\{C_6\,|\,00\frac{1}{2}\}$, the characters are thus $e^{i\kappa \cdot \tau_R}\,e^{2\pi i l/6}$ where $l = 0$ is called A, $l = 3$ is called B, $l = 1$ and 5 form E_1, and $l = 2$ and 4 form E_2 in the point group notation. For $\kappa = 0(\Gamma)$, $l = 1$ and 5 are complex conjugate, as are $l = 2$ and 4; hence the twofold degeneracies. In the interior (Δ), this complex conjugacy breaks down, and the E_1 and E_2 branches will split as shown in the detailed calculations for iodoform by Neto, Oehler, and Hexter.[4] But at the boundary (A), the factor $e^{i\kappa \cdot \tau_{C_6}} = e^{i\kappa_3/2} = e^{i\pi/2} = i$, and new complex conjugate pairs prevail. If we follow Neto, Oehler, and Hexter in designating $l = 1$ and $l = 5$ as E_1' and E_1'', $l = 2$ and $l = 4$ as E_2' and E_2'', then by use of the formula $i\,e^{i\pi l/3}$ we easily discover that $l = 0, 3$; $l = 1, 2$; and $l = 4, 5$ are now complex conjugate, so that the branches which begin at Γ as A and B, E_1' and E_2', and E_1'' and E_2'' become degenerate at the zone boundary point A; this is clearly illustrated in Fig. 6-8(a).

$D_{3d}^6(R\bar{3}c)$

Our final example describes the calcite structure. The symmorphic reciprocal space group is D_{3d}^5 with sites (a) $= D_{3d}(\Gamma)$, (b) $= D_{3d}(Z)$, (c) $= C_{3v}(\Lambda)$, (d) $= C_{2h}(L)$, (e) $= C_{2h}(F)$, as illustrated in Fig. IX-4. At $Z = (\frac{1}{2}\frac{1}{2}\frac{1}{2})$ in rhombohedral coordinates, $\kappa_R = (\pi\pi\pi)$ for R not in C_{3v}. Since $\tau_S = (\frac{1}{2}\frac{1}{2}\frac{1}{2})$ for S not in S_6, a maximal site, the test $e^{i\kappa_R \cdot \tau_S} = 1$ fails, and as shown in BC, the representations must be found amongst those of the abstract group G_{24}^1 and are all doubly degenerate. Moreover, two of these are complex conjugate, so the energies will be fourfold and twofold degenerate. The correlation from Γ to Z via Λ is shown in Table X-3.

As noted in Section 6-3(e), ν_1 in calcite gives rise to species $A_{1g} + A_{1u}$ at Γ; from Table X-3, one sees that the two branches $A_1 + A_2$ in the zone interior coalesce at Z; similarly ν_2 yields $A_{2g} + A_{2u}$ at Γ and goes to E at Z. The molecularly degenerate modes ν_3 and ν_4 each yield $E_g + E_u$ at Γ and become fourfold degenerate at Z because of the time reversal degeneracy of ${}^1F$ and ${}^2F$.

The two other high symmetry points on the zone boundary are L and F. Of these, the F point [rhombohedral coordinates $(\frac{1}{2}\frac{1}{2}0)$] satisfies the $e^{i\kappa_R \cdot \tau_S} = 1$ test because the only non-vanishing $\kappa_R = (\bar{\pi}\bar{\pi}0)$ and hence yields unity with $\tau_S = (\frac{1}{2}\frac{1}{2}\frac{1}{2})$. Thus the symmetry at F may be described with the aid of the point group C_{2h}. However, at $L = (0\frac{1}{2}0)$, κ_R takes on the value $(0\bar{\pi}0)$ when $R = i$ and the $e^{i\kappa_R \cdot \tau_S} = 1$ test is not satisfied. Turning to BC (pages 362 and 228), one finds that the only representation at L is doubly degenerate, time reversal having no additional effect.

4. N. Neto, O. Oehler, and R. M. Hexter, *J. Chem. Phys.*, **58,** 5661 (1973). See Table 6-15 and Fig. 6-8(a).

Table X-3 Correlation of symmetry species along the threefold axis in D_{3d}^{6}

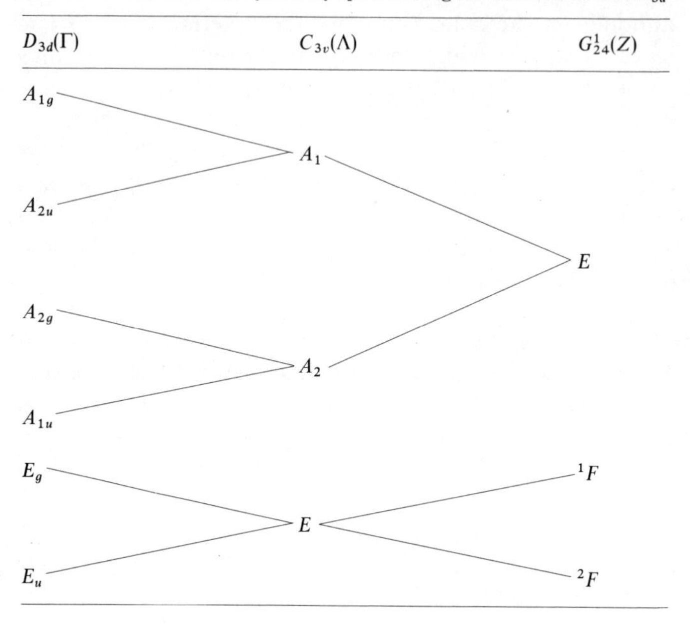

APPENDIX

XI

EVALUATION OF LATTICE SUMS

In this appendix we outline the two kinds of lattice summation techniques which have been used in evaluating dipole wave sums, such as the field propagation tensor of Section 5-7. In this outline we discuss the physical basis of each technique, as well as their relationship to each other. The older of the two techniques is the so-called Ewald–Kornfeld procedure, first developed by Ewald.[1] While this procedure has enjoyed a number of expositions[2,3,4] we primarily follow here that of Slater.[5]

The fundamental problem in carrying out a dipole sum over a three-dimensional lattice is that this sum is *conditionally convergent*—its value depends upon the *order* of the summation: different results are obtained depending on which limit, $k \to 0$ or $R \to \infty$, is approached first. Some of the difficulties imposed by these convergence problems have been discussed by Hexter.[6]

Ewald's method was originally devised to solve the Madelung problem—that is, to evaluate the constant α in the Coulomb energy of an ionic lattice,

$$\frac{\alpha}{a} = \sum_{l}' \frac{1}{|\mathbf{l}|} - \sum_{l} \frac{1}{|\mathbf{l} - \mathbf{r}|}$$

1. P. P. Ewald, *Dissertation,* München, 1912; *Ann. Phys.,* **54,** 519, 557 (1917); **64,** 253 (1921); *Nachr. Ges. Wiss. Göttingen, N. F. II,* **3,** 55 (1938).
2. M. Born and K. Huang, *Dynamical Theory of Crystal Lattices,* Oxford University Press, 1954, section 30 and Appendices II and III.
3. C. Kittel, *Introduction to Solid State Physics,* John Wiley & Sons, Inc., New York, 1953, Appendix B.
4. J. M. Ziman, *Principles of the Theory of Solids,* Cambridge University Press, New York, 1964.
5. J. C. Slater, *Insulators, Semiconductors and Metals,* McGraw-Hill, 1967, Section 9-4.
6. R. M. Hexter, *J. Chem. Phys.,* **33,** 1833 (1960).

where a is the lattice constant and $\mathbf{r}$ is the distance between adjacent ions. However, we are interested in a lattice of, say, unit dipoles oriented parallel to, say, the z-axis of a crystal. We seek the electric field due to this lattice at a general site in it. We first calculate the *potential* at an arbitrary point, $\mathbf{r}$, of this lattice. This potential is equivalent to the derivative $-\partial/\partial z$ of the potential ϕ at the same point $\mathbf{r}$ of a lattice of unit, positive charges. The z component of the electric field is then equal to the second derivative, $\partial^2\phi/\partial z^2$. This extension of the Ewald procedure to dipole lattices was originally carried out by Kornfeld.[7]

In more than one dimension, this sum is also conditionally convergent, and a counting scheme must be devised, such as that due to Evjen.[8] In general, however, a direct sum of Coulomb terms $1/r$ is slow to converge. Alternatively, the potential may be found by the solution of Poisson's equation, $\nabla^2\phi = -4\pi\rho$. In doing so, both the potential ϕ and the charge density ρ are written as a Fourier series, such as

$$\rho = \sum_{\mathbf{k}} \rho_{\mathbf{k}}\, e^{i\mathbf{k}\cdot\mathbf{r}} \tag{XI-1}$$

where $\mathbf{k}$ is a vector of the reciprocal lattice. Substitution in Poisson's equation gives

$$\phi = 4\pi \sum_{\mathbf{k}\neq 0} \rho_{\mathbf{k}}\, e^{i\mathbf{k}\cdot\mathbf{r}}/k^2 \tag{XI-2}$$

The omission of the $\mathbf{k} = 0$ term is necessary to prevent divergence of this sum; it must likewise be absent from (XI-1). Since

$$\rho_{\mathbf{k}} = \frac{1}{\mathrm{v}} \int_{\text{one unit cell}} \rho(\mathbf{r})\, e^{-i\mathbf{k}\cdot\mathbf{r}}\, d\mathrm{v} \tag{XI-3}$$

$\rho_0 = 0$ means that the average charge of a unit cell vanishes, as it must. Equation (XI-3) can also be used to find the other $\rho_{\mathbf{k}}$, provided that we have a model for $\rho(\mathbf{r})$.

Assume first that the positive charge distribution is represented by a delta function centered at the origin, $\mathbf{r} = 0$, but that the negative ion distribution is uniform throughout the unit cell. The total charge of the cell, as we have said, vanishes. Since the Fourier transform of a delta function is unity, $\rho_{\mathbf{k}\neq 0} = \mathrm{v}^{-1}$. The negative ion charge density makes no contribution to $\rho_{\mathbf{k}\neq 0}$, since

$$\frac{\rho(r = \text{const.})}{\mathrm{v}} \int_{\text{one unit cell}} e^{-i\mathbf{k}\cdot\mathbf{r}}\, d\mathrm{v} = 0$$

At $\mathbf{k} = 0$, the negative ion charge density is equal and opposite to that of the delta function, in order for $\rho_0 = 0$. Since $\rho_{\mathbf{k}} = \mathrm{v}^{-1}$, Eq. (XI-2) may converge only because of its $\mathbf{k}^{-2}$ term. But if (XI-2) is replaced by an integral, we will soon see that each spherical shell of thickness $d\mathbf{k}$ makes the same contribution to ϕ; hence, the potential does not converge with this model for $\rho(\mathbf{r})$. We therefore replace the delta

7. H. Kornfeld, *Z. Phys.*, **22**, 27 (1924).
8. H. M. Evjen, *Phys. Rev.*, **39**, 675 (1932).

function by the Gaussian distribution

$$\rho(\mathbf{r}) = \left(\frac{\varepsilon^3}{\pi^{3/2}}\right) e^{-\varepsilon^2 \mathbf{r}^2} \tag{XI-4}$$

where ε is a width parameter. If (XI-3) is written as a product of integrals in terms of the cartesian components of **r** and **k**, it is easily found that*

$$\rho_k = \frac{1}{V} e^{-k^2/4\varepsilon^2}$$

and

$$\phi = \frac{4\pi}{V} \sum_{\mathbf{k} \neq 0} \frac{e^{i\mathbf{k}\cdot\mathbf{r} - k^2/4\varepsilon^2}}{k^2} \tag{XI-5}$$

The factor $e^{-k^2/4\varepsilon^2}$, which is the result of replacing the point (delta function) charges with a Gaussian distribution, insures the convergence of ϕ.

In fact, the potential due to the delta functions centered at the origin of each unit cell, plus a uniform negative charge, is almost what is desired. We take the potential of (XI-5) and add another to it, one which consists of the potential due to point charges at each lattice point plus negative Gaussian distributions centered at the same points. The latter distribution *exactly* cancels the positive Gaussians assumed in (XI-4), and the result is the potential due to the desired distribution.

The additional potential may be found by integration of Poisson's equation. (Recall that each Gaussian is spherically symmetric.) The similarity of this problem to the Debye–Hückel problem can be recognized. The potential consists of three parts:

1. That due to the point positive charge, of the form r^{-1},
2. A term of the form

$$-\frac{1}{R}\int_0^R e^{-\varepsilon^2 r^2}\, d\mathbf{r}$$

which is the potential of the negative charge distribution within the arbitrary radius R, all concentrated at the origin, and

3. A term of the form

$$-\int_R^\infty \frac{e^{-\varepsilon^2 r^2}}{r}\, d\mathbf{r}$$

which represents the potential due to the remainder of the negative charge distribution.

* While the tail of any one Gaussian extends outside a unit cell, the tails of all the Gaussians of all other cells extend into the origin cell, so the integral over one cell may be replaced by another over all space.

The three terms are additive, and the result is that the additional potential is

$$\phi = \frac{1 - \operatorname{erf}(\varepsilon r)}{r} \tag{XI-6}$$

where

$$\operatorname{erf} x = \frac{2}{\pi^{1/2}} \int_0^x e^{-x^2}\, dx$$

Contributions like (XI-6), one for each lattice point, are added to (XI-5) to obtain the total potential.

For the potential at a lattice point, we can write (from (XI-5) and (XI-6))

$$\phi = \frac{4\pi}{V} \sum_{k \neq 0} \frac{e^{-k^2/4\varepsilon^2}}{k^2} + \sum_t \frac{1}{t} G(\varepsilon t) \tag{XI-7}$$

where

$$G(\varepsilon t) = \frac{2}{\pi^{1/2}} \int_t^\infty e^{-y^2}\, dy = 1 - \operatorname{erf}(\varepsilon t)$$

Notice that the first sum in (XI-7) is over reciprocal lattice vectors, while the second is over the vectors of the direct lattice. In each case, however, the width parameter ε plays a prominent role. The trick of the Ewald procedure is to choose ε such that both sums converge rapidly; however, the result is independent of the value of ε chosen.

The potential of a similar array of dipoles is found by differentiating (XI-7). The part deriving from the first part (XI-5) is

$$\phi_{p1} = -\frac{\partial \phi_1}{\partial z} = -\frac{4\pi i}{V} \sum_{\mathbf{k}} \frac{k_z}{k^2} e^{i\mathbf{k}\cdot\mathbf{r} - k^2/4\varepsilon^2}$$

and the z component of the resultant dipole field is

$$E^z_{p1}(\mathbf{r}) = \frac{\partial^2 \phi_1}{\partial z^2} = -\frac{4\pi}{V} \sum_{\mathbf{k}} \left(\frac{k_z}{k}\right)^2 e^{i\mathbf{k}\cdot\mathbf{r} - k^2/4\varepsilon^2}$$

The field at a lattice point ($\mathbf{r} = 0$) is then

$$E^z_{p1}(0) = -\frac{4\pi}{V} \sum_{\mathbf{k}} \left(\frac{k_z}{k}\right)^2 e^{-k^2/4\varepsilon^2} \tag{XI-8}$$

Similarly, by taking first and second derivatives of the second sum in (XI-7) we obtain

$$E^z_{p2} = -\sum_t \left[\left(\frac{1}{t^3} - \frac{3z^2}{t^5}\right) G(\varepsilon t) + \frac{2\varepsilon}{\pi^{1/2}} \left(\frac{1}{t^2} - \frac{3z^2}{t^4}\right) e^{-\varepsilon^2 t^2} - \frac{4\varepsilon^3}{\pi^{1/2}} \frac{z^2}{t^2} e^{-\varepsilon^2 t^2}\right] \tag{XI-9}$$

The total z component of the dipole field is $E^z_{p1} + E^z_{p2}$. By equating the arguments of the exponential factors in (XI-8) and (XI-9) which control the convergence for the lattice points on the z-axis, it is found that[9]

$$\varepsilon^4 = k^2/4t^2$$

The Ewald–Kornfeld method has been used by a number of authors to obtain dipole sums for particular crystals,[10–13] some of which have been described in earlier chapters (see Sections 5-7 and 6-5).

The second technique of lattice summation we shall describe is often referred to as the method of *planewise summation.* It was originally proposed in a series of articles by Nijboer and deWette[14] and has been subsequently further developed by deWette and Schacher.[15] In this method, instead of carrying out the dipole sum over an infinite lattice, it is first taken over a finite crystal of a given shape. The limit of the infinite crystal is then approached with conservation of shape. Thus a procedure for obtaining an infinite sum is provided.

We have previously observed that the convergence of sums like those of Coulomb terms over a lattice (as well as dipole sums) is slow. Rapid convergence may be obtained by the construction of an *auxiliary function.* Suppose a smooth function $f(x) \rightarrow 0$ slowly for $x \rightarrow \infty$. Then the sum

$$S = \sum f(n)$$

converges slowly. However, an auxiliary function $\mathscr{F}(x)$, which does approach zero quickly as $x \rightarrow \infty$, may be found. Since

$$S = \sum f(n)\mathscr{F}(n) + \sum f(n)[1 - \mathscr{F}(n)]$$

the first sum will now have good convergence properties, and the second will be no worse than the original sum. Because the Fourier transform of a smooth function is sharp, we can seek the transform of the second sum, and sum it in reciprocal space, where it should converge quickly. This is the principle of the planewise summation method.

In the case of the Coulomb problem, Nijboer and deWette use an integral expression of the Madelung constant

$$\alpha = \int \frac{\omega(\mathbf{r})}{\mathbf{r}}\, d\mathbf{r}$$

where

$$\omega(\mathbf{r}) = \sum_{l}{}' e^{i\mathbf{k}_{1/2}\cdot\mathbf{r}}\, \delta(\mathbf{r} - \mathbf{r}_l)$$

9. R. E. Frech and J. C. Decius, *J. Chem. Phys.*, **51,** 1536 (1969).
10. L. W. McKeehan, *Phys. Rev.*, **43,** 1022, 1025 (1933).
11. H. Mueller, *Phys. Rev.*, **47,** 947 (1935); **50,** 547 (1936).
12. J. M. Luttinger and L. Tisza, *Phys. Rev.*, **70,** 95 (1946); **72,** 257 (1947).
13. D. P. Craig and J. Walsh, *J. Chem. Soc. (London)*, (1958) 1613.
14. B. R. A. Nijboer and F. W. deWette, *Physica*, **23,** 309 (1957); **24,** 422, 1105 (1958).
15. F. W. deWette and G. E. Schacher, *Phys. Rev.*, **137,** A78 (1965).

$\mathbf{k}_{1/2} = a^{-1}(\frac{1}{2}\frac{1}{2}\frac{1}{2})$, so that the exponential furnishes the charge alternation as the lattice is traversed. The delta function representation for the charge distribution is the same as that used by Ewald. The complementary error function $\Phi(\pi^{1/2}r) = 1 - \text{erf}\,(\pi^{1/2}r)$ is then used as the auxiliary function, thus:

$$\alpha = \int \frac{\omega(\mathbf{r})}{r} \Phi(\pi^{1/2}r)\, d\mathbf{r} + \int \frac{\omega(\mathbf{r})}{r} [1 - \Phi(\pi^{1/2}r)]\, d\mathbf{r}$$

Parseval's formula* is then used, together with the Fourier transforms

$$\int \frac{\text{erf}\,(\pi^{1/2}r)}{r} e^{i\mathbf{k}\cdot\mathbf{r}}\, d\mathbf{r} = 4\pi \frac{e^{-k^2}}{k^2}$$

and

$$\int \omega(\mathbf{r})\, e^{i\mathbf{k}\cdot\mathbf{r}}\, d\mathbf{r} = \text{v}^{-1} \sum_{\bar{\mathbf{k}}} \delta[\mathbf{k} - (\bar{\mathbf{k}} - \mathbf{k}_{1/2})] - 1$$

(where $\bar{\mathbf{k}}$ is a vector of the reciprocal lattice) to write the second integral as

$$\int \frac{\omega(\mathbf{r})}{r} [1 - \Phi(\pi^{1/2}r)]\, d\mathbf{r} = \int \left\{ \sum_{\bar{\mathbf{k}}} \frac{4\pi}{\text{V}} \delta[\mathbf{k} - (\bar{\mathbf{k}} - \mathbf{k}_{1/2})] - 1 \right\} \frac{e^{-k^2}}{k^2}\, d\mathbf{k}$$

$$= \frac{4\pi}{\text{V}} \sum_{\bar{\mathbf{k}}} \frac{e^{-(\bar{\mathbf{k}} - \mathbf{k}_{1/2})^2}}{(\bar{\mathbf{k}} - \mathbf{k}_{1/2})^2}$$

Thus an expression for the Madelung constant is obtained which is equivalent to that derived using the Ewald procedure, with the Ewald parameter ε^2 equal to π.

Because the derivatives of the Coulomb potential can be expanded in spherical harmonics, the coefficients of which are the nth-order multipoles of the charge distribution,[16] the field at the origin due to that multipole can be written as

$$S'_{l,m}(0 \mid \mathbf{k}, n) = \sum_t{}' \frac{Y_{lm}(\theta, \varphi)\, e^{i\mathbf{k}\cdot\mathbf{r}_t}}{r_t^{2n+l}}$$

Each such field has its own auxiliary function. That for the dipole field $S'_{2,0}(0 \mid \mathbf{0}, \frac{1}{2})$ is the incomplete gamma function

$$\Gamma(n, x) = \int_x^\infty e^{-t} t^{n-1}\, dt$$

with $n = 5/2$, $x = \pi r_t^2/a^2$. The sum is again factored into two parts, the first of which is over the direct lattice, and the second (after taking the Fourier

* If $F(\mathbf{k})$ and $G(\mathbf{k})$ are the three-dimensional Fourier transforms of $f(\mathbf{r})$ and $g(\mathbf{r})$, respectively, then Parseval's formula is

$$\int F^*(\mathbf{k}) G(\mathbf{k})\, d\mathbf{k} = \int f^*(\mathbf{r}) g(\mathbf{r})\, d\mathbf{r}$$

16. J. A. Stratton, *Electromagnetic Theory,* McGraw-Hill, New York, 1941, chapter III.

transform) is over the reciprocal lattice, as in the Coulomb problem. The second sum, which is independent of the crystal structure, is identical with the tensor $(4\pi/3a^3)(\delta^{\mu\nu}) + (4\pi/a^3)(k^{\mu\nu}/k^2)$ of (5-7-9) for an infinite crystal, and with the $\boldsymbol{S}$ tensor of (5-7-14) for a finite crystal.

In cubic crystals, the first sum vanishes identically, but for all others the planewise summation technique has practical advantage. We consider two separate contributions to the first sum. S_{I}' gives the dipole field due to all dipoles in the lattices except those in the plane $z = 0$; S_{II}' gives the remainder. S_{I}' is summed first over the lattice vectors in the xy plane, and only after that over the planes in the z direction. The two-dimensional sum is Fourier transformed into an equivalent sum over the two-dimensional reciprocal lattice. S_{II}' is summed using the original procedure, with the auxiliary function $\Gamma(\frac{3}{2}, \pi\sigma^2)/\Gamma(\frac{3}{2})$ where σ is the radius vector in the $z = 0$ plane. A computer program utilizing this technique has been developed by Dickmann.[17] Good convergence of S_{I}' is effectively achieved after summation over a very few planes in the z direction. S_{II}' converges well if it is carried out over approximately 400 unit cells.

It should be noted that in this formulation of the dipole summation, the mathematical recipe effectively assumes a slab-shaped, albeit infinite, crystal. The convergence forced by the auxiliary functions preserves this shape in proceeding to a crystal of finite dimension—the actual domain over which convergence is achieved. Results for other shapes are then found, very simply, by the addition of the $\boldsymbol{S}$ tensor.

17. D. B. Dickmann, M.S. Thesis, US Naval Postgraduate School, 1966.

APPENDIX

XII

ROTATIONAL MOTION IN CRYSTALS

Most known molecular crystals, with the exception of H_2, can be treated within the framework of the methods developed in Chapter 6, in particular by regarding both the internal and external (rigid) modes as vibrations.[1] However, for hydrogen, the rotational motion is so far from being vibrational that it is much better to regard these degrees of freedom as a slightly perturbed free rotation.

The uniqueness of crystalline H_2 arises in part from the fact that H_2 has the smallest moment of inertia of any molecule, and in part from the weakness of the intermolecular forces—largely consisting, so far as rotation is concerned, of quadrupole interactions. One must also take into account the different nuclear spin states: pure para-hydrogen consists only of molecules which have a resultant nuclear spin of zero (opposite spins of the two nuclei) and, as is well known, exists only in even J rotational states. Thus at temperatures of a few degrees Kelvin, para-hydrogen will have a ground state in which all molecules have $J = 0$, and excitations in which no change in nuclear spin is allowed can only occur in which one or more molecules go from $J = 0$ to $J = 2, 4$, etc.

Pure ortho-hydrogen, on the other hand, has a ground state in which all molecules are in the $J = 1$ state, and its excitations must consist of transitions in which one or more molecules are excited to $J = 3, 5$, etc., or if $J = 1$ is split, to higher M states.

To build crystal wavefunctions for the rotational motion, we must use the exciton method, since as shown in Section 6-4, one can only use the phonon

1. *Para-H_2*: J. Van Kranendonk and G. Karl, *Rev. Mod. Phys.*, **40,** 531 (1968). *Ortho-H_2 and para-D_2*: S. Homma, K. Okada, and H. Matsuda, *Progr. Theoret. Phys.*, **36,** 1310 (1966); **38,** 767 (1967). H. Veyama and T. Matsubara, *Progr. Theoret. Phys.*, **36,** 784 (1967). J. C. Raich and R. D. Etters, *Phys. Rev.*, **168,** 425 (1968).

method for the oscillator problem (quadratic potential function). To avoid cumbersome subscripts, it is convenient to designate the one-molecule free rotation wavefunctions by the analogous symbols of atomic spectroscopy, i.e., for $J = 0$ we use s; for $J = 1$, p; for $J = 2$, d; etc. The ground state is represented by

$$\psi_0 = \prod_{\tau} s(\tau) \tag{XII-1}$$

which is exact in the limit of no intermolecular forces: this expression is limited to the case of one molecule per unit cell.

If the unit cell is multiply occupied, it will be convenient to use $\Delta\tau$ to represent the fractional lattice vector positions of the several molecules. In the case of two molecules per unit cell, one might choose $\Delta\tau = 0$ for the first molecule, and write

$$\psi_0 = \prod_{\tau} s(\tau)s(\tau + \Delta\tau) \tag{XII-2}$$

At very low temperatures, para-hydrogen is believed to exist in the hexagonal close-packed D_{6h}^4 structure with two molecules per unit cell on Wyckoff sites (b) of symmetry D_{3h}. To investigate the symmetry of excited rotational states, i.e., *rotons,* we first find the species of the one-molecule wavefunctions at the sites, and then use correlation between D_{3h} and D_{6h} to obtain the unit cell (factor group) symmetries. Since para-hydrogen (and o-D_2) exists only in even J states, we shall only give the symmetry analysis for even J in Table XII-1.

Although correlation between species A_1' of D_{3h} and D_{6h} yields $A_{1g} + B_{2u}$, one notes that the $J = 0$ wavefunctions yield only a single state under D_{6h}, which is the ground state and of course is totally symmetric.* The one-molecule first excited states in para-hydrogen will have all molecules except one still in the s state: the excited molecules can be in any of the five d_α states where α stands for z^2, $x^2 - y^2$, etc. The wavefunction can be written (see Section 6-4)

$$\psi_{2\alpha}^{\pm}(\boldsymbol{\kappa}) = \psi_0 \sum_{\tau} \left[\frac{d_\alpha(\tau)}{s(\tau)} \pm \frac{d_\alpha(\tau + \Delta\tau)}{s(\tau + \Delta\tau)} \right] e^{-i\boldsymbol{\kappa}\cdot\boldsymbol{\tau}} \tag{XII-3}$$

Table XII-1 Symmetry species of linear molecule rotational wavefunctions in the hexagonal (D_{6h}^4) structure

J	M	Wavefunction	D_{3h}	D_{6h}
0		s	A_1'	A_{1g}
2	0	d_{z^2}	A_1'	$A_{1g} + B_{2u}$
	± 1	$d_{x^2-y^2}, d_{xy}$	E'	$E_{2g} + E_{1u}$
	± 2	d_{xy}, d_{yz}	E''	$E_{1g} + E_{2u}$

* The A_{1g} and B_{2u} states differ only in the sign attributed to the molecular wavefunctions at the A and B positions in the hexagonal unit cell, i.e., $A_{1g} = \prod_{\tau} s(\tau, A)s(\tau, B)$ and $B_{2u} = \prod_{\tau} s(\tau, A)[-s(\tau, B)]$, and hence are not distinct states.

The selection rules under D_{6h} symmetry permit infrared transitions from A_{1g} to A_{2u} or E_{1u}, and Raman activity from A_{1g} to A_{1g}, E_{1g}, or E_{2g}. Thus for the $J = 0 \rightarrow J = 2$ transition, one expects a single infrared line (to the E_{1u} state) and three Raman lines, in agreement with experiment.

Note that this rotational analysis yields very different predictions from those which would have been made in the librational limit. In the latter case, the small vibrational coordinate for libration at the D_{3h} site would be of species E'', yielding by correlation librational normal coordinates of species $E_{1g} + E_{2u}$; of these, the E_{1g} mode would be Raman active and infrared inactive. Of course, in this limit the observed frequency would be much higher than the gas phase $J = 0 \rightarrow J = 2$ frequency.

The rotational analysis can, of course, be carried out on the basis of other assumed crystal symmetries. For example, an ordered structure with an arrangement like cubic close packing belongs to the space group T_h^6 (compare CO_2) with four molecules on $C_{3i} = S_6$ sites. This cubic structure consists in the ordered form exclusively of o-H_2 (or p-D_2) molecules, which in the free rotation limit can only be in odd J states. Here, as in the hexagonal case, the correlation of the ground molecular state (in this case $J = 1$) from site to unit cell symmetry yields more than one crystal state, in particular $A_u + F_u$ states under T_h because of the tetramolecular unit cell. Again, however, these crystal states expressed as products of molecular states are not in fact different, because the F_u states would differ from the A_u state only with respect to the signs of the p_z wavefunctions at the several molecules in the unit cell. For this reason, we suppress the F_u which would normally appear under T_h in the first row of Table XII-2.

Table XII-2 Symmetry species of linear molecule rotational wavefunctions in the cubic (T_h^6) structure

J	M	Wavefunction	S_6	T_h
1	0	p_z	A_u	A_u
	± 1	p_x, p_y	E_u	$E_u + 2F_u$

The Raman spectrum should allow scattering corresponding to a new class of excitation of the rotational energy; at the low temperature of the experiment, the molecules may be expected to be in the $J = 1$ ground state (for o-H_2 and p-D_2), but in contrast to the earlier case of p-H_2, one should be able to observe a $\Delta J = 0$ excitation between the A_u ground state corresponding to p_z and the excited $E_u + 2F_u$ states associated with p_x and p_y molecular wavefunctions. These excitations, which are called *librons*, have been observed by Hardy, Silvera, and McTague.[2]

2. W. N. Hardy, I. F. Silvera, and J. P. McTague, *Phys. Rev. Lett.*, **26**, 127 (1971).

AUTHOR INDEX

Allen, H. C., Jr., 52
Aller, L. H., 69
Allin, E. J., 233
Andermann, G., 243, 264
Anderson, A., 235
Aung, S., 287

Balslev, I., 271
Barron, T. H. K., 174, 287
Bass, J. L., 192
Batterman, B. W., 204
Bernstein, E. R., 300
Berreman, D. W., 192
Bickermann, A., 108
Blackman, M., 134
Bloom, M., 13
Blount, C. E., 305
Bonadeo, H., 262
Bonn, R., 310
Born, M., 77, 170, 190, 277, 371
Bosanquist, C. H., 204
Bouckaert, L. P., 111
Bowers, M. T., 305
Bradley, C. J., 105, 128, 327, 362, 363, 364
Bragg, W. L., 204
Brillouin, L., 111, 360
Brockhouse, B. N., 5, 81, 132, 144, 174
Brüesch, P., 186
Bryant, J. I., 238, 309
Buchanan, R. A., 282
Burbank, R. D., 250

Cabana, A., 305
Cahill, J. E., 235
Cardona, M., 272
Carlson, R. E., 264, 265
Casher, A., 327
Casimir, H. B. G., 212
Caspers, H. H., 282
Chadwick, B. M., 303
Charles, S. W., 303
Charles, W., 305
Charneau, R., 306
Cheesman, L. E., 178
Cheung, H., 236
Claus, R., 178
Cochran, W., 132, 173, 174, 277
Cohen, M. H., 174, 202, 287
Cole, H., 204
Cooke, R. D., 299
Coulson, C. A., 197
Couture, L., 188, 195
Couture-Mathieu, L., 195
Cowley, E. R., 287
Cowley, R. A., 132, 173
Cracknell, A. P., 105, 128, 164, 327, 362, 363, 364
Craig, D. P., 375
Crawford, B. L., Jr., 53
Cross, P. C., 15, 52, 219, 326
Cundell, M. A., 311
Curl, Jr., R. F., 289
Cyvin, S. J., 45, 197

Damany, N., 304
Damen, T. C., 169, 178, 198, 239
Darwin, C. G., 203
Decius, J. C., 15, 45, 96, 186, 197, 210, 219, 238, 251, 264, 265, 287, 291, 292, 300, 309, 326, 375

deLaunay, J., 102, 142, 148
deWette, F. W., 207, 375
Diamant, Y., 289, 305
Dicke, R. H., 274
Dickmann, D. B., 207, 257, 281, 377
Dolling, G., 5
Downs, A. J., 303
Dows, D. A., 235, 243, 251, 255, 262, 264, 280, 281, 287, 303
Dressler, K., 304, 305
Dubost, H., 306

Elkin, E. L., 265
Etters, R. D., 378
Evjen, H. M., 372
Ewald, P. P., 103, 371
Ewing, G. E., 305

Fahrenfort, J., 188
Falk, M., 300, 302
Fano, U., 211
Fine, P. C., 134
Flygare, W. H., 305
Fox, D., 202, 251, 263, 265, 287
Francia, M. D., 300
Frech, R. E., 186, 192, 264, 287, 375
Fredrickson, L. R., 231, 238
Friedrich, H. B., 234, 299

Gilat, G., 166
Gillis, N. S., 108
Giordmaine, J. A., 198, 239
Glück, M., 327
Goldfarb, T. D., 306
Gordon, D. J., 309
Gordon, R. G., 52, 73
Gorelik, V. S., 280
Gouterman, M., 306
Günthard, H. H., 282
Gunier, A., 204
Gur, Y., 327

Haas, C., 309
Halford, R. S., 119, 156, 266, 306
Hallam, H. E., 303
Hamermesh, M., 30
Hamilton, W. C., 3
Hardy, W. N., 380
Hare, W. F. J., 233
Harrick, N. J., 192
Harrison, W. A., 159, 160
Hartwig, C. M., 241, 264
Harvey, S. R., 303
Heller, W. R., 174
Hellwege, K. H., 240
Henry, C. H., 178
Henry, N. F. M., 119, 168, 345
Herley, G. I., 305
Hermann, T. S., 303
Herring, C., 164, 365
Herzberg, G., 52, 313, 327
Herzfeld, K. F., 174
Hexter, R. M., 166, 202, 225, 236, 246, 250, 251, 263, 265, 282, 287, 289, 305, 369, 371
Hiebert, G. L., 298, 299, 305
Hirsch, P. B., 203
Hisatsune, I. C., 309
Hoffman, R. E., 223
Hollenberg, J. L., 262
Homma, S., 378
Hopfield, J. J., 178, 210
Hopkins, H. P., Jr., 289
Hornig, D. F., 95, 118, 223, 280, 298, 299, 305, 310
Hoshino, S., 108
Hrostowski, H. J., 298
Huang, K., 170, 174, 175, 199, 277, 371
Hughes, A. E., 298

Ibers, J. A., 3
Ipatova, I. P., 102, 134, 294

Jacobi, N., 235, 281
Jacobson, J. L., 309
Jacox, M. E., 303, 305
James, R. W., 204
Johnson, F. A., 166, 277
Jones, G. O., 306
Jordan, T. H., 281

Kanney, L. B., 108
Karl, G., 378
Kauzmann, W., 69
Keffer, F., 174, 202, 287,
Kellerman, E. W., 174, 287
Ketelaar, J. A. A., 188, 309
Khare, B. N., 306
Kingsley, J. D., 265
Kinsey, E. L., 282
Kittel, C., 3, 371
Klein, M. V., 181, 205
Knop, O., 300, 302
Kopelman, R., 301
Kornfeld, H., 372
Kovalev, O. V., 327
Kramers, H. A., 243
Krause, P. F., 234, 299
Kravitz, L. C., 265
Krimm, S., 97
Kronig, R. de L., 243
Kubo, R., 52, 216, 274
Kursunoglu, B., 210

Lax, M., 292
Lederman, W., 77
Lee, J. W., 212
Lee, K. O., 305
Leibfried, G., 102, 134, 142
Leighton, R. B., 134
Leroi, G. E., 235, 305

Lesch, W., 240
Liang, C. Y., 97
Linevsky, M. L., 304
Lipscomb, W. N., 281
Lonsdale, K., 119, 168, 345
Lorentz, H. A., 205
Loudon, R., 155, 166, 195, 296, 313
Love, W. F., 128, 327
Ludwig, W., 102, 134, 142
Luttinger, J. M., 206, 207, 375
Lüty, F., 308
Lyubarskii, G. Ya., 327
Lyddane, R. H., 173, 174

Maccoll, A., 197
MacDonald, R. E., 233
Maeda, S., 193
Mahan, G. D., 211, 265
Maker, P. D., 45
Maki, A., 309
Malan, O. G., 292
Mara, R. T., 282
Maradudin, A. A., 102, 134, 278, 287, 294, 295, 296
Marcus, A., 174
Martin, T. P., 209
Marusak, G. V., 305
Marzocchi, M. P., 262
Maslakowez, I., 310
Mathieu, J. P., 188, 195, 326
Matsubara, T., 378
Matsuda, H., 378
McClellan, A. L., 266
McDonald, R. S., 309
McKeehan, L. W., 375
McTague, J. P., 380
Melveger, A. J., 299
Merrithew, R. B., 305
Merzbacher, E., 70
Meserole, F., 265
Metselaar, R., 310
Meyer, B., 306
Midwinter, J. E., 197
Miller, S. C., 128, 327
Milligan, D. E., 303, 305
Mills, I. M., 71
Mitin, G. G., 280
Montroll, E. W., 102, 134, 294
Morgan, H. W., 309
Morse, M., 165
Mosteller, L. P., 186
Motegi, H., 108
Mueller, H., 207, 375
Mulliken, R. S., 313

Nagamiya, T., 216
Narayanamurti, V., 289, 308, 310
Neto, N., 166, 236, 282, 287, 369
Newman, R., 306
Niimura, N., 108
Nijboer, B. R. A., 375
Nixon, E. R., 300
Norman, I., 303

Oden, L. L., 300
Oehler, O., 166, 236, 282, 287, 369
Okada, K., 378
Overend, J., 309
Oxton, I. A., 300, 302

Parmenter, R. H., 163, 164
Peake, S. C., 303
Peiser, H. S., 204, 304
Penner, S. S., 73
Pezolet, M., 234
Phillips, J. C., 161
Philpott, M. R., 212
Pilar, F. L., 45
Pimentel, G. C., 266, 298, 303, 305
Pitzer, K. S., 289
Placzek, G., 313
Plihal, M., 240
Pohl, R. O., 308, 310
Polder, D., 212
Pople, J. A., 233
Porter, G., 303
Porto, S. P. S., 169, 178, 198, 239, 241, 264
Poulet, H., 195, 197, 326
Price, W. C., 244, 307, 308, 310

Raich, J. C., 108, 378
Ramachandran, G. N., 203
Raubenheimer, L. J., 166
Rauch, J. E., 45, 197
Rayleigh, Lord, 60, 291
Redington, R. L., 305
Robinson, T. S., 244
Rochkind, M. M., 306
Romand, J., 304
Ron, A., 280
Roncin, J., 304
Rosenstock, H. B., 174, 287
Rousseau, D. L., 241, 264
Rush, J. J., 299

Sachs, R. G., 173
Sakisaka, Y., 204
Sandhu, H. S., 13
Savage, C. M., 45
Savitsky, G. B., 305
Savoie, R., 234
Schaack, G., 240
Schacher, G. E., 207, 375
Schachtschneider, J. H., 36
Schatz, P. N., 193
Schettino, V., 235, 280, 281, 282, 287, 309
Schnepp, O., 233, 235, 281, 289, 304, 305
Schroeder, L. W., 299
Schrötter, H. W., 178
Scott, J. P., 178
Seitz, F., 3, 95, 109, 134, 135, 356

Seward, W. D., 289, 310
Sherman, W. F., 307, 308, 309, 310, 311
Shimaoka, K., 108
Silvera, I. F., 380
Slater, J. C., 3, 112, 114, 134, 327, 371
Smith, D. F., Jr., 309, 310
Smith, H. W., 281
Smoluchowski, R., 111
Smulovitch, P. P., 310
Snyder, R. G., 97
Speiser, A., 16, 85, 301
Staats, P. A., 309
Stephen, M. J., 212
Stratton, J. A., 376
Strauss, H. L., 287
Streib, W. E., 281
Suarez, N. H., 309
Sushchinskii, M. M., 280
Sutherland, G. B. B. M., 97, 282
Sutton, L. E., 197

Taddei, G., 262
Tell, B., 169, 178
Teller, E., 173
Terhune, R. W., 45
Thompson, H. W., 292
Thyagaragan, G., 193
Tisza, L., 206, 207, 375
Tobin, M. C., 95
Turrell, G. C., 309

van der Elsken, J., 309, 310
Van Hove, L., 161
Van Kranendonk, J., 378
Vedder, W., 188, 310
Veyama, H., 378
von Kármán, Th., 77

Wallis, R. F., 278, 295, 296
Walmsley, S. H., 233, 235
Walsh, J., 375
Warren, J. L., 102, 133, 313, 327
Weiss, G. H., 102, 134, 158, 287, 294
Wells, A. J., 53
Welsh, H. L., 233
Wenzel, R. G., 133
Whittle, E., 303
Wickersheim, K. A., 282
Wigner, E. P., 111, 134, 135, 356, 365
Wilczek, T., 258, 261
Wilkinson, G. R., 307, 308, 309, 310
Williams, D. E., 261
Wilson, E. B., Jr., 15, 53, 54, 219, 326
Winston, H., 119, 156, 266
Wittke, J. P., 274
Wolf, E., 190
Woodfine, J. M., 306
Woods, A. D. B., 132, 174
Wooten, F., 186
Wyckoff, R. W. G., 101, 118, 222, 230, 345

Yacoby, Y., 272
Yarnell, J. L., 133
Yust, S., 272

Zachariasen, W. H., 106, 119
Zak, J., 327
Zernike, F., 197
Ziman, J. M., 274, 371

SUBJECT INDEX

Absorption, Einstein coefficient for induced, 51
Absorption cross-section, 68
Accidental degeneracies, 166
Acentric crystals, 14
Acoustic branches, 84, 102
Activation of ω_L, 295
Activity representations, 47, 49
α-quartz, 178
α-tungsten, 122
Analysis of multiplets, 249–265
Analytic critical point, 161
Anharmonicity, 65–67, 273–280
Annihilation operators, 322–325
Atom-atom interaction theory, 254–257
Auxiliary function, 375

B, effective field matrix, 216
B, matrix of transformation from cartesian to internal coordinates, 36
Band, two-phonon, 154
Base-centered monoclinic lattice, 106
Basis, 85
Berreman conditions, 209
Betti numbers, 165
BH_4^- in alkali halide matrices, 310
Biexcitons, bound, 280
BO_2^- in alkali halide matrices, 309
Body-centered cubic, secular determinant as a function of $\boldsymbol{\kappa}$ in, 143
Body-centered lattice, symmetry species at $\boldsymbol{\kappa} \neq 0$, 129
Border groups, 85
Boron nitride, 101
Brillouin zone, 110
Brillouin zones and special points, 361–363
Bouckaert, Smoluchowski and Wigner (BSW) points, 111

C_1 (line), 93
C_i (line), 86
C_6^6 ($P6_3$), irreducible representations in, 368
Calcite ($CaCO_3$), 10, 123, 229, 238–241, 264, 363
CaF_2, symmetry species at $\boldsymbol{\kappa} = 0$, 121–122
Calcite structure:
  irreducible representations in, 369
  series of compounds with, 229, 264
$Ca(OH)_2$, 282
CH_4 in Xe, Kr and Ar matrices, 305
C_6H_6 in C_6D_6, 300
Character, 30
  of axial vector, 320
Character tables, 32, 312–317
CH_3Cl, 248–260
CHI_3, 227, 236, 282–287, 363, 369
Class:
  crystal, 108
  of operations, 21
ClO_4^-, 311
CN^- in alkali halides, 308
CO, 9
  in Ar matrix, 305
CO_2, 9, 38, 223–227, 234–235, 281, 362, 367
Combinations and overtones involving degenerate states, 50
Combination transitions, 51

Compensation, 166
Computer sampling techniques to obtain density of state functions, 166
Conditional convergence of dipole sums, 257, 371
Continuum, uniformly polarized, 205
Correlation between symmetry species of groups, 58, 118, 326–335, 359
Correlation diagram, 120
  for CaF_2, 121
  for calcite, 229
  for CH_3Cl, 250
  for CHI_3, 227
  for CO_2, 224
  for diamond, 120
  for NaCl, 121
  for $NaClO_3$, 230
  for NaN_3, 228
  for TiO_2, 168
Correlation field coupling, 8, 251
Correlation function, 52, 74, 274
Correlation tables, 58, 130, 326–335
Correlation theorem, 59
Cosets, 87, 301, 342
  double, 301
Covalent crystals, 5
Covering operation, 16
Creation operators, 322–325
Critical points (c.p.'s), 158
  analytic, 161
  and impurity spectra, 295–298
  fluted, 164
  ordinary, 163
  singular, 161, 163, 165
Crystal imperfection, 204
Crystallographic space group, 105, 109, 343
Crystals, optically active, 198
Crystals containing complex ions, 6, 307–311
  (*See also* specific crystals or ions)
CsCl, 149–153
Cubic and quartic terms in the potential energy, 67, 278–280
Cubic forms of ZnS, 118
Cyanate ion, NCO^-, 59

$D_{3d}^6(R\overline{3}c)$ irreducible representations in, 369–370
Damping constant γ_j, 179
$\mathbf{D}_{ij}$, field propagation tensor between molecules, i and j, 201
$\mathbf{D}_{\tau m,\tau' m'}$, field propagation tensor between molecule m in unit cell $\boldsymbol{\tau}$ and molecule m' in unit cell $\boldsymbol{\tau}'$, 215
DBr (*see* HBr)
DCl (*see* HCl)
Decomposition formula, 32, 58
Density of modes, 84
Density of one- and two-phonon states, 271–273
Density of states and critical points, 156–167
Diamond, symmetry species at $\boldsymbol{\kappa} = 0$, 113–120
Diamond lattice, 123
Diastereoisomers, 302
Diatomic linear lattice:
  with symmetry, 86–93
  without symmetry, 93–94
Dielectric constant, 179
  and electrical susceptibility, 273
  and mechanical anharmonicity, 278
  and overtone or combination transitions, 276
  and refractive index, 182
  imaginary part of, 182
  microscopic theory of, 214–217
  real part of, 182
Dielectric tensor, 179
Dihedral groups, 19
Dimension of a representation, 31
Dipole autocorrelation function, 52, 74, 274
Dipole-coupling theory, 257–267
Dipole derivatives, 51, 71, 263–265
Dipole sum, conditionally convergent, 257, 357
Dipole-wave sum, 201, 371
Directional and polarization properties in single crystals, 167–170
Direct sums, 47
Dispersion curves:
  and joint density-of-states, 270
  for diamond, 133
  for KBr, 132
  for linear lattices, 80, 83, 90
  of the polariton, 211–213
Distinct orientations as physically inequivalent, 302
DNA, 95
Domain size, 204
Doppler half-width b_D, 72
Double coset, 301
Dynamic matrix:
  in the FG form, 134
    for body-centered cubic lattice, 143
    for face-centered cubic lattice, 140
    for lattice with a basis; CsCl, 149
    for simple cubic lattice, 138
  for coupled internal modes, 221, 222–231

Effective inverse mass matrix, **G**, 36
Effective field, 215
Electrical anharmonicity, 269
Electrical susceptibility χ, 273
Emission, Einstein coefficients for spontaneous and induced, 51, 67
Energy bands, 4
Energy transfer and relaxation, 14
Equivalent representation, 31
Errors in relative intensities for polarized Raman spectra, 198
Evaluation of lattice sums, 371–377
Ewald-Kornfeld summation procedure, 202, 371–375
Ewald parameter, 373, 376

Exciton and phonon methods of analyzing intermolecular couplings, 246–249
Exciton:
treatment of rotation of crystal H_2, 378
vibrational, 246, 252
Expansion of the electric dipole, 275
in CO_2, 235
in CHI_3, 237
Extraordinary wave, 182

F, force constant matrix, 36
Face-centered cubic lattice, secular determinant as a function of $\boldsymbol{\kappa}$, 140
Factor group, 343
of space group, 87
Factor group coupling, 251
Factor group theorems, 110, 124, 342–344
Fermi Golden Rule, 69, 70, 279
Fermi resonance, 66
in CO_2, 235, 280
Field propagation tensor, 201, 215
Fluoroapatite, $Ca_{10}(PO_4)_6F_2$, 265
Fluted critical point, 164
Free rotation, 13
Frequency distribution functions, 158
in the neighborhood of an analytical critical point, 161–162
Frequency product rule, 53
Functions, symmetry adapted, 33
Fundamental excitations, 50

GaP, 178
Gas-phase vibrational spectroscopy, 1
Gerade representations, 32
Giant molecule approach, 246
Glide reflection, 109
G, effective inverse mass matrix, 36
G matrix elements, 54
Graphite, 101
Green's function:
and anharmonicity, 280
and impurities, 294
Group of the wave vector, 343

Harmonic oscillator state functions, 322–325
HBr, mixed crystals with DBr, 299
HCl, 9
crystal structure of, 107
in Ar matrix, 305
mixed crystals with DCl, 299
HCN, 9, 222, 233, 362
mixed crystals with DCN, 299
Heisenberg representation, 73, 274
HF in rare gas matrices, 305
HF_2^- in alkali halide matrices, 309
Hindered rotation, 13
H_2O in Ar matrix, 305
Hooke's Law, 76, 170
Hot bands, 66
Hydrogen-bonded crystals, 6

Hyperpolarizability, 44, 197
Hyper-Raman effect, 44

Improper subgroups, 342
Improper symmetry operations, 319
Impurity mode, 295
Impurity theory, 290–298
InAs, 6
$InBO_3$, 263
Induced emission, Einstein coefficient for, 51
Inequlvalent representations, 31
Inert gas matrices, 303–306
Intermolecular forces, 11
Intermolecular potentials, quantitative theories of, 249–267
Internal and external coordinates in molecular crystals, 218–221
Internal coordinates, 35
Internal modes:
in CO_2, 235
in CHI_3, 236
International notation, 22
International symbols for the space groups, 110, 336–341
Inter-site vector, 119
Invariants, transition matrix elements as, 45
Invariant subgroup, 124, 342
Ionic crystals, 5
Ionic matrices, 307–311
i(pa)s, 169, 241
Irreducible representations, 28
in non-symmorphic groups, 364–370
Isotope effects, 53–60
magnitude of frequency shifts, 59
Isotopic solid solutions, 298–303
of DBr in HBr, 299
of DCl in HCl, 299
of DCN in HCN, 299
of deuterated ammonium ion, 300
of deuterated benzene, 300
Isotopic substitution:
with change of symmetry, 57
without change of symmetry, 56

Joint density-of-states, function for, 270
Joint density of summation states, 158, 270

k, integer defining irreducible representation for one-dimensional translation group, 78
k, wave vector in three dimensions, 103
$\mathbf{k}_i$, wave vector of incident radiation, 195
$\mathbf{k}_p$, wave vector of phonon, 195
$\mathbf{k}_s$, wave vector of scattered radiation, 195
$\boldsymbol{\kappa}$, vector of pure numbers, κ_i, representing the wave vector, 104
$\bar{\boldsymbol{\kappa}}_R$, $\bar{\boldsymbol{\kappa}}_S$, vectors in reciprocal space associated with point group operations, *R*, *S*, 125, 126
KDP, 7
KHF_2, 7

Kinks, 166
K_2SO_4, 265

Lattice, 105, 143
  secular determinant as a function of $\boldsymbol{\kappa}$ for:
    body-centered cubic, 143
    face-centered cubic, 140
    simple cubic, 138
Lattice modes of N_2, CO and CO_2 crystals, 281
Lattice sum, 201–210, 257
  evaluation of, 371–377
Lattice types, 106
Lattice with a basis, secular determinant as a function of $\boldsymbol{\kappa}$ for, 149
Librons, 380
$LiIO_3$, 178
$LiNbO_3$, 178
Line and band shapes, widths and intensities, 68–74
Linear lattice, vibrations of, 75–100
  continuum, 76
  diatomic, 85–94
  monatomic, 77–84
Linear polymers, vibrations of, 94–100
Line group, 85, 95
$LiNO_3$, 264
Lithium hydroxide, 122, 263
Local field, 251
Localized vibration, 295
Longitudinal acoustic (LA) phonon, 84
Longitudinal and transverse modes, 170–178
Longitudinal modes, 78
Longitudinal optic (LO) branch, 131
Lorentz-Lorenz field, 217
Lyddane, Sachs, Teller (LST) relation, 173

Macroscopic second order susceptibility, 197
Madelung problem, 371
Matrix:
  defining the kinetic energy, 36
  defining the potential energy, 35
  representing a symmetry transformation, 16
Matrix isolation, 303
Matrix representation, 24
Maugin-Hermann symbols for crystal symmetry, 22
Maximal site, 123, 359
Mechanically anharmonic terms, effect on intensities of two-phonon bands, 279
Mechanically harmonic system, 269
Metallic crystals, 4
Metals, 4
$MgCO_3$, 264
$Mg(OH)_2$, 263
Microscopic theory (*see* Dielectric Constant or Transverse and Longitudinal Modes)
Modulation spectroscopy, 271–273
Molecular ions, 11
Mosaics, single crystals composed of, 204
Multiphonon spectra, 281
Multiphonon transitions, selection rules for, 154–156
Multiplets:
  relative intensities of, 266–267
  splitting due to atom-atom potential, 251, 254–257
  splitting due to dipole-dipole potential, 251, 257–263

N_2, 9
N_3^- in alkali halide matrices, 309
$NaClO_3$, 10, 230, 241–245, 264, 362
NaN_3, 228, 237, 363
$NaNO_3$, 10, 229, 264
NCO^-, cyanate ion, 59
  in alkali halide matrices, 309
NH_3, in Ar matrix, 305
NH_4^+ in alkali halide matrices, 310
NH_4Cl, 232
NH_4NCS, 300
NH_4NO_3, 306
NH_4ReO_4, 302
NMR line-narrowing, 13
NO_2^- and NO_3^- in alkali halide matrices, 310
N_2O, 9
N_2O_5, 11
Non-equivalent irreducible representations, 31, 365
Non-primitive translation, 109
Non-symmorphic space groups, 109, 364
Normal coordinates, 33–35
Normal incidence, reflectivity at, 184
Normal modes of simple symmetric molecules, 60–65
Normal modes of vibration, 33–43

O_2, 9
$O_h^7(Fd3m)$, irreducible representations in, 366
Oblique incidence, reflectivity at, 186
OH^- in alkali halide matrices, 308
One-dimensional, diatomic crystals, frequency distribution functions of, 158
One molecule per unit cell, exciton and phonon methods of analyzing intermolecular couplings with, 246–248
One-phonon transitions, selection rules for, 154
One-site excitons, 248
Operator:
  annihilation, 322
  covering, 16
  creation, 322
  Hamiltonian, 16
  symmetry, 17–18
  time reversal, 92, 365
  Wigner Projection, 33, 79, 87, 91, 104
Optical branches, 102
Optical mode, 91
Optically active crystals, 198
Order, of a group, 19
Order-disorder transitions, ferro-electricity, 13
Ordinary critical point, 163

Ordinary wave, 181
Orientation:
  for a certain κ, 129
  in CaF_2, 149
  of degenerate symmetry cordinates, 135
Orientationally ordered crystal, 300
Oriented gas model, 266
Orthogonality:
  of characters, 32
  of matrix elements for irreducible representations, 30
Ortho-hydrogen, 378
Oscillator strength, 71
Overlapping modes, reflectivity of crystals with, 188
Overtones of non-degenerate modes, 50
Overtone transitions, 51

Para-hydrogen, 378
Partner functions, 33
Pb, 81
PBr_5, 11
PCl_5, 11
Perturbation approach to the one-dimensional impurity problem, 291
Perturbation method, 8
Phonon, 84
PH_3, 300
Plane wave propagation in anisotropic crystals, 181
Planewise summation, method of, 375
Planar XY_3, example of, 40–43
Plastic crystals, 13
Point groups, 15, 16
Polariton, 177, 210–213
Polarizability:
  and transition strengths, 215
  derivatives, 49, 312–317
    for rutile, 169
  in relation to Raman transitions, 45
  molecular, 44
Polarization of Raman scattering, 193–200
Polyethylene, 95, 97–99
Potassium acid fluoride, 299
Potential energy:
  cubic and quartic terms in, 278
  expressed in internal coordinates, 35
  expressed in normal coordinates, 34
  intermolecular terms in, 220
Pressure broadening, 52, 72
Primitive lattice, 106
Primitive unit cell, 115
Proper subgroups, 342
Proper symmetry operations, 319
Product rule, 53, 57
Pseudo-vector, 29, 320
Pure rotational operations, 19, 319

$\mathbf{q}_{k\alpha}(\boldsymbol{\tau}, m)$, molecular normal coordinate for molecule m in unit cell $\boldsymbol{\tau}$, 220

$Q_{k\alpha}(\boldsymbol{\kappa}, m)$, phonon coordinate of wave vector $\boldsymbol{\kappa}$ for molecules of type m, 221
Q, vector whose components are molecular normal coordinates, 37
Quadrupole-induced dipoles, 233
Quantum mechanics of the normal modes, 35

R tensors, 207
Radiant power, 68
Raman effect, 1, 44
Raman scattering:
  at other angles, 199
  at right angles, 193
  polarization of, 193–198
  relative intensities of longitudinal and transverse modes, 195–198
  selection rules for, 44–49
Raman spectrum:
  of calcite, 239
  of NH_4Cl, 231
  second order, 272
Rayleigh's theorem, 59, 291, 294
Reciprocal lattice, 103, 111
Reciprocal lattices and space groups, 356
Reciprocity of sites and special points, 356–360
Redundancy, 40
Reflectivity:
  normal incidence on a principal plane, 184
  normal incidence on an arbitrary face, 188
  oblique incidence on a principal plane, 186
  TE polarization, 186
  TM polarization, 186
Refractive index, 181
  separation into real and imaginary parts, 182
Representations:
  activity, 47, 49
  direct product, 45
  irreducible, 28
  multiplier, 128, 283
  of non-symmorphic space groups, 364–370
  of point groups, 23
  small, 365
  symmetrized direct product, 45
  transition, 47
Retarded dipole interaction, 211
Retarded van der Waals interaction, 212
Right-angle scattering, 193
Rock salt, symmetry species at $\boldsymbol{\kappa} = 0$, 121–122
Rotation-reflection axis, 19
Rotation of molecules in solids, 12
Rotational motion in crystals, 378–380
Rutile (TiO_2), 167

Sampling techniques for Brillouin zone, 166
Scattering geometry in a Raman experiment, 177
Schöflies symbols, 22
Schrödinger representation, 74
Screw-rotation, 109
Second-order Raman spectra, 272

Secular determinant:
as a function of $\boldsymbol{\kappa}$, 133–152
for a linear diatomic lattice, 89
separation of internal and external modes, 219
Selection rules, 44–53, 152–167
Selection rules and intensities in molecular crystals:
combinations and overtones, 267–270
fundamentals, 231–245
Selection rules for impurity-activated phonon absorption, 296–298
Selenium, 101
Separation of high and low frequencies, 219
Several molecules per unit cell, exciton and phonon methods of analyzing intramolecular couplings with, 248
Shape tensor **S**, 206
SiF_4, 287
Silicon, 101
Similarity transformation, 27
Simple cubic lattice, secular determinant as a function of $\boldsymbol{\kappa}$ in, 138
Singular critical point, 161, 165
Site, 8, 119, 359
Sites in the 230 space groups, 345–355
Site field, 251
Site groups, 119, 359
SO_4^{2-}, 311
Sodium azide, 228, 237
Sodium chloride, symmetry species at $\boldsymbol{\kappa} = 0$ in, 120
Space group, crystallographic, 105
complete listing, with generating elements, 336–341
non-symmorphic, 109, 364
operations $\{R \mid \tau\}$, 109
symmorphic, 109
Special points, 359
Splittings, 8, 249–267
Spontaneous emission, 67
Static field, 251
Star of $\boldsymbol{\kappa}$, 123, 359
Structure of a reducible representation, 32
Subgroups, 342
Surface polarization charge, 206
Surfaces of constant energy, 160
Susceptibility: (*see* Dielectric constant)
second order, macroscopic, 197
Symmetrically equivalent nucleus, 16
Symmetries of the $\boldsymbol{\kappa} = 0$ modes of $NaClO_3$ ($T^4 = P2_13$), 230
Symmetrization of coordinates under $H(\boldsymbol{\kappa})$, 129
Symmetrized direct product, 45
Symmetry adapted functions, 33
Symmetry analysis in cartesian coordinates, 318–321
Symmetry analysis of polyethylene, 97–99
Symmetry and internal coordinates, 38–40
Symmetry coordinates, 33
Symmetry element, 17
Symmetry of the $\boldsymbol{\kappa} = 0$ modes in CO_2, space group T_h^6, 224
Symmetry of the vibrations of the calcite lattice, 229
Symmetry of the vibrations in rocksalt, diamond, and fluorite, 131
Symmetry restrictions on transitions allowed by anharmonicity, 66
Symmetry species at $\boldsymbol{\kappa} = 0$, 113
Symmetry species at $\boldsymbol{\kappa} \neq 0$, 123
Symmetry species at $\boldsymbol{\kappa} = 0$ for iodoform, space group $C_6^6 = P6_3$, 227
Symmetry species in molecular crystals by correlation, 221–231
Symmetry species of the $\boldsymbol{\kappa} = 0$ vibrations of NaN_3, space group D_{3d}^5, 228
Symmetry species of translation, rotation and forms of the dipole ($\boldsymbol{\mu}' = \partial\boldsymbol{\mu}/\partial Q$) and polarizability ($\boldsymbol{\alpha}' = \partial\boldsymbol{\alpha}/\partial Q$) derivatives for the crystallographic point groups, 312–317
Symmetry transformations, distinction from symmetry elements, 18
Symmorphic space group, 109, 122, 359
System, crystal, 105, 106, 108

τ, vector of pure integers representing lattice translation, 103
τ_R, (fractional) translation associated with the point group operation, R, 124
τ_σ, fractional vector labeling one of a set of equivalent sites, 123
T_d unit cell, 194
T_h^6 ($Pa3$), irreducible representations in, 367
TE (electric vector transverse to the plane of incidence), 186
Time reversal, 92, 364–369
Time reversal symmetry, 367
TM (magnetic vector transverse to the plane of incidence), 186
Topographical methods, 204
Totally symmetric representation, 32
Trace of a matrix, 30
Transformation between internal and normal coordinates, 35
Transformation of cartesian coordinates, 318
Transformation of translationally symmetrized coordinates, 123
Transition probability, 67
Transition representation, 47
Transition strength, $S_j^{(\sigma)}$, 179
Transitions, selection rules for:
combination, 51
flundamental, 51
overtone, 51
Translation, rotation, and forms of the dipole ($\boldsymbol{\mu}' = \partial\boldsymbol{\mu}/\partial Q$) and polarizability ($\boldsymbol{\alpha}' = \partial\boldsymbol{\alpha}/\partial Q$) derivatives for the crystallographic point groups, 312–317
Translation group, 102, 105, 109, 343
Translational symmetry group, 77

Transmission through a slab, 190–193
Transverse acoustic (TA) phonon, 84
Transverse and longitudinal modes, 131
  and wave propagation in anisotropic crystals, 181
  in the Raman spectrum, 195–198
  macroscopic theory, 170–178
  microscopic theory, 200–214
Transverse modes, 78, 81
Transverse optic (TO) branch, 131
Trimethylamine, 306
Twinning, 198
Two-parameter lattice, 86
Two-phonon transitions, 154–156
Two-site exciton, 248

U-center, 298
Ungerade representations, 32
Unitary matrix, 34
Unit cell group, 359

Van der Waals crystals, 5
Vector **k**, 103
Vector of the reciprocal lattice, 111
Vibrational exciton, 246
Vibration-rotation branches, 67

Wave functions for phonons, relation between real and complex, 325
Wigner projection operator, 33, 79, 87, 91, 104

XO_4^{-n}, oxyanions in alkali halide matrices, 311
XY_2 molecule, 60, 61
$XY_3(C_{3v})$ molecule, 62
$XY_3(D_{3h})$ molecule, 62
XY_3Z molecule, 64
XY_4 molecule, 64
XYZ molecule, 61
XYZ_2 molecule, 63

ZnO, 178
ZnS, cubic form of, 118